THE ROUTLEDGE HANDBOOK OF SMART BUILT ENVIRONMENT

The primary aim of this edited volume is to document the current theories, best practices, and technological advancements in the move towards a Smart Built Environment (SBE). The needs to accelerate towards the SBE are numerous and include:

- Increasing complexities and the need for interconnectivity within the built environment (e.g. mega infrastructure projects)
- Data-driven decision-making resulting in higher demand from clients (e.g. smart design, construction, operation, and end of life [EOL])
- High requirements from stakeholders (e.g. system efficiency, environmental performance, green procurement)
- Fast-paced technological advancement and integration
- Natural disaster resilience of the built environment (e.g. prediction, smart control of building component)
- Sustainability issues around the built environment

In this book, the interrelationships among the various lifecycle stages: design, construction, operation, and EOL; the collective benefit of synergy at building level, multi-infrastructure level, and city-level, as well as the ultimate goals in relation to the deployment of smart technologies in the industry are addressed. Part I covers smart design and construction, Part II smart living, and operation, and Part III broadens the scope to the whole smart city. Chapters examine:

- How smart technologies can improve the effectiveness, productivity, and efficiency of the built environment
- An overview of theories and practices that are enabled by innovations and technologies for developing the SBE
- The basis for new research agendas, new concepts, and frameworks for future development.

This handbook documents the current theories, practices, and technologies and develops a holistic approach for research and practice by adopting a multidimensional outlook for the SBE. It is an essential reference work for all built environment stakeholders, from academia through to the professions.

Professor Weisheng (Wilson) Lu is Chair Professor (Digital Construction) and Head of Department of Real Estate and Construction, Faculty of Architecture at The University of Hong Kong.

Professor Chimay J. Anumba is Dean of the College of Design, Construction and Planning at the University of Florida, USA.

Designed cover image: Martha Willette Lewis

First published 2025
by Routledge
4 Park Square, Milton Park, Abingdon, Oxon OX14 4RN

and by Routledge
605 Third Avenue, New York, NY 10158

Routledge is an imprint of the Taylor & Francis Group, an informa business

British Library Cataloguing-in-Publication Data
A catalogue record for this book is available from the British Library

ISBN: 978-1-032-46208-0 (hbk)
ISBN: 978-1-032-46929-4 (pbk)
ISBN: 978-1-003-38384-0 (ebk)

DOI: 10.1201/9781003383840

Typeset in Times New Roman
by Apex CoVantage, LLC

THE ROUTLEDG
HANDBOOK OF SMA
BUILT ENVIRONMEN

Edited by Weisheng Lu and Chimay J. Anumba

Routledge
Taylor & Francis Group
LONDON AND NEW YORK

CONTENTS

Foreword *viii*
Faye Bowser

Introduction 1
Chimay J. Anumba and Weisheng Lu

1 Digital twins for smart decision-making 4
Chady Elias, Isabelle Southern and Raja R.A. Issa

2 From particles to spaces: The intelligent planning unit (IPU) theory to simplify the built environment 17
Suyash Padhye and Makarand Hastak

3 Generative design for excellence (DfX) for a smart built environment: From rule-based imitation to data-driven exploration 35
Ziyu Peng, Yi Zhang, Weisheng Lu, Junjie Chen, Liupengfei Wu and Jinying Xu

4 Project delivery methods to digital fabrication in architecture: A comparative case study from a modularity perspective 46
Tan Tan, Daniel Hall, Eleni Papadonikolaki, Grant Mills and Konrad Graser

5 Unlocking the power of language models for smart configuration in the AEC domain using spoken language understanding 64
Yinyi Wei, Xiao Li, Chengke Wu and Frank Petzold

6 Physical fatigue assessment (PFA) for construction workers with AI and sensory techniques 83
Yantao Yu and Zirui Li

7 Smart or not? Perspectives on digital fabrication and digital construction 104
Qian Chen, Ming Shan Ng and Jürgen Hackl

8 Methodical review of exoskeleton use in the construction industry 118
Abiola Akanmu, Anthony Yusuf and Nihar Gonsalves

9 Ontological models for construction scheduling and resource planning 133
Zigeng Fang, Songbo Hu, Junlin Wang, Yihai Fang and Qiuchen Lu

10 Iterative digital twin development for high-technology healthcare buildings: The case of a new ophthalmology diagnostic hub 162
Grant Mills, Duncan Wilson, Anne Symons, Irinie Roufaeel, Peter Scully, David Shanks, Cephas Bhaskar, Ian Eames, Hari Jayaram, Sobha Sivaprasad and Paul Foster

11 A theoretical framework for integrating digital twins in building lifecycle management 174
Karla Saldana Ochoa, Yasin Delavar, Deepak Balakrishnan, Chady Elias, Ravi S. Srinivasan and Chimay J. Anumba

12 Digital twin-based prototype system for healthcare facilities management 190
Obinna C. Madubuike and Chimay J. Anumba

13 Activity-based models for smart and sustainable urban environment 220
Qian-Cheng Wang and Li Wan

14 Sustainable urban built environment: Principles, tools, and applications 243
Phillip Jones and Jianxiang Huang

15 Exploring the potentials and obstacles of city information modelling 258
Junhao Wu, Feng Qiu and Ling Ma

16 Mobility-as-a-Service (MaaS) and the built environment 271
Qiao Si, Anthony G.O. Yeh, Guan Huang and Yang Yue

17 Integrated smart curb management in urban built environments addressing human dynamics and disruptive technologies 284
Yan Wang

18 Digital technologies for climate-adaptive smart urban transportation systems 296
Jinying Xu, Kristen MacAskill, Asaf Tzachor and Catherine Richards

19 Blockchain for smart built environment: Current status, challenges, and prospects 314
Liupengfei Wu, Weisheng Lu, Jinying Xu, Jinfeng Lou and Liang Yuan

20 Effects of signage color on human psychophysiological states and wayfinding performance in urban underground spaces with different noise levels: A combined EEG and VR experiment 326
Isabelle Y.S. Chan, Hao Chen and Pei Ma

Conclusion 339
Weisheng Lu and Chimay J. Anumba

Index *341*

FOREWORD

Accelerating the smart built environment

Today, the digital world influences and shapes the buildings where we live, work, and learn as profoundly as our everyday lives. Not surprisingly, according to the U.S. Environmental Protection Agency (EPA), as much as 90% of our lives is spent indoors. Working for a technology company, I understand how the built environment is key to productivity and success, as well as being healthy, comfortable, safe, secure, and efficient. As a company, Siemens believes in creating technology that transforms the everyday, for everyone. Personally, I am fascinated by the rate of innovation and how technology can play a role in creating an environment where people can thrive.

The smart built environment can exist at a variety of scales – campuses, neighborhoods, districts, cities, regions, and beyond. Using the campus of higher education institutions as an example, it is evident that these institutions have been at the forefront of digitalization and embraced the benefits that a smart built environment brings to the campus experience. Students today are younger than Smart Technology itself and more digital savvy: they expect a seamless, intuitive experience and often base their choice of institution on their perception of the campus experience, as well as the institution's purpose and value. Likewise, faculty and staff have their own expectations for their workspaces to focus on their core purpose of teaching or research, as well as promoting the mission of the institution. And alumni, who want to celebrate their alma mater and often contribute financially, have expectations of the campus evolving to remain competitive and relevant.

In many ways campuses operate like small, self-contained cities, each a microcosm of society as a whole. A campus faces the same challenges and needs that a small city would have for energy, security, operability, and livability. But if higher education's approach is any indicator, then creating smart built infrastructure for cities of all sizes across the country – as this book describes – can and should be done. Higher education institutions, as early adopters of both technology and sustainability initiatives, serve as living laboratories that demonstrate what is technically and economically feasible today while accelerating the pace of innovation by testing and proving new concepts.

Cities face challenges with facilities and infrastructure that have been developed over decades, often resulting in a mix of building systems that lack communication and vary in digital readiness. The proactive adoption of technology not only identifies which solutions can best meet users' needs but also provides a blueprint for systematically increasing digital maturity over time. The cities at the forefront of transitioning to being a smart built environment stand to reap considerable benefits and can be an exemplar for other cities contemplating the transition.

Today's leaders know it is vitally important to deliver the optimal experience of the built environment for its population. I have experienced the development and integration of smart technology for cities around the world and understand the importance of having clarity in that vision and how to create a roadmap for success. Today's cities are at a critical turning point in adopting digitalization and smart buildings to be more efficient and sustainable. They must learn from the example of early adopters by using a proven methodology to be digital-ready and embrace the technology available today and in the next decade. This book helps illuminate the way forward for cities and infrastructure.

In a way, a smart built environment mirrors the human nervous system, operating seamlessly to inform, react, and safeguard, carrying messages, sharing and coordinating actions, and reacting appropriately – and even learning from its experiences. With AI and automation, what the smart built environment learns it can build upon so that every year more people access an ever-more advanced digital experience.

Innovative technologies, from data-driven design to digital twin to electric transportation to smart buildings to smart cities – among many other topics discussed in this book – can be applied to nearly every aspect of the urban community, with the goal of not just sustainability but much enhanced living, which both make for a much better human experience.

If we take the right steps today, we can move forward and transform our urban infrastructure so that it is smarter, safer, more sustainable, and even more resilient and truly enhance our everyday life, while enjoying the benefits of enhanced energy efficiency, improved comfort, health and safety, and lower operational costs.

For me, this is exciting. Really exciting!

Faye Bowser
Vice President Higher Education Vertical Market
Siemens Smart Infrastructure

INTRODUCTION

Chimay J. Anumba and Weisheng Lu

Over the last few decades, there has been a tremendous growth in the integration of technologies into the built environment, with the goal of making it smarter. The primary objective is to develop a smart built environment, which leverages emerging technologies to improve the built environment and the quality of life for the broader society that live, work and recreate within this environment. While efforts to develop the smart built environment continue apace, this book is an effort to capture the current state of the art and includes contributions from researchers and professionals from around the world. It is made up of 19 chapters divided into three main sections: Smart Design and Construction, Smart Living and Operation of Facility, and Smart Cities.

Part I: **Smart Design and Construction** consists of nine chapters that address a variety of topics:

Chapter 1 by Elias et al. discusses digital twins (DT) for smart decision-making. It describes how DTs leverage that data from a physical asset to unlock value and support smart decision-making based on the integration of artificial intelligence (AI) and Internet of Things (IoT).

In Chapter 2, Padhye and Hastak explore the use of intelligent planning unit (IPU) theory to address the complexity in the built environment, which is made up of a complex system of systems. Using appropriate case scenarios, they detail how IPUs can be combined with emerging technologies such as artificial intelligence (AI) and blockchains to the evolution of the built environment.

The focus of Chapter 3 by Peng et al. is generative design for excellence (DfX) for a smart built environment. It advocates for a data-driven generative design paradigm that overcomes the limitations of rule-based approaches to the development of a smart built environment.

Chapter 4 examines the challenges in the delivery methods of projects with varying degrees of digital fabrication. Using a comparative case study, Tan et al. identify three key elements that need to be considered: the adoption of modular products and processes, adoption of an integrated project delivery method, and the role of informal relationships.

Wei et al., in Chapter 5, investigate how the power of large language models can be unlocked for smart configuration and user needs understanding in the architecture, engineering, and construction (AEC) domain. They identify the opportunities and challenges associated with the use of language models in this domain.

DOI: 10.1201/9781003383840-1

Physical fatigue assessment of construction workers using AI and sensory techniques is the subject of Chapter 6 by Yu and Li. They propose a heuristics-based approach for the management of risk categories and fuzzy reasoning for mapping joint-level risk levels to whole-body risk levels.

In Chapter 7, Chen et al. describe the implementation framework for digital fabrication and construction with the aim of improving data and material flows, thereby reducing rework and improving project quality. Case studies are utilized to illustrate the key concepts and their potential benefits.

Chapter 8 by Akanmu et al. provides a methodological review of the use of exoskeletons in the construction industry. It provides a taxonomy of exoskeletons, presents a comprehensive summary of current research efforts to develop exoskeletons within and outside the construction industry, and outlines research gaps in facilitating the adoption of exoskeletons in construction projects.

Fang et al. focus on ontological models for construction scheduling and resource planning in Chapter 9. They review recent research in this area and highlight both the benefits and challenges of adopting ontological models. Using a case study from the dynamic scheduling of crane operations, they present a novel approach to harnessing the benefits and addressing the challenges.

Part II: **Smart Living and Operation of Facility**, is comprised of three chapters:

Chapter 10 by Mills et al. presents a digital twin for ophthalmology building design. It details the integration of cyber-physical data and information in the delivery of an ophthalmology clinic. It includes a case study that addresses the complexities associated with integrating data from a variety of Internet of Things (IoT) and other sensors and various visual representations, to create an advanced ophthalmology building design.

In Chapter 11, Saldana-Ochoa et al. describe a theoretical framework for integrating digital twins in various aspects of the building lifecycle – design, construction, operation, and maintenance. It describes the transition path from building information modelling (BIM) to digital twins (DTs) and includes a comparative analysis of the different platforms for developing DTs. Future developments in DTs are also outlined with a focus on the integration of AI and other smart technologies.

Madubuike and Anumba's work in Chapter 12 is concerned with a digital twin-based prototype system for healthcare facilities management. It identifies key healthcare facilities management problems and presents four DT-based scenarios that were vetted with practitioners. The developed prototype system is also presented including details of its development process, operation, and evaluation.

Part III: **Smart Cities** includes eight chapters that cover different aspects of the topic:

Activity-based models for smart and sustainable urban environments are covered in Chapter 13 by Wang and Wan. They explore the role of these models in capturing the interaction between urban environments and residents' activity patterns and potentially contributing to city digital twins.

Jones and Huang, in Chapter 14, discuss how to change conventional approaches to a sustainable urban environment using new theories, tools, and strategies. The novel methods are expected to result in enhanced occupant comfort and health.

In Chapter 15, Wu et al. explore the potential benefits and obstacles of city information modeling (CIM), which they consider critical to the evolution of smart cities. They discuss how CIM facilitates great connectivity between multi-dimensional and multi-scale data generated from a variety of sources to enable informed decision-making in a smart city.

Chapter 16, by Si et al., examines Mobility-as-a-Service (MaaS) and its potential to transform transportation systems by facilitating public transit and reducing car ownership and driving. It discusses how MaaS can contribute to the achievement of a smart urban built environment with improved accessibility, energy consumption, carbon emissions, and reduced need for public parking.

Wang presents, in Chapter 17, the concept of integrated smart curb management in urban built environments. The chapter identifies the key challenges of the curb environment that demand smart solutions and presents the essential components of a smart curb environment that can maximize equitable and convenient access while minimizing greenhouse gas emissions, relieving congestion, and boosting livability for residents.

Digital technologies for climate-adaptive smart urban transportation systems (UTS) are covered in Chapter 18 by Xu et al. The chapter discusses the complexity of UTSs and the growing focus for them to be more sustainable and resilient to climate-change-induced hazards. It maps the value of different digital technologies – such IoT, digital twins, big data, and artificial intelligence (AI) – against the requirements of a climate-adaptive smart UTS.

In Chapter 19, Wu et al. discuss the current status, challenges, and prospects of the use of blockchain technology for the smart built environment. They identify seven application areas, including smart cities, and they outline the key technical and non-technical challenges to blockchain implementation. They state that the prospects for blockchains in the smart built environment are mainly in smart contracts, token-based incentives, and integration with other digital technologies.

Chan et al., in Chapter 20, discuss the effects of signage color on human psychophysiological states and wayfinding performance by focusing on urban underground spaces. They conducted an electroencephalography (EEG)-virtual reality (VR) experiment and found out that participants in the black-and-white wayfinding system demonstrated better wayfinding performance than their counterparts in the color-coded system under both low and medium noise levels.

1

DIGITAL TWINS FOR SMART DECISION-MAKING

Chady Elias[1], Isabelle Southern[2] and Raja R.A. Issa[3]

[1]PhD Student, M.E. Rinker, Sr. School of Construction Management, University of Florida, email: chadyelias@ufl.edu

[2]MS Student, M.E. Rinker, Sr. School of Construction Management, University of Florida, email: isouthern@ufl.edu

[3]UF Distinguished Professor, M.E. Rinker, Sr. School of Construction Management, University of Florida, email: raymond-issa@ufl.edu

1 Introduction

Digital twins (DTs) are quickly becoming synonymous with smart cities because they create the ability to interact virtually with all physical smart devices in the built environment at all levels of the smart city. DTs allow us to integrate in a physical environment data and information on what is happening in that environment thus converting that physical environment into a virtual one that can be used in real-time – or in aggregate – to facilitate the analyses of physical spaces.

In the architecture, engineering, construction, and operations (AECO) industry, the focus on DTs has increased due to the proliferation of digitalization and integration processes. For example, there has been a marked growth in the development of building information modeling (BIM), however, while BIM has instigated the digitalization and integration of design and construction information, its utility for smart decision-making in the post-construction stage is limited as the data and information captured in BIM outputs are static. Consequently, the need for technology that enables dynamic optimization of BIM data and operational data has grown. DTs have shown early promise in the AECO industry to elevate BIM from static to actionable and dynamic virtual models.

2 Background

The path from BIM to DTs contained within smart cities involves the integration of multiple data sources (see Figure 1.1).

2.1 Building information modeling

BIM as a digital technology continues to stimulate new workflows in the AECO industry. The capabilities of BIM to generate three-dimensional visualizations and data-rich models have resulted in its extension to facility lifecycle activities including facility management (FM), operations and maintenance (O&M), commissioning and close-out, energy management, space management, and all key aspects of FM (Becerik-Gerber et al. 2012). Major progress has been made in the efficient

DOI: 10.1201/9781003383840-2

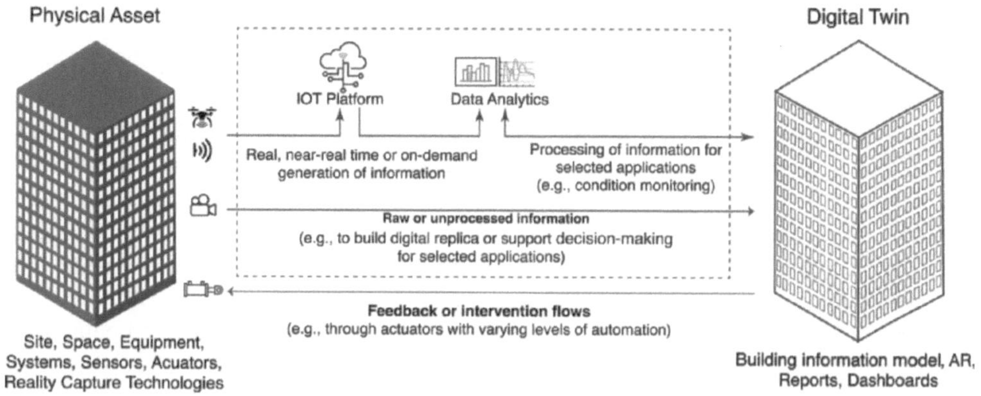

Figure 1.1 Essential components to create a building DT
Source: adapted from Khajavi et al. 2019.

Table 1.1 AIA E202 LODs (AIA 2022)

§	*Levels of Development (LOD)*
4.2	100: The Model Element may be graphically represented in the Model with a symbol or other generic representation but does not satisfy the requirements for LOD 200. Information related to the Model Element (e.g., cost per square foot, tonnage of HVAC, etc.) can be delivered from other Model Elements.
4.3	200: The Model Element is generically and graphically represented within the Model with approximate quantity, size, shape, location, and orientation.
4.4	300: The Model Element, as designed, is graphically represented within the Model such that its quantity, size, shape, location, and orientation can be measured.
4.4.1	350: The Model Element, as designed, is graphically represented within the Model such that its quantity, size, shape, location, orientation, and interfaces with adjacent or dependent Model Elements can be measured.
4.5	400: The Model Element is graphically represented within the Model with detail sufficient for fabrication, assembly, and installation.
4.6	500: The Model Element is a graphic representation of an existing or as-constructed condition developed through a combination of observation, field verification, or interpolation. The level of accuracy shall be noted or attached to the Model Element.

transfer of design and construction data to FM systems, e.g., computerized maintenance management systems (CMMS), by using open standards, e.g., construction operation building information exchange (COBie). Identification and specification of the required data during the facility design stage and requiring appropriate BIM deliverables are essential to developing models that are beneficial to facility managers. The data allows facility managers to analyze operational data while allowing owners a complete view of their assets (Asare et al. 2021).

Selecting the proper BIM level of development (LOD) is crucial in successfully developing a DT. The LODs for sharing building information models with project participants, as prescribed in the AIA E202 contract (AIA 2022), range from LOD 100 to LOD 500 (see Table 1.1). BIM, specifically the LOD 500 model, is the foundation of the Existence DT. The next step in the integration of BIM, FM, and O&M lies in the development of DTs.

2.2 Cyber-physical systems

Cyber-physical systems (CPS) are a set of embedded physical devices, objects, and equipment that interact with cyberspace through a communication network (Baheti and Gill 2011). These embedded systems and sensors are wirelessly connected with each other and the internet leading to the integration of the physical world and cyberspace in the form of CPS. Six common CPS characteristics are: hybrid physical and logical systems, hybrid analytical and measurement methods, control, component classes, time, and trustworthiness (Greer et al. 2019). Communication between physical and cyber elements is fundamental, since understanding the physical and the computational components as well as their interaction is essential.

CPS have a high degree of complexity and require a high level of networked communication integration between physical and cyber elements. The trend is for CPS to depend less on human decision-making and more on computational intelligence as they continue to evolve. They already have a major impact on health and medical equipment, telecommunications, transportation, user electronics, manufacturing, smart grids, and intelligent buildings. CPS are changing AECO's attitudes toward contemplating a more adaptable and flexible environment. A major challenge to deploying CPS has been the need to design systems that are reliable, cyber secure, and safe (NIST 2013).

2.3 Internet of things

The Internet of Things (IoT) presents us with opportunities for transforming work and everyday life. The IoT is at the intersection of the physical and digital worlds, with all kinds of devices harnessing the power of interconnectivity to provide seamless experiences for businesses and consumers alike. To reach its full potential, IoT has to shift from continuing to provide incremental value amid siloed clusters to unlocking its vast potential value as a fully interconnected IoT ecosystem. This will require an integrated IoT network within and across all industries. The main obstacle to be confronted is the cybersecurity risk which detrimentally impacts the trust needed to integrate IoT applications and networks. For smart cities, as with other applications, the expected solution lies in the merging of IoT and cybersecurity to form a new, integrated system (Greer et al. 2019).

2.4 Cyber-physical systems and Internet of Things convergence

Connecting the physical and logical worlds is a main characteristic attributed to both CPS and IoT in most recent definitions (Greer et al. 2019). This connectivity provides the basis for a unified components and interactions model applicable to both CPS and IoT. This unified model reflects the convergence of CPS and IoT definitions where IoT and CPS comprise interacting physical, logical, transducer, and human components designed for functioning through integrated logic and physics which allows for interchangeably labeling them (Greer et al. 2019).

The DT, which is the virtual representation of a real-world entity, i.e., the physical twin (PT), is a crucial facilitator for the advances promised by CPS in the interaction between the virtual and physical worlds. The key to the success of DTs is the establishment of practical reference architectures to be used by different firms and disciplines. This aspect includes the need to develop service-based and real-time enabled infrastructures for vertical and horizontal integration.

2.5 *Digital twins*

A DT is comprised of three principal parts: a physical system in real space, the PT; a virtual system in cyberspace, the DT; and the connection between real and cyber space for transferring data and information using CPS/IoT. A DT creates an accurate digital model of the physical system in cyberspace that can accurately replicate and simulate the behaviour of the PT. According to Tao et al. (2018), a DT can also provide a digital footprint of products by integrating geometry, structure, behaviour, rules, and functional properties. Salvador Palau et al. (2019) noted that DTs can also be considered as intelligent agents with prediction, communication, and data preprocessing capabilities.

Kritzinger et al. (2018) distinguished the various digital forms and categorized them as digital model, digital shadow, and DT based on the automated dataflow between them (see Figure 1.2). The digital model is a digital representation of an existing or planned physical object that does not use automated data exchange between the physical object and the digital object. Changes in the state of the physical object have no direct impact on the digital object and vice versa. The digital shadow is derived from the digital model and represents the one-way data flow between the state of an existing physical object and a digital object. A change in state of the physical object leads to a change of state in the digital object but not vice versa. The DT is characterized by automated bi-directional data flow between the physical and digital objects, which possess intelligence and decision-making capabilities that enable the automated feedback loop to the physical entity. A change in the state of the physical object directly results in a change in state of the digital object and vice versa.

The combination of the physical system – the PT – and its corresponding DT is the fundamental building block of fully connected and flexible systems that can learn and adapt to new demands. Some of the DT roles include remote monitoring, predictive analytics, simulating future behaviour, and optimization. To fulfil these roles, DTs have to rely on certain capabilities that exist across all of them. These required DT capabilities are summarized as follows (Redelinghuys et al. 2020):

- Acquire PT state: the DT must be able to acquire data from a variety of sensor types (e.g., temperature, pressure and vibration sensors, and counters or PLC registers) from the PT. The PT sensor data collected is refined and enriched (e.g., through combination and adding context) into information sets that describe the state of the PT.

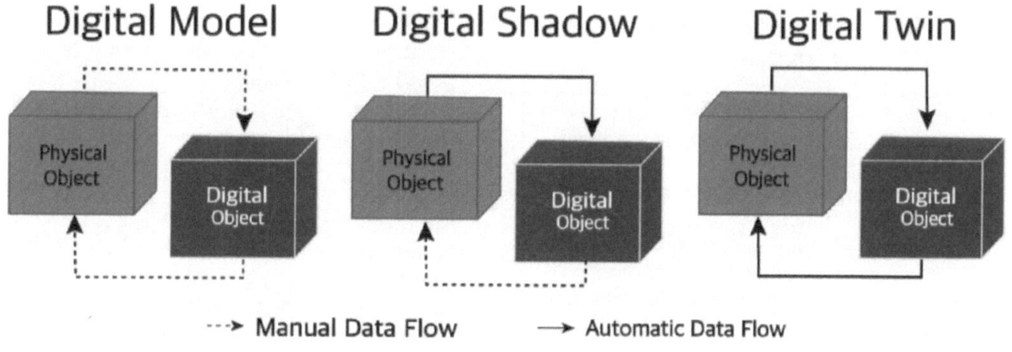

Figure 1.2 Evolution from digital forms to digital twins

Source: adapted from Kritzinger et al. 2018.

- Maintain information repository: the state information obtained from the PT sensors is stored for easy access through the internet. This repository typically relies on cloud-based storage, since large volumes of data may be stored for long periods of time.
- Simulate operation: the simulation of the PT's operation, i.e., predicting its future behavior from a given starting state and selected set of conditions, is required for some of the envisioned roles of the DT including the evaluation of new processes and different operation schedules.
- Emulate operation: using emulation to imitate and visually represent or reproduce the action or function of the PT in real time using feedback from embedded sensors.

CPS and DTs are similar in their description of cyber-physical integration, which is comprised of physical and cyber/digital parts. CPS and DTs share similarities but also have differences. CPS are more foundational as they do not directly reference implementation strategies or specific applications, whereas the DT is related to an engineering category (Tao et al. 2019).

Changes in the physical process will impact the digital world through the feedback of real-time embedded sensors and actuators. Using this data feedback, digital models can be used to interpret the behavior of machines or systems and predict future states from real-time and historical data, as well as experience and knowledge. The core elements of a DT are the models and data. CPS – and the technologies required for developing CPS – are considered as a necessary foundation for implementing DTs.

Some of the major benefits of adopting DTs in terms of optimization, holistic oversight, collaboration and transparency, and predictive analysis are shown in Figure 1.3 (SWAN 2022). DT technology provides visually rich data analysis across different user interfaces, enabling the streamlining of the FM process. Yan et al. (2021) developed a DT system for multidisciplinary stakeholder engagement in an office building complex to help improve work order responses to maintenance needs and requests. Their study found the multidisciplinary integration of the DT

Overall Operational Oversight	Collaboration and Transparency
• Single comprehensive view • Improved asset management • Reduced cost of operations • Reduced maintenance costs • Reduced unplanned outages • Anomaly detection • Analysis of multiple what-if scenarios • Reduced compliance issues	• Transparency across sectors and departments • Increased interdepartmental collaboration • Knowledge retention for new workforce • Proactive workforce • Understand historical engineering and operational decisions
Optimization	**Predictive Analytics**
• Multi-objective simulations for optimal cross-sector design • Experimentation in digital form to implementation • Improved decision making • Increased stakeholder collaboration • Reduced construction and maintenance conflicts • Continuous operational optimization	• Proactive operation instead of reactive, preventing future missed moments or failures • Near real-time holistic connection between the physical and the digital world • Ability to run what-if analysis at any time • Ability to adapt and respond to rapidly changing scenarios and disturbance

Figure 1.3 DT benefits (SWAN 2022)

improved the work order response rate by 38%, providing FM staff with the baseline knowledge and virtual interface to effectively understand, communicate, and solve maintenance needs. Similarly, Cheng et al. (2020) studied a data-driven framework for predictive maintenance at a higher education institution that utilized IoT sensor networks for mechanical, electrical, and plumbing systems management and found that the choice of prediction method and machine learning impacted the level of accuracy and processing time for monitored data.

Major challenges to adopting DTs include global connectivity, data integration and interoperability, data standardization, security and integrity, real-time performance and reliability, as well as barriers to its implementation and legacy system transformation (Attaran and Celik 2023). These challenges play a fundamental role in the development of DTs, as the connections between the PT and its corresponding DT typically rely on internet-enabled connectivity.

3 Classifications and levels of maturity

There is not one single definition for what a DT is or the capabilities it provides. There are numerous types of DTs and levels of functionality based on the needs of the organization or project and the maturity of the data available. The development of a DT is a continuum, with the model evolving with the addition of new data and capabilities; the following classifications were developed by KPMG (2022) to indicate the level of functionality of the DT based on the types and level of data and capabilities that the system provides (see Figure 1.4):

- Existence DT: furnishes principal project information, e.g., details on asset location and properties, enabling a single source of truth for asset data across the project. Traditional CAD and BIM systems are examples of an existence twin. This DT definition differs from that of the Kritzinger et al. (2018) classification of the evolution from digital forms to DTs
- Status DT: provides information on the status and condition of assets as detected by embedded IoT sensors. This can provide important insights into construction quality and progress, as well as asset health status over its lifecycle, and it enables prediction of future performance based on the data collected
- Operational DT: allows for a real-time view of the project and the operational asset. This view can provide critical insights into real-time performance and risks, both during construction and operations, and it enables more informed decision-making

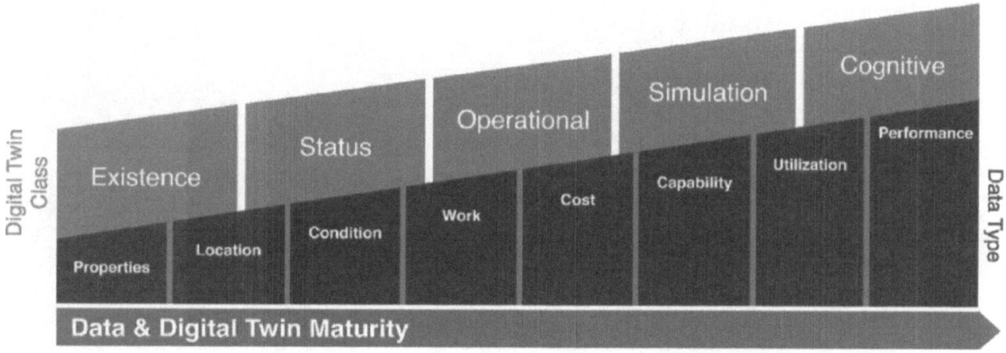

Figure 1.4 Data and DT maturity

Source: adapted from KPMG 2022.

- Simulation DT: enables teams to assess the impact of different design, construction, and operational decisions, allowing optimization of improvements in cost, performance, and risk. A simulation twin enables better and more thorough planning and can help minimize the risk of costly design and construction errors and faulty operational control changes.
- Cognitive DT: uses AI and real-time data collection to analyze data, make decisions, and optimize operational performance. This enables refinements made in real time based on live data to adapt to existing conditions. The ability to take a set of behavioral observations and convert that set into key performance indicators for FM decision-making is the functionality that distinguishes between cognitive and other DTs.

These classifications imply that an organization does not need to develop a highly advanced and complex model to see value from investing in DT technologies. For example, existence and status DTs can provide important project management insights on the physical configuration, properties, budget and cost, and as-built condition of a project and its component assets.

3.1 Gemini Principles

The Gemini Principles (see Figure 1.5) developed by the Centre for Digital Built Britain (CDBB), prescribe the values that should guide the development of DTs (Bolton et al. 2018). They emphasize that access to data ensures decisions are made based on the best available timely information. Using the Gemini Principles assures data integrity in the DT-driven decision-making process.

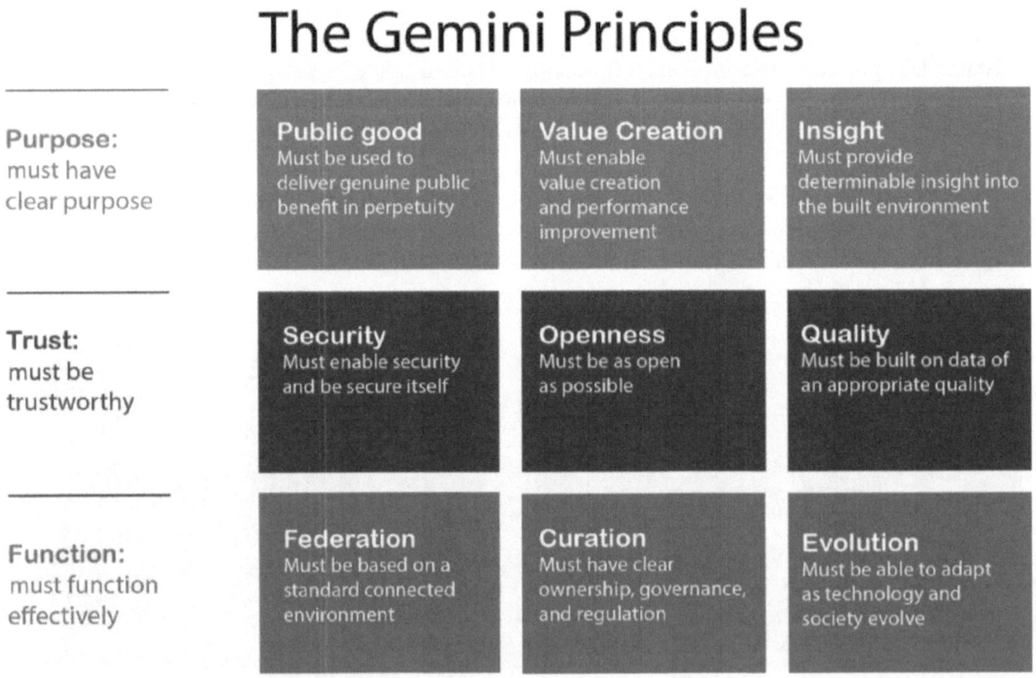

Figure 1.5 The Gemini Value Principles

Source: adapted from Bolton et al. 2018.

3.2 Differences between building information modeling and digital twins

Figure 1.6 shows the main differences and scopes of building information models and DT (Delgado and Oyedele 2021). The building information model focuses on replicating the physical asset throughout the asset's lifecycle, while the DT replicates and enables a connection with the physical asset during operations. In BIM, there is a total split between the digital and the physical asset, while for DTs this split is distorted due to asset instrumentation and data synchronization between the physical and the digital asset. BIM is used in the three main phases of the built asset's lifecycle, design, build, and operations (Brilakis et al. 2019), while the DT is focused primarily on operations. As-designed building information models are used for BIM Levels 1 and 2 and as-built models for BIM Level 3, with their corresponding LOD specifications (see Figure 1.6). These models have varying degrees of detail for their specific use cases, i.e., built asset design, design-construction coordination, optimal asset delivery, and FM. For DTs, there are no standard specifications for model detail or fidelity available. BIM has limited support for asset monitoring and control and for asset performance simulations during operations, while the DT does not consider discipline coordination for built asset delivery (Delgado and Oyedele 2021).

3.3 Digital twins and smart cities

As listed in Table 2, there is a different understanding of DTs within the context of smart cities. Tao et al. (2019) summarized and analyzed the existing understandings of DTs and proposed the ideal characteristics of DTs in the model, data, connection, service/function, and physical dimensions

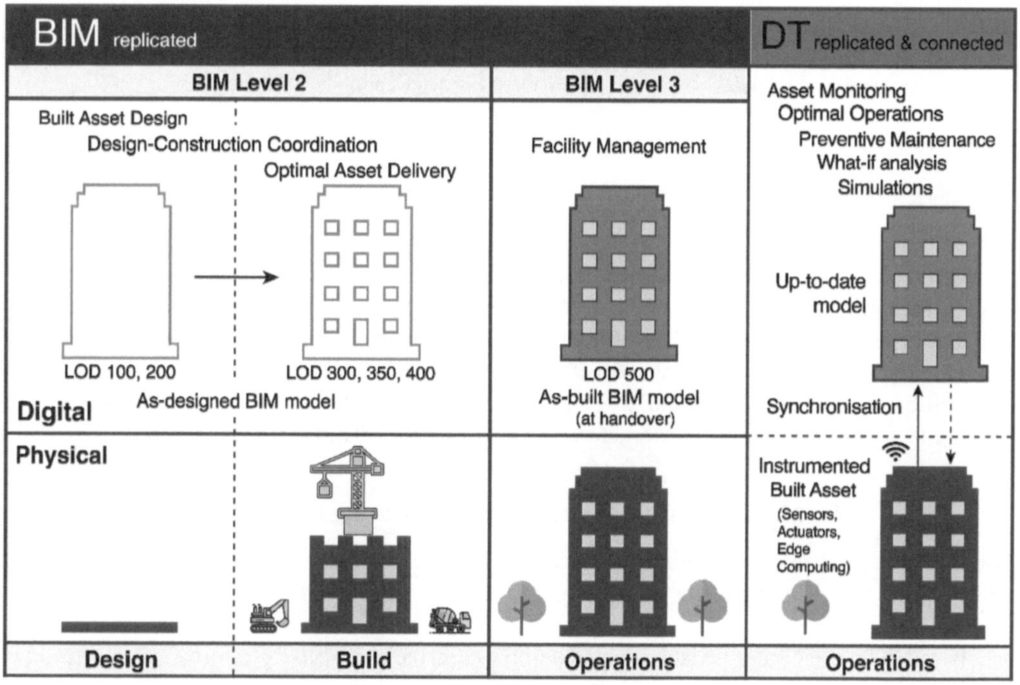

Figure 1.6 Differences between building information models and DTs

Source: adapted from Delgado and Oyedele 2021.

Table 1.2 Different understandings of DT in smart cities

Part of Understanding	*Ideal Characteristics*	*DT Cities*	*Dimensions*
Digital twins are 3D models.	Multi-dimensional/multi-space-time/multi-scale.		Model
Digital twins are copies of physical entities.	Dynamic/evolutive/interactive.		
Digital twins are virtual prototypes.	High fidelity/highly reliable/high precision.		
Digital twins are data/big data.	Total factor/all-service/complete flow scheme/full life circle.	Accurate mapping	Data
Digital twins are product lifecycle management (PLM).	Virtual-real fusion/multi-source fusion/heterogeneous integration.		
Digital twins are digital thread.	Real-time update/real-time interact/real-time respond.		
Digital twins are Digital shadow.			
Digital twins are physical union platforms.	Bi-directional connection/interaction/driving.	Virtual-real interaction	Connection
Digital twins are industrial Internet platforms.	Cross-agreement/interface/platform.		
Digital twins are simulation.	Model driven + data driven simulation verification/visualization/control/predict/optimize.	Software definition	Services/functions
Digital twins are virtual verification.			
Digital twins are visualization			
Digital twins are pure digital representation or virtual bodies.	Models vary from object to object/data vary by feature/services and functions vary according to needs.	Intelligent feedback	Physical
Digital twins are irrelevant to entities.			

Source: adapted from Deren et al. 2021.

(as shown in Table 1.2). They noted that, because DTs at different stages exhibit different characteristics, the understanding and practice of DTs cannot be separated from specific objects, specific applications, and specific needs. Therefore, the actual application does not necessarily require the established DT to have all the ideal characteristics in Table 1.2, and it is acceptable for it to meet the specific needs of users (Tao et al. 2019). As an extension of the DT concept in the urban field, the DT city also exhibits specific characteristics based on the ideal DT characteristics.

4 Methods for designing and implementing digital twins

Based on data and implementation characteristics, there are three methods for designing and implementing DTs (Adamenko et al. 2020b):

Data-based DTs: the design is data-focused and requires the use of IoT platforms. This method is best suited for data-driven applications and performance monitoring.

System-based DTs: the design is more focused on the actual physical object. If the system's characteristics are mapped using a consistent asset information model, the simulation models can retrieve the information needed in a standardized manner. This method is considerably more complex than the data-based structure, since exact information and specifications are required from the manufacturer to be able to generate a model that is as accurate as possible.

Hybrid method (building up method): This method combines the data-based and system-based methods. AI can be used with this method to analyze and visualize data on the current state or performance of the system.

4.1 Platforms for implementing digital twins

There are three categories of data platforms available for implementing DTs (Adamenko et al. 2020a):

- IoT platforms: they provide data connectivity between the real and virtual world. Typically equipped with resources that establish connections between networked devices and the applications that process and/or visualize the data. Examples include Azure DTs and IoT, Amazon Web Services IoT TwinMaker, and Siemens MindSphere. DT design with such tools is more data-based. They provide a user interface (UI) for data modeling
- Gaming engines: These platforms facilitate the development of executable video game-like applications. Their high-end visualization capabilities can be combined with IoT Platforms to achieve user-friendly DT applications. Examples include Unreal Engine and Unity 3D. They require extensive programming to model DT data
- Commercial modeling and simulation platforms: These tools typically support the design and implementation of system-based DTs. Examples include ANSYS, Autodesk Tandem, NVIDIA Omniverse, Microsoft Azure, and Unreal Engine. They each provide a user interface (UI) for data modeling, e.g., Digital Twin Definition Language (DTDL) based on JSON-LD for Azure models and the Universal Scene Description (USD) language for the Omniverse platform.

4.2 Cognitive digital twins

Cognitive digital twins (CDTs) allow data obtained from building sensors and actuators to be transformed into usable knowledge. The overall concept of CDTs is to maximize FM efficiency by connecting the physical world to the virtual world, utilizing multidimensional coordination and cognitive data analytics afforded by virtual intelligence. The process of developing CDTs from data collection to FM decision-making was found to require the following functionalities (Al Faruque et al. 2021; El Mokhtari et al. 2022):

- Perception: building components are monitored by IoT sensor networks and the building automation systems (BAS) that collect performance data from components like HVAC and other utility systems. Data is obtained and is continuously steamed to a virtual model that replicates the physical asset, forming usable representations of the data in the CDT (El Mokhtari et al. 2022)
- Attention: after receiving the ingested data, the CDT system filters out invalid or irrelevant data with the help of AI and machine learning algorithms that determine what data warrants attention (El Mokhtari et al. 2022). This functionality allows the CDT system to focus on certain levels of performance driven by externally defined signals, e.g., HVAC air flow rates below a threshold of cost efficiency and HVAC anomaly prediction conducted with a machine learning framework (Villa et al. 2022)

- Memory: memory encompasses the storage of historic and present behavioral data collected by the CDT system, serving as a reservoir of performance data that is tracked and stored for visualization in the model and learned by AI and machine learning algorithms for further predictive analysis (El Mokhtari et al. 2022). The process of memory allows information to be encoded, timestamped, stored, maintained, and retrieved (El Mokhtari et al. 2022) for FM decision-making when system anomalies or failures are detected by sensor networks
- Reasoning and problem solving: reasoning relies on data and detected anomalies stored in memory to extract solutions for resolving building issues. This functionality is what distinguishes CDTs from conventional DTs by taking a set of behavioral observations and translating them into key performance indicators that aid decision-making (El Mokhtari et al. 2022). For predictive maintenance, machine learning algorithms analyze data in real time, relate it to stored performance patterns, and identify early degradation/deviations from normal patterns through inductive or deductive reasoning (El Mokhtari et al. 2022)
- Learning: once behavioral data is collected, attention is allocated to specific performance indicators and stored in memory, and analysis of the data is conducted by way of virtual intelligence. The conclusions made from reasoning are recorded as learned experiences (El Mokhtari et al. 2022). Learning allows the system to critically analyze results from reasoning and predict future degradation ahead of time for lifetime extension and the minimization of inefficient performance (Cheng et al. 2020). As building components perform along their lifecycle, the building CDT detects potential anomalies based on learned experiences and predicted patterns, allowing FM staff to develop strategies ahead of time for maintenance plans and resource assembly. D'Amico et al. (2022) considered the CDT as an approach to improve FM and concluded that CDTs not only enable failure prediction and proactive maintenance but also enhance the interconnectedness between more reliable systems and more informed management

5 Conclusion

DTs provide a new outlook for smart decision-making in the AECO industry. From an operations and maintenance perspective, DTs can provide a dynamic view of facility status, enable operational control, support scenario planning and testing, and afford overall operational intelligence. To achieve high-performing DTs for smart decision-making, it is important to develop high-quality BIM outputs from the design and construction process and complement them with high-fidelity data. This requires a clear understanding of the different types of DTs, the tools for developing them, as well as knowing how and where to apply DTs. This chapter has presented a roadmap to transforming BIM to DTs. As research and industrial traction on DTs increases, it is important to address the issues of standardization of DT design and implementation. This includes testing implementation tools and methods towards identifying how best they can facilitate the creation of DTs that truly make decision-making in the AECO industry smarter. Future research on building DTs as components of smart cities should explore developing standardized data integration methods, advanced data and DT visualization, and integration of DTs from diverse domains.

References

Adamenko, D., Kunnen, S., and Nagarajah, A. (2020a). Comparative Analysis of Platforms for Designing a Digital Twin. In: Ivanov, V., Trojanowska, J., Pavlenko, I., Zajac, J., Peraković, D. (eds.), *Advances in Design, Simulation and Manufacturing III. DSMIE 2020. Lecture Notes in Mechanical Engineering*. Springer, Cham. https://doi.org/10.1007/978-3-030-50794-7_1.

Adamenko, D., Kunnen, S., Pluhnau, R., Loibl, A., and Nagarajah, A. (2020b). Review and Comparison of the Methods of Designing the Digital Twin. *Procedia CIRP*, 91, 27–32 (Elsevier B.V.). https://doi.org/10.1016/j.procir.2020.02.146.

Al Faruque, M. A., Muthirayan, D., Yu, S. Y., and Khargonekar, P. P. (2021). Cognitive Digital Twin for Manufacturing Systems. *Design, Automation & Test in Europe Conference & Exhibition (IEEE)*, 440–445.

American Institute of Architects (AIA). (2022). AIA Document E202™-2022 BIM Exhibit for Sharing Models with Project Participants. *AIA*. Retrieved June 1, 2023, from https://zdassets.aiacontracts.org/ctrzdweb02/zdpdfs/Preview_E202-2022.pdf?_gl=1*2c1he*_ga*NjgyNTkxMjE3LjE2NjQyNzY3Mzk.*_ga_E3VWY9JRK3*MTY2NDI3NjczOS4xLjEuMTY2NDI3NjgwNC42MC4wLjA.

Asare, K. A. B., Issa, R. R. A., Rui, L., and Anumba, C. (2021). BIM for Facilities Management: Potential Legal Issues and Opportunities. *Journal of Legal Affairs and Dispute Resolution in Engineering and Construction*, 13(4). https://doi.org/10.1061/(ASCE)LA.1943-4170.0000502.

Attaran, A., and Celik, B. G. (2023). Digital Twin: Benefits, Use Cases, Challenges, and Opportunities. *Decision Analytics Journal*, 6, 100165. https://doi.org/10.1016/j.dajour.2023.100165.

Baheti, R., and Gill, H. (2011). Cyber-Physical Systems. *The Impact of Control Technology*, 12(1), 161–166.

Becerik-Gerber, B., Jazizadeh, F., Li, N., and Calis, G. (2012). Application Areas and Data Requirements for BIM-Enabled Facilities Management. *Journal of Construction Engineering and Management*, 138(3), 431–442. https://doi.org/10.1061/(ASCE)CO.1943-7862.0000433.

Bolton, A., Butler, L., Dabson, I., Enzer, M., Evans, M., et al. (2018). *Gemini Principles*. Centre for Digital Built Britain (CDBB). https://doi.org/10.17863/CAM.32260.

Brilakis, I., Pan, Y., Borrmann, A., Mayer, H.-G., Rhein, F., Vos, C., Pettinato, E., and Wagner, S. (2019). Built Environment Digital Twining. *International Workshop on Built Environment Digital Twinning presented by TUM Institute for Advanced Study and Siemens AG*. https://doi.org/10.17863/CAM.65445.

Cheng, J. C. P., Chen, W., Chen, K., and Wang, Q. (2020). Data-Driven Predictive Maintenance Planning Framework for MEP Components Based on BIM and IoT Using Machine Learning Algorithms. *Automation in Construction*, 112, 103087. https://doi.org/10.1016/j.autcon.2020.103087.

D'Amico, R. D., Erkoyuncu, J. A., Addepalli, S., and Penver, S. (2022). Cognitive Digital Twin: An Approach to Improve the Maintenance Management. *CIRP Journal of Manufacturing Science and Technology*, 38, 613–630. https://doi.org/10.1016/j.cirpj.2022.06.004.

Delgado, J. M. D., and Oyedele, L. (2021). Digital Twins for the Built Environment: Learning from Conceptual and Process Models in Manufacturing. *Advanced Engineering Informatics*, 49, 101332. https://doi.org/10.1016/j.aei.2021.101332.

Deren, L., Wenbo, Y., and Zhenfeng, S. (2021). Smart City Based on Digital Twins. *Computational Urban Science*, 1, 4. https://doi.org/10.1007/s43762-021-00005-y.

El Mokhtari, K., Panushev, I., and McArthur, J. J. (2022). Development of a Cognitive Digital Twin for Building Management and Operations. *Frontiers in Built Environment*, 8, 856873. https://doi.org/10.3389/fbuil.2022.856873.

Greer, C., Burns, M., Wollman, D., and Griffor, E. (2019). *Cyber-Physical Systems and Internet of Things*. National Institute of Standards and Technology (NIST) Special Publication 1900-202. https://doi.org/10.6028/NIST.SP1900-202.

Khajavi, S. H., Motlagh, N. H., Jaribion, A., Werner, L. C., and Holmstrom, J. (2019). Digital Twin: Vision, Benefits, Boundaries, and Creation for Buildings. *IEEE Access*, 7, 147406–147419. https://doi.org/10.1109/ACCESS.2019.2946515.

KPMG. (2022). Insight Report: Innovation and R&D in Construction. *KPMG*. Retrieved June 1, 2023, from https://www.cca-acc.com/wp-content/uploads/2022/06/Insight-report-June-2022.pdf.

Kritzinger, W., Karner, M., Traar, G., Henjes, J., and Sihn, W. (2018). Digital Twin in Manufacturing: A Categorical Literature Review and Classification. *IFAC-PapersOnLine*, 51(11), 1016–1022. https://doi.org/10.1016/j.ifacol.2018.08.474.

National Institute of Standards and Technology (NIST). (2013). Foundations for Innovation in Cyber-Physical Systems. *National Institute of Standards and Technology: Workshop Report*. Retrieved June 1, 2023, from https://www.nist.gov/document/cps-workshopreport-1-30-13-finalpdf.

Redelinghuys, A. J. H., Basson, A. H., and Kruger, K. (2020). A Six-Layer Architecture for the Digital Twin: A Manufacturing Case Study Implementation. *Journal of Intelligent Manufacturing*, 31, 1383–1402. https://doi.org/10.1007/s10845-019-01516-6.

Salvador Palau, A., Dhada, M. H., and Parlikad, A. K. (2019). Multiagent System Architectures for Collaborative Prognostics. *Journal of Intelligent Manufacturing*, 30(8), 2999–3013. https://doi.org/10.1007/s10845-019-01478-9.

SWAN. (2022). *Digital Twin Readiness Guide*. Retrieved June 1, 2023, from https://swan-forum.com/wp-content/uploads/2022/05/SWAN-Digital-Twin-White-Paper_Final.pdf.

Tao, F., Cheng, J., Qi, Q., Zhang, M., Zhang, H., and Sui, F. (2018). Digital Twin-Driven Product Design, Manufacturing and Service with Big Data. *International Journal of Advanced Manufacturing Technology*, 94(9–12), 3563–3576. https://doi.org/10.1007/s00170-017-0233-1.

Tao, F., Qi, Q., Wang, L., and Nee, A. Y. C. (2019). Digital Twins and Cyber – Physical Systems Toward Smart Manufacturing and Industry 4.0: Correlation and Comparison. *Engineering*, 5(4), 653–661. https://doi.org/10.1016/j.eng.2019.01.014.

Villa, V., Bruno, G., Aliev, K., Piantanida, P., Corneli, A., and Antonelli, D. (2022). Machine Learning Framework for the Sustainable Maintenance of Building Facilities. *Sustainability*, 14(2), 681. https://doi.org/10.3390/su14020681.

Yan, J., Yin, K., Lu, Q., and Shen, X. (2022). Developing a Web-Based BIM Asset and Facility Management System of Building Digital Twins. In: Issa, R. (ed.), *Proceedings Computing in Civil Engineering*, 2021, 490–497. https://doi.org/10.1061/9780784483893.061.

2

FROM PARTICLES TO SPACES

The intelligent planning unit (IPU) theory to simplify the built environment

Suyash Padhye[1] *and Makarand Hastak*[2]

[1]PhD Student, Lyles School of Civil Engineering, Purdue University, 550 Stadium Mall Drive West Lafayette, IN 47907-2051, USA. Email: spadhye@purdue.edu

[2]Dernlan Family Head of Construction Engineering and Management and Professor of Civil Engineering, Lyles School of Civil Engineering, Purdue University, 550 Stadium Mall Drive West Lafayette, IN 47907-2051, USA. Email: hastak@purdue.edu

2.1 Introduction

Our built environment is a system of systems with complex layers of infrastructure, multiple processes, and diverse stakeholders. With its dynamic nature and complexity, establishing a streamlined flow (e.g., optimizing construction processes, the flow of information, minimizing waste and emissions, etc.) throughout the entire system becomes challenging (Kirchherr et al., 2017; Cui et al., 2010; Moffatt and Kohler, 2008). Most construction projects, based on the current state of the art, are non-replicable, i.e., unique designs, specifications, and other lifecycle operations. Furthermore, resource optimization and standardization pose a challenge, such as planning for time, cost, labour, supply chain, quality, and equipment (Jeon et al., 2022; Cho and Hastak, 2013). With such dynamics, it is difficult to target continuous improvement in products and processes of the built environment.

To address these concerns, researchers have proposed building information modelling (BIM) as a tool for promoting replicability (Cao et al., 2015). Industry Foundation Classes (IFC codes) are developed as a standardized and digital description of the built assets. Also, innovative project delivery methods such as advanced work packaging (AWP), front-end planning, and modularization are proposed to support substantial project performance improvements and to involve the stakeholders right from the planning stage of a construction project (CII, 2016, 2009). These advances are expected to benefit users to collaborate and break down silos of information throughout the lifecycle of various assets.

Despite such research efforts directed to improve project objectives such as time, cost, quality, and safety, client expectations are evolving to fulfil additional targets such as remote monitoring capabilities, net-zero emissions, lean principles, and circular economy models (CII, 2023; Jeon et al., 2022). Therefore, to accommodate such dynamic expectations over a project lifecycle, there is a significant need to reduce the complexity of the built environment system. Furthermore, it is also critical to centralize the valuable data that could be collected in this system of systems. Such information could be analysed to support data-driven decisions like net-zero goal planning, detour planning (Padhye et al., 2021; Morshedi et al., 2023), disaster preparedness, and energy planning and utility management (Bhattacharyya and Hastak, 2023).

DOI: 10.1201/9781003383840-3

2.2 Research background

Past researchers have conducted studies to address process and network improvement-related aspects to break down the built environment into manageable units. According to Niu et al. (2016), the concept of smart construction objects is introduced as the foundational components for future construction. This approach aims to tackle the difficulties posed by large and complex construction projects, such as their unique and non-replicable nature, as well as the presence of diverse workflows based on specific project characteristics.

Albreshne et al. (2014) proposed an ontology-based framework intending to improve services within residential environments. By integrating smart objects into complex scenarios and effectively controlling various home devices, this framework enables enhanced energy savings and improved user comfort. Material passports illustrated by Benachio et al. (2020) and Sauter et al. (2019) focus on the storage of important data generated from the building components throughout their lifecycle. Such innovative planning strategies can be useful to achieve targets associated with net-zero carbon cities and circular economies. Additionally, such smart construction objects can be integrated with blockchain technologies to enable transparency and improve credibility across the built environment (Lu et al., 2021).

There is an emerging need for connected infrastructure or a connected built environment. Shi (2020) proposed methods to facilitate vehicle-infrastructure integration from the perspective of managing traffic conditions on risk-prone highways in cold climate regions. They combined smart sensing and mobility applications to propose alternative routes to road users in real time based on the highway conditions. The next-generation highways can be designed in alignment with such objectives. Similarly, Adegoke et al. (2019) developed applications for positioning autonomous vehicles by implementing the concept of connected infrastructure. Other studies like Min and Bjornsson (2008), Grau et al. (2009), and Akhavian and Behzadan (2012) have developed solutions for real-time inventory management, weekly production, site material tracking, remote monitoring, and remote planning for construction operations.

Advanced methodologies like data-driven discrete event simulation, agent-based modelling, and artificial intelligence models are used to solve challenges due to complexity in the construction sector as well as in other industries. However, no study has ever addressed different levels of physical entities and strategies to standardize them while optimizing for relevant key performance indicators (KPIs) to reduce the complexity in our built environment system. In this chapter, we present the theory of intelligent planning units (IPUs) along with two case scenarios to present proposed applications. The chapter is organized in the following way: Section 3 outlines the concept, characteristics, definition, and benefits of IPUs; Section 4 describes proposed application cases for (a) automated structural health monitoring of bridges, (b) optimized electricity demand management across residential, commercial, industrial, and transportation sectors, and (c) designing a sports centre; and in Section 5 we present the challenges and future research avenues.

2.3 Concept of the proposed intelligent planning unit (IPU)

Hastak and Koo (2017) proposed an innovative theory, the intelligent planning unit (IPU), to standardize complex entities at a modular scale in the built environment. The built environment is a complex system of interrelated infrastructure systems in facilities, in design, development, and operation, similar to the intricacy of a human biological system.

The built environment encompasses scalable units that span from the nanoscale to vast cities similar to how multiple cells within a human body function together to form a digestive system or a respiratory system. The built environment entails factors like energy, temperature, humidity, illuminance, and noise that can be characterized at the nanoscale level by utilizing sensors. Specialized bricks can be designed by optimizing such nanoscale properties along with the project objectives (e.g., safety, cost, schedule). Furthermore, such specialized bricks that are optimized for the project objectives as well as embedded with sensing capabilities, when strategically placed, can give rise to intelligent rooms. These intelligent rooms can scale up to intelligent buildings or campuses with specific purposes (e.g., net-zero target, user comfort, maintenance planning). Expanding further, these intelligent structures can scale up into intelligent cities (e.g., circular economy, self-sufficient energy). This scalability demonstrates how multiple nanoscale IPUs can come together to form composite IPUs that are capable of achieving higher objectives.

In terms of planning, IPUs are meticulously designed and optimized to enhance project safety and productivity. For example, in masonry work planning, considerations include crew size, safety equipment, optimal working spaces, and scaffold positioning. These IPUs can be replicated throughout construction sites, continuously monitored, and refined for improved decision-making. The next section outlines the definition, characteristics, benefits, and process of implementing IPUs in the real world.

2.3.1 Definition

An intelligent planning unit (IPU) is a well-defined planning unit at a predetermined scale with specific purposes and functions that are optimized with respect to the desired KPIs including cost, schedule, quality, safety, and sustainability. IPUs are replicable, scalable, and self-refining while integrating knowledge in real time.

In the context of Figure 2.1, the theory of IPUs sheds light on the influence of nanoscale units, enabling the utilization of their unique properties to design and develop an intelligent built

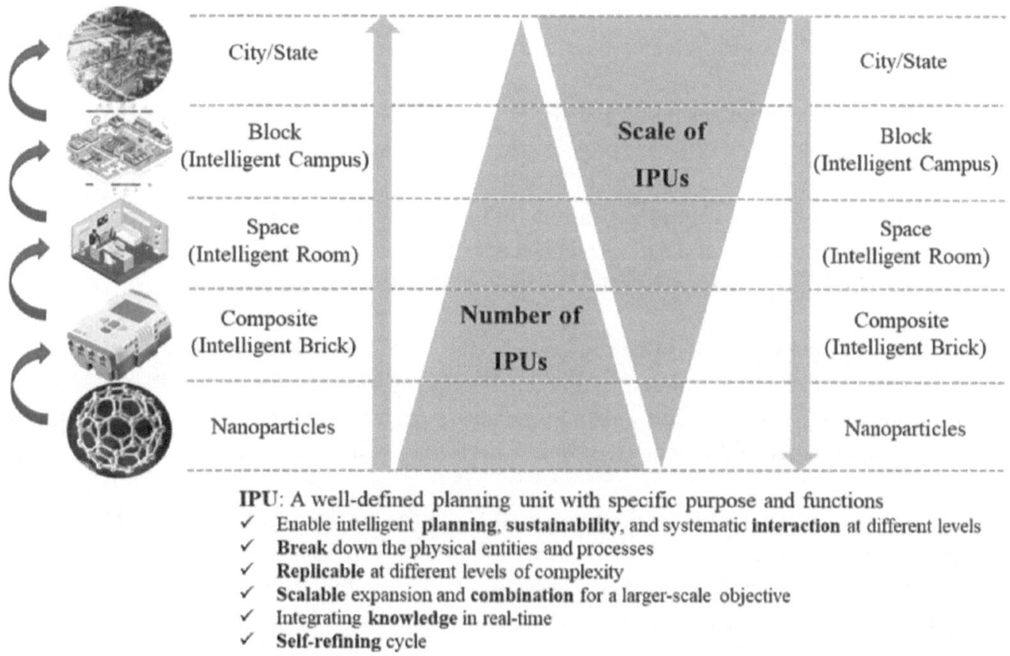

Figure 2.1 Scale and number of IPU entities in the complex built environment

environment. Figure 2.1 illustrates how the number of IPUs decreases as the IPUs scale up across the built environment (e.g., from bricks to a city). The IPUs spanning each scale are combined to design the succeeding-tier IPU (i.e., multiple IPU bricks to an IPU room and multiple IPU rooms to an IPU building). These properties provide a unique opportunity to meticulously plan the built environment ranging from a nanoscale level to the city level. The theory of IPU thus helps to standardize the complex physical entities and processes at a modular scale, to accumulate knowledge at a central level, to enhance infrastructure performance, and finally for robust data-driven decision-making.

To summarize, IPUs can be useful in bringing together the principles of process improvement and network improvement along with smart construction technologies such as smart construction objects, digital twins, and modular construction.

Main characteristics of an IPU

- Well-defined planning unit for specific purposes, including economic, environmental, and social goals
- Breaks down the built environment into measurable, optimized units, including nanoscale
- Replicable at various complexity levels, with composite IPUs for higher objectives
- Scales from nanoscale particles to large structures like buildings and cities
- Uses Internet of Things (IoT) for real-time integration and predictive analytics in a cloud-based knowledge centre
- Enhances performance with an integrated system and self-organized refinement via IoT.

Potential advantages

- Strategic approach for cost effectiveness, resource management, energy efficiency, and human health
- Granular approach optimizes performance at various levels
- Scalable and adaptable to different project sizes, integrating multiple IPUs for comprehensive results
- Covers the full spectrum of physical entities, aiding intelligent planning and operations
- Enhances communication and data-driven decisions through IoT and stakeholder collaboration
- Optimizes performance with real-time data and intelligent communication.

IPUs are designed to break down the complexity of the built environment into replicable and scalable units that have interactional and self-refining capabilities. Such IPUs, when replicated throughout the project, can enhance the entire project lifecycle with respect to appropriate key performance indicators (KPIs) and help the project stakeholders make informed decisions.

To facilitate the design and development of IPUs in the real world, Hastak and Koo (2017) have proposed a comprehensive three-fold implementation process that consists of: (a) IPU Planning, (b) IPU Application, and (c) IPU Network. The following section provides detailed information about the steps involved in each of these phases and lists their associated critical activities.

IPU planning

- IPU design:
    - Define the aim, objectives, and functions of an IPU (WHY)
    - Specify associated properties and specifications (WHAT)
    - Identify appropriate methodologies and technologies (HOW)
- IPU strategy:
    - Determine the required physical scale/level of the IPU (WHERE)
    - Identify the IPU life cycle (WHEN)
    - Identify stakeholders and decision-makers (WHO).

IPU application

- IPU replication:
    - Optimize and replicate a well-planned IPU at similar levels
    - Develop composite IPUs by integrating several types for larger objectives
- IPU physical entities:
    - Replicate IPUs at different levels, from nano- to large-scale (brick, building, city)
    - Design IPUs to interact with each other across various levels.

IPU network

- IPU interaction:
    - Store IPU operational data in cloud-based repositories
    - Use predictive and prescriptive analytics for better decision-making
- IPU refinement:
    - Refine IPUs based on deviations between as-planned and as-built for future applications
    - Explore self-organized refinement mechanisms for IPUs

The next section will expand on each of these phases with detailed steps and examples.

2.3.2 Phase 1: IPU planning

The planning phase of an IPU involves two key aspects: IPU design and IPU strategy. The following considerations need to be addressed: (a) WHY focuses on the objective and functions; (b) WHAT involves the properties and specifications; (c) HOW entails the methodologies and technologies; (d) WHERE pertains to the physical entities; (e) WHEN relates to the life cycle processes; and (f) WHO encompasses the stakeholders who act as decision-makers.

2.3.2.1 IPU design

To design an intelligent planning unit (IPU) effectively, three key aspects must be clearly defined: WHY, WHAT, and HOW. The planning and design of an IPU can take place in the virtual or digital world, with a well-defined objective, identified properties, associated specifications, and suitable methodologies and technologies. All these characteristics are optimized with respect to the KPIs such as cost, schedule, safety, and sustainability. Furthermore, IPUs can be designed for a process or a product/element with an essential property that will be replicated 'n' times in that particular project. For example, an IPU wall designed to optimize the indoor environment can be replicated in all the rooms of that building project. Nanotechnologies with unique characteristics, such as self-cleaning abilities, lighter and stronger products, and improved physical quality, can be utilized to achieve specific design needs. Computer-aided methodologies like data mining, geographic information systems, building information modelling, optimization, agent-based modelling, and stochastic analysis can also contribute to IPU design. An IPU can take the form of a brick containing relevant nanosensors, made of synthetic material. It can also be defined as a room that incorporates these special bricks and other IPUs, forming a composite IPU.

For example, suppose a client wants to build a K–12 school in the suburbs close to an airport. The IPU theory can be applied to design this school, with one objective of maintaining indoor environmental quality in the classrooms and addressing noise and vibration issues from the external environment. Nanosensors, embedded in a brick that is strategically placed throughout the room, can continuously monitor and control factors like noise, temperature, humidity, and illuminance according to the classroom's specific requirements. Such an optimized room forms a composite IPU, i.e., an integration of multiple component IPUs (e.g., noise dampers, sensor bricks, HVAC systems, layout design for purpose, etc.). Depending on the requirement of classrooms in the project, this composite IPU can be replicated (with minor adjustments for orientation, size, etc.) and implemented in the project.

2.3.2.2 IPU strategy

To establish a successful IPU strategy, three additional aspects must be clearly defined: WHERE, WHEN, and WHO. First, in terms of WHERE, it is important to determine the level of physical entities in the complex built environment system where the IPU will be implemented (e.g., nanoparticles, components, elements, buildings, bridges, roads, transportation systems, cities, or even countries). The specific level at which the IPU will operate should align with the defined objectives. Second, in terms of WHEN, the IPU can be implemented at any stage of the project life cycle, whether it be during planning, design, construction, or operation and maintenance. The goal is to maximize services to decision-makers throughout the entire life cycle process. Defining the life cycle stage is essential to the IPU's implementation. Third, in terms of WHO, the IPU should be planned and designed with the end user in mind. The IPUs should possess the capability to interact with one another and effectively integrate with external environments at any time and in any location.

The requirements associated with WHY-WHAT-HOW and WHERE-WHEN-WHO of IPUs are explained with an example of designing a classroom as presented in Figure 2.2. IPU #A can be defined to optimize and monitor the desired indoor environment inside the classroom. A well-defined IPU #A can be replicated throughout the planning phase of the project. Similarly, IPU #B can be designed to manage the external environment.

The project manager can act as a stakeholder until the end of the construction phase and then hand over the control system to other stakeholders. Similarly, this classroom can be further developed with additional IPUs (IPUs #C, #D, #E, #F) essential to optimize its lifecycle. For example, an IPU #C smart meter can be installed during the operation and maintenance phase to optimize the energy demand of the classroom. After this classroom is optimized with all the KPIs, it can be replicated across the building as explained in the IPU application phase.

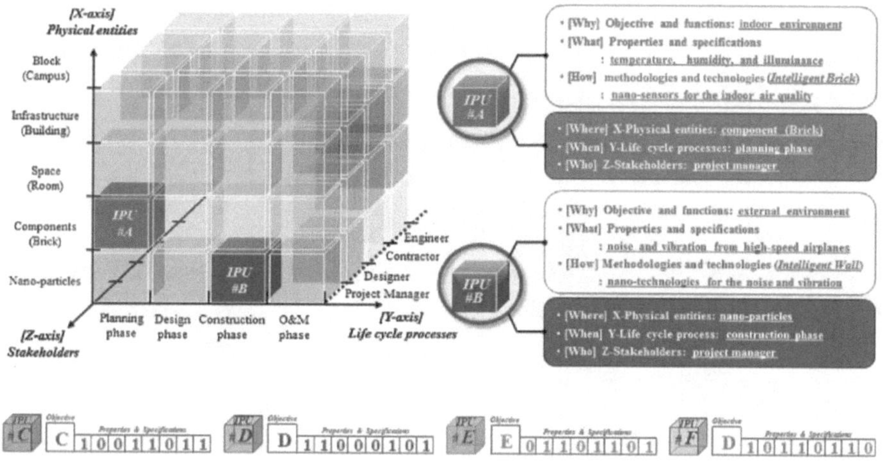

Figure 2.2 IPU planning phase

Source: adopted from Hastak and Koo, 2017.

2.3.3 *Phase 2: IPU application*

2.3.3.1 *IPU replication*

The concept of an IPU introduces a novel and innovative perspective to the complex built environment system by viewing it as an object-oriented planning unit. This perspective involves the process of abstraction, allowing us to extract essential characteristics and create a new framework for planning and design. Each IPU possesses a unique combination of binary scales, signifying its independent nature.

Figure 2.3 displays four distinct types of IPUs (IPUs #C, #D, #E, and #F) with varying objectives using binary scales (for instance, IPU #D could be represented as 10110110 in binary form). These binary scales reflect the exact properties, specifications, and technology associated with that specific IPU. Consequently, IPUs can be replicated for real-world applications to deliver improved performance. Moreover, composite IPUs can be developed by integrating multiple nano IPUs defined at lower scales, contributing to higher-level objectives within the intricate built environment system. As illustrated in Figure 2.3, a composite IPU #1 can be a combination of IPUs #C, #D, #E, and #F. Similarly, composite IPUs #2, #3, #4, and so on can be defined based on the project characteristics. The implementation of an IPU in the real world necessitates defining its frequency and location within a space, be it a room or a broader area like a collection of rooms, facilities, or even an entire campus. Based on these characteristics, a range of IPUs or composite IPUs can be replicated and effectively deployed at various levels of physical entities in real-world scenarios.

2.3.3.2 *IPU physical entities*

Just as a nanoparticle is an incredibly tiny particle, a brick can be considered a small element of a building. Building upon this idea, a building can be viewed as a small component of a city. By implementing science, engineering, and technology at the brick level or higher, we can achieve the objectives defined for an IPU and its associated properties. With these characteristics in mind, the theory of an IPU can be applied to plan, build, operate, and connect different types of infrastructure. The third step, the IPU network, introduces the concept of a connected built environment.

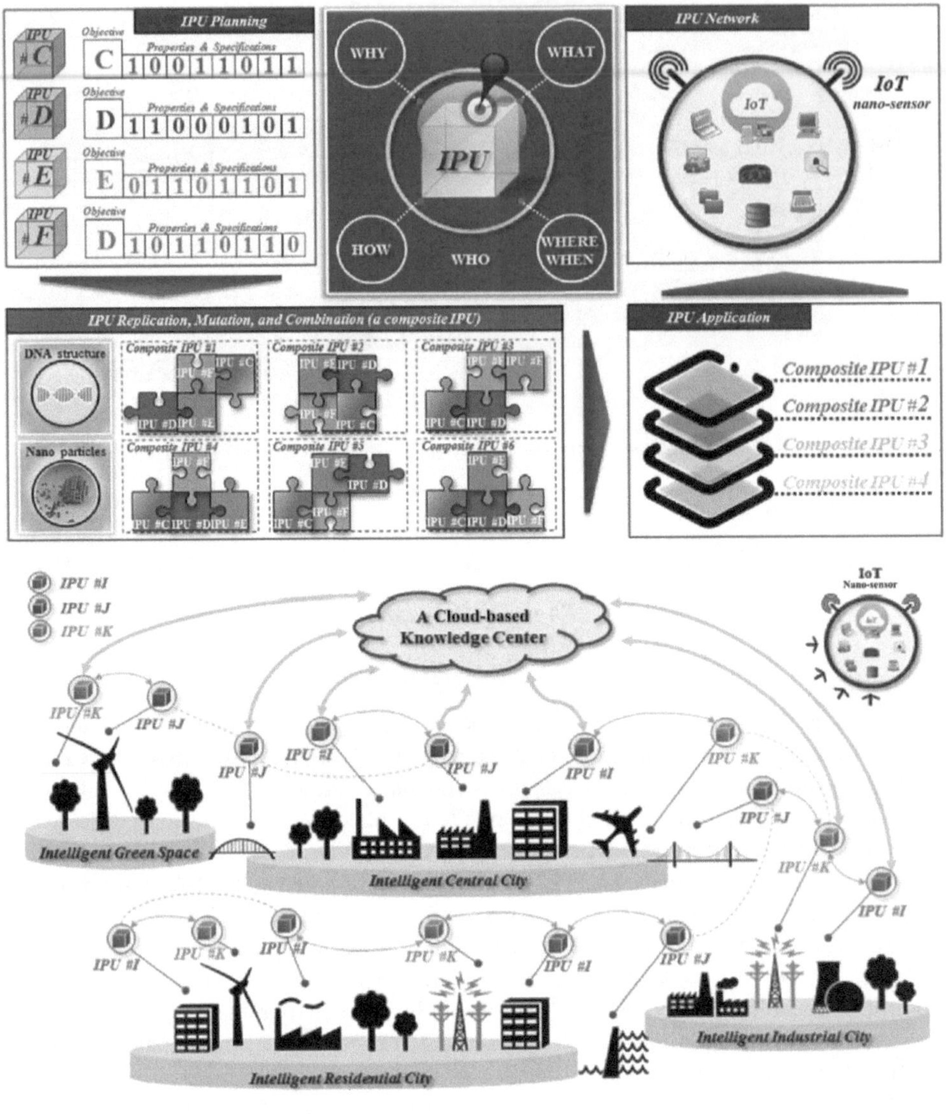

Figure 2.3 Composite IPUs and IPU network

Source: adopted from Hastak and Koo, 2017.

2.3.4 Phase 3: IPU network

2.3.4.1 IPU interaction

Real-time integration of all IPU information can be achieved through a cloud-based knowledge centre. The Internet of Things (IoT) technology is particularly valuable in this context, supporting seamless communication and data exchange between IPUs. Figure 2.3 demonstrates the authors' concept of an IPU City, wherein various physical entities in the built environment are designed for their objectives but also can interact with their external environment to achieve greater objectives.

For example, IPU #J can be designed to monitor the structural health of a bridge. It can be replicated throughout all the bridges in the city with a higher-level objective such as creating traffic management plans, disaster preparedness plans, or planning for rehabilitation. IPU #Js can interact with other IPUs #K and #I in real time with IoT-based technologies. Furthermore, IPU #I can be designed to optimize user comfort (illuminance, temperature, and humidity) and can be replicated to form intelligent residential space, intelligent commercial and industrial space, and intelligent parks. All these IPUs can also interact with the external environment throughout their lifecycle processes and are controlled by designated stakeholders.

The interaction among the IPUs operates in three main parts: (a) the input part, where information is received from the external environment; (b) the internal part, responsible for collecting, classifying, storing, evaluating, and disseminating information based on the needs of the decision-maker; and (c) the output part, which sends information back to the external environments at the appropriate scale of physical entities. For example, flood protection measures designed for an IPU industrial space may be triggered based on its interaction with IPU #J which monitors the structural health of the bridges connected to that space.

2.3.4.2 IPU refinement

Various factors can lead to differences between the planned and actual performance of an IPU, such as incorrect application, lack of expertise, user behaviour, environmental variations, marketing strategies, administrative boundaries, and more. To bridge this gap and ensure that an IPU delivers its intended or improved performance, it is crucial to establish a feedback process and a refinement mechanism within the IPU. Such a mechanism will help the stakeholders to refine their IPUs for future projects. Also, depending on how the IPU is installed in the real world, the existing IPUs can either be replaced with the newly refined IPUs or can be retrofitted to cater to the current needs.

2.4 IPU case scenario

This section presents scenarios to illustrate the planning, application, and network phases of IPU implementation. IPUs can be defined for design, process, or network improvement.

2.4.1 Case I. Structural health monitoring (SHM) of bridges

Suppose a DOT is planning to construct a new suspension bridge over a river. This bridge will complement the existing bridge to support the increasing traffic demand in the city. The DOT expects this bridge to self-report its structural health status (Cardno, 2020). This will help the DOT transition from a traditional maintenance-as-scheduled approach to an innovative maintenance-on-demand functionality. This bridge can be designed and managed using the IPU theory. A suspension bridge consists of structural components like the deck, girders, towers, cables, anchorages, and piers. Each of these structural components can be designed as IPUs with the end goal of providing automated structural health monitoring functionality (e.g., IPU Bridge Deck).

2.4.1.1 IPU planning (design and strategy)

As outlined in Section 3, the IPU planning phase comprises design (WHY-HOW-WHAT) and strategy (WHERE-WHEN-WHO). Figure 2.4 provides an overview of how the decision-makers can proceed to plan IPUs for the automated structural health monitoring (SHM) of bridges. IPUs

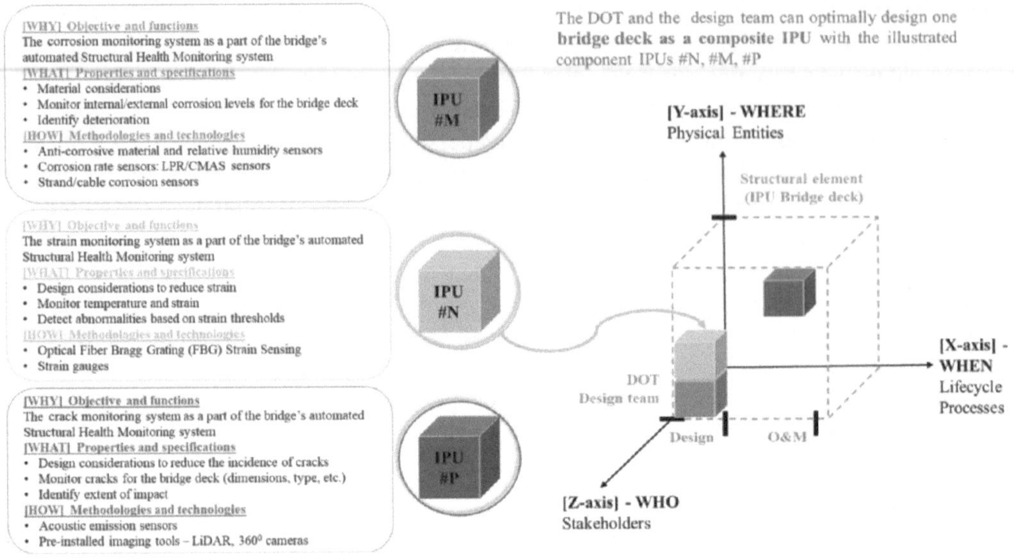

Figure 2.4 IPU planning for bridge SHM

#M, #N, and #P are defined to minimize and monitor corrosion, strain, and cracks in various structural elements of a bridge.

In the **IPU design** stage, at first, the decision-makers need to clearly define the objective and functions (WHY) of the IPU. This would provide clarity in terms of the reason to implement the IPU theory in a real-world scenario. In this example, the need to implement an automated structural health monitoring (SHM) system is to design, construct, and maintain bridges safely and sustainably throughout the lifecycle of the bridge. This will help the respective State DOTs shift from traditional maintenance-as-scheduled to innovative maintenance-on-demand activities to support their bridge infrastructure. Second, the specifications and properties (WHAT) that should be incorporated or embedded, measured, and managed in the IPU for achieving the WHY must be identified. In the case of the automated SHM system, the inclusion of resilient material and the ability to record and transmit structural data are the two most critical requirements. Case studies can be performed to further refine the exact material and technological requirements for various structural components of a bridge (e.g., deck, girder, pier, cable). Further, the advanced and emerging methodologies (HOW) should be surveyed to realize the required specifications and properties. In this case, anti-corrosive reinforcements, regenerative material, fibre optic cable sensors, acoustic sensors, etc. are some of the advancements that could be considered to develop the IPUs. Figure 2.4 illustrates examples of corrosion, strain, and crack monitoring IPUs based on the methods and technologies used by Song and Saraswathy (2007), Xu et al. (2016), and Enckell et al. (2011). These researchers have used a combination of sensors and materials to autonomously monitor the structural health of bridges. The actual implementation of these sensors could vary based on the characteristics (e.g., type, span, location) of the bridge.

In the **IPU strategy** phase, from the perspective of WHERE, the decision-makers need to establish the hierarchical levels or the physical entity levels to implement an IPU. When these IPUs are strategically placed within each structural component of the bridge, every bridge across the cities will become an IPU capable of optimizing its structural health (i.e., various nanoscale IPUs scale up to form composite IPUs). To address the WHEN step, these IPUs can be implemented right from the planning stage for new projects to maximize their potential benefits. This entire IPU framework should be strategized from the perspective of the stakeholders (WHO, viz., DOT, design team, contractor, maintenance service providers, and academic researchers).

2.4.1.2 IPU application (replication and physical entities)

The second phase, the IPU application phase, involves IPU replication (mutation and combination) and IPU physical entities (scaling up to achieve higher-level objectives). Well-defined IPUs can be replicated with their existing or modified properties to achieve their objectives. Figure 2.5 demonstrates various types of component IPUs that can be developed to monitor structural health based on factors like load, displacement, corrosion, cracks, strain, and so on. As illustrated in Figure 2.5, IPUs can be replicated as-is or with subset properties (e.g., IPUs #L1–#L4, IPUs #M1–#M4, IPUs #N1–#Q2).

In this case, once the DOT and designers have defined one IPU (e.g., IPU Bridge Deck), the same IPU can be replicated throughout all upcoming bridge projects. Depending on the specific requirement of the structural element, different types of nanoscale IPUs can be replicated and integrated to form composite IPUs (e.g., Composite IPU Bridge Deck can be designed with a combination of IPUs #L4, #M3, #N1, #P1, and #Q1). This composite IPU Bridge Deck can then be replicated based on the specific number of panels required for the span of the suspension bridge. This IPU will be capable of transmitting its SHM-related data to a central cloud-based knowledge centre. This data will be processed in real-time to schedule appropriate maintenance activities. This enables the stakeholders to transition to an innovative maintenance-on-demand strategy over the traditional maintenance-as-scheduled approach. When it scales up, such composite IPUs can assist in achieving higher-level objectives (e.g., disaster response strategies, carbon emission tracking, etc.).

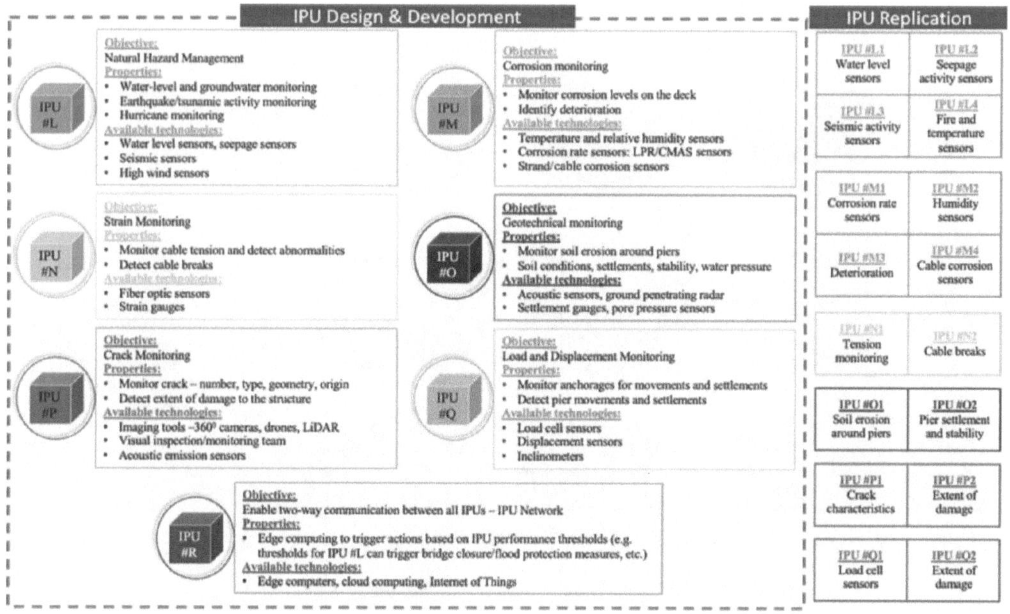

Figure 2.5 IPU application for bridge SHM

2.4.1.3 IPU network (interaction and refinement)

The last phase, the IPU network, enables the interaction and refinement of principles. With the use of smart features (viz. IoT, digital twins, edge computing, cloud storage), information related to SHM can be autonomously connected and processed to monitor the structural health of respective structural components and schedule maintenance-on-demand activities. Figure 2.6 presents a detailed overview of how IPUs can scale up to achieve higher-level objectives.

In this case, as a composite IPU, an IPU girder (IPU #M1 + #N2 + #P1 + #R) can be applied to the bridge to reduce the lifecycle deterioration and consistently monitor aspects like corrosion, cracks, stress, and strain. Similarly, such structural IPU components (e.g., IPU pier, IPU cable, IPU deck) can scale up to form an IPU bridge with a well-defined and planned purpose. At an even larger scale, such IPU bridges can be used to plan IPU cities. Performance-based thresholds can be computed with a cloud-based knowledge centre and real-time decisions could be undertaken specific to the IPU in consideration (e.g., scheduling maintenance, notifying relevant suppliers for repair, etc.). IPUs can be refined if their as-built performance deviates from expectations. Installed IPUs can be retrofitted with modifications and new IPUs can be redesigned from the planning stages. This allows for a continuous process improvement cycle in the IPUs.

2.4.2 Case II. Optimized electricity demand management

The electricity grid system in the state of Indiana comprises five main utilities: Northern Indiana Public Services Co. (NIPSCO), Vectren Corporation (VECTREN), Indiana Michigan Power (I&M), Duke Energy (DUKE), and Indianapolis Power and Light Company (IPL). These utilities are responsible for supplying electricity to the residential, commercial, and industrial sectors within their respective service territories. Within each sector, there are individual units consisting of various electrical and electronic equipment, thereby transforming them into smart systems.

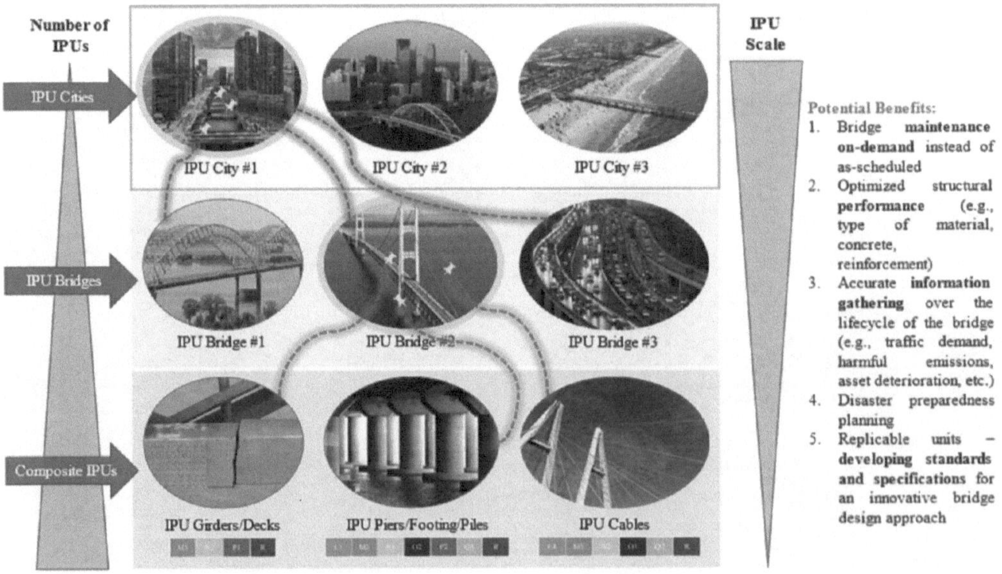

Figure 2.6 IPU network for bridge SHM

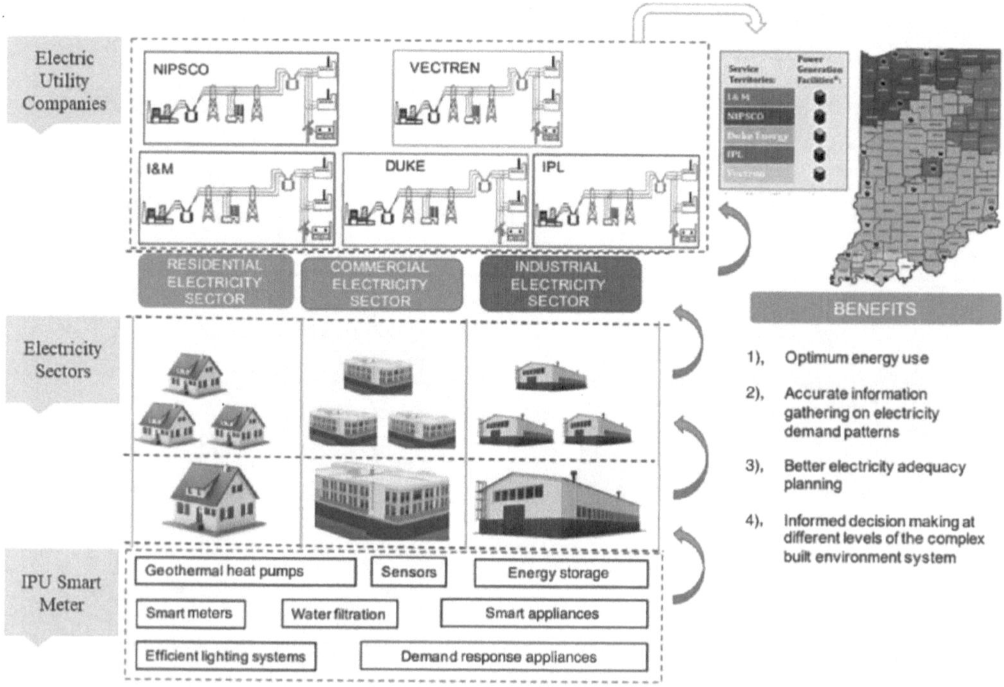

Figure 2.7 IPUs for optimizing energy efficiency
Source: adopted from Mukherjee et al., 2017.

These units encompass a range of technologies such as geothermal heat pumps, sensors, energy storage systems, smart meters, water filtration systems, smart appliances, efficient lighting systems, and demand response appliances. Figure 2.7 illustrates how the IPU concept can be implemented for optimizing energy performance across residential, commercial, and industrial sectors in the State of Indiana.

As discussed in the previous example, the decision-makers are expected to iterate through the IPU planning, IPU application, and IPU network phases to implement the proposed smart meter in a real-world scenario. As outlined in Figure 2.7, a standardized IPU Smart Meter can be applied to each unit in the electricity sectors (e.g., residential, commercial, industrial). Potential stakeholders can be electric utility companies and state utility agencies. Collectively, the expected benefits could be improved planning for electricity demand, analysing consumption patterns, detecting anomalies, and such informed decision-making at different levels of complexity in the built environment.

2.4.3 Case III. Sports centre design

Kim et al. (2021) proposed to implement IPUs to design a sports centre facility as illustrated in Figure 2.8. First, standardized rooms are designed as IPUs (e.g., IPU #A for lecture room, IPU #C for gymnasium, IPU #E for laboratory). Then, these IPUs are used to develop floor plans, design and specifications, schedule, inventory, and such project attributes. Energy, environment, and cost were established as primary objectives and each IPU was developed to optimize these objectives over the lifecycle of the project.

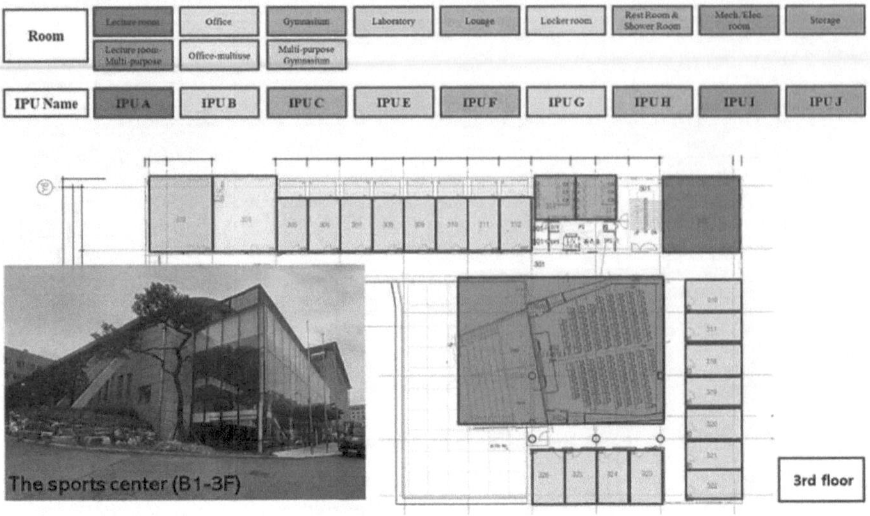

Figure 2.8 IPU sports centre
Source: adapted from Kim et al., 2021.

The third-floor plan in Figure 2.9 is designed with IPUs #A, #B, #E, #G, and #H, i.e., lecture room, office space, laboratory, locker rooms, and restrooms. For example, IPU #A (lecture room), is a composite IPU consisting of nanoscale IPUs that are optimized for (a) planning-centric factors such as design, material selection, labour requirement, and optimum schedule for construction along with (b) user-centric factors such as temperature, air, humidity, technology requirement, and overall comfort. After this IPU configuration is approved by the associated stakeholders, it can be replicated throughout the sports centre. Similarly, office spaces, locker rooms, and other facilities can be optimized and replicated throughout the building. In a real-world scenario, an architecture/engineering/construction (AEC) firm that specializes in developing sports centres can have a portfolio of IPU rooms with various properties and specifications. Clients can configure their unique sports centres replicating, combining, and refining the IPU rooms offered by the AEC firm's portfolio. Such a setup is expected to fast-track overall project delivery along with enhancing the comfort of the end users.

2.5 Challenges and prospects

This chapter introduced the theory of intelligent planning units (IPUs) and its three-fold implementation process, i.e., IPU planning, IPU application, and IPU strategy. However, certain challenges and limitations need to be addressed to enhance the performance of the proposed IPU theory. The next section explains these challenges in IPU implementation and prospects for future researchers and industry professionals.

2.5.1 IPU rating system

After the successful implementation of IPUs in the real world, best practices could be proposed to improve the as-built performance of an IPU. This will promote maintaining a continuous improvement cycle for the IPUs. To this end, a standardized inspection and rating system will assist in benchmarking the performance improvements of an IPU.

Example: consider an IPU girder designed to reduce dead load and enable autonomous SHM of the girder elements of a bridge.

Let's say the current benchmarks for this IPU girder are as follows:

- Dead-load = 1 ton/meter and SHM sensing depth = 6"

The project team proposes another IPU girder with the following capabilities:

- Dead-load = 0.8 ton/meter and SHM sensing depth = 9"

The IPU rating for the dead-load can be 1.25 (i.e., 1/0.8) and the IPU rating for SHM sensing depth can be 1.5 (i.e., 9/6). This indicates a 25% increase in dead-load performance and a 50% increase in SHM sensing depth performance respectively.

Such rating mechanisms can be useful for managing client expectations in a better way by breaking down complex construction processes into manageable IPUs.

2.5.2 Developing robust decision-support tools

Appropriate decision support tools can be developed to facilitate IPU planning by considering the project uncertainties, expected performance, simulation results, and the associated stakeholders. An IPU implemented through a robust planning process might be useful to provide both physical and virtual connectivity within the complex infrastructure layers of the built environment. All the operating data collected from the nanoscale IPUs can be used to generate insightful information about the current state of various physical entities (e.g., transportation systems, residential and commercial units, educational institutions, and so on). Challenges associated with traffic, energy, disaster preparedness/response, and such complex tasks can be managed in a better way by having replicable and connected IPUs throughout the built environment.

2.5.3 Real-world application scenarios

This study demonstrated how IPUs can be implemented in the real world with the help of three examples: (a) enabling automated structural health monitoring of bridges to shift from a traditional maintenance-as-scheduled to an innovative maintenance on-demand solution, (b) implementing IPU smart meters for designing residential, commercial, and industrial buildings to optimize the energy efficiency of these sectors, and (c) designing a sports centre by implementing IPU rooms to optimize cost, schedule, and supply chain during the construction phase and to enable intra-centre communication between IPUs using IoT tools. However, all the examples are hypothetical applications of the IPU theory. Therefore, the authors encourage industry practitioners to use these hypothetical applications to explore real-world implementation scenarios. The theory of IPU can potentially be implemented in various scenarios like project controls, safety management, transportation planning, sustainability management, process improvement, and so on. For example, stakeholders associated with data centres can explore designing future projects with the IPU theory to compress the schedule, reduce cost, and optimize the carbon footprint of their data centres.

2.5.4 Addressing credibility and cybersecurity concerns

The IPU theory proposed technological developments like sensors embedded in structural components, intelligent construction objects, edge computing, IoT, and a cloud-based infrastructure knowledge centre. However, as discussed by Mantha and de Soto (2019), such dynamic interactions between several interconnected infrastructure projects and stakeholders pose significant security risks. Such risks can manipulate the performance of IPUs in the form of cyber-attacks. However, advances in fields like blockchain technologies and internet security protocols have promising prospects to protect the credibility of data generated through IPUs. As highlighted by Padhye and Hastak (2023), any data should be adequately screened for credibility before it is used for any decision-making tasks. The proposed IPU rating system can be complemented with a credibility score to rate the reliability of data generated by the corresponding IPUs.

2.5.5 Path forward

Along with addressing the challenges that were presented earlier, future research can focus on developing a range of applications for the IPU theory. A few proposed areas of development are: (a) intelligent nanotechnology (e.g., lighter and durable elements, self-repairing components), (b) intelligent city planning (e.g., urban heat islands, noise maps, air index, etc. can be improved by implementing appropriate IPUs), (c) intelligent energy (e.g., photovoltaic systems, solar/wind setup, smart grids), (d) intelligent disaster preparedness (e.g. autonomous structural health monitoring, early hazard detection, connected infrastructure response), and so on. Furthermore, by consistently collecting data through all these nanoscale IPUs, disparities can be examined between the expected vs. planned performance of the IPUs, and appropriate solutions can be developed to enable continuous improvement of the built environment system. Han and Golparvar-Fard (2017) have demonstrated how data collected through multiple point clouds on a construction site can be modelled for management insights. Exploring this field holds significant potential for the prospective evolution of the constructed environment system, encompassing structures, infrastructure, power generation, and transportation, along with their associated sectors such as energy and the environment, on a global scale. Despite prior endeavours to address the complexities of the built environment system-of-systems, there was insufficient understanding of the fundamentals associated with intelligently planning this system-of-systems at a modular scale. To address this challenge, the theory of IPUs at a different level of complexities (nanoscale, structure, infrastructure system, city) was proposed by Hastak and Koo (2017). This chapter elaborated on the three-fold implementation process along with application case scenarios to ensure that potential readers can better understand the IPU theory. Furthermore, this study has also highlighted potential challenges and future avenues of research to expand this theory into real-world application scenarios. Ultimately, the goal of this study was to facilitate addressing the complexity of the built environment at a manageable scale and satisfying the objectives of the stakeholders (both directly and indirectly associated with the project).

References

Adegoke, E. I., Zidane, J., Kampert, E., Ford, C. R., Birrell, S. A., & Higgins, M. D. (2019). Infrastructure Wi-Fi for connected autonomous vehicle positioning: A review of the state-of-the-art. *Vehicular Communications*, 20, 100185.

Akhavian, R., & Behzadan, A. H. (2012). An integrated data collection and analysis framework for remote monitoring and planning of construction operations. *Advanced Engineering Informatics*, 26(4), 749–761.

Albreshne, A., Lahcen, A. A., & Pasquier, J. (2014). Using a residential environment domain ontology for discovering and integrating smart objects in complex scenarios. *Procedia Computer Science*, 32, 997–1002.

Benachio, G. L. F., Freitas, M. D. C. D., & Tavares, S. F. (2020). Circular economy in the construction industry: A systematic literature review. *Journal of Cleaner Production*, 260, 121046.

Bhattacharyya, A., & Hastak, M. (2023). A data-driven approach to quantify disparities in power outages. *Scientific Reports*, 13(1), 7247.

Cao, D., Wang, G., Li, H., Skitmore, M., Huang, T., & Zhang, W. (2015). Practices and effectiveness of building information modelling in construction projects in China. *Automation in Construction*, 49, 113–122.

Cardno, C. A. (2020). Smart bridges, evolved. *Civil Engineering Magazine Archive*, 90(6), 42–49.

Cho, K., & Hastak, M. (2013). Time and cost-optimized decision support model for fast-track projects. *Journal of Construction Engineering and Management*, 139(1), 90–101.

CII. (2009). *Front End Planning for Renovation/Revamp Projects (Best Practice). RT-242 Topic Summary.* Austin, TX: Construction Industry Institute. https://www.construction-institute.org/resources/knowledgebase/best-practices/front-end-planning/topics/rt-242 (Last Accessed on 7/17/2023).

CII. (2016). *Transforming the Industry: Making the Case for AWP as a Standard (Best) Practice. RT-319 Topic Summary.* Austin, TX: Construction Industry Institute. https://www.construction-institute.org/resources/knowledgebase/best-practices/advanced-work-packaging/topics/rt-319 (Last Accessed on 7/17/2023).

CII. (2023). *Defining Capital Efficiency Metrics for Downstream and Chemicals (DCC) Projects. RT-399 Topic Summary.* Austin, TX: Construction Industry Institute. https://www.construction-institute.org/resources/knowledgebase/knowledge-areas/result/topics/rt-399 (Last Accessed on 7/17/2023).

Cui, Q., Hastak, M., & Halpin, D. (2010). Systems analysis of project cash flow management strategies. *Construction Management and Economics*, 28(4), 361–376.

Enckell, M., Glisic, B., Myrvoll, F., & Bergstrand, B. (2011). Evaluation of a large-scale bridge strain, temperature and crack monitoring with distributed fibre optic sensors. *Journal of Civil Structural Health Monitoring*, 1, 37–46.

Grau, D., Caldas, C. H., Haas, C. T., Goodrum, P. M., & Gong, J. (2009). Assessing the impact of materials tracking technologies on construction craft productivity. *Automation in Construction*, 18(7), 903–911.

Han, K. K., & Golparvar-Fard, M. (2017). Potential of big visual data and building information modeling for construction performance analytics: An exploratory study. *Automation in Construction*, 73, 184–198.

Hastak, M., & Koo, C. (2017). Theory of an intelligent planning unit for the complex-built environment. *Journal of Management in Engineering*, 33(3), 04016046.

Jeon, J., Padhye, S., Bhattacharyya, A., Cai, H., & Hastak, M. (2022). Impact of COVID-19 on the US construction industry as revealed in the purdue index for construction. *Journal of Management in Engineering*, 38(1), 04021082.

Kim, J., Hong, T., Hastak, M., & Jeong, K. (2021). Intelligent planning unit for the artificial intelligent based built environment focusing on human-building interaction. *Journal of Asian Architecture and Building Engineering*, 20(6), 729–746.

Kirchherr, J., Reike, D., & Hekkert, M. (2017). Conceptualizing the circular economy: An analysis of 114 definitions. *Resources, Conservation and Recycling*, 127, 221–232.

Lu, W., Li, X., Xue, F., Zhao, R., Wu, L., & Yeh, A. G. (2021). Exploring smart construction objects as blockchain oracles in construction supply chain management. *Automation in Construction*, 129, 103816.

Mantha, B. R., & de Soto, B. G. (2019). Cyber security challenges and vulnerability assessment in the construction industry. In *Creative Construction Conference 2019* (pp. 29–37). Budapest University of Technology and Economics.

Min, J. U., & Bjornsson, H. C. (2008). Agent-based construction supply chain simulator (CS 2) for measuring the value of real-time information sharing in construction. *Journal of Management in Engineering*, 24(4), 245–254.

Moffatt, S., & Kohler, N. (2008). Conceptualizing the built environment as a social – Ecological system. *Building Research & Information*, 36(3), 248–268.

Morshedi, M., Padhye, S., Mwamba, I., Kang, K., Labi, S., & Hastak, M. (2023). Suitability assessment of detour routes for road construction projects: Framework and case studies. *Journal of Management in Engineering*, 39(2), 04022077.

Mukherjee, S., Yoon, S., & Hastak, M. (2017). Implementing intelligent planning unit (ipu) concept for optimized electricity demand management in the complex built environment. Leadership in Sustainable Infrastructure 2017 in Vancouver, Canada.

Niu, Y., Lu, W., Chen, K., Huang, G. G., & Anumba, C. (2016). Smart construction objects. *Journal of Computing in Civil Engineering*, 30(4), 04015070.

Padhye, S., & Hastak, M. (2023). Forthcoming. A framework to evaluate information and source credibility: International construction decision-making. *Journal of Management in Engineering*. https://doi.org/10.1061/JMENEA/MEENG-5533.

Padhye, S., Mwamba, I., Kang, K., Labi, S., & Hastak, M. (2021). Safety, mobility, and cost benefits of closing one direction of the interstate in rural areas during construction work (*Joint Transportation Research Program*, Publication No. FHWA/IN/JTRP-2021/21). West Lafayette, IN: Purdue University. https://doi.org/10.5703/1288284317345.

Sauter, E., Lemmens, R., & Pauwels, P. (2019). CEO and CAMO ontologies: A circulation medium for materials in the construction industry. In *6th International Symposium on Life-Cycle Civil Engineering (IALCCE)* (pp. 1645–1652). CRC Press.

Shi, X. (2020). More than smart pavements: Connected infrastructure paves the way for enhanced winter safety and mobility on highways. *Journal of Infrastructure Preservation and Resilience*, 1(1), 1–12.

Song, H. W., & Saraswathy, V. (2007). Corrosion monitoring of reinforced concrete structures-A. *International Journal of Electrochemical Science*, 2(1), 1–28.

Xu, J., Dong, Y., Zhang, Z., Li, S., He, S., & Li, H. (2016). Full scale strain monitoring of a suspension bridge using high performance distributed fiber optic sensors. *Measurement Science and Technology*, 27(12), 124017.

3

GENERATIVE DESIGN FOR EXCELLENCE (DFX) FOR A SMART BUILT ENVIRONMENT

From rule-based imitation to data-driven exploration

Ziyu Peng[1], *Yi Zhang*[2], *Weisheng Lu*[3], *Junjie Chen*[4], *Liupengfei Wu*[5] *and Jinying Xu*[6]

[1]PhD Candidate, Department of Real Estate and Construction, Faculty of Architecture, The University of Hong Kong, Pokfulam Road, Hong Kong SAR. Email: ziyupeng@connect.hku.hk

[2]PhD Student, Department of Real Estate and Construction, Faculty of Architecture, The University of Hong Kong, Pokfulam Road, Hong Kong SAR. Email: zhengyi0709@connect.hku.hk

[3]Professor, Department of Real Estate and Construction, Faculty of Architecture, The University of Hong Kong, Pokfulam Road, Hong Kong SAR. Email: wilsonlu@hku.hk

[4]Research Assistant Professor, Department, Department of Real Estate and Construction, Faculty of Architecture, The University of Hong Kong, Pokfulam Road, Hong Kong SAR. Email: chenjj10@hku.hk

[5]Post-doctoral Fellow, Department of Real Estate and Construction, Faculty of Architecture, The University of Hong Kong, Pokfulam Road, Hong Kong SAR. Email: liupengfeiwu@connect.hku.hk

[6]Marie Curie Future Roads Fellow, Department of Engineering, University of Cambridge, UK. Email: jx314@cam.ac.uk

3.1 Introduction

Generative design is an iterative process that produces various design alternatives (Krish, 2011). Since its workflow is linear and repetitive, it is expected to do so more efficiently than traditional design processes. Although the concept is not new – Spanish architect Antonio Gaudi used shape grammar, a traditional generative design method, to create his masterpieces, the Church of Colònia Güell and the Sagrada Família (Peng et al., 2023) – recent advancements in digitalization and simulation techniques have opened up new possibilities (Hao et al., 2023).

There are several reasons for the contemporary flourishing of generative design. One is the rapid development of digital tools. With the increasing availability of affordable software and open-source building information modeling (BIM) libraries (Gan, 2022), generative design tools have become more accessible to designers. Another reason is the growing complexity of design

DOI: 10.1201/9781003383840-4

challenges in the modern built environment, including aesthetics (Muehlbauer, 2018), performance (Shea et al., 2005), cost (Hassan et al., 2022), buildability (Ma et al., 2021), thermal comfort (Peng et al., 2023), energy efficiency (Caldas, 2008), and building codes (Zhang et al., 2021). In the presence of such factors and constraints, computer-aided generative design offers an efficient approach to exploring a broader design space and producing and evaluating more design alternatives. Last, amid our urgent need to address climate change and resource scarcity, there is a growing emphasis on design that minimizes energy consumption (Caldas, 2008) and reduces waste (Laovisutthichai et al., 2022b; Wu et al., 2023). Generative design, with its ability to consider performance factors during the design stage, is a valuable tool in creating more sustainable built environments.

In conjunction with design for excellence (DfX), an umbrella term for design that optimizes functionalities, generative design can encompass both aesthetics and practicality for architects and engineers (Laovisutthichai et al., 2022a). By considering a wide range of excellence variables, generative design and DfX principles combine to augment designers' abilities and lead to the creation of more sustainable buildings.

Against this backdrop, this chapter aims to explore the application of generative DfX in the SBE context. SBE will enable the built environment industry to transition into a digital realm where information is effortlessly exchanged, leading to the development of innovative business models, improved quality, reduced expenses, and enhanced profitability for society, companies, and users. In current generative design literature, both rule-based and data-driven paradigms exist. The former is proficient in representing a designer's will by making rules to accelerate and dominate the design process, while the latter focuses on deploying nascent machine learning techniques to explore various design patterns in previous masterpieces and use them to guide current design. The workflows and applications of these two paradigms in the SBE are yet to be properly discussed in the literature.

3.2 Generative DfX: an emerging paradigm

The promotion of DfX arises from a wide range of challenges in the architecture, engineering, and construction (AEC) industry, including issues such as low productivity, inadequate quality, an aging workforce, unsustainable development, and unsafe working conditions (Laovisutthichai et al., 2021). Originating in the product design and manufacturing industry, DfX is a design philosophy and integrated approach that aims in AEC to incorporate all desirable characteristics throughout the entire lifespan of a building, from inception to demolition (Garvin, 1984). "X" or "excellence" can be substituted to specify a favorable attribute, such as manufacture and assembly (DfMA), environment (DfE), recyclability (DfR), logistics (DfL), life cycle (DfLC), safety (DfS), maintainability (DfMt), and testability (DfT) (Huang & Mak, 1999; Laovisutthichai et al., 2021).

Generative DfX is a revolutionary concept that has the potential to transform the AEC industry. A design exploration process that uses algorithms to generate and evaluate multiple design alternatives based on parameters from the user (Krish, 2011), it can take many "X" factors into account based on generative design principles. For instance, machine learning-enabled generative DfX has been used to transform building codes into a domain-specific language to explore interior space arrangement alternatives by (Sydora & Stroulia, 2020). The AEC industry can reap multiple advantages from the implementation of generative DfX. A key benefit is cost savings and improved overall profitability, achieved through optimized design and reducing materials (Laovisutthichai & Lu, 2022). Generative DfX can also lead to more efficient design and construction processes, resulting in reduced project time (Laovisutthichai & Lu, 2022), and it also has the potential to enhance sustainability in the construction industry by reducing material waste and improving energy efficiency (Laovisutthichai et al., 2022b). Additional advantages might encompass inventive designs and enhanced safety (Laovisutthichai et al., 2021). Therefore, it is necessary to comprehend the initial conceptual design process, understand the interplay of numerous factors in DfX, and integrate implicit experiential knowledge in order to enhance computational intelligence for generative DfX.

3.3 Rule-based generative DfX for SBE

3.3.1 The workflow and application of rule-based generative design

Rule-based generative design is a methodology that uses predefined rules for building shape generation to automate the design process (Sydora & Stroulia, 2020). Some of the rules are rooted in aesthetic conventions such as the L-system, whereby designers imitate tree growth to achieve naturalism in building forms (Rian & Sassone, 2014). Other rules are created to ensure the generated design layouts satisfy structural requirements (Liao et al., 2022) or other man-made regulations (Sydora & Stroulia, 2020) governing the construction of the subject facility.

As explained in Figure 3.1, a typical process of rule-based generative design involves three stages, namely rule specification, variable definition, and design production. To begin, designers describe a set of rules and constraints using the language of mathematical formulas and/or parametric algorithms. In the past, these regulations were established and computed by hand, as demonstrated by Gaudi's use of the funicular system when designing the Church of Colònia Güell (Huerta, 2006). The mainstream approach now is to incorporate architecture rules into parametric programs using software such as Grasshopper and Unity, allowing a more efficient exploration of shape variants (Khan & Awan, 2018). Rule-based generative design also demonstrates integration potential with building performance assessment tools that simulate energy consumption, thermal comfort, and view, so as to achieve generative design for building performance assessment (Oxman, 2009). At the variable definition stage, the start condition is input into the system in order to regulate the constraint variables for building shape generation and their values, including site condition, building codes, and client requests. Finally, the design production stage follows the rules and start conditions to automatically generate a series of design alternatives.

This three-stage rule-based workflow enables designers to explore a vast number of design possibilities and optimize performance parameters with the help of modern computing instruments. The rules also reflect the will of the designers, including their attitudes toward aesthetics, expectations of building performance, and other excellence factors. While the generated options are the result of algorithms, expert knowledge is still needed to refine them to ensure their feasibility and alignment with project requirements.

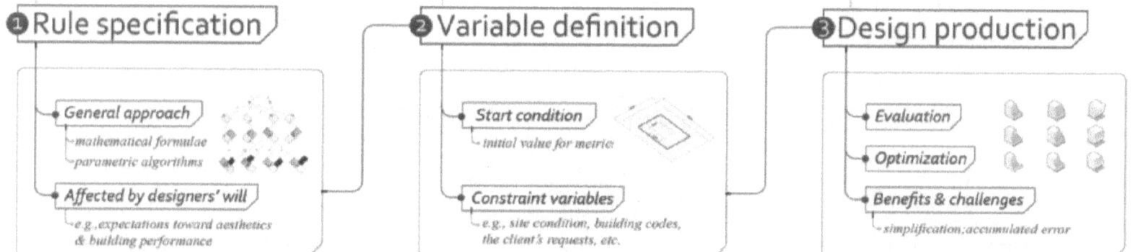

Figure 3.1 The overall workflow of rule-based generative design

The rule-based generative DfX workflow has been widely applied in developing the SBE, particularly for design problems involving intensive trial and error. Design of building structures is a typical case. By integrating rules such as load distribution, material properties, and safety regulations, engineers can generate design variants that ensure the structural integrity of the target building (Li et al., 2023). The rule-based workflow can also be used to reduce the environmental footprint of a structure by encoding rules to define the relationships between different elements of the structure and their environmental parameters (Gan et al., 2019). Another typical scenario is energy-efficient building design. By inputting relational rules that indicate the impacts of architectural elements on energy efficiency, such as the impact of window-to-wall ratio on natural lighting and building orientation on thermal insulation, designers can generate design options with minimized energy consumption and carbon footprints.

3.3.2 The advantages and challenges of rule-based generative design

Rule-based generative design has emerged as a promising methodology for automating and optimizing the design process in the SBE context. Compared with conventional design dependent on craftsmanship, rule-based generative design allows exploration of a greater number of design possibilities within a limited timeframe. Ever-increasing computing power gives designers time to try out different rules and examine the design solutions (Hornby & Pollack, 2001). It also facilitates the optimization of various performance parameters, including those concerning sustainability (Alsakka et al., 2023) through consideration of factors such as energy efficiency and environmental footprint during the rule specification stage. The rule-based workflow can therefore promote the creation of more sustainable buildings that minimize resource consumption and reduce the overall carbon footprint.

Despite the benefits mentioned earlier, rule-based generative DfX faces several challenges. First is the need to explicitly decode past design knowledge and restructure it into machine-readable rules. The quality of generated solutions relies on the accuracy and completeness of the rules defined (Jiang et al., 2022), since the rule-based workflow is a one-way onward process where an error in the start condition will only amplify throughout the generation process. Additionally, while this workflow can generate design options, human expertise is still required to interpret and refine them (Dogan et al., 2019). Last but not least, confining the designs to what we already know might constrain them to what is essentially an imitation of what we already have. As human society is poised to construct increasingly complex mega structures within ever-challenging constraints, this rule-based generative design paradigm may be ill-equipped to tackle the design challenges envisaged.

To conclude, rule-based generative design offers benefits in terms of exploring design possibilities and integrating sustainability principles. However, challenges related to rule specification, validation, and adaptability to complex design problems may hinder its implementation. These challenges have led designers to tap into data-driven generative DfX.

3.4 Data-driven generative DfX for SBE

Data-driven generative design uses large amounts of data to inform the creation of new design alternatives. Unlike rule-based generative design, it uses existing data (e.g., design layouts) as learning samples to train a model that automatically produces design variants without pre-set rules, on the basis that this data contains knowledge that can be learned and applied to produce new design patterns.

3.4.1 Big data and SBE

The Internet and technological innovations have brought about an exponential growth in the volume and variety of data in existence. From this has emerged the notion of big data, a term used to describe the vast and intricate datasets that pose challenges for conventional processing techniques (Hilbert & López, 2011). Big data is revolutionizing scientific research, empowering scientists to scrutinize unprecedentedly extensive and diverse data from multiple sources, thereby discerning previously imperceptible correlations and trends. Moreover, with the emergence of ground-breaking technologies (e.g., deep learning), organizations have come to realize the immense potential of mining data for business intelligence to gain a competitive edge (Bajari et al., 2019). Therefore, data-driven technologies have garnered widespread acceptance, enabled extraction of valuable insights, and facilitated data-driven decision-making.

Data-driven technologies (i.e., AI) have played a pivotal role in propelling industrial modernization and digital transformation in the AEC industry. This is particularly the case in the context of SBE, where from the perspective of either smart cities (e.g., transportation systems, population distribution) or smart buildings (e.g., energy usage, occupant experience), data is gathered from interconnected systems, sensors, and IoT devices and subsequently analyzed to modify the system (Bessis & Dobre, 2014). This leads to a positive feedback cycle of data collection, analysis, and prediction. However, it is noteworthy that many applications of data-driven approaches in SBE often focus on measuring building performance rather than addressing fundamental design issues. This contrasts with evidence in the field of sustainability demonstrating that passive energy-saving methods (i.e., relying on building design ontology to conserve energy) are promoted to a greater extent than active methods (Shao et al., 2020). Therefore, it is intriguing to consider how data-driven technologies can revolutionize the design phase to help achieve an SBE.

3.4.2 Redefining the rules by harnessing big data

As stated in the preceding section, data-driven AI has been utilized predominantly in statistic-centric domains. Its limited application in architecture has focused primarily on post-design stages (e.g., construction). However, the advent of generative AI models (e.g., DALL-E 2, MidJourney) in late 2022 and ChatGPT in early 2023 signal the inception of the AI era. We have come to realize that AI – or predictive models derived from extensive data training (Zhuang et al., 2017) – can disrupt and transform human-centric creative industries, including architectural design. This illustrates how data can be used in design (Loyola, 2018): big data enables architects to collect design conditions (e.g., height restriction), performance (e.g., energy consumption), and experiential outcome labels (e.g., layout, massing) and make faster and better-informed design decisions.

The reason data can be designed is that it has allowed the process of auto-encoding information to be perfected (Kramer, 1992). Over the years, designers have dedicated themselves to encoding design knowledge, progressing from traditional design theory (e.g., typology and morphology), to adaptive design theorems (e.g., Plug-in City, Metabolism), and then to rule-based algorithmic computation. These approaches aim to describe high-dimensional architectures with low-dimensional features (e.g., bubble diagrams) through progressively more expressive and automated approaches (Wang et al., 2016). However, since the rules for feature extraction are manually specified, these methods inevitably suffer from information loss, making it difficult to comprehensively describe complex systems such as architecture (e.g., style, construction drawing). In contrast, the latest developments in deep learning technology can maximize the preservation of information comprehensiveness during the process of dimension reduction, effectively redefining the rules (Vahdat &

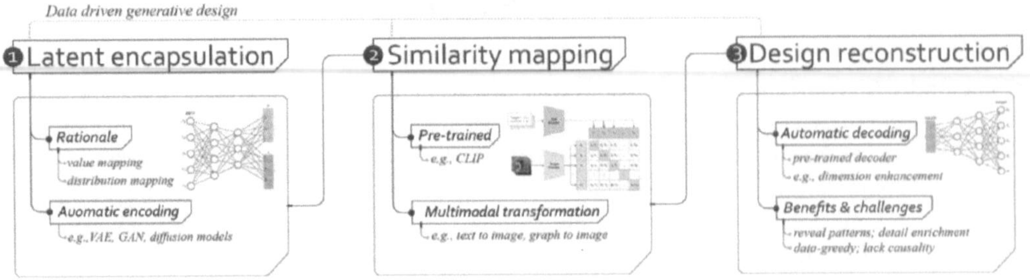

Figure 3.2 The overall workflow of data-driven generative design

Kautz, 2020). This is realized through the capability of deep learning to concentrate exclusively on preeminent features, a learning process that closely resembles that of human beings.

As shown in Figure 3.2, the workflow of data-driven generative design comprises three stages, namely, latent encapsulation, similarity mapping, and design reconstruction. First, rather than relying on manually defined rules, data-driven approaches attain feature extraction through auto-encoders, whose rationale lies in two aspects: (a) mapping the features into latent space vectors and (b) establishing a correspondence between the probability distribution of real-world images and the latent vector distribution (usually modeled as a Gaussian). For instance, in the Variational Autoencoder (VAE), 2D images are cast into a set of vectors using an encoder-decoder architecture; either generative adversarial network (GAN) or diffusion model sampling random noise from a normal distribution and transforming it into a same-dimensional image through respective network structures. Once the latent spaces are established, the similarity between latent vectors can be calculated through pre-trained models (e.g., CLIP), empowering various functionalities. For example, Lin et al. (2024) transformed graphs into modular building layouts, demonstrating how latent similarity enables multimodal transformation.

Last, the neural networks reparametrize the latent features into high-dimensional design solutions utilizing auto-decoders. From the aforementioned, it is evident that the accuracy of encoding is of utmost importance, and the quantity and richness of data determine it. When the data is abundant (e.g., residential floor plans), neural networks can accurately capture diverse features imperceptible to humans, revealing intricacies during prediction and even exhibiting emergence, thereby revealing the underlying causes. However, the reliance on data in data-driven generative design means that it performs poorly with limited data as an accurate encoding process cannot be realized, lacks causal explanation since features are latent, and rarely challenges cognitive boundaries with the new, unlike rule-based generative design.

In summary, data-driven generative design has demonstrated the potential to liberate architects from mundane design tasks enabling aesthetic exploration, notably in virtual reality and post-concrete materiality contexts.

3.4.3 Data-driven generative DfX: challenges and prospects

In architecture, the core challenge for data-driven generative design lies in the scarcity of labeled 3D data. Taking building style as an example, generating the style of Corbusier presents difficulties in defining the Le Corbusier style, discerning whether other buildings resemble this style (since Corbusier's work is limited), and finding sufficient corresponding 3D models. These problems are challenging to solve and, consequently, it is difficult to build a massive labeled dataset upon which supervised learning can be conducted.

Current data-driven endeavors have focused on addressing manual labeling through unsupervised deep-learning techniques. Given an ample amount of data, these methods aim to automatically identify intrinsic connections among similar data. Notable instruments developed based on GAN and diffusion models include ArchiGAN, TxtImg2Img (Liao et al., 2022), and D3M (Su et al., 2024). Additionally, XGBoost is derived from the Random Forest algorithm and has shown promising performance in specific research scenarios (Yan et al., 2022). Nevertheless, these research efforts focus on generating 2D layout plans and graphical renderings, not 3D representation and dataset richness. Thus, mature algorithms for 3D building morphology generation have yet to be developed, although there are some studies on 3D encoding that have shown potential, such as NeRF rebuilding, 2D layout slicing, and 3D CLIP (Wang et al., 2021; Hegde et al., 2023). Additionally, model distillation (Polino et al., 2018) is a method used to derive smaller models from pre-trained large models, such as ChatGPT, to effectively handle subtasks.

Numerous studies have shown that data-driven generative design techniques are deeply embedded in the DfX framework. By leveraging data and analytics, the design process can be optimized across various disciplines. For example:

- Data-driven design can investigate data for manufacture (e.g., production records) to identify potential design improvements (e.g., reduce complexity). If there are certain components consistently weighing down previously streamlined production processes, big data analysts could identify them and inform designers so that they can make modifications (Tao et al., 2018)
- Similarly, big data incorporates part interaction and tolerance information, which can be utilized to search for optimum assembly sequences (i.e., minimum assembly steps and errors) in the design stage (Boninsegna et al., 2018)
- Data-driven approaches handle environmental and sustainable design well. Research works on ventilation (Gan et al., 2022) and thermal comfort (Peng et al., 2023) have demonstrated the vast capability that environmental data (e.g., life cycle assessment, energy consumption) affords designers
- More issues regarding building management, such as reliability (e.g., failure rates, field performance) and serviceability (e.g., repair records, customer feedback), can be handled within the design stage with the assistance of data analysis (Maddalena et al., 2020).

In summary, data-driven generative design plays a positive role in achieving sustainability and prolonging building lifespans. This process is commonly realized through a "collection-train-optimization" workflow: as in a modular integrated construction (MiC) project, data related to the complex processes can first be obtained from either physical assembly or computer simulation. Then the dependence between each module assembly sequence and various objectives can be learned using deep regression models (e.g., XGBoost). Last, if the model performs well on the test set, users can subsequently obtain the optimal solution whether by algebra in simpler cases or meta-heuristic algorithms in more complicated cases. Therefore, data-driven generative design empowers designers to make evidence-based decisions and optimize designs across various aspects of DfX.

3.5 Discussion and conclusion

The built environment, as the collective sum of our construction-transformed surroundings, reflects the glory of civilization. Design activities have long followed rules explicitly specified in technical treatises (e.g., the Yingzao Fashi) or implicitly encoded in experience (e.g., the golden rectangle).

Even with the advent of computers, this rule-based approach persists. However, relying solely on established knowledge confines the design space within the boundaries of what we already know, resulting in designs that are essentially imitations of the past. This approach no longer aligns with our present and future need and ambition to construct increasingly complex mega structures within ever-challenging constraints. The situation calls for an innovative but implementable approach to SBE design.

Data-driven approaches have the potential to fundamentally transform existing practices. It is not new to exploit data available to train AI engines. However, most of these AI use predictive engines to correlate data or make decisions in mundane tasks such as regression and classification. The rise of generative AI such as ChatGPT and other large language models demonstrates the feasibility of machines learning hidden knowledge from data for creative tasks. It is especially suitable for open-ended domains like architectural design where the problem is not well defined. Utterly different from rule-driven practice, such data-driven approaches do not require a formalism of existing domain knowledge (architecture-wise, structure-wise, inter alia) to guide the generative design. Rather, the knowledge is automatically learned and implicitly encoded into weights and parameters of the AI networks. As such, the approach could uncover design patterns that have not been formalized or even noticed by humans, which is conducive to generating unconstrained designs that are open for exploration.

The data-driven approach promises to catalyze a paradigm shift in how design work is conducted in an SBE. As more data is fed into the system, the machine becomes increasingly creative and capable of generating design alternatives that were previously unthinkable. At a micro level, this capability can assist architects, engineers, and designers in kick-starting their work with highly plausible machine-generated options. On a macro level, data-driven machine designers act as bionic brains, expanding the human imagination. These blue-sky designs challenge design professionals to rethink existing architectural design theory, genres, and conventions. However, the final decision on whether a design is good or bad still rests with humans. By harnessing the machine's advantage in pattern learning and combining it with human adaptability and creativity, a new paradigm of human–robot collaborative design is emerging. The two parties can enhance each other's strengths and compensate for their respective weaknesses. Through this mutual enhancement, new discoveries and insights can be unveiled, ultimately enriching the collective body of design knowledge.

While the prospects appear exciting, the path forward is not without challenges. The first concern is the potential scarcity of data. Generative AI is data-greedy in essence. It needs a vast volume of past design data to fuel the process of training and weight refinement. While information technology has significantly lowered the cost of data production and communication, in a profession-based sphere like architecture design, it remains unclear whether domain-specific data such as drawings and footprints can be generated and shared with the same level of speed and openness.

Even if sufficient data are solicited, a question that follows is how to ensure the quality and a balanced distribution of the data. Machine learning is limited by what is encoded in data. In the profession of building design, this "garbage in, garbage out" mechanism can have enormous implications. For example, if the training set is dominated by Western Baroque-style castles, the model may not generate designs that align with the local natural and cultural landscape, such as an Eastern pagoda-like building. To ensure generative DfX serves the general societal good, it is critical to guarantee the inclusiveness of its training data.

Another challenge lies in standardization of the data. Building design data can be manifested in various formats (paper-based, digital files, or vectorized). Even though modern AI instruments have achieved great progress in generalizing across diverse data modalities, it is still difficult to learn design patterns from such heterogeneous data. To facilitate machine learning, a formal and generative design-oriented representation of the data is needed.

References

Alsakka, F., Haddad, A., Ezzedine, F., Salami, G., Dabaghi, M., & Hamzeh, F. (2023). Generative design for more economical and environmentally sustainable reinforced concrete structures. *Journal of Cleaner Production, 387*, 135829. https://doi.org/10.1016/j.jclepro.2022.135829

Bajari, P., Chernozhukov, V., Hortaçsu, A., & Suzuki, J. (2019). The impact of big data on firm performance: An empirical investigation. In *AEA papers and proceedings* (Vol. 109, pp. 33–37). Nashville, TN: American Economic Association. https://doi.org/10.1257/pandp.20191000

Bessis, N., & Dobre, C. (Eds.). (2014). *Big data and internet of things: A roadmap for smart environments* (Vol. 546). Basel: Springer International Publishing. https://doi.org/10.1007/978-3-319-05029-4

Boninsegna, L., Banisch, R., & Clementi, C. (2018). A data-driven perspective on the hierarchical assembly of molecular structures. *Journal of Chemical Theory and Computation, 14*(1), 453–460. https://doi.org/10.1021/acs.jctc.7b00990

Caldas, L. (2008). Generation of energy-efficient architecture solutions applying GENE_ARCH: An evolution-based generative design system. *Advanced Engineering Informatics, 22*(1), 59–70. https://doi.org/10.1016/j.aei.2007.08.012

Dogan, K. M., Suzuki, H., Gunpinar, E., & Kim, M. S. (2019). A generative sampling system for profile designs with shape constraints and user evaluation. *Computer-Aided Design, 111*, 93–112. https://doi.org/10.1016/j.cad.2019.02.002

Gan, V. J. (2022). BIM-based graph data model for automatic generative design of modular buildings. *Automation in Construction, 134*, 104062. https://doi.org/10.1016/j.autcon.2021.104062

Gan, V. J., Wang, B., Chan, C. M., Weerasuriya, A. U., & Cheng, J. C. (2022). Physics-based, data-driven approach for predicting natural ventilation of residential high-rise buildings. In *Building simulation* (Vol. 15, pp. 129–148). Tsinghua University Press. https://doi.org/10.1007/s12273-021-0784-9

Gan, V. J., Wong, C. L., Tse, K. T., Cheng, J. C., Lo, I. M., & Chan, C. M. (2019). Parametric modelling and evolutionary optimization for cost-optimal and low-carbon design of high-rise reinforced concrete buildings. *Advanced Engineering Informatics, 42*, 100962. https://doi.org/10.1016/j.aei.2019.100962

Garvin, D. A. (1984). Product quality: An important strategic weapon. *Business Horizons, 27*(3), 40–43. https://doi.org/10.1016/0007-6813(84)90024-7

Hao, T., Huang, J., He, X., Li, L., & Jones, P. (2023). A machine learning-enhanced design optimizer for urban cooling. *Indoor and Built Environment, 32*(2), 355–374. https://doi.org/10.1177/1420326X221112857

Hassan, S. R., Megahed, N. A., Eleinen, O. M. A., & Hassan, A. M. (2022). Toward a national life cycle assessment tool: Generative design for early decision support. *Energy and Buildings, 267*, 112144. https://doi.org/10.1016/j.enbuild.2022.112144

Hegde, D., Valanarasu, J. M. J., & Patel, V. (2023). CLIP goes 3D: Leveraging prompt tuning for language grounded 3D recognition. In *Proceedings of the IEEE/CVF international conference on computer vision* (pp. 2028–2038). https://doi.org/10.48550/arXiv.2303.11313

Hilbert, M., & López, P. (2011). The world's technological capacity to store, communicate, and compute information. *Science, 332*(6025), 60–65. https://doi.org/10.1126/science.1200970

Hornby, G. S., & Pollack, J. B. (2001, May). The advantages of generative grammatical encodings for physical design. In *Proceedings of the 2001 congress on evolutionary computation (IEEE Cat. No. 01th8546)* (Vol. 1, pp. 600–607). IEEE. https://doi.org/10.1109/CEC.2001.934446

Huang, G. Q., & Mak, K. L. (1999). Design for manufacture and assembly on the internet. *Computers in Industry, 38*(1), 17–30. https://doi.org/10.1016/S0166-3615(98)00105-5

Huerta, S. (2006). Structural design in the work of Gaudi. *Architectural Science Review, 49*(4), 324–339. https://doi.org/10.3763/asre.2006.4943

Jiang, Z., Wen, H., Han, F., Tang, Y., & Xiong, Y. (2022). Data-driven generative design for mass customization: A case study. *Advanced Engineering Informatics, 54*, 101786. https://doi.org/10.1016/j.aei.2022.101786

Khan, S., & Awan, M. J. (2018). A generative design technique for exploring shape variations. *Advanced Engineering Informatics, 38*, 712–724.

Kramer, M. A. (1992). Autoassociative neural networks (Computers & Chemical Engineering). *Neutral Network Applications in Chemical Engineering, 16*(4), 313–328. https://doi.org/10.1016/0098-1354(92)80051-A

Krish, S. (2011). A practical generative design method. *Computer-Aided Design, 43*(1), 88–100. https://doi.org/10.1016/j.cad.2010.09.009

Laovisutthichai, V., & Lu, W. (2022). Design for Manufacture and Assembly (DfMA) in architectural design meetings: From a case study to knowledge-to-action framework. *Smart and Sustainable Built Environment*. Advance online publication. https://doi.org/10.1108/SASBE-07-2022-0136

Laovisutthichai, V., Lu, W., & Bao, Z. (2022a). Design for construction waste minimization: Guidelines and practice. *Architectural Engineering and Design Management, 18*(3), 279–298. https://doi.org/10.1080/17452007.2020.1862043

Laovisutthichai, V., Lu, W., & Lau, S. S. Y. (2022b). Design for excellence in architecture, engineering, and construction: A multi-stakeholder model. In Guo, H., Fang, D., Lu, W., & Peng, Y. (Eds.), *Proceedings of the 26th international symposium on advancement of construction management and real estate. CRIOCM 2021. Lecture notes in operations research.* Singapore: Springer. https://doi.org/10.1007/978-981-19-5256-2_58

Li, M., Liu, Y., Wong, B. C., Gan, V. J., & Cheng, J. C. (2023). Automated structural design optimization of steel reinforcement using graph neural network and exploratory genetic algorithms. *Automation in Construction, 146*, 104677. https://doi.org/10.1016/j.autcon.2022.104677

Liao, W., Huang, Y., Zheng, Z., & Lu, X. (2022). Intelligent generative structural design method for shear wall building based on "fused-text-image-to-image" generative adversarial networks. *Expert Systems with Applications, 210*, 118530. https://doi.org/10.1016/j.eswa.2022.118530

Lin, X., Chen, J., Lu, W., & Guo, H. (2024). An edge-weighted graph triumvirate to represent modular building layouts. *Automation in Construction, 157*, 105140. https://doi.org/10.1016/j.autcon.2023.105140

Loyola, M. (2018). Big data in building design: A review. *Journal of Information Technology in Construction, 23*(2018), 259–284. https://www.itcon.org/2018/13

Ma, W., Wang, X., Wang, J., Xiang, X., & Sun, J. (2021). Generative design in building information modelling (BIM): Approaches and requirements. *Sensors, 21*(16), 5439. https://doi.org/10.3390/s21165439

Maddalena, E. T., Lian, Y., & Jones, C. N. (2020). Data-driven methods for building control – A review and promising future directions. *Control Engineering Practice, 95*, 104211. https://doi.org/10.1016/j.conengprac.2019.104211

Muehlbauer, M. (2018). Towards typogenetic tools for generative urban aesthetics. *Smart and Sustainable Built Environment, 7*(1), 20–32. https://doi.org/10.1108/SASBE-10-2017-0052

Oxman, R. (2009). Performative design: A performance-based model of digital architectural design. *Environment and Planning B: Planning and Design, 36*(6), 1026–1037. https://doi.org/10.1068/b34149

Peng, Z., Lu, W., Hao, T., Tang, X., Huang, J., & Webster, C. (2023). Cost-aware generative design for urban 'cool spots': A random forest-principal component analysis-augmented combinatorial optimization approach. *Energy and Buildings*, 113317. https://doi.org/10.1016/j.enbuild.2023.113317

Polino, A., Pascanu, R., & Alistarh, D. (2018). Model compression via distillation and quantization. *ArXiv Preprint*. Advance online publication. https://doi.org/10.48550/arXiv.1802.05668

Rian, I. M., & Sassone, M. (2014). Tree-inspired dendriforms and fractal-like branching structures in architecture: A brief historical overview. *Frontiers of Architectural Research, 3*(3), 298–323. https://doi.org/10.1016/j.foar.2014.03.006

Shao, T., Zheng, W., & Jin, H. (2020). Analysis of the indoor thermal environment and passive energy-saving optimization design of rural dwellings in Zhalantun, Inner Mongolia, China. *Sustainability, 12*(3), 1103. https://doi.org/10.3390/su12031103

Shea, K., Aish, R., & Gourtovaia, M. (2005). Towards integrated performance-driven generative design tools. *Automation in Construction, 14*(2), 253–264. https://doi.org/10.1016/j.autcon.2004.07.002

Su, P., Lu, W., Chen, J., & Hong, S. (2024). Floor plan graph learning for generative design of residential buildings: A discrete denoising diffusion model. *Building Research & Information, 52*(6), 627–643. https://doi.org/10.1080/09613218.2023.2288097

Sydora, C., & Stroulia, E. (2020). Rule-based compliance checking and generative design for building interiors using BIM. *Automation in Construction, 120*, 103368. https://doi.org/10.1016/j.autcon.2020.103368

Tao, F., Qi, Q., Liu, A., & Kusiak, A. (2018). Data-driven smart manufacturing. *Journal of Manufacturing Systems, 48*, 157–169. https://doi.org/10.1016/j.jmsy.2018.01.006

Vahdat, A., & Kautz, J. (2020). NVAE: A deep hierarchical variational autoencoder. *Advances in Neural Information Processing Systems*, *33*, 19667–19679. https://doi.org/10.48550/arXiv.2007.03898

Wang, Y., Yao, H., & Zhao, S. (2016). Auto-encoder based dimensionality reduction. *Neurocomputing*, *184*(1), 232–242. https://doi.org/10.1016/j.neucom.2015.08.104

Wang, Z., Wu, S., Xie, W., Chen, M., & Prisacariu, V. A. (2021). NeRF – Neural radiance fields without known camera parameters. *ArXiv Preprint*. Advance online publication. https://doi.org/10.48550/arXiv.2102.07064

Wu, L., Lu, W., Peng, Z., & Webster, C. (2023). A blockchain non-fungible token-enabled 'passport' for construction waste material cross-jurisdictional trading. *Automation in Construction*, *149*, 104783. https://doi.org/10.1016/j.autcon.2023.104783

Yan, H., Yan, K., & Ji, G. (2022). Optimization and prediction in the early design stage of office buildings using genetic and XGBoost algorithms. *Building and Environment*, *218*, 109081. https://doi.org/10.1016/j.buildenv.2022.109081

Zhang, J., Liu, N., & Wang, S. (2021). Generative design and performance optimization of residential buildings based on parametric algorithm. *Energy and Buildings*, *244*, 111033. https://doi.org/10.1016/j.enbuild.2021.111033

Zhuang, Y. T., Wu, F., Chen, C., & Pan, Y. H. (2017). Challenges and opportunities: From big data to knowledge in AI 2.0. *Frontiers of Information Technology & Electronic Engineering*, *18*, 3–14. https://doi.org/10.1631/FITEE.1601883

4

PROJECT DELIVERY METHODS TO DIGITAL FABRICATION IN ARCHITECTURE

A comparative case study from a modularity perspective

Tan Tan[1], *Daniel Hall*[2], *Eleni Papadonikolaki*[3], *Grant Mills*[4] *and Konrad Graser*[5]

[1]Department of Real Estate and Construction, The University of Hong Kong, Hong Kong SAR. Email: tant@hku.hk.

[2]Faculty of Architecture and the Built Environment, Delft University of Technology, Netherlands. Email: d.m.hall@tudelft.nl

[3]Faculty of Civil Engineering and Geosciences, Delft University of Technology, Netherlands. Email: e.papadonikolaki@tudelft.nl.

[4]Bartlett School of Sustainable Construction, University College London, United Kingdom. Email: g.mills@ucl.ac.uk.

[5]School of Architecture, Design and Civil Engineering, Zurich University of Applied Sciences, Switzerland. Email: konrad.graser@zhaw.ch.

4.1 Introduction

Integrating digital fabrication (DFAB) in the architectural engineering and construction (AEC) industry is a complex design and construction process. DFAB refers to data-driven production, where the generated workflow and data enable numerically controlled manufacturing equipment to fabricate parts or products (Bock and Linner, 2015; Ng et al., 2021). Empirical investigations have revealed an increasing number of digital instruments, such as the processes of 3D printing and robotic manufacturing and assembly, which have significantly enriched the practice of DFAB (Agustí-Juan and Habert, 2017). The potential and capabilities of DFAB are transformative in nature, particularly in how structures are conceived and materialised (Pawar et al., 2017). Nevertheless, the adoption rate of DFAB within the AEC industry remains low (Ng et al., 2022; Ng et al., 2021). The application of DFAB in the AEC industry faces complex challenges, including technical expertise requirements, high implementation costs, regulatory constraints, and significant shifts in traditional design and construction paradigms (Tan et al., 2023).

A significant barrier to the implementation of DFAB is the project delivery methods. DFAB necessitates an elevated level of coordination to ensure that all parties involved can efficaciously

DOI: 10.1201/9781003383840-5

collaborate throughout the project lifecycle (Ng et al., 2022). During the process of project delivery, seamlessly integrating various DFAB technologies and tools constitutes a significant challenge. Substantial integration can also consume a considerable amount of time and cost, leading to an extension in the implementation timeline of DFAB. Therefore, in addition to integration, modularity approaches are required as well (Graser et al., 2021). Furthermore, the AEC industry is characterised by stringent regulations and standards, whilst DFAB is a new field. Despite successful delivery instances for some unique cases, DFAB lacks project delivery method research for wider implementation and standardised adoption.

Advancements in digitally enabled project delivery are poised to augment the implementation efficiency of DFAB. A research gap resides in the relationship between project delivery methods and DFAB. DFAB comprises various techniques, such as 3D printing, robotics, and computer numerical control cutting. The extent of DFAB implementation varies across different projects, which impacts the considerations for adopting project delivery methods. For instance, certain projects may only incorporate DFAB for partial components, whilst others might embrace DFAB across the entire building. Presently, there is no standardised metric to gauge the degree of DFAB implementation, which also challenges understanding the relationship between DFAB and project delivery methods.

This study aims to explore the implementation of DFAB in architecture from the project delivery perspective through a comparative case study. The study initially delineates four levels of implementation of DFAB, four types of project delivery methods, and project delivery strategies from a modularity perspective in Section Two. Subsequently, Section Three introduces the research methods. Section Four describes the implementation of project delivery methods and DFAB technologies in two specific projects. Finally, the study discusses the advancements, insights, challenges, and future research directions in this field.

4.2 Project delivery methods to DFAB through a lens of modularity

4.2.1 Implementation of DFAB

There has been no research concerning the degree of DFAB implementation. Based on the four-level model of modular and offsite construction concepts proposed by Pan (2019), DFAB could also be categorised to facilitate understanding its implementation. The degree of DFAB implementation in a project can broadly be encapsulated in four levels (see Figure 4.1). The full DFAB level involves an extensive application of DFAB technologies in most or all processes, representing a degree of implementation between 60% and 100%. Hybrid DFAB is a level where DFAB coexists with traditional manufacturing techniques, contributing to a 30% to 60% degree of implementation. Partial DFAB, on the other hand, indicates a project where a fraction of processes involve DFAB technologies. While less dominant, this 15% to 30% degree of implementation still influences the project's success. Finally, the minimal DFAB is the stage with the lowest degree of implementation, where DFAB technologies are sparingly used, often in very limited steps, indicating a 0% to 15% degree of implementation. Each level signifies a different extent of DFAB implementation, offering a flexible approach depending on a project's specific characteristics and needs. The degree of DFAB implementation could be quantified through a formula encompassing three key parameters: the proportion of DFAB data, the proportion of DFAB equipment usage, and the proportion of workflow steps involving DFAB. The degree can be computed as the weighted average of these parameters.

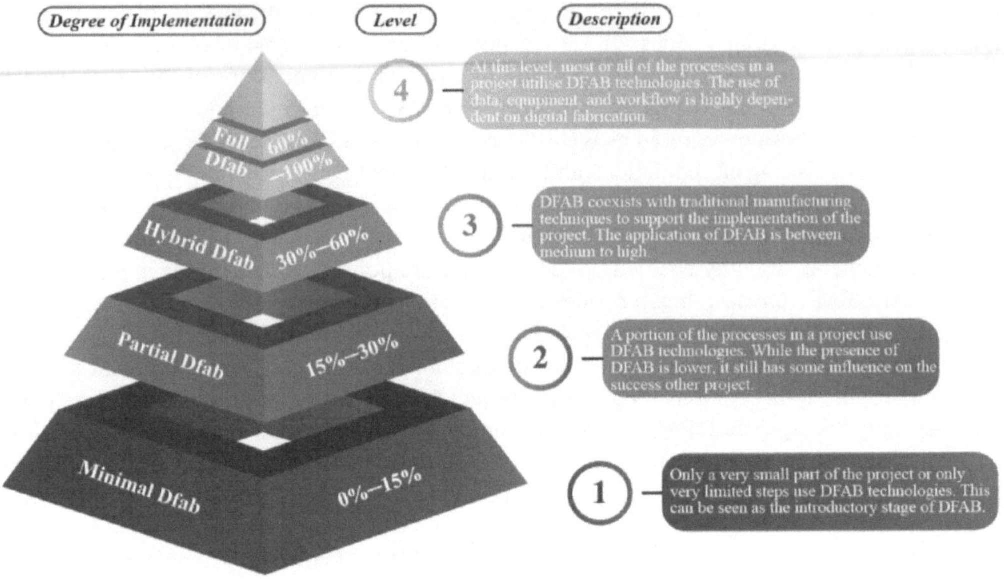

Figure 4.1 Implementation degree of DFAB in architecture

4.2.2 Supply chain modularity and project delivery methods

A project delivery method is a system employed by project owners or managers to systematically arrange and fund a project or facility's design, construction, operations, and maintenance aspects (Al Khalil, 2002). This involves forming contractual agreements with one or more parties involved in the project. There are several commonly used methods for project delivery. These include design-bid-build (D-B-B), Design-Build (D-B), construction manager at risk, construction management multi-prime, public-private partnership, and integrated project delivery (IPD). Voordijk et al. (2006) established the connection between project delivery methods and supply chain modularity. Modularity refers to a hierarchical system structure consisting of smaller sub-systems that can be designed independently but operate as a holistic system (Baldwin et al., 2000; Ulrich, 1995). Modularity is a relative system attribute (Baldwin et al., 2000). Every system is somewhat modular (Campagnolo and Camuffo, 2010), or integrated. Modularity research in the supply chain is an emerging area (Salvador et al., 2002; Voordijk et al., 2006) and is spreading in construction (Doran and Giannakis, 2011; Voordijk et al., 2006). The trend to modularise the supply chains facilitates the transformation and reorganisation of value creation within supply chains (Doran and Roome, 2003; Collins et al., 1997). Supply chain modularity refers to whether certain supply functions or tasks are conducted by a single supplier or not and whether they can be explicitly distinguished from others (Wolters, 2002), thus aiming to mitigate the complexity within supply chain coordination. Supply chain modularity focuses on the division of labour within a supply chain network for specific supply chain functions and tasks and how companies interact, which is for a relatively flexible and interchangeable relationship among suppliers, customers, and partners (Fine et al., 2005). By analogy with modular products and processes, a modular supply chain responds to the changing demands on functionality and performance of different supply chain variants by cultivating alternative capabilities to deal with different versions of functional components (Voordijk et al., 2006).

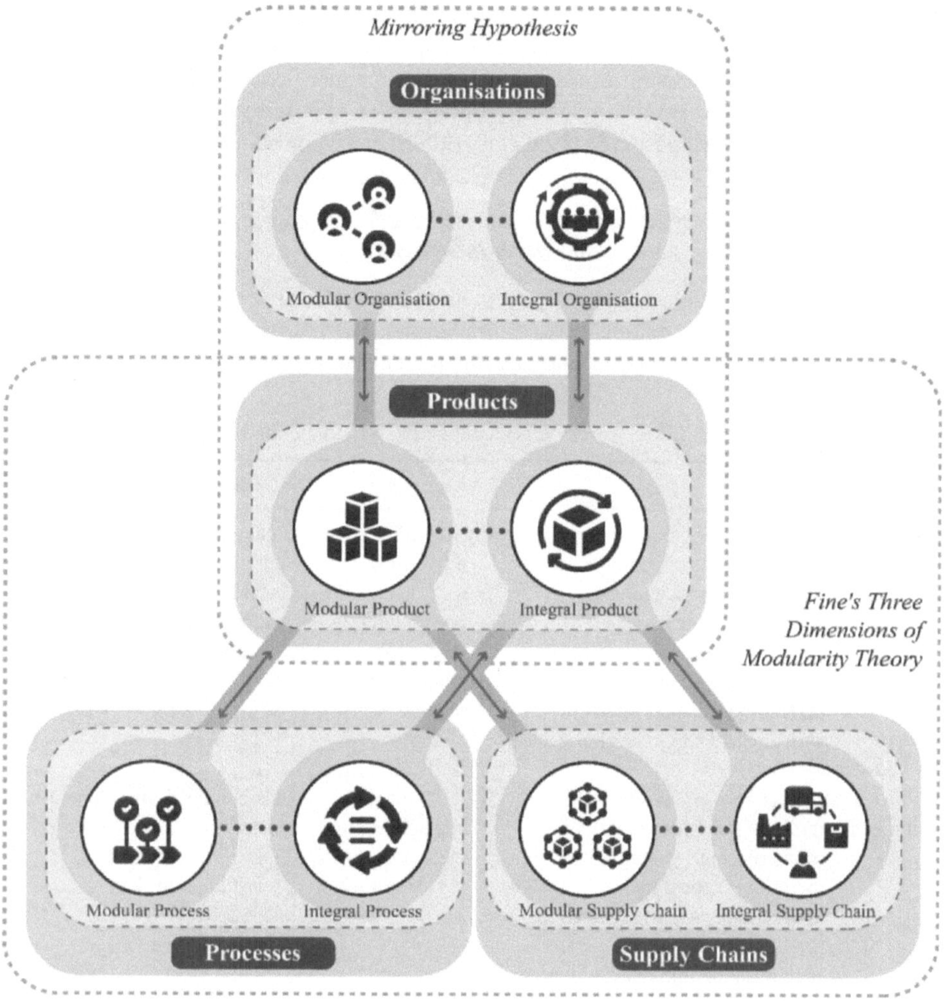

Figure 4.2 Relationships between multiple dimensions of modularity

In the AEC industry, Voordijk et al. (2006) claim supply chain modularity is assessed based on the separation or integration of design and execution responsibilities within organisational models. When design and execution are separated, there is usually higher modularity; for example, in the traditional model, components and responsibilities are distinctly allocated through multiple contracts. However, in integrated models such as the D-B approach, one organisation handles both design and construction, leading to less modularity. The least modular is the brochure plan model, where a single dominant entity controls all supply chain stages, including assembly and manufacturing (Voordijk et al., 2006). As shown in Figure 4.2, Fine et al. (2005) predict that a modular product tends to be designed by a modular process and modular supply chain. Voordijk et al. (2006) further test and describe this proposition in construction by adopting the variations of project delivery methods as different modularity strategies. However, Tee et al. (2019) assert that modular designs may mitigate coordination problems by decreasing the interdependencies between modules. These designs could also impede collaboration due to the increased focus on specialisation within the modules. This elucidates that modularity and integration, typically seen as contradictory, can also function in a complementary manner. Hence, there is a potential for better flexibility and coordination through misalignment, as opposed to the sole presence of the alignment relationship, as suggested by Fine et al. (2005) and Voordijk et al. (2006).

4.2.3 Four types of project delivery methods categorised based on modularity lens

Based on the modularity lens, project delivery methods could be divided into four types through two dimensions: vertical integration, vertical modularity, horizontal integration, and horizontal modularity (see Table 4.1). The table categorises project delivery methods through vertical and

Table 4.1 Four types of project delivery methods categorised based on the modularity theory

Category	*Definition*	*Advantages*	*Disadvantages*	*Examples*
Vertical Integration	A project delivery method where multiple stages or components (e.g., design, procurement, construction) are managed by a single entity or closely aligned team. This typically involves a centralised decision-making structure and a high level of control.	Increased efficiency, reduced communication delays, and clear lines of responsibility and decision-making.	Potential for less flexibility and can be less adaptable to changes.	Engineering, procurement, and construction (EPC), design-build.
Vertical Modularity	A project delivery method where different stages or components of a project are managed within their own vertical hierarchies but are relatively separate across the project. This allows for greater specialisation and independence within different stages of a project.	Allows for specialisation, clear phase responsibilities.	This can result in communication barriers, coordination challenges between phases.	Design-bid-build
Horizontal Integration	A project delivery method where various stages and components of a project collaborate and share information and resources on the same level. This typically involves cross-functional teams and a highly collaborative work environment.	Enhanced collaboration, greater flexibility and adaptability, conducive to innovation.	Potentially more complex decision-making, may require more time for consensus building.	Integrated project delivery (IPD)
Horizontal Modularity	A project delivery method where different stages or components of a project operate as relatively independent modules on the same horizontal level. This allows each module to have a degree of autonomy while coordinating and collaborating as needed.	Flexibility, modular collaboration as needed.	Potential for inconsistencies across modules, coordination can still be a challenge.	Construction management multi-prime

horizontal dimensions of modularity and integration. Vertical integration, exemplified by the D-B model, features a hierarchical structure with a single entity managing multiple stages, ensuring streamlined communication. Vertical modularity, as seen in the D-B-B method, involves dividing the project into stages managed by different entities within a hierarchical structure, providing flexibility but possibly disjointed communication. Horizontal integration, demonstrated by IPD, emphasises collaboration across different teams at the same level within the organisation, fostering resource sharing and collaborative decision-making. Last, horizontal modularity, represented by the construction management procurement route, involves segmenting the project into independent modules managed by different teams at the same organisational level, promoting flexibility and specialisation but potentially facing coordination challenges.

In project delivery methods, applying modularity theory helps in understanding how to organise and optimise resources, responsibilities, and communication in project management processes. The distinction between vertical and horizontal further describes how modularity and integration occur across different levels within an organisation (see Figure 4.3). Vertical integration and modularity primarily deal with hierarchical structures, where responsibilities are either consolidated under one entity (integration) or divided among different entities in a tiered manner (modularity). On the other hand, horizontal integration and modularity focus on collaboration and division of tasks, respectively, across teams or units at the same organisational level, emphasising peer-level coordination and resource sharing. In summary, incorporating the ideas of modularity theory provides more structural insights and flexibility for the delivery of construction and engineering projects.

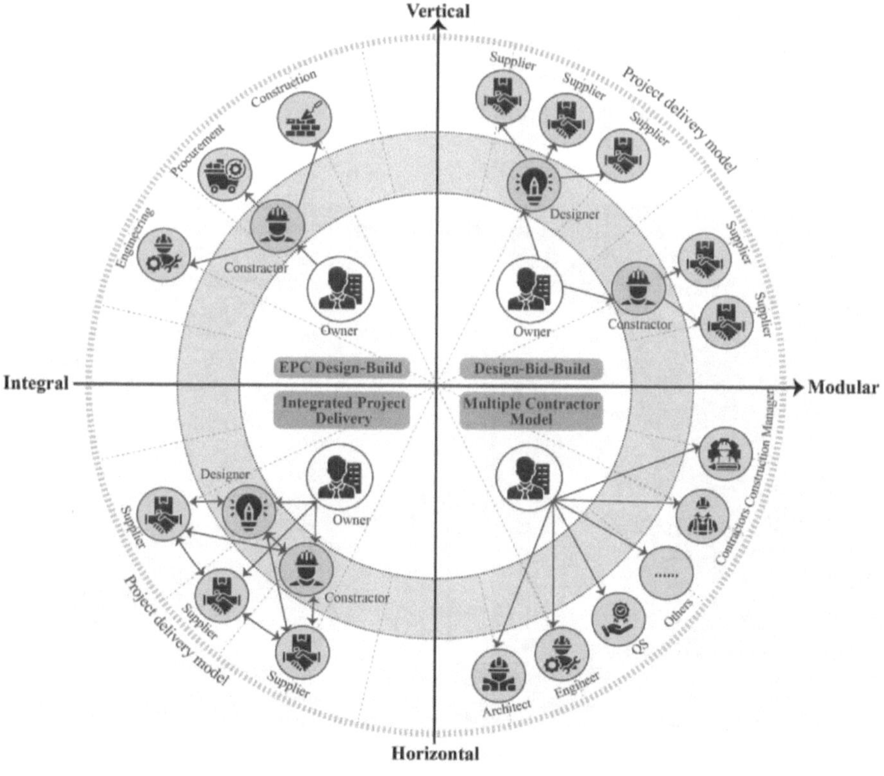

Figure 4.3 Integration-modularity in project delivery methods

4.3 Methodology

A comparative and exploratory case study was conducted to analyse and compare the relationships between DFAB implementation and project delivery methods. The case study is a well-established approach to reveal and explore emerging context-based phenomena (Yin, 2009). DFAB is new to the wide industry implementation. Thus, an exploratory study would be beneficial to understanding its project delivery. Some of the authors were personally involved in the design and construction processes of these two DFAB projects. This study adopted an analysis strategy to compare the levels of DFAB implementation and integration-modularity in project delivery methods. The framing and categorisation methods for these two comparative units have been clarified in the literature review sections. There are four types of DFAB implementation levels, including full, hybrid, partial, and minimal DFAB. Four types of project delivery methods through the lens of modularity include vertical integration, vertical modularity, horizontal integration, and horizontal modularity. The relationships between multiple dimensions of modularity (see Figure 4.2) are used as the overarching lens to see the relationships.

The case selection strategy aims to identify cases with different DFAB implementation levels and investigate if and how their implementation levels impact the project delivery method selections. Second, more importantly, is to investigate if and how the project delivery methods have impacts on the DFAB implementation. A significant exploration is to see if Fine's three dimensions of modularity theory can be applied to analyse and explain the relationship between DFAB implementation levels and project delivery methods. For example, to see if there is an integral-modular alignment relationship across product, process, and supply chain. This study collected interview, observation, and archival data from two cases and conducted an interpretivism-based analysis of the case data.

This study identifies two cases. The DFAB House (i.e. Case A, see Figure 4.4), erected from 2016 to 2018, serves as a forward-looking residential model in Switzerland, demonstrating state-of-the-art DFAB methods like 3D printing and robotics. This initiative is a joint endeavour

Figure 4.4 Case A: DFAB house

Source: NCCR dfab.

Figure 4.5 Case B: Tujia Pan-Museum

Source: the author.

among the Swiss National Centre of Competence in Research (NCCR) DFAB, architects, construction firms, and scientists showcased on the NEST building of Empa and Eawag. This project has indeed elevated the benchmark for environmentally friendly and digitally fabricated architecture. Another case B is the Chinese Tujia Pan-Museum (See Figure 4.5). Case B is a museum construction project with a 40,000 m^2 gross floor area for the Tujia minority in Enshi, Hubei province, China. The design was from December 2017 to December 2018, and the construction was from February 2019 to July 2021. The Chinese Tujia Pan-Museum is an innovative approach to poverty reduction and cultural promotion among the isolated Tujia people. Honouring traditional architecture yet adopting a modern design, the building utilises sustainable, prefabricated timber, reducing costs, environmental impact, and potential indoor pollution.

4.4 Case studies

4.4.1 Case A: DFAB house

4.4.1.1 DFAB implementation to case A

The DFAB house (see Figure 4.4) integrates six novel digital building processes (see Table 4.2), including the In situ Fabricator, an autonomous construction robot, and the Mesh Mould, a robotic process for steel-reinforced concrete structures. The Smart Slab, a 3D-printed formwork for integrated ceiling slabs and Spatial Timber Assemblies, a robotically fabricated timber structure, were also utilised. In the DFAB house project, the implementation of all these technologies is optimised and controlled through digital models and algorithms to achieve maximum efficiency and precision. The project demonstrates the potential of DFAB techniques in architectural design and production, capable of transforming how buildings are produced and how we understand architectural form and function. Regarding the implementation levels, the DFAB house can be regarded as a full DFAB.

4.4.2 Project delivery to case A

The DFAB house project, an experimental, non-commercial initiative, was designed to showcase innovative research outcomes and required a significant transformation of conventional project delivery and roles. The project's unique approach combined D-B and IPD to synergise the efforts of planners, designers, and contractors, blurring the boundaries between planning and execution. It entailed adopting novel platforms like building information modelling and DFAB house plug-ins, prompting designers and engineers to assume roles like DFAB managers while also dealing with decreased

Table 4.2 DFAB techniques in DFAB house

Technology	*Description*	*Benefits*	*Photos*
In situ Fabricator	A versatile autonomous on-site construction robot.	Allows for increased efficiency and precision in construction. Can operate independently, reducing the need for human labour in potentially dangerous environments.	
Mesh Mould	A formwork-free, robotic process for steel-reinforced concrete structures.	Eliminates the need for formwork, reducing material waste and labour costs. The process can create complex and bespoke shapes with reinforced concrete.	
Smart Dynamic Casting	An automated concrete slip-forming process.	This technique allows for the rapid construction of vertical concrete structures, saving time and reducing labour requirements. It also allows for the creation of unique and complex forms.	
Smart Slab	Integrated ceiling slabs fabricated with 3D-printed formwork.	The 3D printing process allows for the creation of complex and bespoke shapes, which can be optimised for material efficiency and performance.	
Spatial Timber Assemblies	A robotically fabricated timber structure.	The use of robotics allows for precise assembly of timber structures, reducing waste and increasing efficiency. The process can accommodate complex and unique designs.	
Lightweight Translucent Facade	A membrane skin filled with translucent thermal insulation.	This facade technology provides insulation while allowing for natural light penetration, enhancing energy efficiency. It also offers a unique aesthetic due to its translucency.	

Source: NCCR dfab.

personal involvement due to automation. Despite these novel adaptations, flexibility challenges arose within project relationships. Although formal contractual obligations existed between stakeholders, the exploratory nature of the DFAB house necessitated less formal, personal agreements and fostered a climate of interdisciplinary problem-solving. This necessitated more informal, organic collaboration mechanisms beyond formal established project delivery methods, often reflected in self-organised meetings among various experts such as roboticists, structural engineers, and material scientists. The project delivery also hinged on the integration of key actors early in the project timeline – especially during the conceptual stage – and the strategic co-location of the project team. This co-location fostered continuous interaction and collaboration, enhancing the collective body of knowledge. The industrial partners maintained a strong presence throughout the project stages, enabling direct involvement in planning, research, and industry decision-making processes. This proximity to production allowed stakeholders to move beyond remote judgements and gain a deeper comprehension of the project's needs, thus facilitating the exploration and realisation of DFAB's full potential.

The DFAB house project demonstrates a unique interplay between bottom-up self-organisation and top-down controls. This balance ensures that autonomous exploration aligns with the realities of project delivery. While research-led development is crucial, a more structured approach becomes necessary to prevent unending explorations from impeding project completion; as a contractor noted, "If you would have let them do this indefinitely, there still wouldn't be a finished building." Addressing this issue required modularity and integration in the project structure. Process modularity allowed each DFAB application to be independently and concurrently developed within its self-managed, highly integrated organisation module. Compared to software development by a project investigator, this approach decoupled operative project management from technology development and focused on integration, thus shielding researchers from the project's "managerial side." Such a strategy insulated the complex and uncertain aspects of DFAB development from interface interactions. However, coordinating interdependencies among multiple parallel DFAB developments posed challenges. A CEO highlighted the need for a seamless view of interfaces without sharp delineations of responsibility. Properly coordinated interfaces at module boundaries fostered the integration of different DFAB applications. A sense of collective responsibility was encouraged – "You are just a part of the project, you are not alone. . . . Everyone depends on the others." In this context, it's pertinent to discuss the role of "bridge function" actors. These individuals, possessing a broad array of skills and experience, were pivotal in managing information exchange across module interfaces. Their role was instrumental in the successful organisation of the project, illustrating the importance of roles that straddle both technical proficiency and cross-functional coordination in complex, exploratory projects like DFAB House.

4.4.3 Case B: Tujia Pan museum

4.4.3.1 DFAB implementation to case B

In light of the considerable task volume inherent to timber processing, this endeavour employed robotic fabrication for the prefabricated timber rooftop assembly, as delineated in Figure 4.6. The technological groundwork for DFAB was provided by NURBS parametric modelling, specifically

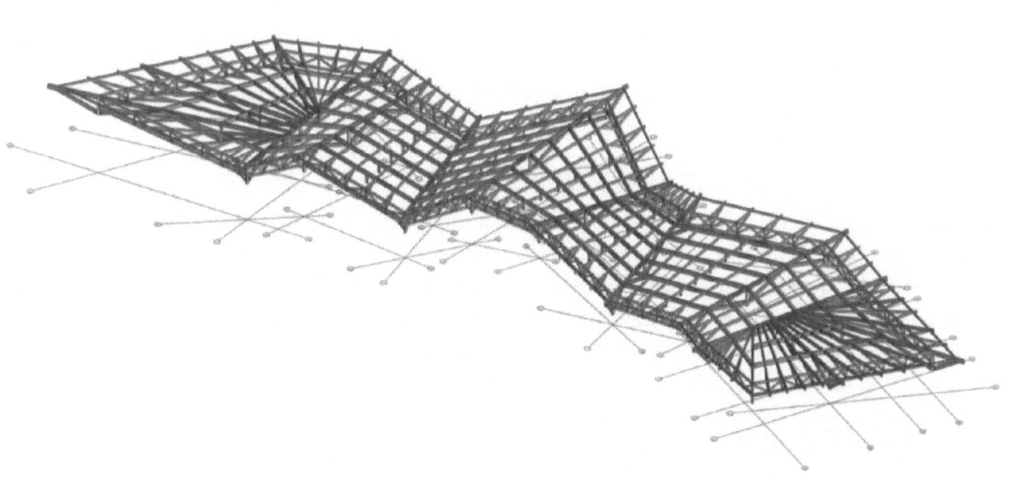

Figure 4.6 Prefabricated timber rooftop structure through DFAB

Table 4.3 DFAB techniques in Tujia Pan-Museum

Technology	*Description*	*Benefits*	*Photos*
Robotic arm cutting	A robotic arm equipped with cutting tools to perform precise and automated cutting of wood materials.	Contributes to increased efficiency, cost savings, and quality enhancement in manufacturing processes.	

Source: the author.

the Rhinoceros 3D software. The entity that manufactured the robotic appendages teamed up with a subcontractor to devise a timber structure processing plug-in, thereby enabling robotic fabrication, as detailed in Table 4.3. A virtual replication and generation of the physical robot arm's real-time operation were made possible within the 3D software environment, facilitating the simulation of all modular component fabrication and manufacturing procedures. The enhancement of the robotic arm's project-specific adaptability was a result of personalised software development and maintenance services. The integration of robotic fabrication with digital-enabled design has significantly bolstered design quality and production efficiency while concurrently mitigating labour requirements. Design information interchange at the corresponding phase is propelled by automated production implements.

However, from a holistic project scope, automation in construction constitutes merely a minor fraction. The ultimate architectural product is presently not factory-manufactured, indicating that a substantial journey remains before complete automation dependency materialises for the final construction. The deployment of robotic arms is accompanied by numerous obstacles. Equipment for processing has its limitations, encompassing component size, tool size and shape, and processing direction. These attributes invoke novel requisites for designers, necessitating intimate communication between designers and manufacturers to comprehend the characteristics of materials and processing and manufacturing capabilities, thereby achieving Design for Manufacture and Assembly (DfMA). Taking into account the DFAB conditions within this project, it can be perceived as a partial DFAB. Specifically, 15–30% of the construction is reliant on the DFAB process and associated technologies, such as the wooden roof structure. However, conventional construction techniques are still employed for aspects like the building's foundations, walls, and curtain walls, and they were subsequently amalgamated with the DFAB-produced wooden roof structure.

4.4.3.2 Project delivery to case B

In terms of project delivery methods, this project adopted the EPC model in the contract formation. However, in this project, the implementation of DFAB does not dictate the choice of project delivery method. Instead, the project delivery method was established first, followed by the decision on DFAB implementation. The substantial relationships of stakeholders within the project are not solely influenced by the contract-based project delivery method but, more importantly, depend on the informal relationships established through trust among the parties. These informal relationships are the primary factors affecting risk and responsibility allocation among all stakeholders in the project and essentially constitute the project delivery method rather than being fully determined by the contractual model. Although in theoretical EPC the main contractor takes the lead, in this project, the chief architect, as the actual system integrator, played a leading role in promoting the implementation of DFAB and convinced the client to adopt DFAB. The conflict resolution and system integration issues in the implementation of DFAB are also aggregated to the chief architect. This informal project cooperation relationship, hidden under the formal EPC model, is a substantial factor that affects DFAB.

In this case, the integration and DfMA relied on the collaboration between the three parties' architects and the main contractor's project manager. The principal architect from the design firm took a major role in design changes, design optimisation, and DfMA process integration. The complexity of design activities is decomposed through the modular process. For example, engineers and manufacturers' design activities were "hidden" and not directly involved in cross-organisation communication; they were led by the architects of their firms. The interface in the design process represents the rules of interaction and how different groups' design activities interact. For example, in this case, design firms prefer not to share all information or 3D models with other stakeholders because contracts have no such requirements. For example, the single-line model is usually a design task conducted by civil engineers. In this case, architects took the role of this design task and confirmed the section dimension size with civil engineers. Because civil engineers were not responsible for the appearance of forms and the aesthetics of the structural system, the architects allocated these tasks to themselves to keep design quality and make the OSC efficient. The change of design tasks was a reconfiguration of the design process. Relatively simpler information documents (i.e. single-line model) were exchanged as an interface for different design tasks between architects and civil engineers, representing the modularity in the design process.

4.4.4 Cross-case analysis

The comparative case study has revealed some similarities and differences in implementing DFAB. The extent of DFAB implementation affects the degree of change in the project organisation at the technical, procedural, and design levels. It can be seen in case B that the extent of DFAB implementation is far less than in case A. However, this does not mean that the implementation challenges faced by DFAB are reduced. In case A, due to the deep implementation of DFAB, all project stakeholders have a greater consensus on the acceptance and understanding of DFAB. Additionally, as a non-commercial research-oriented demonstrator, all parties are more willing to embrace this change's interdisciplinary challenges, providing a foundation of trust and cooperation. On the other hand, in case B, the implementation of DFAB is only executed in one part of the building (the roof), and communication and negotiation with multiple stakeholders involved in traditional construction are necessary. This situation presents not only interdisciplinary knowledge issues but also problems of trust and consensus on interests. Therefore, a lower degree of DFAB implementation does not necessarily mean lower implementation difficulties. Instead, at the level of project delivery, it may also face many challenges. These depend on the degree of correlation that the part implementing DFAB will have with other parts, i.e., the degree of independence or dependence of its DFAB component.

The shared insights distilled from the two case studies are as follows: (a) the adoption of modular products and processes; (b) the adoption of an integral type of project delivery method; and (c) the significant role of informal relationships. Whether it is the EPC, IPD, or DB, it can be seen that their inherent project delivery methods lean towards a more integral approach (see Figure 4.3). The real underlying stakeholder relationships within project delivery, which are established through informal relationships, also feature the commonality in both cases of assisting the design and workflow of the project to become more integrated, thereby resolving the challenges and problems in the engineering process. The type of informal relationship presents an inconsistency with the formal relationships among stakeholders established by contractual project delivery methods. However, this inconsistency does not necessarily have a purely negative effect. In the two cases selected for this study, these informal relationships have promoted the adoption and implementation of DFAB. Due to the establishment of formal contractual relationships, which are influenced and restricted by various policies, environmental factors, legal issues, and project backgrounds, informal relationships provide flexibility at the organisational and supply chain levels to implement DFAB. This relationship flexibility can help better implement relevant technologies and strategies in the early stages of DFAB introduction, compensating for the lack of an established and compatible DFAB project delivery method.

4.5 Discussion

4.5.1 Reflecting on the demarcation of modularity and integration in architecture

This comparative case study invites us to reconsider the prevailing definitions of modularity and integration within the field of architecture. Pan (2019) delineated modular construction as the apex of the offsite construction spectrum, wherein 60–90% of building components are fabricated offsite. According to this definition, the DFAB house described in case A could be exemplified as a quintessential modular building. Nevertheless, a salient observation is the presence of multiple integrative measures in Case A, such as integrated project delivery solutions. DFAB is also regarded as a typical practice of integrative design, entailing a design-to-manufacture integration driven by design data. If integration and modularity represent the two extremities of a spectrum, how is this paradoxical phenomenon, the simultaneous presence of integration and modularity, realised? This question triggers researchers and practitioners to reconsider the definitions of these terms.

This study argues that modularity and integration can coexist due to inconsistencies among different system dimensions and hierarchical levels within an object. Therefore, the object under examination must be explicitly defined when discussing modularity and integration. For instance, in Figure 4.7, both A and B in the two illustrations are highly integrated internally due to their mutual dependencies and connections. The interface between the two modules is changed based on the reconfiguration of the two groups' design activities. With the change of boundaries (i.e. dashed boxes), the interface between two groups is changed from one connection to two connections, which is activated by the move of "C." The left side grouping way is relatively more modular than the right side. In addition, two modules formed by A and B are linked by Interface C, presenting A and B as highly independent submodules with modular characteristics from this perspective. This discrepancy results from scrutinising different levels of the modular-integrative spectrum. As illustrated in Figure 4.2, modularity and integration can exist in different dimensions (i.e., product, process, organisation, supply chain), and both Case A and Case B reflect a misalignment relationship between different dimensions of modularity. Specifically, both cases demonstrate modular products and processes, yet they simultaneously possess integral supply chain features.

4.5.2 Reflecting on Fine's three dimensions of modularity theory

The conclusion diverges from Fine's three dimensions of modularity theory (Fine, 2010) but further builds on and develops Tee et al.'s (2019) arguments on the benefits of misalignment and complementarities between modularity and integration. In this comparative case study, the modular products of the two cases are formed by modular processes and integrated supply chains,

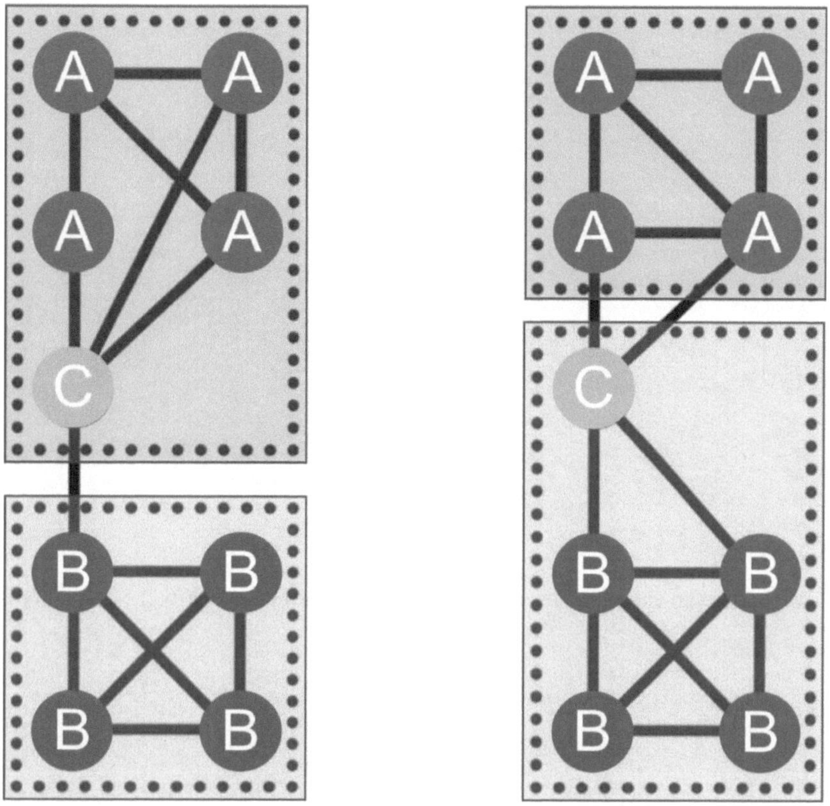

Figure 4.7 Reconfiguration of key design activities

representing a type of misalignment relationships amongst the product, process, and supply chain. "Misalignment relationships" in multi-dimensional modularity denote inconsistencies or unclear correspondence in the degree of modularity across different dimensions such as product, process, organisation, and supply chain. Theoretically, the degree of modularity in these different dimensions should be aligned, that is, modular products should be produced by modular processes, modular processes should be conducted within modular organisations, and modular organisations should be supported by modular supply chains. For instance, Voordijk et al. (2006) carried out research on the application of Fine's three dimensions of modularity theory in the construction industry and examined this alignment relationship. However, in practice, this alignment may not always be present, potentially leading to the so-called "misalignment."

Increasing the degree of process modularity and adopting modular processes may be attributable to the aim of reducing complexity during the design process and simultaneously enhancing the task flexibility of interdisciplinary teams, mitigating risks, and promoting system integration (see Figure 4.8). Within the context of DFAB, modular processes enhance project adaptability and efficiency. Independent module design, production, and assembly allow for swift adjustments amidst demand changes or technological innovations. Digital optimisation of a module can be immediately replicated across all relevant projects, significantly boosting efficiency. Quality control is also reinforced, as digitally controlled environments enable independent production and testing of modules, and any defective module can be rectified without disrupting the entire project.

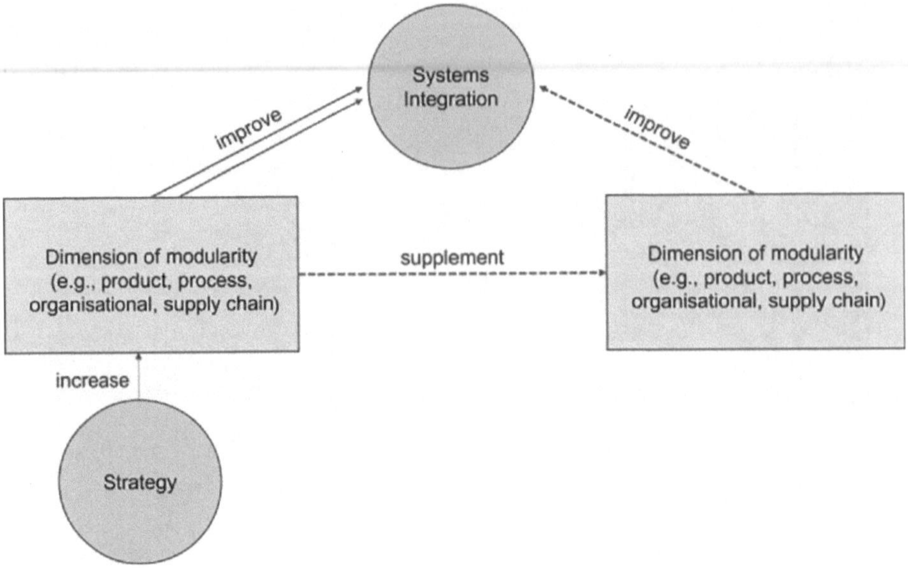

Figure 4.8 Modular complement relationship through the increase of modularity (i.e. addition complement)

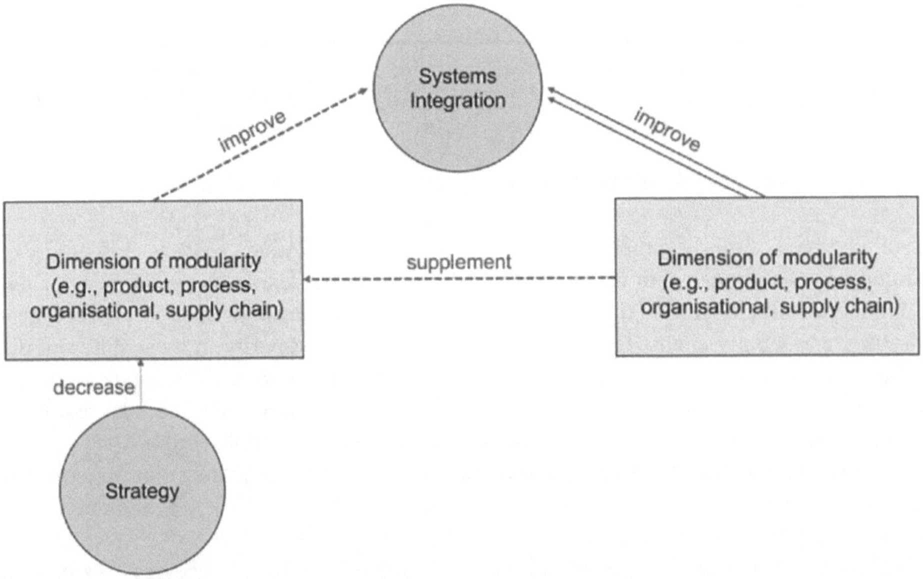

Figure 4.9 Modular complement relationship through the decrease of modularity (i.e. subtraction complement)

Reducing the modularity scale of the supply chain and adopting an integrated supply chain could potentially be a strategy to address the complexity challenges inherent in implementing DFAB. This degree of misalignment may facilitate system integration by tackling the multifaceted issues associated with DFAB execution (see Figure 4.9). IPD and an integrated supply chain, used to enhance collaboration and quality control, can address the complexity challenges in DFAB. IPD encourages tighter collaboration among stakeholders, improving coordination, reducing errors, and raising project quality. An integrated supply chain allows for comprehensive quality control and traceability of each component. The full involvement of all parties across project stages often results in higher satisfaction with project outcomes.

4.5.3 Reflecting on project delivery to DFAB

Understanding and managing modularity misalignment provides valuable insights for devising solutions suited for DFAB project delivery, especially under varying degrees of DFAB implementation. First, DFAB projects necessitate an integrated perspective on products, processes, organisations, and supply chains. This calls for comprehensive understanding and coordination to ensure effective collaboration. For instance, when using modular design and manufacturing processes, the supply chain and organisational structure must support this modularity. Higher DFAB implementation complexity may involve more automation and intricate data management.

Second, the potential unpredictability caused by misalignment demands flexible and adaptive project delivery solutions. This may imply utilising tools and methods that accommodate change, like agile development or planning for possible changes from the project's inception. Projects with lower DFAB implementation might require stronger flexibility and adaptability due to increased manual involvement and frequent alterations. Third, misalignment represents an opportunity for innovation and improvement in our project delivery solutions. This could include adopting new technologies or methods, improving processes or organisational structures, or finding better ways to coordinate our supply chains. In projects with lower DFAB implementation, innovation might depend more on human resources and knowledge, while technology and data might be key in higher implementation projects.

Although addressing misalignment might increase project complexity and difficulty, the long-term benefits are substantial. A highly modular and integrated project delivery solution enhances efficiency, quality, and satisfaction, boosting long-term project success. Different investments and strategies might be required for projects with varying implementation degrees for optimal results. Overall, understanding and managing the misalignment of modularity across various dimensions can enhance the design and implementation of DFAB project delivery solutions, improving project success rates. Strategies and methods must be tailored based on the degree of DFAB implementation to effectively tackle potential challenges and opportunities.

4.6 Conclusions

This study conducts a comparative analysis of two cases implementing DFAB at varying degrees (full and partial) to identify relevant project delivery strategic patterns. As the adoption of DFAB in the broader field of architecture and the exploration of associated project delivery methods remain emergent topics, this study attempts to fill a gap in the existing body of knowledge through its exploratory case study approach. A key finding is that the degree of DFAB implementation significantly influences the interrelationship between DFAB and project delivery solutions. The level of DFAB implementation doesn't necessarily correlate with the degree of difficulty or challenges faced; regardless of its intensity, DFAB implementation will impact project delivery solutions, but the manner of these impacts will vary. Projects with a high level of DFAB adoption necessitate sophisticated data management systems due to increased automation and interconnectivity, demanding a higher degree of collaboration. These projects face potential technological risks and require significant initial investments. However, they also present abundant opportunities for process optimisation and efficiency gains through data analytics and automation. Conversely, projects with a lower degree of DFAB adoption predominantly depend on human-centric coordination and traditional construction knowledge. While these projects may experience a greater likelihood of human error, they also leverage human creativity and experience for innovation. Initial investments for these projects tend to be smaller but may incur higher operational costs due to increased manual labour. Thus, project delivery solutions must consider these dynamics, tailoring strategies to suit the specific demands and conditions inherent in the degree of DFAB implementation.

This study invites a reconsideration of modularity and integration in architecture, using a comparative case study to show that these two concepts can coexist due to inconsistencies among different system dimensions and hierarchical levels. In the context of DFAB, modular processes increase adaptability and efficiency, while integrated supply chains tackle complexity challenges inherent in DFAB implementation. Notably, this study diverges from Fine's three dimensions of modularity theory, revealing misalignment relationships among product, process, and supply chain. Understanding and managing this modularity misalignment is crucial for devising DFAB project delivery solutions. It necessitates an integrated perspective on products, processes, organisations, and supply chains. The unpredictability triggered by misalignment requires adaptive project delivery solutions and provides a platform for innovation. Despite the complexity of addressing misalignment, the potential for improved efficiency, quality, and satisfaction boosts long-term project success. In conclusion, managing modularity misalignment across various dimensions can optimise DFAB project delivery solutions, enhancing project success rates. Tailored strategies based on the DFAB implementation degree are required to effectively address potential challenges and opportunities. This comparative case study on DFAB project delivery has limitations. The research findings might be bound by the confines of theoretical models, such as the modularity theory, which have inherent limitations that affect outcomes. Future research could proceed in several directions:

- Comparing architectural projects and industries across different geographical contexts to better understand the role of DFAB in project delivery.
- Identifying measurable relationships and trends, complemented by longitudinal studies to uncover the long-term impacts of DFAB implementation
- Examining the influence of evolving DFAB technologies, including AI, robotics, and automation, on project delivery mechanisms.

4.7 Acknowledgement

This chapter is based on the work of DFAB by the authors. Some contents of Sections Four and Five have been published in peer-reviewed journals, including Tan et al. (2022, 2024) and Graser et al. (2021). For copyright issues, please contact the publishing editor of the book.

References

Agustí-Juan, I. & Habert, G. 2017. Environmental design guidelines for digital fabrication. *Journal of cleaner production*, 142, 2780–2791.

Al Khalil, M. I. 2002. Selecting the appropriate project delivery method using AHP. *International journal of project management*, 20, 469–474.

Baldwin, C. Y., Clark, K. B. & Clark, K. B. 2000. *Design rules: The power of modularity*, MIT Press.

Bock, T. & Linner, T. 2015. *Robot oriented design*, Cambridge University Press.

Campagnolo, D. & Camuffo, A. 2010. The concept of modularity in management studies: A literature review. *International journal of management reviews*, 12, 259–283.

Collins, R., Bechler, K. & Pires, S. 1997. Outsourcing in the automotive industry: From JIT to modular consortia. *European management journal*, 15, 498–508.
Doran, D. & Giannakis, M. 2011. An examination of a modular supply chain: A construction sector perspective. *Supply chain management: An international journal*, 16(4), 260–270.
Doran, D. & Roome, R. 2003. An evaluation of value-transfer within a modular supply chain. *Proceedings of the institution of mechanical engineers, part D: Journal of automobile engineering*, 217, 521–527.
Fine, C. H. 2010. *Clockspeed: Winning industry control in the age of temporary advantage*, ReadHowYouWant.com.
Fine, C. H., Golany, B. & Naseraldin, H. 2005. Modeling tradeoffs in three-dimensional concurrent engineering: A goal programming approach. *Journal of operations management*, 23, 389–403.
Graser, K., Kahlert, A. & Hall, D. M. 2021. Dfab house: Implications of a building-scale demonstrator for adoption of digital fabrication in AEC. *Construction management and economics*, 39, 853–873.
Ng, M. S., Chen, Q., Hall, D. M., Hackl, J. & Adey, B. T. 2022. Designing for digital fabrication: An empirical study of industry needs, perceived benefits, and strategies for adoption. *Journal of management in engineering*, 38, 04022052.
Ng, M. S., Graser, K. & Hall, D. M. 2021. Digital fabrication, BIM and early contractor involvement in design in construction projects: A comparative case study. *Architectural engineering and design management*, 19, 39–55.
Pan, W. 2019. Modular integrated construction: Innovating higher. *The Hong Kong Institution of Engineers (HKIE) Building Division 18th Annual Seminar: Innovation and Technology for Productivity Enhancement.*
Pawar, V. M., Stuart-Smith, R. & Scully, P. 2017. Toward autonomous architecture: The convergence of digital design, robotics, and the built environment. *Science robotics*, 2, eaan3686.
Salvador, F., Forza, C. & Rungtusanatham, M. 2002. Modularity, product variety, production volume, and component sourcing: Theorizing beyond generic prescriptions. *Journal of operations management*, 20, 549–575.
Tan, T., Mills, G. & Papadonikolaki, E. 2024. Exploring multidimensional modularity: Strategies to reduce complexity in design activities. *Journal of management in engineering*, 40, 05024002.
Tan, T., Mills, G., Papadonikolaki, E., Li, B. & Huang, J. 2022. Digital-enabled Design for Manufacture and Assembly (DfMA) in offsite construction: A modularity perspective for the product and process integration. *Architectural engineering and design management*, 1–16.
Tan, T., Ng, M. S. & Hall, D. 2023. Demystifying barriers to digital fabrication in architecture. *Engineering project organization conference*, Berlin, Germany.
Tee, R., Davies, A. & Whyte, J. 2019. Modular designs and integrating practices: Managing collaboration through coordination and cooperation. *Research policy*, 48, 51–61.
Ulrich, K. 1995. The role of product architecture in the manufacturing firm. *Research policy*, 24, 419–440.
Voordijk, H., Meijboom, B. & De Haan, J. 2006. Modularity in supply chains: A multiple case study in the construction industry. *International journal of operations & production management*, 26(6), 600–618.
Wolters, M. 2002. *The business of modularity and the modularity of business* (No. ERIM PhD Series; EPS-2002-011-LIS).
Yin, R. K. 2009. *Case study research: Design and methods*, Sage.

5

UNLOCKING THE POWER OF LANGUAGE MODELS FOR SMART CONFIGURATION IN THE AEC DOMAIN USING SPOKEN LANGUAGE UNDERSTANDING

Yinyi Wei[1], Xiao Li[2], Chengke Wu[3] and Frank Petzold[4]

[1]DEPARTMENT OF CIVIL ENGINEERING, THE UNIVERSITY OF HONG KONG,
EMAIL: WYYY@CONNECT.HKU.HK.

[2]DEPARTMENT OF CIVIL ENGINEERING, THE UNIVERSITY OF HONG KONG,
CORRESPONDING AUTHOR, EMAIL: SHELL.X.LI@HKU.HK.

[3]SHENZHEN INSTITUTE OF ADVANCED TECHNOLOGY, CHINESE ACADEMY OF SCIENCE,
EMAIL: CK.WU@SIAT.AC.CN.

[4]TUM SCHOOL OF ENGINEERING AND DESIGN, TECHNICAL UNIVERSITY OF MUNICH,
EMAIL: PETZOLD@TUM.DE.

5.1 Introduction

Undergoing three industrial revolutions, the current fourth-generation revolution, known as Industry 4.0, represents a notable paradigm shift that aims to integrate advanced technologies into industrial systems, fostering intelligence, efficiency, and interconnectedness. This transformation also extends its profound impact to the architecture, engineering, and construction (AEC) domain, where the utilization of big data and artificial intelligence plays a pivotal role in facilitating decision-making processes and enhancing production efficiency to align with the dynamic and intricate nature of project demands (Yan, et al., 2020). Industry 4.0 and smart construction highlight the significance of human-machine collaboration, wherein humans and machines work in synergy, capitalizing on each other's strengths. Consequently, machines must comprehend human-generated data and instructions and respond accordingly, while humans must appropriately guide the machines toward more meaningful directions. In practical terms of construction, although a small portion of data is structured, over 80% of data remains unstructured (Wu, et al., 2022). Moreover, since language and voice serve as the most accessible and economical means of human communication, a substantial portion of this unstructured data is text-based (Gharehchopogh & Khalifelu, 2011). Therefore, it becomes crucial to effectively convert unstructured textual data into structured information throughout the life cycle of a project, enabling further analysis and response (Wu, et al., 2022).

To meet the aforementioned requirements, natural language processing (NLP) has emerged as a prominent tool in the AEC domain. NLP, considered as the jewel in the artificial intelligence crown,

DOI: 10.1201/9781003383840-6

encompasses two primary tasks: natural language understanding (NLU) and natural language generation (NLG). NLU primarily focuses on comprehending textual information and extracting pertinent contents, such as information extraction (Grishman, 2015). On the other hand, NLG involves generating necessary knowledge or responses for users, as exemplified by chat systems (Chen, et al., 2017). In the AEC domain, NLP is commonly referred to as NLU, which is employed to interpret unstructured texts generated throughout the project life cycle. This usually entails transforming a given task into either a text classification task or a sequence labeling task, allowing for the classification of documents into predefined types or the extraction of desired information from a given sequence.

One direct application of NLP lies in smart configuration in the context of mass customization, which involves constant interaction between humans and building information modeling (BIM) for decision-making about the design and construction of modular buildings. Presently, prevalent approaches rely heavily on heuristic or traditional machine learning methods to parse and understand user needs (Shin & Issa, 2021; Nabavi, et al., 2023). However, the conventional NLP techniques employed in smart configuration encounter challenges in handling volatile and complex user-generated texts, impeding effective decision-making and the fulfillment of individual preferences. These challenges in the AEC domain stem from diverse educational and disciplinary backgrounds among stakeholders (Lee & Jeong, 2012), leading to disparities in stakeholder and disciplinary terminologies and the formation of isolated knowledge silos. For instance, variations in understanding terms as common as stakeholder and community can lead to misunderstandings and conflicting expectations (Colvin, et al., 2016). Therefore, it is crucial to utilize more precise techniques for understanding generated texts from various stakeholders. Failure to do so can result in discordance, inconsistency, and misunderstandings among stakeholders (Kaiya, et al., 2005), leading to significant errors and inaccuracies throughout the process.

To adapt to the evolving landscape of smart construction and meet the demands of smart configuration for mass customization, this chapter introduces a novel perspective by integrating language models into user needs understanding. Language models are artificial intelligence systems specifically designed to comprehend and generate human languages. Leveraging the established framework of spoken language understanding (SLU) in the field of NLU, we propose to conceptualize the process of user needs understanding as an SLU task. During the experimental design process, we recognize two crucial roles of bridle-wise language models, namely well-informed knowledge bases and powerful backbone models. To elaborate further, we illustrate their significance through the example of understanding user needs that pertain to attribute changes and queries about building elements. Specifically, as knowledge bases, language models facilitate the acquisition of professional and practical knowledge about modular buildings and assist in curating a complicated dataset. On the other hand, backbone models are utilized to address specific user needs and generate structured information for subsequent stages of configuration. Through extensive experiments and comprehensive analyses, we reveal the opportunities and challenges associated with the integration of language models in this domain, providing inspiration for further research and practical applications in the field of smart configuration and smart construction.

5.2 Background research

5.2.1 Natural language processing in smart construction

Within the context of smart construction, natural language processing (NLP) techniques are predominantly employed to handle a diverse range of documents generated throughout the life cycle of a project (Wu, et al., 2022), e.g., safety reports (Li, et al., 2021), collaboration texts (Lee &

Yi, 2017), constraint management (Wu, et al., 2023), and other documents from online sources (Zheng, et al., 2020). The tasks associated with smart construction can be regarded as natural language understanding (NLU) tasks, typically falling under the paradigms of either text classification or sequence labeling. Each paradigm undergoes the following technical process: (a) text pre-processing involves denoising and cleaning raw texts by employing tokenization, stop word removal, and stemming. (b) Text feature engineering indicates capturing relevant features from the provided texts. Syntactic features, obtained through syntactic analysis and semantic features, acquired from pre-trained word embeddings, are commonly utilized. (c) Text understanding is accomplished through rule-based methods or machine learning approaches, thus providing evidence for the analysis of new data. (d) Performance evaluation, which uses suitable metrics to assess the effectiveness of the designed methods, ensuring their suitability for real-world practical processes. By following this systematic scheme, NLP techniques are applied to a diverse range of construction documents. Smart construction can benefit from enhanced document management and knowledge extraction, improving efficiency and effectiveness.

For the paradigm of text classification, it can be generally divided into two distinct senses: narrow sense and broad sense. Classification under the narrow sense assigns a category label to each text, under which all texts share common features or express the same subject matter. This enables the automated organization of documents from various sources, such as injury narratives (Marucci-Wellman, et al., 2017; Fang, et al., 2020), construction accidents (Goh & Ubeynarayana, 2017; Zhang, et al., 2019), and building quality complaints and control (Zhong, et al., 2019; Zhang, et al., 2022). On the other hand, in the broad sense, classification involves the vectorization of a text, followed by retrieval or matching in the vector space to obtain similar samples, thus establishing a historical foundation for judgment and decision-making. Commonly encountered application scenarios include information retrieval in the BIM environment (Wu, et al., 2019; Wang, et al., 2022) and case-based reasoning utilizing historical documents (Zou, et al., 2017; Ayhan & Tokdemir, 2019).

Unlike text classification, which operates on the text as a whole, the paradigm of sequence labeling assigns a label to each token in a given text. In the context of smart construction, it generally involves identifying keywords in a text for further action (Nabavi, et al., 2023; Tian, et al., 2021) or recognizing entities and determining their intrinsic relations (Zhong, et al., 2020; Zheng, et al., 2022). Such tasks primarily aim to extract structured information from unstructured texts, facilitating the conditions for subsequent batch processing.

Through the utilization of NLP techniques, data can be effectively filtered, organized, and analyzed, facilitating the extraction and generation of valuable knowledge from vast quantities of information. The implementation of NLP enables various applications, including automated compliance checking and risk prediction, to significantly enhance safety and operational efficiency within the construction process.

5.2.2 Mass customization and smart configuration

Modular construction is a pioneering construction methodology that emphasizes the creation of three-dimensional and self-contained building modules in a controlled environment, followed by their subsequent transportation to the construction site for assembly (Langston & Zhang, 2021; Tan, et al., 2023). These modules are pre-engineered and pre-fabricated to encompass a diverse array of interior and exterior elements, such as structural, mechanical, electrical, and decorative components. This integrative strategy effectively encapsulates an entire section of a building. Compared with conventional construction, modular construction possesses notable advantages, including heightened productivity, environmental sustainability, and enhanced quality control.

In pursuit of enhanced flexibility and personalized design and construction, the application of mass customization from the manufacturing industry has been introduced into modular construction, which emphasizes the delicate balance between mass production and the fulfillment of individual preferences (Da Silveira, et al., 2001; Ghannad & Lee, 2023). The overarching objective of this syncretic concept is to achieve large-scale production of kit-of-parts that satisfy specific user requirements. The term kit-of-parts in this context refers to a curated selection of pre-engineered building components that are carefully designed to facilitate their assembly into diverse configurations, resulting in a fully realized modular building (Howe, et al., 1999).

With the concept of mass customization in mind, it is crucial to empower the planning, design, and construction process of modular construction by incorporating big data and advanced technologies (Cao, et al., 2021; Bakhshi, et al., 2022). This integration aims to cater to specific user needs and foster effective human-machine collaboration within the smart configuration process (Norouzi, et al., 2015). Configuration driven by artificial intelligence can be broadly categorized into two parts: computer vision-based (CV-based) and NLP-based. The former provides users with visually satisfying outcomes through automated optimization and generation techniques within the design processes, such as floor planning and layout design (Ghannad & Lee, 2022; Benjamin, et al., 2022). The latter focuses on voice- and text-based interactions between humans and configuration systems with the goal of acquiring structured information that can be used for subsequent modification or retrieval of kit-of-parts (Shin & Issa, 2021; Wang, et al., 2020; Wang, et al., 2022).

5.2.3 Language models

To derive meaningful and interpretable features, traditional approaches in NLP depend on manually crafted feature design, requiring domain expertise and substantial human efforts. After acquiring relevant features, these approaches utilize traditional machine learning techniques, such as support vector machine (Hearst, et al., 1998), convolutional neural networks (Gu, et al., 2018), and recurrent neural networks (Salehinejad, et al., 2017), to characterize the curated features and produce the outcomes. However, the advent of Transformers, an encoder-decoder architecture based on self-attention mechanisms (Vaswani, et al., 2017), has kicked off the paradigm shift in NLP.

The current scope of language models is primarily built upon the Transformer architecture, which can be broadly categorized into three main types: (a) Encoder-only architecture, such as BERT (Devlin, et al., 2018); (b) decoder-only architecture, exemplified by the GPT family (Brown, et al., 2020; Ouyang, et al., 2022); and (c) encoder-decoder architecture, as seen in T5 (Raffel, et al., 2020). These language models are trained using carefully designed pre-training tasks on extensive amounts of textual data, granting them the transformative ability to predict the likelihood of word sequences or generate new texts based on given inputs. The prevailing paradigm during this period involves pre-training and fine-tuning, wherein the knowledge embedded in language models is harnessed to enhance the performance of downstream tasks. However, small-scale language models also encounter numerous challenges, such as their limited ability to comprehensively capture intricate linguistic phenomena, as well as their inherent limitations in zero- and few-shot learning scenarios.

In recent times, there has been notable interest directed towards large language models (LLMs). To facilitate alignment with their smaller counterparts mentioned earlier, we denote large language models, whose parameter counts exceed one hundred billion, as large-scale language models in this chapter. Research has shown that increasing the model size not only enhances its performance in low-resource scenarios but also unveils emergent capabilities that are not observed in smaller models. This scalability facilitates strong generalizability across unseen tasks (Wei, et al., 2022).

Currently, ChatGPT[1] stands as one of the most popular applications of large-scale language models. It follows the training methodology of the GPT family and employs reinforcement learning from human feedback (RLHF) (Christiano, et al., 2017) on high-quality conversational-style datasets. In addition to its unparalleled dialogue capabilities, researchers have made extensive efforts to uncover other embedded capabilities in ChatGPT, such as its performance in various general NLP tasks (Li, et al., 2023; Pan, et al., 2023; Chang, et al., 2023). Moreover, investigations are underway to explore the potential impact of ChatGPT in diverse domains, such as the education sector (Basic, et al., 2023) and the medical realm (Tu, et al., 2023). Rather than relying on fine-tuning with labeled data from downstream tasks, large-scale language models can achieve satisfactory results through careful prompt engineering, which involves utilizing only a few samples or just task descriptions.

This chapter aims to investigate the capabilities of multi-scale language models for understanding user needs in the context of smart configuration, revealing the opportunities and challenges that may arise.

5.2.4 Spoken language understanding

Existing efforts to characterize user needs have predominantly relied on keyword searching and expansion [6, 7]. However, there is a lack of a unified definition and systematic framework to comprehensively encompass the entire process, from initial comprehension to subsequent disposal. To address this gap, this chapter proposes the formulation of user needs understanding as a spoken language task for normalization.

Spoken language understanding (SLU), also referred to as dialogue understanding, is a fundamental component within task-oriented dialogue systems (Qin, et al., 2021; Weld, et al., 2022). Its primary objective is to extract structured semantic frames from input texts, which may originate from either voice or raw text sources. In the case of voice inputs, automatic speech recognition is employed prior to the processing stage. SLU comprises two main sub-tasks: intent detection and slot filling. For a user query, the intent represents the user's desired outcome or goal, indicating their underlying intention. On the other hand, slots are the associated arguments that align with the expressed intention. Usually, the slots manifest as individual words or spans of words within the input. However, in certain cases, the slots may not be explicitly evident in the input. Figure 5.1 illustrates examples from two domains, movie and smart configuration, to provide further clarity.

To solve the SLU task, intent detection is commonly approached as a text classification task, aiming to leverage the whole semantics of a given text to identify the underlying intent. On the other hand, slot-filling is typically modeled as a sequence labeling task, wherein a label is assigned to each individual token using the BIO-format. To mitigate potential issues related to error

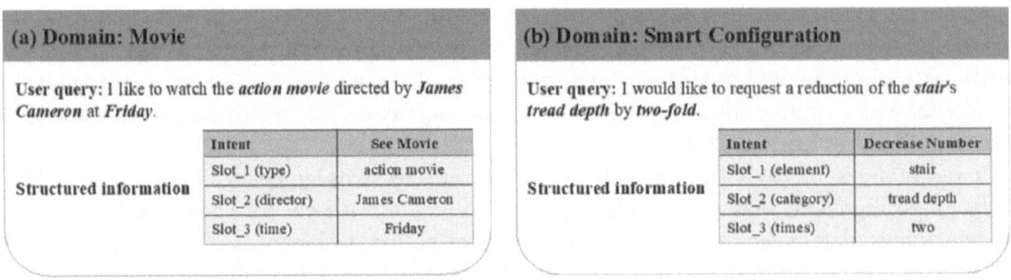

Figure 5.1 Examples of spoken language understanding (SLU) specifically focus on intent detection and slot filling within two domains: movie and smart configuration

propagation and cascading, many existing studies adopt a joint modeling approach to simultaneously address both intent detection and slot-filling sub-tasks (Cui, et al., 2023).

In this chapter, we detail the dual roles of language models through the experimental process of the transformed user needs understanding task, namely knowledge bases and backbone models. To address the scarcity of data, we adopt ChatGPT as a knowledge base and leverage its capabilities to curate a dataset in the form of an SLU task. Additionally, we utilize both small-scale and large-scale language models as backbone models to effectively handle user queries in a target manner.

5.3 Language models as knowledge bases

As the concept of NLP-based smart configuration in the AEC domain is still emerging, there is currently a lack of open-source datasets available in this domain. Consequently, this chapter aims to ameliorate this gap by creating a dataset from scratch that pertains to attribute changes and queries about building elements. The entire process is illustrated in Figure 5.2.

The first step in the process involves the identification of building elements and their respective attributes and properties. To accomplish this, one effective approach is to solicit answers directly from ChatGPT. Since the fact that ChatGPT is trained on a plethora of textual data across various domains, it can generate domain-specific knowledge through appropriate prompt-based guidance. Nevertheless, it is important to note that generative models like ChatGPT are inherently uncontrollable, which introduces the possibility of generating harmful or factually incorrect content. To mitigate the potential risks associated with this issue, we employed external knowledge sources to pilot and rectify the knowledge acquisition process. In particular, we used two external references: the Industry Foundation Classes (IFC) specification and the buildingSMART Data Dictionary (bSDD) compiled by buildingSMART,[2] along with the Building Element Ontology developed by Pieter Pauwels.[3]

For the technical process in this chapter, with the assistance of ChatGPT and other external references, we carefully picked 11 building elements in the scope of smart configuration: beam,

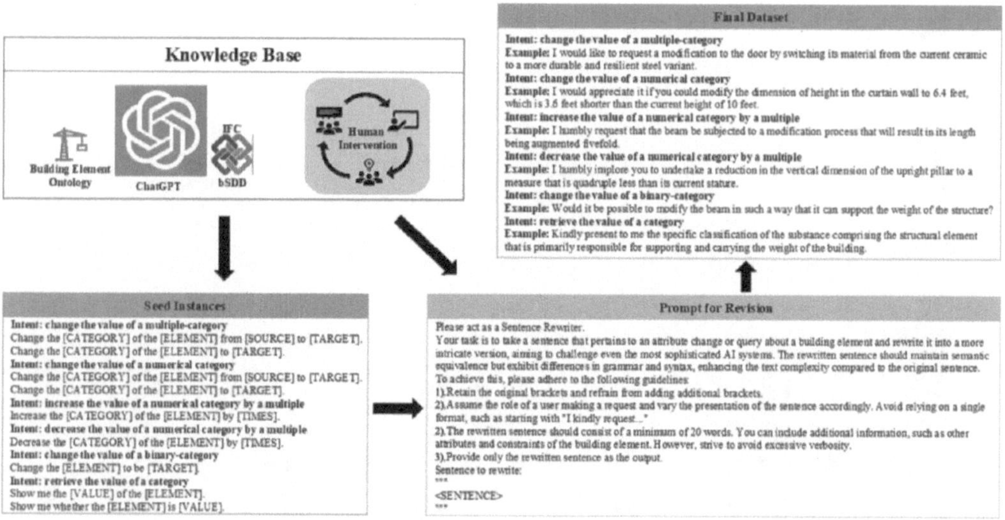

Figure 5.2 Utilizing ChatGPT as a knowledge base to acquire professional knowledge in the AEC domain and employing it to refine seed instances into final queries in line with real-life scenarios

chimney, column, covering, curtain wall, door, railing, roof, stair, wall, and window. The categorization of attributes and properties for each building element is structured into three distinct categories. These categories were designed primarily for evaluating the ability of language models to understand diverse user needs and do not ensure the completeness of the taxonomy. Specifically, the three categories are as follows: (a) multiple category, encompassing properties with three or more options; (b) numerical category, represented by specific numerical values; and (c) binary category, featuring properties with two mutually exclusive options. For instance, when considering the building element of a window, the frame material falls under multiple category, offering choices such as wood, aluminum, steel, vinyl, glulam, stone, and concrete. On the other hand, the width of the window corresponds to the numerical category, while the ventilated attribute is under the binary category, indicating whether the window should be ventilated or not.

To obtain the fundamental user queries, we manually compiled a set of seed instances, each corresponding to one of the six distinct intents: (a) change the value of a multiple category; (b) change the value of a numerical category; (c) increase the value of a numerical category by a multiple; (d) decrease the value of a numerical category by a multiple; (e) change the value of a binary category; and (f) retrieve the value of a category. These seed instances serve as templates, with placeholders denoted by brackets, which would be automatically populated in the subsequent steps. The identified intents cover the basic needs of potential users for NLP-based smart configuration in the AEC domain. We then filled in the placeholders with building elements, their respective attributes and properties, in a predetermined proportion, yielding the fundamental user queries.

The data acquired in this way possessed an oversimplified grammar and syntax, rendering it inadequate and unqualified for real-life scenarios where the user queries may be variable, intricate, and ambiguous. To address this limitation, we proceeded with ChatGPT to rephrase the fundamental user queries into more complicated versions that maintain semantic equivalence but exhibit differences in grammar and syntax. The specific prompt employed for this purpose is illustrated in Figure 5.2.

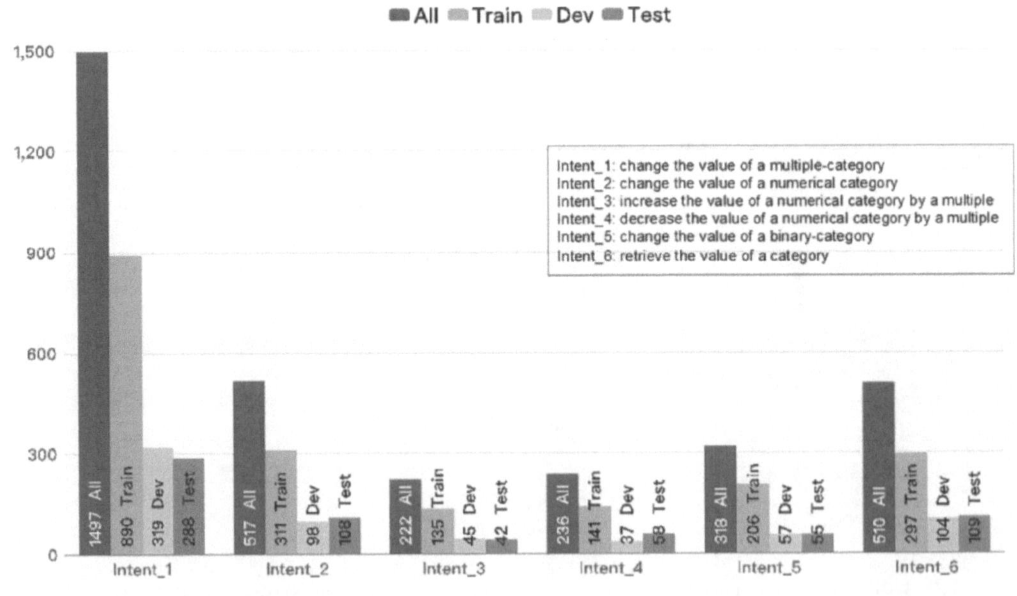

Figure 5.3 Statistics of the curated dataset

We produced 300 queries for each building element, resulting in a total of 3,300 queries. To determine the composition of the queries for each intent associated with a building element, we employed weight-based sampling with the following proportion: 8:5:5:5:4:6. The whole dataset was then partitioned into a training set, development set, and test set, with a ratio of 6:2:2. The generated dataset fit the real-life scenarios well with the nature of imbalanced data distribution. The statistical details of the curated dataset are presented in Figure 5.3.

This section presents an approach for creating a comprehensive dataset leveraging ChatGPT as a knowledge base to provide domain-specific knowledge in the AEC domain and to be a user-informed rewriter to enhance the complexity of plain queries. Throughout this process, human intervention is essential. To ensure smooth and successful operations, language models require guidance from humans, including prompt engineering, validation, and correction of generated results, to carry out specific tasks effectively.

5.4 Language models as backbone models

When a user query arrives, the identification of the object and corresponding operations has traditionally relied on keyword extraction in previous methods (Shin & Issa, 2021; Wu, et al., 2019; Wang, et al., 2022). This process typically involves analyzing the query from a syntactic perspective, determining whether a word serves as a keyword based on its syntactic role in the query. Based on empirical and heuristic observations, keywords, which encompass users' intents, are considered to be located within the noun nodes or noun phrases that are connected along the path from the root to the boundary nodes in the syntactic trees. Once the keywords have been extracted, further steps are taken to access standardized terms, including keyword expansion and mapping using domain ontologies. However, these methods face challenges when dealing with complex user queries that involve implicit intents and multiple corresponding slots. Such queries may consist of irregular multi-token keywords and possess more than one noun and noun phrase, making them difficult to handle using these simplistic approaches.

Considering the unprecedented capabilities of language models in NLU tasks, the study presents three prospective methods to address the aforementioned challenges by leveraging diverse

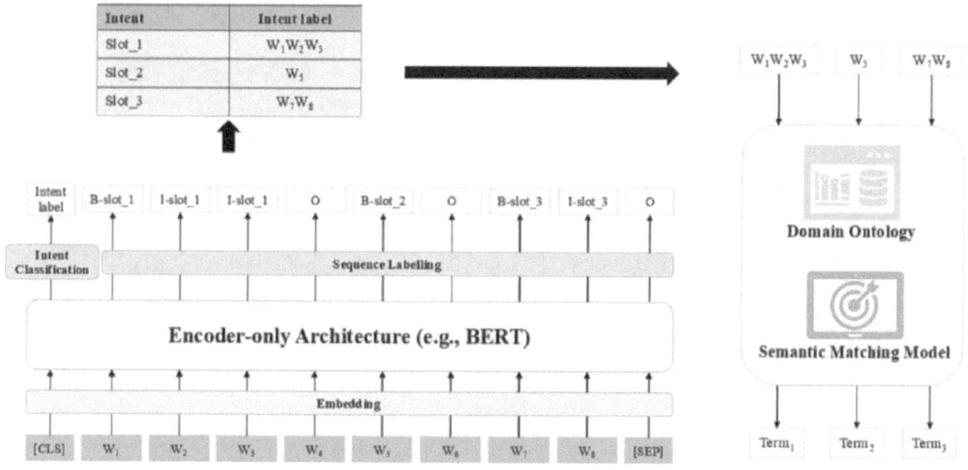

Figure 5.4 Utilizing encoder-only architectures (e.g., BERT) for user needs understanding

Transformer-based architectures as backbone models. These prospective backbone models are described here:

(a) Encoder-only architectures for user needs understanding

Based on the underlying architecture of the Transformer's encoder, a pre-trained language model exhibits a comprehensive understanding of user queries, thus enabling the provision of meaningful representations for input tokens. To address downstream tasks, classification heads can be seamlessly incorporated atop the encoder, facilitating efficient handling of these tasks.

The process of utilizing encoder-only architectures for user needs understanding is depicted in Figure 5.4. Specifically, the task of user needs understanding can be formulated as a combination of two tasks: intent detection and slot filling. Initially, the user query is tokenized into a sequence of tokens. Subsequently, the encoder-only architecture, such as BERT, is employed to encode the tokens into representations within a vector space. To determine the intent of the input query, a text classification head is applied to the representation of the [CLS] token, which can be considered as the comprehensive representation of the input query. This intent classification task is akin to a general text classification problem. On the other hand, slot information is obtained through a sequence labeling task utilizing BIO-format, where each token is labeled as B-Slot (Begin Slot), I-Slot (Inside Slot), or O (Others). To concurrently model these two tasks in a joint mode, the cross-entropy losses from both tasks are combined to form the final optimization objective.

As with traditional approaches, the aforementioned process also needs to map the extracted slots to standardized terms. To this end, a domain ontology comprising the standardized terms and a semantic matching model for term-matching are utilized to address the challenge of aligning slots with their corresponding standardized terms.

Despite the application of language models, the presence of implicit slots in input queries still poses challenges. For instance, consider the query "Make the wall beige." In this case, the intent of the query is "change the value of a multiple category," and the corresponding slot "multiple category" is "color," which, however, is not explicitly mentioned in the query. This issue highlights a limitation of encoder-only architectures, as they can only capture individual words or spans of words within the input. Consequently, we do not implement this approach in our subsequent experiments.

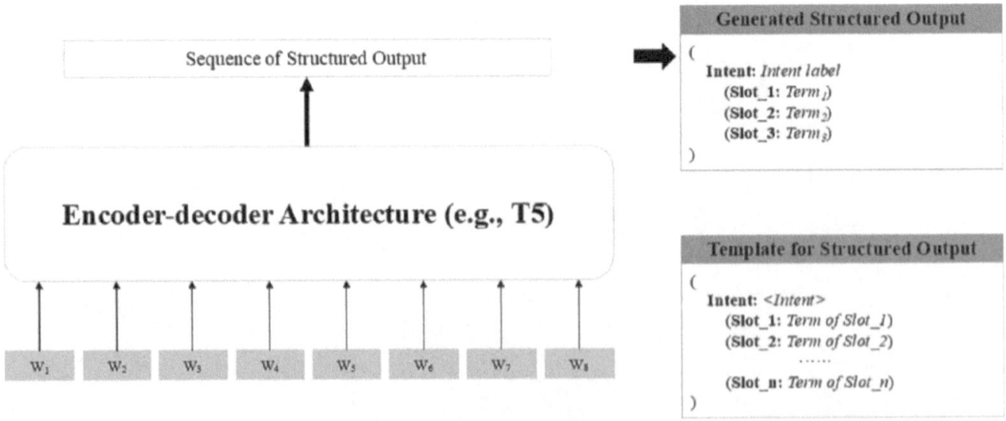

Figure 5.5 Utilizing encoder-decoder architectures (e.g., T5) for user needs understanding

(b) Encoder-decoder architectures for user needs understanding

The incorporation of both an encoder and decoder within the Transformer architecture offers remarkable potential in comprehending input queries and generating the desired output. Consequently, we propose addressing the user needs understanding task through the utilization of an encoder-decoder architecture, as depicted in Figure 5.5.

In order to utilize a language model with an encoder-decoder architecture for a general natural language understanding task, it is crucial to devise the structure of the output text, thereby facilitating the conversion into a text-to-text task. As shown in Figure 5.5, we design a template for structured output, incorporating the intent of an input query and its associated slots in a unified manner. This method efficiently addresses the user needs understanding task by transforming it into a text-to-structure task, eliminating the need for additional term-matching procedures employed in encoder-only architectures.

During the proposed process, the input query is first tokenized into individual tokens, and then an encoder-decoder architecture, such as T5, is utilized to encode the input and decode it into a sequence of structured output. The resulting output adheres to the predefined format of the designed structured output. After appropriate training, the model becomes capable of generating the intent and slots for a user query in a structured output format, consisting of standardized terms. Moreover, the model exhibits the ability to automatically deduce implicit slots based on the semantic information in the user query. To yield the desired results, the final step involves decoding the structured output into specific intents and slots. This process can be effectively executed through the utilization of regular expressions in conjunction with manually designed rules.

(c) Decoder-only architectures for user needs understanding

Language models built on the Transformer's decoder architecture are trained using extensive text corpora with the objective to predict the next token based on the existing token sequence. These models, known as autoregressive language models, were initially utilized for natural language generation

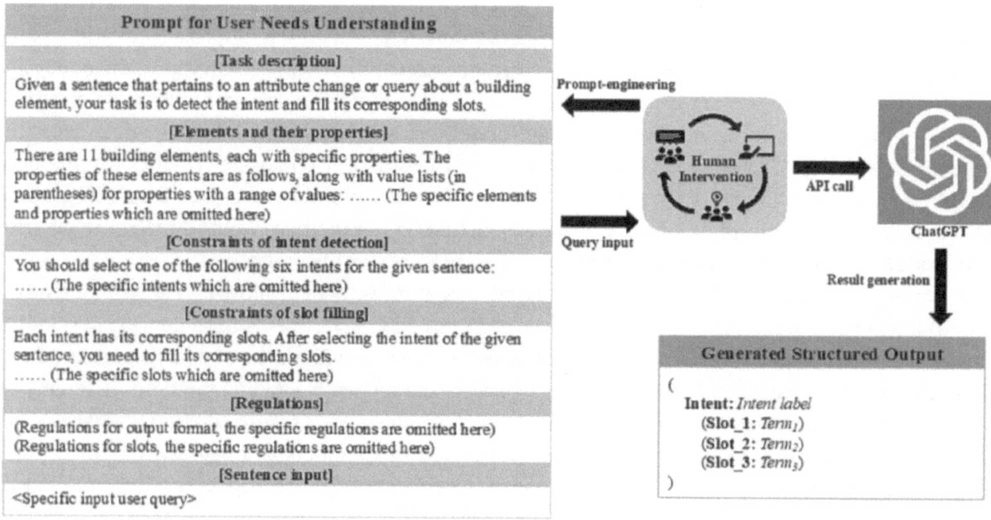

Figure 5.6 Utilizing decoder-only architectures, specifically large-scale language models like ChatGPT, for user needs understanding

purposes. Nevertheless, recent studies have demonstrated that, by scaling up their size, decoder-only architectures can effectively tackle natural language understanding tasks in low-resource scenarios with only a few samples or just task descriptions. This ability is considered one of the advantageous features of large-scale language models. Presently, owing to the interplay between architectures and parameters, large-scale language models consistently adopt decoder-only architectures.

In consideration of the significant workload and cost associated with fine-tuning large-scale language models for downstream tasks, we exclusively harness their capabilities for the user needs understanding task, utilizing only the task description and pertinent details. The entire process is illustrated in Figure 5.6. In order to effectively utilize large-scale language models like ChatGPT, prompt engineering emerges as a pivotal step. A well-constructed prompt necessitates careful compilation and thorough revision prior to its utilization. As with the encoder-decoder architectures, the final step in this regard also relies on a decoding procedure to extract intents and slots from structured outputs.

5.5 Experiment and evaluation

In the previous sections, we have discussed the two roles played by language models in the user needs understanding task. First, we utilized large-scale language models as knowledge bases to construct a dataset pertaining to attribute changes or queries about building elements. Second, we employed language models as backbone models to devise concrete methods for identifying user intents and populating the relevant slots within an input query. In this section, we conducted experiments and evaluations to further elaborate on the opportunities and challenges associated with the use of language models in NLP-based smart configuration in the AEC domain.

5.5.1 Experimental setup

Given the limitations of encoder-only architectures in handling implicit slots, we only focused on the implementation of encoder-decoder architectures and decoder-only architectures for effective user needs understanding. As an exemplar of encoder-decoder architectures, T5 (Raffel, et al., 2020) presents a text-to-text framework that fosters a coherent and unified approach for addressing diverse NLP tasks, resulting in cutting-edge performance on numerous benchmarks. Consequently, we employed T5 in various sizes as small-scale language models of the encoder-decoder architecture. Additionally, for a large-scale language model, we adopted ChatGPT as the decoder-only architecture. Notably, ChatGPT was exclusively used in zero-shot scenarios, where only the task description and pertinent details are provided. It is worth mentioning that we used ChatGPT during both the dataset curation and prediction processes, which is justified as we used the same version of the model for both processes, mitigating potential risks of data leakage.

We evaluated our proposed methods on our curated dataset, using the following metrics for assessment: (a) intent accuracy, quantified as the ratio of correctly identified intents of sentences; (b) slot F1 scores, representing the harmonic mean scores between precisions and recalls of exact slots extraction and matching with standardized terms; and (c) overall accuracy, calculated as the ratio of sentences in which both intents and slots are correctly predicted. The three metrics were adopted to evaluate the performance of intent detection, slot filling, and overall user need understanding, respectively.

In our implementation, we selected three specific variants of the T5 model, namely T5-small, T5-base, and T5-large, based on their parameter count. To optimize the models, we employed the AdamW optimizer with a learning rate of 3e-4 and a weight decay of 5e-5. The training process was conducted for a maximum of 20 epochs, with a batch size of 8. To ensure reliable results, we

Table 5.1 Experimental results

Model	*Intent Acc. (%)*	*Slot F1. (%)*	*Overall Acc. (%)*	*Parameter Count*
Full data training				
T5-small	99.49 (± 0.08)	98.51 (± 0.13)	96.12 (± 0.37)	≈ 60 million
T5-base	99.94 (± 0.07)	99.02 (± 0.04)	96.97 (± 0.13)	≈ 220 million
T5-large	99.82 (± 0.15)	99.02 (± 0.14)	96.88 (± 0.48)	≈ 770 million
Zero-shot Prediction				
ChatGPT	96.36	94.17	82.42	Not Given

Note: for variants of the T5 model, the mean values and standard deviations are shown. The parameter count of each model is also presented.

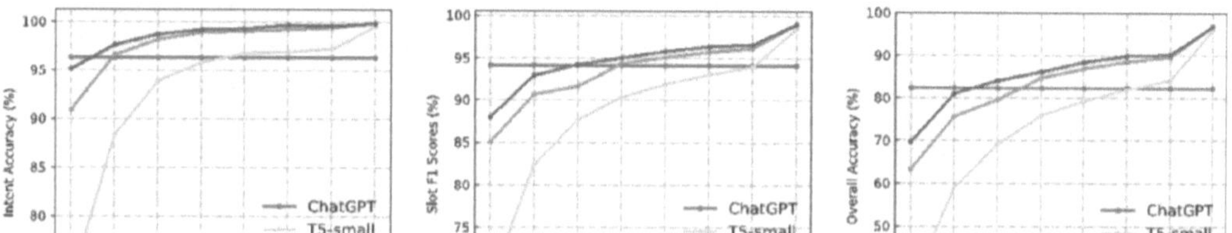

Figure 5.7 Results of k-shot experiments with variants of the T5 model, in comparison with zero-shot prediction of ChatGPT

performed five independent model runs on five different random seeds and averaged the outcomes. For the experiments involving ChatGPT, we utilized the API of the GPT-3.5-turbo model in its May 24 version.

5.5.2 Experimental results

The experimental results evaluating the effectiveness of language models as backbone models for understanding user needs are presented in Table 5.1, including full training with different variants of the T5 model, as well as zero-shot prediction using ChatGPT. From Table 5.1, we can observe that (a) in terms of encoder-decoder architectures, the performance generally improved as the model size was scaled up. However, T5-large performed worse than T5-base, likely due to both models being trained with the same arguments, which leads to insufficient training for the much larger T5-large model; (b) the large-scale language model ChatGPT produced satisfactory results solely based on the task description and relevant details. This demonstrated the remarkable capabilities of such models in the user needs understanding task.

As discussed in the preceding sections, the curated dataset exhibits an imbalanced distribution, thereby exerting pressure on the training procedure. To assess the performance of the T5-model in low-resource scenarios and demonstrate the zero-shot prediction capabilities of ChatGPT, we randomly picked k samples for each intent to train different variants of the T5 model and evaluated them on the complete test set. The results are shown in Figure 5.7. Notably, it required approximately 70, 40, and 30 samples, respectively, for each intent of T5-small, T5-base, and T5-large to achieve comparable performance to ChatGPT's zero-shot prediction outcomes across all evaluation metrics, highlighting the impressive prowess of ChatGPT.

5.5.3 Error analysis

Based on the experimental process, we analyzed the errors and limitations of different kinds of language models through concrete cases.

5.5.3.1 Limitations of the T5 models

(a) Semantics misunderstanding of user needs.

Due to the capacity constraints of small-scale language models, they might misunderstand the semantics of user needs to some extent. For example, the case, "Could you present to me the type of substance that constitutes the safety barrier installed along the perimeter?" refers to a query to retrieve the material of the railing. The desired slot, denoted as the "retrieved value," should be identified as the "material." However, due to the presence of the word "type," the T5-model mistakenly recognized the slot as "type" instead. This observation highlights the ongoing challenges faced by small-scale language models in comprehending the precise local semantics of user needs.

(b) Lack of association abilities.

Previously, we mentioned one merit of encoder-decoder architectures: their ability to generate outputs directly without resorting to additional term-matching procedures. However, in situations where the expression is highly complex and lacks corresponding training data, these architectures are unable to effectively address the issue of association. For instance, consider the following case: "may I be presented with the composition constituting the vertical barrier serving as a separation within the structure?" The desired label for the slot "building element" is "wall," as indicated by the expression "vertical barrier serving as a separation within the structure," which, however, was wrongly identified as "railing" by the T5 model.

(c) Highly dependent on sufficient training data.

In this chapter, the user needs understanding task is addressed by transforming it into a text-to-structure task using encoder-decoder architectures. However, to ensure accurate generation of the desired output format, these models require a substantial amount of training data. In cases where the training data is limited, the models are prone to making mistakes. For instance, consider the following user inquiry: "I would like to inquire if it is feasible to reduce the pitch angle of the roofing system by a factor of 9, as this modification would greatly improve the overall design of the building while maintaining its structural integrity." The corresponding label for the "numerical category" slot is "slope," which is indicated by the term "pitch angle." Unfortunately, the T5 model tended to use the term "pitch angle" instead of the standardized term "slope." This discrepancy arose due to the scarcity of training samples that contain the term "slope," accounting for only 0.9% of the entire training dataset. Hence, insufficient training on this specific term and its many different synonyms led to inaccuracies in the generated outputs.

(d) Insensitivity of numbers.

Due to the inherent limitations of small-scale language models compared to their large-scale counterparts, they encounter difficulties in effectively converting numerical values from digital forms to word forms and vice versa. To demonstrate this challenge, consider the following example: "could you please alter the vertical elevation of the smokestack to a measure of nine thousand seven hundred and thirty-nine millimeters?" The corresponding representation for the "target value" slot is expressed as "9,739 mm" in word form. Unfortunately, the T5 model fell short of successfully transitioning from the word form to the digital form.

5.5.3.2 Limitation of ChatGPT

In our experimental process, we only employed ChatGPT for zero-shot prediction, wherein only the task description and relevant details are available. The results demonstrated the superiority of directly applying ChatGPT. Nevertheless, it is imperative to acknowledge the presence of certain limitations, primarily manifested in two aspects.

First, ChatGPT exhibits restrictions in effectively managing domain-specific terminologies, resulting in imprecise classification. Second, ChatGPT's inconsistent adherence to all instructions provided in the prompt introduces the possibility of generating unexpected outputs. These outputs may deviate from the desired format or contain irrelevant content, thereby bringing substantial errors to the decoding procedure.

5.6 Discussion

In this chapter, we conducted a comprehensive research process to gain insights into user needs understanding for smart configuration in the AEC domain. Our approach involved the development of appropriate approaches and the design of experiments, leveraging the two roles of language models: knowledge bases and backbone models. By exploring the diverse architectures and scales of language models, we revealed their superiority and limitations. In this section, we delved deeper into the potential opportunities and challenges associated with NLP-based smart configuration in the AEC domain.

5.6.1 Opportunities

5.6.1.1 Domain-specific language model

In our process of curating the dataset and conducting experiments, we identified the limitations of both small-scale and large-scale language models when handling domain-specific texts within the context of smart configuration, which are replete with a multitude of professional terminologies. These challenges are further exacerbated in the broader AEC domain, wherein diverse terminology guidelines are adopted for various projects and specifications, thereby introducing substantial complexities in comprehending texts in the AEC domain, particularly in low-resource scenarios. Recognizing this gap, Zheng et al. (2022) introduced ARCBERT, a BERT-based language model designed specifically for the AEC domain. While large-scale language models have gained considerable prominence in various fields, such as ChatLaw (Cui, et al., 2023) in the legal domain and BloombergGPT (Wu, et al., 2023) in the financial domain, there is currently a lack of similar large-scale language models in the AEC domain to facilitate its digital transformation. Consequently, there is an urgent need to develop large-scale language models tailored for specific scenarios in this domain.

5.6.1.2 Integration with domain ontology

In the AEC domain, most approaches involve utilizing domain ontologies to manage diverse concepts and definitions across the entire life cycle of a project (Li, et al., 2023; Li, et al., 2022). In order to leverage the efficiency and comprehensiveness of these domain ontologies and to enhance the interpretability and validity of large-scale language models in the AEC domain, additional efforts can be undertaken to integrate language models with pre-existing domain ontologies.

5.6.1.3 Utilization of external tools

Without precise guidance (e.g., prompt engineering), large-scale language models are challenged when handling specific tasks, particularly those featured with abundant instructions and constraints. To streamline the utilization process and unleash the full potential of these models, external tools have been devised. For example, Auto-GPT[4] enables users to achieve comprehensive objectives by completing partial and disassembled goals. Similarly, LangChain[5] enhances the capabilities of large-scale language models by combining them with additional sources of computation or knowledge, resulting in a truly formidable application. Given the abundant documentation and repositories available in the AEC domain, it becomes imperative to devise strategies for seamless integration and effective utilization of these invaluable knowledge sources with external tools capable of harnessing large-scale language models. Through judicious employment of external tools, considerable progress can be achieved in augmenting the efficacy and efficiency of large-scale language models within practical scenarios. This integration holds promising potential for advancing the state of the art in smart design and construction, enabling novel and sophisticated applications in the industry.

5.6.1.4 Multi-modal models

Despite the substantial presence of textual data in the AEC domain, the significance of visual data cannot be overlooked, such as in the scenarios of floor planning and layout design. To harness the synergistic potential of both textual and visual information, the utilization of multi-modal models becomes essential. Advanced models like GPT-4 exhibit the ability to comprehend and process both textual and visual data, facilitating a seamless transition between the two modalities. This capability holds significant promise in enhancing the efficacy of smart design and construction practices.

5.6.1.5 Human-in-the-loop methods

Large-scale language models are commonly trained on extensive corpora, which may consist of high-quality and low-quality data. Consequently, these language models are prone to generating toxic, biased, or potentially harmful content. To align these models with human values (e.g., helpful, honest, and harmless), it is essential to incorporate human-in-the-loop methods. A widely utilized technique, Reinforcement Learning with Human Feedback (RLHF), exemplifies this approach. On the other hand, applying large-scale language models necessitates human intervention in compiling prompts, as evidenced by our research process. In the AEC domain, this is particularly pronounced due to the intricate requirements and complex production environment. To advance toward an automated future, it holds significant potential to devise supplementary approaches to optimize the utilization of large-scale language models with human-in-the-loop methods.

5.6.2 Challenges

5.6.2.1 Complexity of construction

In the AEC domain, the implementation of language models faces a significant challenge due to the diversity of standards, terms, language, designs, plans, and business cultures. Moreover, the domain-specific knowledge required to comprehend the intricacies of the construction industry can further impede the adoption of language models. Consequently, it is imperative to develop and refine language models that are flexible and adaptable to the diverse and heterogeneous needs and requirements of the construction industry.

5.6.2.2 Demand for high-quality data

One bottleneck in applying large-scale language models in practical scenarios is the acquisition of high-quality data. During the training process, it takes a plethora of textual data alongside human-compiled instructions for pre-training and instruction-tuning procedures. Such data must possess a high degree of quality to facilitate the seamless and successful development of large-scale language models. Furthermore, for their application in a specific domain, it is crucial to employ a rigorous evaluation procedure, which heavily relies on the availability of high-quality evaluation benchmarks. The design and curation of such benchmarks must be extremely careful to ensure optimal performance and minimal errors in downstream tasks. Although the AEC domain possesses a wealth of textual data, the majority of this data is often disordered and of low quality. Therefore, to utilize large-scale language models in real production processes, the data must undergo careful refinement and thorough inspection to meet the necessary standards.

5.6.2.3 Trade-off between cost and effectiveness

In the era of large-scale language models, it is widely acknowledged that model performance improves with increasing model size. However, the practical deployment of large-scale language models in the AEC domain faces significant challenges due to the considerable workload and computing resource requirements. Cost plays a pivotal role when deploying large-scale language models in this domain. Consequently, groups intending to use large-scale language models in this domain must carefully navigate a delicate balance between cost and effectiveness, thoroughly considering various factors. One crucial consideration involves addressing the specific requirements of downstream tasks, aiming to reduce the model size and thereby minimize costs while ensuring that downstream tasks' functionality and performance remain uncompromised.

5.6.2.4 Dynamic project life cycle

The complete life cycle of a project encompasses multiple stages, including planning and design prior to construction, as well as operation and maintenance throughout and after construction. These distinct stages impose a range of requirements on the utilization of language models. Furthermore, the inherent diversity and disparities among projects can introduce difficulties in the application of language models. Consequently, how to effectively customize a particular language model to align with the unique demands of a project throughout its entire life cycle remains a challenge.

5.7 Conclusion

With the advancement of artificial intelligence, integrating artificial intelligence techniques into the architecture, engineering, and construction (AEC) industry has become a prominent topic of discussion. Within the context of smart configuration, a crucial step is to meet individual users' specific requirements and preferences, thus enabling mass customization. Given the prevalence of textual data, utilizing natural language processing (NLP) for smart configuration has garnered considerable attention. A fundamental stage in this process involves understanding user needs, often transforming unstructured user texts into structured information. This chapter introduces language models, including small-scale ones like T5 and large-scale ones like ChatGPT, to enhance the user needs understanding task, formulated as a spoken language understanding task. To accomplish this, we curated a dataset that pertains to attribute changes or queries about building elements, employing language models as knowledge bases. Additionally, we leveraged their role as backbone models to identify user intents and corresponding slots within the texts. Through extensive experiments and analyses, we illustrated the superiority and limitations of language models in this context. Furthermore, we presented a summary of the opportunities and challenges associated with using language models in this domain, aiming to inspire further relevant research and provide project teams with valuable references to enhance the customization of language models in practical applications.

Notes

1 https://chat.openai.com/
2 https://www.buildingsmart.org/
3 https://pi.pauwel.be/voc/buildingelement/index-en.html
4 https://github.com/Significant-Gravitas/Auto-GPT
5 https://github.com/hwchase17/langchain

References

Ayhan BU, Tokdemir OB. Safety assessment in megaprojects using artificial intelligence. Safety Science, 2019, 118: 273–287. https://doi.org/10.1016/j.ssci.2019.05.027.

Bakhshi S, Chenaghlou MR, Rahimian FP, et al. Integrated BIM and DfMA parametric and algorithmic design based collaboration for supporting client engagement within offsite construction. Automation in Construction, 2022, 133: 104015. https://doi.org/10.1016/j.autcon.2021.104015.

Basic Z, Banovac A, Kruzic I, et al. Better by you, better than me, chatgpt3 as writing assistance in students essays. arXiv preprint arXiv:2302.04536, 2023. https://doi.org/10.48550/arXiv.2302.04536.

Benjamin S, Christopher R, Carl H. Feature modeling for configurable and adaptable modular buildings. Advanced Engineering Informatics, 2022, 51: 101514. https://doi.org/10.1016/j.aei.2021.101514.

Brown T, Mann B, Ryder N, et al. Language models are few-shot learners. Advances in Neural Information Processing Systems, 2020, 33: 1877–1901. https://doi.org/10.48550/arXiv.2005.14165.

Cao J, Bucher DF, Hall DM, et al. Cross-phase product configurator for modular buildings using kit-of-parts. Automation in Construction, 2021, 123: 103437. https://doi.org/10.1016/j.autcon.2020.103437.

Chang Y, Wang X, Wang J, et al. A survey on evaluation of large language models. arXiv preprint arXiv:2307.03109, 2023. https://doi.org/10.48550/arXiv.2307.03109.

Chen H, Liu X, Yin D, et al. A survey on dialogue systems: Recent advances and new frontiers. Acm Sigkdd Explorations Newsletter, 2017, 19(2): 25–35. https://doi.org/10.1145/3166054.3166058.

Christiano PF, Leike J, Brown T, et al. Deep reinforcement learning from human preferences. Advances in Neural Information Processing Systems, 2017, 30. https://doi.org/10.48550/arXiv.1706.03741.

Colvin RM, Witt GB, Lacey J. Approaches to identifying stakeholders in environmental management: Insights from practitioners to go beyond the 'usual suspects'. Land Use Policy, 2016. https://doi.org/10.1016/j.landusepol.2015.12.032.

Cui J, Li Z, Yan Y, et al. ChatLaw: Open-source legal large language model with integrated external knowledge bases. arXiv preprint arXiv:2306.16092, 2023. https://doi.org/10.48550/arXiv.2306.16092.

Da Silveira G, Borenstein D, Fogliatto FS. Mass customization: Literature review and research directions. International Journal of Production Economics, 2001, 72(1): 1–13. https://doi.org/10.1016/S0925-5273(00)00079-7.

Devlin J, Chang MW, Lee K, et al. Bert: Pre-training of deep bidirectional transformers for language understanding. arXiv preprint arXiv:1810.04805, 2018. https://doi.org/10.48550/arXiv.1810.04805.

Fang W, Luo H, Xu S, et al. Automated text classification of near-misses from safety reports: An improved deep learning approach. Advanced Engineering Informatics, 2020, 44: 101060. https://doi.org/10.1016/j.aei.2020.101060.

Ghannad P, Lee YC. Automated modular housing design using a module configuration algorithm and a coupled generative adversarial network (CoGAN). Automation in Construction, 2022, 139: 104234. https://doi.org/10.1016/j.autcon.2022.104234.

Ghannad P, Lee YC. Optimizing modularization of residential housing designs for rapid postdisaster mass production of housing. Journal of Construction Engineering and Management, 2023, 149(7): 04023046. https://doi.org/10.1061/(ASCE)CO.1943-7862.0002390.

Gharehchopogh FS, Khalifelu ZA. Analysis and evaluation of unstructured data: Text mining versus natural language processing. In *2011 5th International Conference on Application of Information and Communication Technologies (AICT)*. IEEE, 2011: 1–4. https://doi.org/10.1109/ICAICT.2011.6111017.
Grishman R. Information extraction. IEEE Intelligent Systems, 2015, 30(5): 8–15. https://doi.org/10.1109/MIS.2015.68.
Goh YM, Ubeynarayana CU. Construction accident narrative classification: An evaluation of text mining techniques. Accident Analysis & Prevention, 2017, 108: 122–130. https://doi.org/10.1016/j.aap.2017.08.026.
Gu J, Wang Z, Kuen J, et al. Recent advances in convolutional neural networks. Pattern Recognition, 2018, 77: 354–377. https://doi.org/10.1016/j.patcog.2017.10.013.
Howe AS, Ishii I, Yoshida T. Kit-of-parts: A review of object-oriented construction techniques. In *ISARC'99: International Symposium on Automation and Robotics in Construction (Madrid, 22–24 September 1999)*, 1999: 165–171. https://doi.org/10.22260/isarc1999/0025.
Hearst MA, Dumais ST, Osuna E, et al. Support vector machines. IEEE Intelligent Systems and Their Applications, 1998, 13(4): 18–28. https://doi.org/10.1109/5254.708428.
Kaiya H, Shinbara D, Kawano J, et al. Improving the detection of requirements discordances among stakeholders. Requirements Engineering, 2005. https://doi.org/10.1007/s00766-005-0017-2.
Langston C, Zhang W. DfMA: Towards an integrated strategy for a more productive and sustainable construction industry in Australia. Sustainability, 2021, 13(16): 9219. https://doi.org/10.3390/su13169219.
Lee J, Jeong Y. User-centric knowledge representations based on ontology for AEC design collaboration. Computer-Aided Design, 2012. https://doi.org/10.1016/j.cad.2012.03.011.
Lee JH, Yi JS. Predicting project's uncertainty risk in the bidding process by integrating unstructured text data and structured numerical data using text mining. Applied Sciences, 2017, 7(11): 1141. https://doi.org/10.3390/app7111141.
Li B, Fang G, Yang Y, et al. Evaluating chatgpt's information extraction capabilities: An assessment of performance, explainability, calibration, and faithfulness. arXiv preprint arXiv:2304.11633, 2023. https://doi.org/10.48550/arXiv.2304.11633.
Li X, Wu C, Xue F, et al. Ontology-based mapping approach for automatic work packaging in modular construction. Automation in Construction, 2022. https://doi.org/10.1016/j.autcon.2021.104083.
Li X, Wu C, Yang Z, et al. Knowledge graph-enabled adaptive work packaging approach in modular construction. Knowledge-Based Systems, 2023. https://doi.org/10.1016/j.knosys.2022.110115.
Li X, Zhu R, Ye H, et al. MetaInjury: Meta-learning framework for reusing the risk knowledge of different construction accidents. Safety Science, 2021, 140: 105315. https://doi.org/10.1016/j.ssci.2021.105315.
Marucci-Wellman HR, Corns HL, Lehto MR. Classifying injury narratives of large administrative databases for surveillance – A practical approach combining machine learning ensembles and human review. Accident Analysis & Prevention, 2017, 98: 359–371. https://doi.org/10.1016/j.aap.2016.10.014.
Nabavi A, Ramaji IJ, Sadeghi N, et al. Leveraging natural language processing for automated information inquiry from building information models. Journal of Information Technology in Construction, 2023, 28: 266–285. https://doi.org/10.36680/j.itcon.2023.013.
Norouzi N, Shabak M, Embi MRB, et al. The architect, the client and effective communication in architectural design practice. Procedia-Social and Behavioral Sciences, 2015, 172: 635–642. https://doi.org/10.1016/j.sbspro.2015.01.413.
Ouyang L, Wu J, Jiang X, et al. Training language models to follow instructions with human feedback. Advances in Neural Information Processing Systems, 2022, 35: 27730–27744. https://doi.org/10.48550/arXiv.2203.02155.
Pan W, Chen Q, Xu X, et al. A preliminary evaluation of chatgpt for zero-shot dialogue understanding. arXiv preprint arXiv:2304.04256, 2023. https://doi.org/10.48550/arXiv.2304.04256.
Qin L, Xie T, Che W, et al. A survey on spoken language understanding: Recent advances and new frontiers. arXiv preprint arXiv:2103.03095, 2021. https://doi.org/10.48550/arXiv.2103.03095.
Raffel C, Shazeer N, Roberts A, et al. Exploring the limits of transfer learning with a unified text-to-text transformer. The Journal of Machine Learning Research, 2020, 21(1): 5485–5551. https://dl.acm.org/doi/abs/10.5555/3455716.3455856.
Salehinejad H, Sankar S, Barfett J, et al. Recent advances in recurrent neural networks. arXiv preprint arXiv:1801.01078, 2017. https://doi.org/10.48550/arXiv.1801.01078.
Shin S, Issa RRA. BIMASR: Framework for voice-based BIM information retrieval. Journal of Construction Engineering and Management, 2021, 147(10): 04021124. https://doi.org/10.1061/(ASCE)CO.1943-7862.0002138.

Tan T, Mills G, Papadonikolaki E, et al. Digital-enabled design for manufacture and assembly (DfMA) in off-site construction: A modularity perspective for the product and process integration. Architectural Engineering and Design Management, 2023, 19(3): 267–282. https://doi.org/10.1080/17452007.2022.2104208.

Tian D, Li M, Shi J, et al. On-site text classification and knowledge mining for large-scale projects construction by integrated intelligent approach. Advanced Engineering Informatics, 2021, 49: 101355. https://doi.org/10.1016/j.aei.2021.101355.

Tu R, Ma C, Zhang C. Causal-discovery performance of chatgpt in the context of neuropathic pain diagnosis. arXiv preprint arXiv:2301.13819, 2023. https://doi.org/10.48550/arXiv.2301.13819.

Vaswani A, Shazeer N, Parmar N, et al. Attention is all you need. Advances in Neural Information Processing Systems, 2017, 30. https://doi.org/10.48550/arXiv.1706.03762.

Wu C, Li X, Guo Y, et al. Natural language processing for smart construction: Current status and future directions. Automation in Construction, 2022, 134: 104059. https://doi.org/10.1016/j.autcon.2021.104059.

Wu C, Li X, Jiang R, et al. Graph-based deep learning model for knowledge base completion in constraint management of construction projects. Computer-Aided Civil and Infrastructure Engineering, 2023. https://doi.org/10.1111/mice.12904.

Wu S, Shen Q, Deng Y, et al. Natural-language-based intelligent retrieval engine for BIM object database. Computers in Industry, 2019, 108: 73–88. https://doi.org/10.1016/j.compind.2019.02.016.

Wang H, Meng X, Zhu X. Improving knowledge capture and retrieval in the BIM environment: Combining case-based reasoning and natural language processing. Automation in Construction, 2022, 139: 104317. https://doi.org/10.1016/j.autcon.2022.104317.

Wang Y, Zhao W, Wan WX. Needs-based product configurator design for mass customization using hierarchical attention network. IEEE Transactions on Automation Science and Engineering, 2020, 18(1): 195–204. https://doi.org/10.1109/TASE.2019.2957136.

Wang N, Issa RRA, Anumba CJ. NLP-based query-answering system for information extraction from building information models. Journal of Computing in Civil Engineering, 2022, 36(3): 04022004. https://doi.org/10.1061/(ASCE)CP.1943-5487.0001019.

Wei J, Tay Y, Bommasani R, et al. Emergent abilities of large language models. arXiv preprint arXiv:2206.07682, 2022. https://doi.org/10.48550/arXiv.2206.07682.

Weld H, Huang X, Long S, et al. A survey of joint intent detection and slot filling models in natural language understanding. ACM Computing Surveys, 2022, 55(8): 1–38. https://doi.org/10.1145/3547138.

Wu S, Irsoy O, Lu S, et al. Bloomberggpt: A large language model for finance. arXiv preprint arXiv:2303.17564, 2023. https://doi.org/10.48550/arXiv.2303.17564.

Yan H, Yang N, Peng Y, et al. Data mining in the construction industry: Present status, opportunities, and future trends. Automation in Construction, 2020, 119: 103331. https://doi.org/10.1016/j.autcon.2020.103331.

Zhang D, Li M, Tian D, et al. Intelligent text recognition based on multi-feature channels network for construction quality control. Advanced Engineering Informatics, 2022, 53: 101669. https://doi.org/10.1016/j.aei.2022.101669.

Zhang F, Fleyeh H, Wang X, et al. Construction site accident analysis using text mining and natural language processing techniques. Automation in Construction, 2019, 99: 238–248. https://doi.org/10.1016/j.autcon.2018.12.016.

Zheng J, Wen Q, Qiang M. Understanding demand for project manager competences in the construction industry: Data mining approach. Journal of Construction Engineering and Management, 2020, 146(8): 04020083. https://doi.org/10.1061/(ASCE)CO.1943-7862.0001865.

Zheng Z, Lu XZ, Chen KY, et al. Pretrained domain-specific language model for natural language processing tasks in the AEC domain. Computers in Industry, 2022, 142: 103733. https://doi.org/10.1016/j.compind.2022.103733.

Zhong B, Xing X, Love P, et al. Convolutional neural network: Deep learning-based classification of building quality problems. Advanced Engineering Informatics, 2019, 40: 46–57. https://doi.org/10.1016/j.aei.2019.02.009.

Zhong B, Xing X, Luo H, et al. Deep learning-based extraction of construction procedural constraints from construction regulations. Advanced Engineering Informatics, 2020, 43: 101003. https://doi.org/10.1016/j.aei.2019.101003.

Zou Y, Kiviniemi A, Jones SW. Retrieving similar cases for construction project risk management using natural language processing techniques. Automation in Construction, 2017, 80: 66–76. https://doi.org/10.1016/j.autcon.2017.04.003.

6

PHYSICAL FATIGUE ASSESSMENT (PFA) FOR CONSTRUCTION WORKERS WITH AI AND SENSORY TECHNIQUES

Yantao Yu[1] *and Zirui Li*[2]

[1]Assistant Professor, Department of Civil and Environmental Engineering, the Hong Kong University of Science and Technology, Clear Water Bay, New Territories, Hong Kong SAR of China. Email: ceyantao@ust.hk. (Corresponding author).

[2]PhD Student, Department of Construction and Real Estate, School of Civil Engineering, Southeast University, Nanjing, China. Email: seulzr@126.com.

6.1 Introduction

6.1.1 Fatigue in construction

The global construction industry is currently grappling with substantial occupational health and safety concerns (Umer et al., 2018), with fatigue resulting from physically demanding tasks being a primary cause. Construction workers often endure prolonged work hours without adequate breaks (Aryal et al., 2017), in challenging environmental conditions or confined workspaces (Aryal et al., 2017), thereby substantially increasing the risk of fatigue among them. Persisting in work while fatigued can lead to work-related musculoskeletal disorders (WMSDs), reduced productivity, compromised work quality, and a higher incidence of accidents or falls on construction sites. Consequently, fatigue poses a significant challenge to occupational health and safety within the industry (Wang et al., 2015).

As the construction industry continues to expand and assumes a pivotal role in driving socio-economic development, it places an increasingly substantial burden on its workforce. This, in turn, exacerbates the existing challenges related to occupational health and safety. In summary, fatigue has a detrimental impact on safety, health, and productivity within the global construction sector, owing to heightened workloads, workforce shortages, and an aging demographic (Yu, Yang, et al., 2019). Hence, precise monitoring and prevention of fatigue levels assume paramount importance for construction workers, with the aim to enhancing safety, well-being, and overall productivity performance.

DOI: 10.1201/9781003383840-7

6.1.2 The significance of physical fatigue prevention

Fatigue can be defined as the inability to sustain power output (Fitts, 1994). It arises from the depletion of three essential sources of energy: physical, mental, and emotional. As a result, fatigue can be categorized into three types: physical, mental, and emotional fatigue (Frone & Tidwell, 2015). Mental and emotional fatigue manifest as temporary impairments in cognitive and emotional performance and are influenced by various factors, including elevated stress levels, sleep disorders, fluctuations in mood, and the social environment (Akerstedt et al., 2004; Techera, 2017). These factors are complex to quantify and intertwined with the social cultural atmosphere.

Physical fatigue, also known as muscle fatigue, is biologically characterized by a reduction in the muscle's ability to generate force. It is closely associated with job task characteristics such as awkward working postures, external load, and the duration of operations (Golabchi et al., 2016; Lingard & Wakefield, 2019). Consequently, physical fatigue can be effectively managed through interventions. Additionally, research suggests that construction workers tend to have lower levels of mental health compared to the general population, while their physical health tends to be relatively higher (Maruish, 2004). Recognizing the prevalence of physical fatigue and implementing interventions through construction site management are crucial.

The remainder of this chapter is organized as follows. Section 2 introduces the indicators and collection methods for assessing physical fatigue (PFA), categorized into subjective, physiological, biomechanical, and ergonomic indicators. Subsequently, in Section 3, a comparison is conducted to evaluate the performance of each PFA indicator and data collection method, identifying research gaps. In Section 4, to address the limited applicability of existing ergonomic scales in assessing diverse construction working postures, a heuristic data-driven ergonomic assessment method tailored for such postures is developed. Finally, in Section 5, future research directions are presented.

6.2 Literature review

This chapter provides an overview of major PFA indicators within the construction industry, classified into subjective, physiological, biomechanical, and ergonomic indicators. A comparative analysis is conducted to assess the effectiveness of each PFA indicator and its associated data collection methods. Following this, the chapter systematically identifies research gaps in the field of PFA.

6.2.1 Subjective indicators

Subjective PFA indicators rely on workers' subjective perceptions of physical fatigue and are typically assessed using self-report scales. Two commonly used scales for fatigue assessment are the Borg CR10 scale and the Borg rate of perceived exertion (RPE) scale (Borg, 1998; Jo & Bilodeau, 2021). These scales ask workers to rate their perception of fatigue on a scale from 0.0 to 12.0 or from 6 to 20, with higher scores indicating greater fatigue. However, self-report scales are not practical for measuring physical fatigue among construction workers due to several reasons: (a) reporting may disrupt work activities; (b) the scales only yield a result of the workers' fatigue status, without tracking the development of fatigue; and (c) the reported data may be incorrect or inconsistent due to memory distortion and potential dishonest responses.

In practice, self-report methods are commonly used for manually acquiring subjective data in construction settings. These methods involve workers recalling their postures during construction tasks and responding to questions related to their postures. Questionnaires or interviews are frequently employed to collect self-report data due to their logistical convenience and low initial cost. These studies analyze PFA-related data from various perspectives, such as identifying awkward postures, evaluating manual material handling, and assessing balance stability (Caban-Martinez et al., 2018; DiDomenico et al., 2010; Elders & Burdorf, 2001). However, self-report methods have inherent limitations, particularly in terms of reliability, continuity, and efficiency, despite their widespread utilization.

6.2.2 Physiological indicators

Physiological indicators, such as cardiovascular indicators and electronic indicators, can be used to detect and monitor physical fatigue in construction workers. Cardiovascular indicators, including heart rate, skin temperature, and breathing rate, often increase with high physical strain, indicating the presence of fatigue (Cheng et al., 2013; Gatti et al., 2014; Kirk & Sullman, 2001). These indicators can be measured using wearable sensors attached to workers' bodies, allowing for continuous monitoring of physical fatigue (Gatti et al., 2014; Ghafoori et al., 2023). However, there are limitations associated with the use of these sensors. First, the presence of wearable sensors may impede workers' performance during their work routines. The additional equipment and attachments can restrict movement and hinder their ability to perform tasks efficiently. Moreover, wearable sensors often require frequent charging, typically every few hours, which poses challenges for monitoring physical fatigue over extended periods of time (Yu et al., 2021).

Another electronic indicator commonly used for detecting physical fatigue is surface electromyography (sEMG) (Al-Mulla et al., 2011). Previous studies have utilized sEMG to measure muscle fatigue in construction workers within laboratory settings (Antwi-Afari et al., 2018). However, practical application of surface sEMG on construction sites is limited. The process of attaching sEMG electrodes to the skin of each target muscle group is time-consuming and impractical for measuring whole-body muscle fatigue or overall physical fatigue. Additionally, factors such as sweating and body movement can introduce artifacts to sEMG signals, making it challenging to accurately monitor muscle fatigue under construction site conditions (Jiang et al., 2020).

In addition to sEMG, heart rate (HR) is commonly employed as a physiological measure for monitoring and tracking physical fatigue in construction workers (Anwer et al., 2021). Portable and user-friendly smartwatches allow for real-time acquisition of multiple indicators, including HR. Combined with heart rate variability (HRV), a derived signal from the electrocardiogram (ECG), HR can be used for real-time tracking of individuals' physical fatigue and its associated negative impacts (Ibrahim et al., 2023). Furthermore, in-depth linear and nonlinear analyses of HRV characteristics have been applied to identify and classify physical fatigue in workers (Anwer et al., 2023). Various wearable sensors, such as electroencephalography (EEG) (Wang et al., 2023b), electrocardiography (ECG) (Lee et al., 2022), and electrodermal activity (EDA) (Mehmood et al., 2023), have also been utilized to measure other physiological indicators related to fatigue monitoring [28–30]. For example, Aryal et al. (2017) proposed an approach using wearable sensors, including a heart rate monitor, infrared temperature sensors, and EEG sensors, for real-time detection of physical fatigue. In summary, wearable sensors and devices enable accurate data collection and real-time tracking of physical fatigue. However, due to their expensive cost and intrusiveness, their widespread adoption on construction sites for common applications remains relatively challenging (Yu et al., 2021).

6.2.3 Biomechanical indicators

Biomechanical calculations offer a more detailed and accurate method for assessing subjective physical fatigue by calculating joint forces or torques based on external forces and postures. Software tools like OpenSim (as shown in Figure 6.1), 3DSSPP, and AnyBody Modelling System are commonly used for this purpose (Delp et al., 2007). These tools not only calculate joint forces and torques but also allow for the generation of personalized PFA by comparing the calculated values to individual workers' joint capacities.

In the construction industry, OpenSim and 3DSSPP have been utilized for assessing workers' ergonomic risks and physical fatigue (Seo et al., 2015). However, their practical application is primarily limited to laboratory or virtual environments due to the challenges associated with data collection. For example, to effectively use OpenSim, a comprehensive whole-body motion capture setup is required, involving the placement of 50 markers on various body parts and the use of multiple cameras (Yu, 2020). Such data collection procedures are impractical for monitoring workers' fatigue directly on construction sites. Furthermore, biomechanical analysis relies on the measurement of external forces, which poses significant challenges given the diverse range of forces and movements involved in construction activities. Traditional tools like goniometers and hand dynamometers are commonly used to measure hand forces, but they are limited to straightforward, static, and unnatural postures (Kong & Kim, 2015). The varied hand postures adopted by construction workers call for an ideal hand force measurement tool capable of accommodating the flexibility and continuous variations commonly encountered in the construction industry.

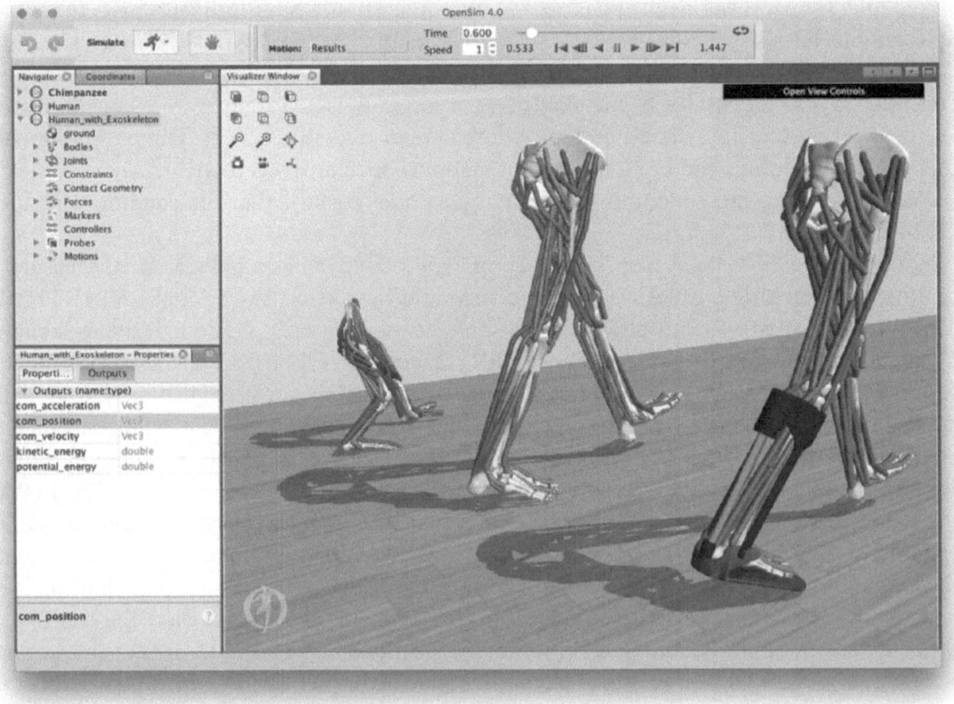

Figure 6.1 User interface of OpenSim (Seth et al., 2018)

Similarly, ground reaction forces (GRF) are typically measured using force platforms, but their application is mostly limited to static cases or small areas (Umer et al., 2018). Wearable plantar pressure sensor systems offer a more versatile approach for accurately and continuously measuring GRF on construction sites (Antwi-Afari et al., 2018; Antwi-Afari et al., 2017; Simeonov et al., 2003). However, it's important to acknowledge that these methods have their measurement range constraints.

The external forces experienced by workers primarily involve forces exerted on the hands and ground reaction forces (Feyen et al., 2000). To enable continuous measurement of hand forces in dynamic scenarios, wearable sensors have been introduced. For instance, surface electromyography (sEMG) sensors have been utilized to quantify the force exerted by the forearm by monitoring muscle activities (Hermens et al., 2000). An example of a sEMG-based measurement system is illustrated in Figure 6.2. However, it is crucial to consider that the accuracy of sEMG sensors may be influenced by various factors such as electrode placement, signal processing, filtering, and individual muscle capabilities (Merletti et al., 2009). Moreover, the

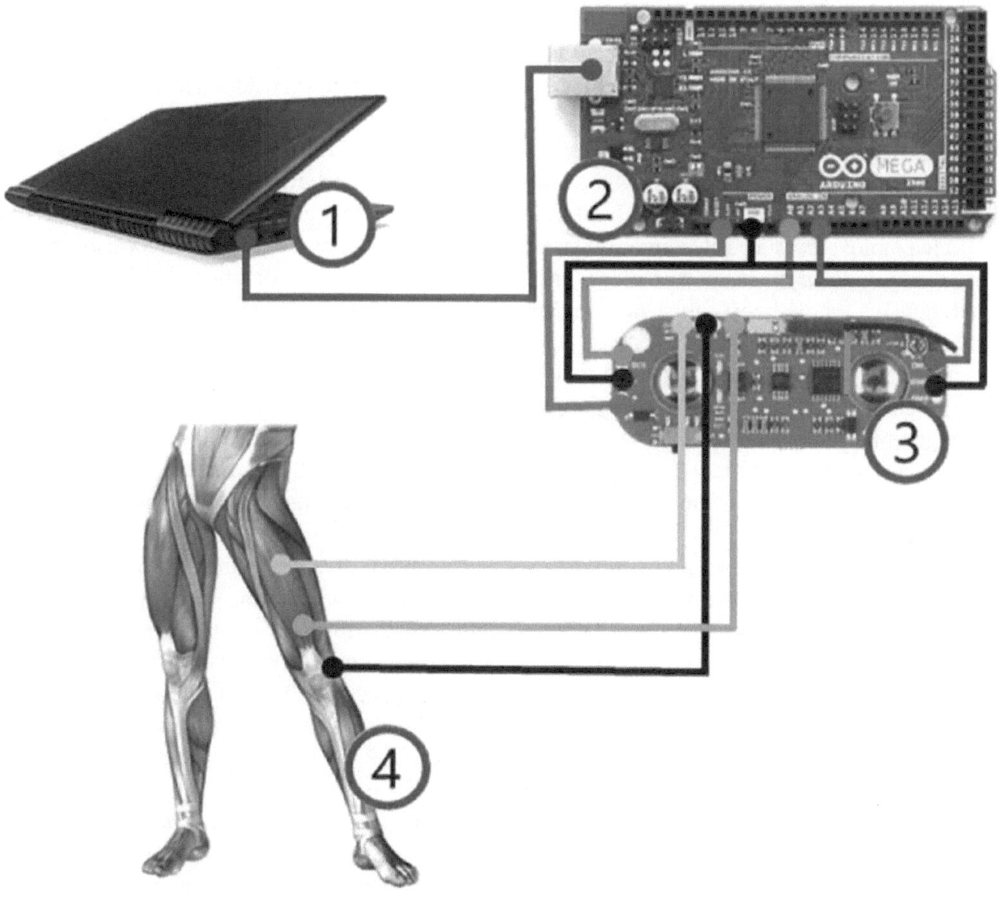

Figure 6.2 A sEMG-based measurement system: (1) personal computer; (2) Arduino MEGA board, (3) surface electromyography (sEMG) sensor by Myoware System, and (4) participant (Toro et al., 2019)

practical utilization of sEMG sensors for construction workers is hindered by the necessity of a dry skin surface for optimal electrode placement and the reliance on wired connections to a signal processor. These requirements present significant challenges in real-world construction environments.

6.2.4 Ergonomic indicators

Ergonomic indicators rely on the analysis of workers' kinematic data to estimate physical fatigue. Numerous studies have employed posture-based methods and models to assess physical fatigue in workers (Hignett & McAtamney, 2000; Nath et al., 2017; Yan, Li, Wang, et al., 2017). The adoption of unergonomic postures significantly contributes to physical fatigue. Previous research has examined physical fatigue by evaluating working postures, often focusing on joint angles (Moriguchi et al., 2013; Yan, Li, Li, et al., 2017; Yan, Li, Wang, et al., 2017). These studies analyze the angles of various body joints in captured working postures, such as neck flexion, extension, and twisting angles. Each joint angle is then categorized into a range of motion and assigned a corresponding physical fatigue score. Another approach involves identifying the working posture through observation and estimating the associated physical fatigue based on the identified posture (Nath et al., 2017).

Several ergonomic scales have been developed, including the Key Item Method (KIM) (Roman-Liu, 2014), the Ovako Working Posture Analysis System (OWAS) (Karhu et al., 1981), the Occupational Repetitive Actions (OCRA) (Occhipinti, 1998), the Rapid Upper Limb Assessment (RULA) (McAtamney & Nigel Corlett, 1993), and the Rapid Entire Body Assessment (REBA) (Hignett & McAtamney, 2000), among others. For instance, the REBA method is an ergonomic scale widely used for postural assessment in healthcare and various industries (Kee, 2021). It employs scoring rules and adjustment rules for six body segments, encompassing the trunk, neck, upper arm, lower arm, legs, and wrists (Hignett & McAtamney, 2000). Due to its comprehensive whole-body assessment and the objectivity derived from scoring rules based on joint angle values [8], the REBA scale has been extensively adopted in the construction industry. These scales incorporate specific rules for coding and recording data, enhancing the accuracy and level of detail in the collected information. However, it is important to acknowledge that subjective judgment errors may arise as the categorization of postures relies on the observer's experience (Golabchi et al., 2016). Moreover, the manual implementation of these methods is notably time consuming and costly (Ghasemi & Mahdavi, 2020), posing challenges in executing efficient and timely interventions.

When it comes to collecting postural data for ergonomic indicators in the construction industry, three types of methods have been employed: systematic observation, contact-sensor-based methods, and non-contact-sensor-based methods. Systematic observation involves a methodical approach to gather reliable and unbiased data on working postures. However, its manual implementation is known to be time consuming and costly (Ghasemi & Mahdavi, 2020). Moreover, subjective judgment errors by experts can occur (Golabchi et al., 2016). Contact-sensor-based methods utilize sensors or markers attached to the body to collect posture-related data of construction workers. Examples include Inertial Measurement Units (IMUs) (Dudek & Jenkin, 2016) and marker-based motion capture systems (Li et al., 2018). These sensors can measure or calculate joint kinematic data, such as joint acceleration, position, and angle. However, when used on construction sites, contact sensors can cause discomfort and intrusiveness. Additionally, there may be additional costs associated with sensor recharging and maintenance (Dudek

& Jenkin, 2016). On the other hand, non-contact-sensor-based methods provide a non-intrusive approach to data collection. These methods typically rely on images or videos captured at construction sites, focusing on visual information related to working postures (Yu, Yang, et al., 2019). Two commonly utilized tools for non-contact sensor-based methods are depth cameras (such as Kinect V1) and Red-Green-Blue (RGB) cameras (Yu et al., 2021). However, depth cameras have limitations such as a restricted measurement range, sensitivity to lighting conditions, and relatively high costs. In contrast, given the widespread installation of low-cost ordinary cameras (RGB) on construction sites (Zhang et al., 2018), RGB video-based PFA strikes an optimal balance between effectiveness and applicability. This approach can be further enhanced by utilizing 3D pose estimation techniques such as OpenPose (Slembrouck et al., 2020) and MediaPipe (Singh et al., 2022).

6.3 Research gaps

Based on the analysis and discussions earlier, the research gaps in PFA are identified as the following two aspects:

6.3.1 Ergonomic indicators facilitate optimal PFA

Building upon the preceding analysis of PFA indicators, this section provides a concise summary and comparison of these indicators, as depicted in Table 6.1. Subjective indicators, relying on manual self-report methods, exhibit limitations, particularly in terms of reliability, continuity, and efficiency, despite their prevalent use on construction sites. Furthermore, both physiological indicators and biomechanical indicators-based PFA collect data through wearable sensors or devices. These solutions facilitate accurate data collection and real-time tracking of physical fatigue. However, their extensive implementation on construction sites has been hindered by high costs and intrusiveness. Additionally, certain methods, such as electronic indicators and software analysis of joint forces, are limited to laboratory settings due to their reliance on complex sensors that are only available in controlled environments.

Distinguished from the aforementioned indicators, ergonomic indicators primarily focus on assessing repetitive working postures, enabling non-intrusive PFA in real-world construction sites through the utilization of pose data from depth or RGB cameras. It is important to note that depth cameras have limitations, including a restricted measurement range, sensitivity to lighting conditions, and relatively high costs. In contrast, low-cost RGB cameras are widely installed on construction sites. Supported by computer vision algorithms such as OpenPose and MediaPipe, recorded RGB images or videos can be used for accurate pose estimation, serving as input for PFA. In conclusion, vision-based ergonomic indicators facilitate optimal PFA in an outdoor and non-intrusive manner.

6.3.2 Limited applicability of ergonomic scales in assessing diversified construction working postures

Ergonomic indicators-based PFA typically involves automated implementation of ergonomic scales, such as REBA (Lee & Lee, 2022) and RULA (Golabchi et al., 2017), using estimated pose data as input. This approach has improved accuracy and cost-effectiveness compared to manual

Table 6.1 The comparison of PFA indicators and data collection methods

PFA Indicators			*Data Collection*	*Environment*	*Intrusiveness*	*Repetitive Work Pattern*
Subjective	Self-report	Self-perception of physical fatigue	Self-repot methods by questionnaires or interviews	Outdoor	√	×
Objective	Physiological indicators	Heart rate, skin temperature, breathing rate, and more.	Wearable sensors or devices	Outdoor	√	×
		sEMG, ECG, EEG, EDA, and more	Wearable sensors or devices	Lab	√	×
	Biomechanical indicators	Joint-level physical fatigue	Wearable sensors and force platforms	Lab	√	×
	Ergonomic indicators	Ergonomic risks due to awkward postures and prolonged or repetitive task patterns	Systematic observation	Outdoor	√	√
			Wearable sensors	Lab	√	√
			Depth and RGB cameras	Outdoor	×	√

expert observations. However, ergonomic scales have limited applicability when assessing construction workers.

Non-standard task procedures, intricate workplace layouts, and individual working patterns collectively contribute to the diversified working postures in construction activities (Yu et al., 2021). In contrast, the manufacturing industry demonstrates relatively fixed working patterns, due to the repetitive tasks performed on the assembly line (Yu, Li, Yang, et al., 2019). This allows for ergonomic assessment under typical manufacturing working scenarios, such as handling and assembling, building upon existing scales and working sampling methods (Brandl et al., 2017). However, for construction workers with unfixed working patterns, defining their typical working scenarios poses challenges. Consequently, the ergonomic scales themselves exhibit two limitations when assessing diversified construction working postures.

Regarding joint-level ergonomic assessment, the discrete boundaries (specified angle ranges or positions) of risk categories may not be applicable to construction working postures. The diversified construction working postures directly contribute to significant movements in joint angles and positions. However, existing ergonomic scales primarily focus on classifying limited ranges of movements observed in repetitive manufacturing tasks. For instance, the discrete boundaries of upper arm flexion in REBA (excluding extension) assign three risk categories to angles below 90°: less than 20°, 20°–45°, and 45°–90°, whereas only one category is provided for angles equal to or greater than 90°. However, certain construction workers, such as barfixers and scaffolders, frequently perform tasks that involve upper arm flexion angles greater than 90°

(Yu, Yang, et al., 2019). Thus, the specified angle ranges for risk categories prove not applicable for construction workers, given their wider range of movements. Additionally, certain modules that determine risk categories by specified positions demonstrate limited applicability on construction sites. For instance, the changes score of the trunk segment in REBA employs an oversimplified binary rule of '+1 if twisting or side flexed' without specifying an angle range for classifying these movements. As a result, this rule is susceptible to subjectivity derived from manual observations, as well as errors introduced by the pose estimation tools. In conclusion, the specified discrete boundaries of risk categories exhibit limited capability in joint-level ergonomic assessment.

In terms of ergonomic assessment results at the body-segment level and the whole-body level, discontinuity is attributed to the discrete boundaries of joint-level risk categories and the reliance on if-then rules for transformation between assessment levels, respectively. When it comes to continuous joint movements, angle variations within a limited range do not lead to synchronous changes in risk categories. This issue stems from the discrete boundaries of specified risk categories. Conversely, angle variations spanning discrete boundaries can lead to sharp changes in risk categories. Therefore, the discrete boundaries of joint-level risk categories ultimately introduce considerable discontinuity and even errors in the outcomes of joint-level and body-segment-level ergonomic assessment. While some attempts have incorporated fuzzy logic to enhance continuity (Golabchi et al., 2017; Wang et al., 2021a; Wang et al., 2023a), the applicability remains uncertain due to numerous subjective settings and assumptions. Furthermore, the if-then rules within existing ergonomic scales, enabling the transformation from the body-segment level to the whole-body level, are not applicable for continuous assessment outputs. For example, the if-then rules extracted from Table A, Table B, and Table C in REBA, such as 'if Table A score is 4 and Table B score is 5, then Table C score is 5', only accommodate integer inputs and result in discrete outputs. Based on the authors' prior study on ergonomic assessment involving six types of construction workers (Yu, Yang, et al., 2019), it is noteworthy that the outcomes of REBA at both levels exhibit discrete score distributions, as illustrated in Figure 6.3. In conclusion, this issue hinders the appropriate feedback on risk information within diversified construction working postures.

6.4 A heuristic data-driven ergonomic assessment method tailored for diversified construction working postures

This study aims to improve the existing ergonomic scales by addressing the aforementioned limitations and to develop an assessment method tailored to diversified construction working postures. Building upon the emphasis of REBA on whole-body assessment, this scale serves as the foundation for the method development. To validate the applicability of the proposed method, on-site experiments are conducted to record operations, while an expert survey is carried out to obtain manual results for comparison purposes. Subsequently, the results of the method development and validation are presented, along with a discussion on the limitations encountered.

6.4.1 Method development

The process of method development comprises three parts. First, to deal with the inapplicable specified angle ranges and positions for joint-level ergonomic assessment, Heuristics Gaussian

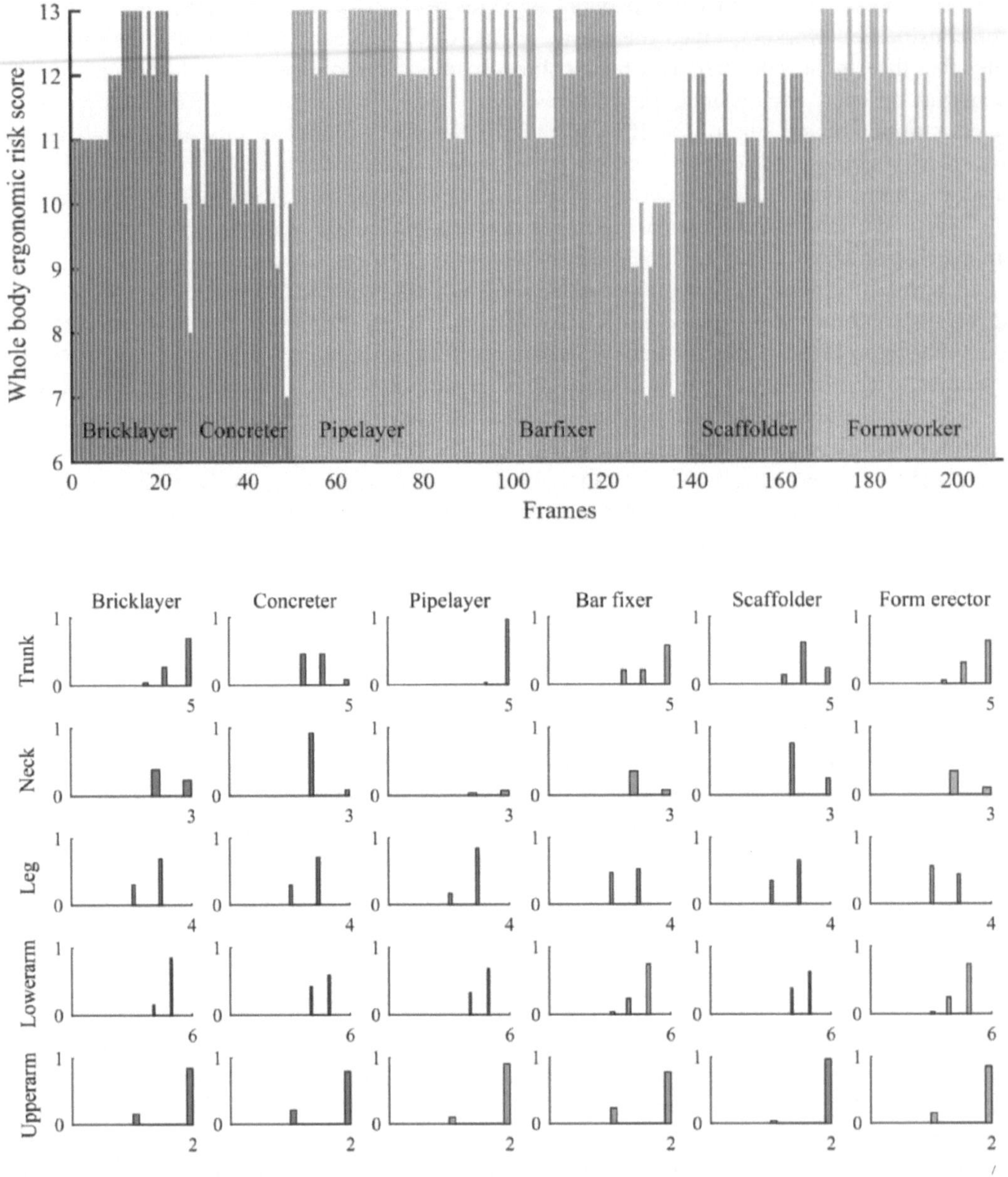

Figure 6.3a, 3b Results of the authors' previous study on ergonomic assessment for six types of construction workers (Yu, Yang, et al., 2019)

Cloud Transformation (H-GCT) is employed. This approach utilizes statistical data from the Construction Motion Data Library (CML) to generate scoring models that align with joint-level modules of REBA. Second, to handle discrete outcomes at the body-segment level, the cloud model (CM) during H-GCT is used to enhance uncertainty representation at the boundaries of adjacent scoring models, leading to transitional boundaries of risk rating scores. In summary, the utilization of H-GCT effectively addresses the inapplicability of specified angle ranges or positions as well as the discrete assessment outcomes at the body-segment level, both of which are consequences of the discrete boundaries specified in REBA. Last, to achieve continuous assessment output at the whole-body level, a fuzzy reasoning mechanism is designed to facilitate the transformation of continuous input across different assessment levels. The specific development process is as follows:

To enable a comprehensive whole-body assessment for construction workers, the method development is based on REBA (Hignett & McAtamney, 2000), which incorporates 15 joint-level assessment modules including five body segments: the trunk, neck, legs, upper arms, and lower arms. Subsequently, statistical data on pose information is extracted from the CML dataset (Tian et al., 2022), which serves as a comprehensive dataset covering major activities within the construction industry. Building upon the REBA and statistical data, H-GCT is implemented to generate 15 sets of scoring models. This data-driven approach, functioning as a robust heuristic and clustering algorithm, allows the generation of a predefined number of clusters from input data samples (Li & Du, 2017). The number of scoring models is determined based on prior knowledge of REBA, aligning with the number of predefined risk categories in each joint-level module. These outcomes replace the inapplicable specified angle ranges or positions for risk categories of REBA and enable continuous output of membership degrees at the joint level.

In addition to addressing the first limitation, the cloud model (CM) during H-GCT is introduced to enhance the representation of uncertainty and achieve transitional boundaries of risk rating scores. This consideration arises from the one-to-many correspondence at the boundaries of adjacent scoring models. Such an approach aligns well with the adaptive cognition, reflecting the anthropomorphic process of judging membership degrees of models (Li & Du, 2017). Consequently, it optimizes the robustness of the assessment. By addressing the one-to-many uncertainty for transitional boundaries, combined with leveraging the one-to-one correspondences at major parts of scoring models, continuous assessment outcomes at the joint level and the body-segment level can be achieved.

Furthermore, to attain continuous assessment outcomes at the whole-body level, a fuzzy reasoning mechanism is designed to enable transformation across assessment levels. The transformation is implemented using 156 one-to-one if-then rules extracted from Table A, Table B, and Table C of REBA (Hignett & McAtamney, 2000). Fuzzy operators, including product t-norm operators (Bobillo & Straccia, 2007) and the Einstein Sum s-norm operator (Azadeh et al., 2014), are selected to handle intersections and unions, respectively. These operators can retain all membership information at the body-segment level during the fuzzy reasoning process. Subsequently, the center of area (CoA) method, known for its ability to synthetically consider membership information, is employed for defuzzification and facilitates continuous assessment outcomes at the whole-body level (Wang et al., 2021a). Ultimately, this mechanism, in conjunction with 15 sets of scoring models, constitutes the ergonomic assessment method.

6.4.2 On-site experiment and validation

To validate the applicability of the proposed data-driven assessment method for construction workers, on-site experiments were conducted at a construction project. For the consideration of portability, an iPhone with a smartphone-based RGB camera was utilized to record videos of workers performing operations. Two workers, a pipelayer and an unloader, were recruited for the experiment, each executing distinct operational procedures. Making use of the 3D pose data estimated by OpenPose, the assessment results were continuously generated by both the proposed method and the conventional REBA (utilized as a baseline method). Subsequently, 11 experts were invited to manually assess the workers' operations based on 15 non-uniformly sampled images and provide feedback on the corresponding risk levels, ranging from level 1 to 5. Using two indicators, namely the average error ($\overline{Error}$) and the root mean square error (RMSE), a comparative analysis was conducted to demonstrate the improved accuracy in risk identification compared to the conventional REBA. This process aims to showcase the applicability of the proposed method.

$$\overline{Error} = \frac{1}{n}\sum_{i=1}^{n}(E_i - L_i)$$

$$RMSE = \sqrt{\frac{\Sigma_{i=1}^{n}(L_i - E_i)^2}{n}}$$

6.4.3 Results and discussion

By utilizing the Heuristics Gaussian Cloud Transformation (H-GCT) based on construction workers' pose data, scoring models for the 15 assessment modules are generated. These scoring models, as illustrated in Figure 6.4, depict the relationships between memberships and angles/indexes. It is important to note that these relationships are not static; instead, this study establishes a heuristic, data-driven, and dynamic assessment method. The membership degrees involved in the assessment process vary within the ranges of persistently unstable systems controlled by three critical parameters: Ex (expectation), En (entropy), and He (hyper entropy). This system effectively captures the uncertainty in the assessment process and exhibits specific anti-interference and adaptive capabilities (Li & Du, 2017). To demonstrate the independence of the generated scoring models, the confusion degree (CD) is computed to reflect the degree of overlap between adjacent models. The results indicate that the generated cloud models (CMs) exhibit excellent independence overall. Specifically, all 52 CD values are below the threshold value of 0.64. It is important to highlight that a CD value exceeding 0.64 indicates the presence of overlap in the core area (Li & Du, 2017), which suggests potential confusion between CMs.

Building upon the risk level outputs provided by the invited experts' manual assessment and the proposed method (Comparison 1), average errors and RMSE values were computed and presented in Table 6.2. Additionally, for reference, the same indicators were computed using the outputs from the conventional REBA and expert scoring (Comparison 2). For the pipelayer, the average errors in the two comparison groups are −3.6% and 17.0%, with corresponding RMSE values of 0.853 and 1.477, respectively. Regarding the unloader, the average errors are 4.2% and 10.9%, accompanied by RMSE values of 0.469 and 1.946. It is evident that both the pipelayer and the unloader consistently exhibit a similar trend in their average errors and RMSE values. Specifically, the absolute values of the output indicators from Comparison 1 (proposed method) are consistently lower than those from Comparison 2 (REBA). This demonstrates that the risk levels identified by the proposed method closely align with those identified by experts within the construction industry, in comparison to the outputs of REBA. As a result, the proposed method can be considered highly applicable for assessing the ergonomic risk levels of construction workers.

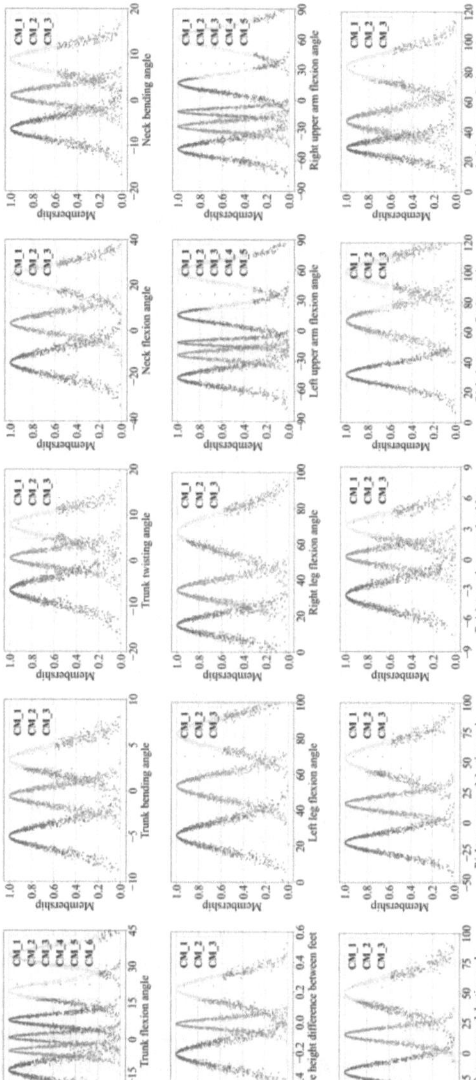

Figure 6.4 Developed scoring models corresponding to the 15 considered assessment modules (Li et al., 2024)

Table 6.2 Calculated $\overline{Error}$ and RMSE for the pipelayer and unloader

Subject	*Indicator*	*CM REBA vs. Experts (Comparison 1)*	*REBAvs. Experts (Comparison 2)*
Pipelayer	$\overline{Error}$	−3.6%	17.0%
	RMSE	0.853	1.477
Unloader	$\overline{Error}$	4.2%	10.9%
	RMSE	0.469	1.946

6.4.4 Limitation of the proposed method

While the method development process is well founded, and its applicability has been thoroughly validated, two primary limitations have emerged: (a) limitations in the dataset and considered assessment modules. The CML dataset provides skeleton data for the 15- or 20-joint system (Tian et al., 2022), resulting in the calculation of certain angles or indexes (e.g., neck twisting) within REBA unavailable. This limitation arises from the constraints of the dataset itself, and it may impact the comprehensiveness of the assessment in capturing all possible contributors for ergonomic risk. (b) limitations in the data-driven approach. To align with the score distribution for each considered module, this study employs a heuristic algorithm-based H-GCT. However, it is crucial to note that this does not guarantee optimal independence among generated scoring models. The independence among the generated scoring models is an important aspect to consider in order to minimize confusion and ensure accurate assessment.

6.5 Potential future research directions for PFA

The authors have attempted to develop a heuristic data-driven ergonomic assessment method tailored to diversified construction working postures. The results contribute to bridging the research gap in the applicability of existing ergonomic scales. To address the two limitations in Section 4.4, the authors have proposed two distinct strategies in Section 5.1. According to the comparison of current research on physical fatigue assessment (PFA), as presented in Table 6.1, it becomes evident that the prior contributions of the proposed method have focused on a subset of ergonomic indicator-based PFA. To further advance this line of inquiry, particularly when utilizing vision-based data collection methods, the authors have outlined additional future research directions in Section 5.2.

6.5.1 Future directions for advancing the proposed ergonomic assessment method

(a) Enhancing and enriching the dataset

The current limitation of the CML dataset is that it only provides skeleton data for the 15 or 20-joint system (Tian et al., 2022), rendering calculation of certain angles in REBA unavailable. For instance, joint-level assessment modules, such as 'neck twisting' and all modules associated with the 'wrist' segment, are absent due to the lack of skeleton data for the hands and head. To overcome this limitation, there are two potential approaches to consider. First, expanding the CML dataset to include a broader range of 3D pose data would directly and effectively address this issue. Another viable method is to obtain more comprehensive posture data through 3D model-based simulations (Li et al., 2019b; Wang et al., 2023a). In addition to enriching the skeleton information, the quality and diversity of the dataset are crucial factors for ergonomic indicator-based PFA. With the prevalence of artificial intelligence algorithms in pose data collection methods, the accuracy and generalization capability of these algorithms are greatly influenced by the quality and diversity of the dataset (Li et al., 2021; Shao et al., 2019). Including diverse workers and activities in the dataset would contribute to the robustness and generalization of the algorithms. Additionally, the absence of standardized content and format for posture-related datasets is a significant challenge (Tian et al., 2022). Previous studies have emphasized the need for standardized datasets, as the lack of standardization hampers the ability to compare accuracy across different studies (Yu et al., 2021). Establishing standards for posture-related datasets would enable researchers to objectively compare the performance of their approaches.

(b) Improving the data-driven approach

It is important to note that the current heuristic approach does not guarantee optimal independence among the CMs generated for certain modules. For instance, some assessment modules, such as trunk flexion, exhibit instances where overlaps exist between the core and basic areas (Li & Du, 2017). To address this issue, one feasible approach is to employ the Adaptive Gaussian Cloud Transformation (A-GCT) (Chen et al., 2015). This approach involves establishing a control condition within the iterative process, where all CDs should be lower than a predefined threshold. By implementing A-GCT, it becomes possible to automate the generation of optimal CMs from the data sample. However, a potential concern arises as this adaptive method might lead to inconsistency between the number of CDs and the corresponding risk categories. In other words, the output membership degrees might render the designed fuzzy reasoning mechanism inapplicable.

6.5.2 Future directions for ergonomic indicator-based PFA

(a) Increasing the accuracy of data collection from non-contact sensors

Monocular RGB cameras are commonly used as non-contact sensors to capture posture data for PFA, thanks to their affordability and non-intrusiveness (Chu et al., 2020). To optimize their performance, the following recommendations can be considered. First, when capturing images from a considerable distance, RGB pictures often have low resolution (Ionescu et al., 2014), which leads to insufficient information for accurate pose estimation (Yu et al., 2021). To address this limitation, one possible approach is to utilize super-resolution algorithms from the field of computer science to improve the resolution of low-resolution images (Wang et al., 2021b). Second, construction sites may have varying lighting conditions, with some areas remaining in darkness. While RGB cameras are limited to well-lit settings, infrared-based depth cameras are ineffective under direct sunlight (Parr et al., 2022). To overcome this challenge, the application of LiDAR and radar for posture recognition and pose estimation on construction sites is worth exploring (Diraco et al., 2013; Sengupta et al., 2020), as they exhibit resilience to varying scene lighting conditions (Petrlík et al., 2021). Furthermore, in dynamic and intricate environments like construction sites, workers may frequently become obscured or concealed in the captured images and video clips (Li & Li, 2022). Therefore, further research is needed to improve and modify existing algorithms for pose estimation and posture recognition under occlusion (Huo & Hendriks, 2012). Additionally, alternative solutions for human pose estimation, such as radio signal-based approaches (Zhao et al., 2018) and the utilization of drone cameras (Ashour et al., 2016; Penmetsa et al., 2014), deserve further exploration and discussion.

(b) Environment awareness integration

Awareness of the surroundings of construction workers, including tools, materials, and environments, provides valuable insights into their behaviors (Yu et al., 2021). This knowledge is crucial for effective site management and identifying unsafe behavior. Continuous monitoring of workers' status, nearby tools, and machinery facilitates the identification of potential safety risks (Wu et al., 2022). Similarly, for ergonomic analysis, identifying the specific tools or materials being used enables the estimation of external loads for biomechanical analysis. Previous studies relied on manual records to estimate these external loads (Yu, Li, Yang, et al., 2019). However, by automatically identifying the tools and materials held by workers and combining it with biomechanical analysis based on 3D posture, a more automated approach can be implemented to estimate external loads (Yu, Li, Umer, et al., 2019). This automation enables more efficient ergonomic analysis when combined with other collected data.

(c) Privacy problems in data collection

The collection of PFA-related data on construction sites raises privacy concerns, as workers may feel uncomfortable being monitored by sensors, cameras, or observers (Park & Brilakis, 2016; Sherafat et al., 2020). Surprisingly, there is a research gap in studying this issue that needs to be addressed. Data ownership also poses challenges in construction projects involving multiple stakeholders, leading to questions about data ownership, sharing permissions, and the need for censorship (Ayodele & Kajimo-Shakantu, 2022). Some studies propose a multi-server information-sharing approach on a private cloud to tackle these challenges in BIM-based projects (Zhang et al., 2017), potentially resolving data ownership and privacy concerns regarding PFA-related data. However, the matter of worker consent remains unresolved and requires further attention.

(d) Integration of different data by multimodal fusion

The data collection methods reviewed in this chapter have primarily relied on single-modality data, resulting in limited optimization of assessment results. To overcome this constraint, the integration of multimodal data through processes like integration, understanding, and reasoning (Majumder et al., 2018) can effectively leverage the wealth of available information and enhance the accuracy and robustness of assessments (Zhang et al., 2020). For example, to address the occlusion issue, instead of solely improving algorithms (Huo & Hendriks, 2012), multimodal fusion enables the integration of advantages derived from various types of technologies (Wang et al., 2020). This technology enhances the robustness and generalizability of trained models (Poria et al., 2017). Therefore, it is crucial to emphasize its applicability in mitigating relevant challenges encountered in adverse conditions.

6.6 Conclusions

This chapter contributes to the existing body of knowledge on physical fatigue assessment (PFA) for construction workers, summarized in the following two aspects.

First, it provides a comprehensive review of studies on PFA in the construction industry, focusing on PFA indicators and data collection methods. The review categorizes PFA indicators into subjective, physiological, biomechanical, and ergonomic indicators. It conducts a comparison to evaluate the performance of each indicator and data collection method. The comparison highlights the advantages and applicability of vision-based ergonomic assessment in construction workplaces.

Second, it highlights the limited applicability of existing ergonomic scales when assessing construction workers. Specifically, it identifies two limitations: the inapplicable discrete boundaries of joint-level risk categories and the reliance on if-then rules for transformation between assessment levels. To address these gaps, the chapter proposes a heuristic data-driven method tailored to diversified construction working postures. This method leverages the Heuristics Gaussian Cloud Transformation (H-GCT), cloud model (CM), and fuzzy reasoning to generate data-driven scoring models that replace discrete boundaries in each considered joint-level module of REBA. This enables appropriate and continuous assessment outputs. Additionally, the chapter outlines future research directions to advance ergonomic indicators-based PFA, particularly when utilizing vision-based data collection methods.

This chapter provides a valuable reference for academics and practitioners interested in PFA for construction workers. It summarizes the performance of indicators and data collection methods in construction management, identifies research gaps, and offers feasible solutions and directions for future research. By presenting a thorough summary and comparison of available PFA indicators and data collection tools, the study enables industrial practitioners to choose the most suitable methods for their specific needs.

However, it is important to acknowledge certain limitations associated with this chapter. First, the exploration of indicators and data collection methods in robotics is limited, with only one study proposing a simulation case (Kurien et al., 2018). Second, the selection of data collection sensors/devices is primarily based on the reviewed studies, which may inadvertently exclude other tools with superior performance that were not included in the review.

References

Akerstedt, T., Knutsson, A., Westerholm, P., Theorell, T., Alfredsson, L., & Kecklund, G. (2004). Mental fatigue, work and sleep. *Journal of Psychosomatic Research, 57*(5), 427–433. https://doi.org/10.1016/j.jpsychores.2003.12.001

Al-Mulla, M. R., Sepulveda, F., & Colley, M. (2011). A review of non-invasive techniques to detect and predict localised muscle fatigue. *Sensors, 11*(4), 3545–3594. https://www.mdpi.com/1424-8220/11/4/3545

Antwi-Afari, M. F., Li, H., Edwards, D. J., Pärn, E. A., Owusu-Manu, D.-G., Seo, J., & Wong, A. Y. L. (2018). Identification of potential biomechanical risk factors for low back disorders during repetitive rebar lifting. *Construction Innovation, 18*(2). https://doi.org/10.1108/CI-05-2017-0048

Antwi-Afari, M. F., Li, H., Edwards, D. J., Pärn, E. A., Seo, J., & Wong, A. (2017). Effects of different weights and lifting postures on balance control following repetitive lifting tasks in construction workers. *International Journal of Building Pathology and Adaptation, 35*(3), 247–263. https://doi.org/10.1108/IJBPA-05-2017-0025

Anwer, S., Li, H., Antwi-Afari, M. F., Umer, W., & Wong, A. Y. L. (2021). Evaluation of physiological metrics as real-time measurement of physical fatigue in construction workers: State-of-the-art review. *Journal of Construction Engineering and Management, 147*(5), 03121001. https://doi.org/10.1061/(ASCE)CO.1943-7862.0002038

Anwer, S., Li, H., Umer, W., Antwi-Afari, M. F., Mehmood, I., Yu, Y., Haas, C., & Wong, A. Y. L. (2023). Identification and classification of physical fatigue in construction workers using linear and nonlinear heart rate variability measurements. *Journal of Construction Engineering and Management, 149*(7), 04023057. https://doi.org/10.1061/JCEMD4.COENG-13100

Aryal, A., Ghahramani, A., & Becerik-Gerber, B. (2017). Monitoring fatigue in construction workers using physiological measurements. *Automation in Construction, 82*, 154–165. https://doi.org/10.1016/j.autcon.2017.03.003.

Ashour, R., Taha, T., Mohamed, F., Hableel, E., Kheil, Y. A., Elsalamouny, M., Kadadha, M., Rangan, K., Dias, J., Seneviratne, L., & Cai, G. (2016). Site inspection drone: A solution for inspecting and regulating construction sites. In *2016 IEEE 59th international midwest symposium on circuits and systems (MWSCAS)*, 16–19 October. Piscataway: IEEE.

Ayodele, T. O., & Kajimo-Shakantu, K. (2022). Challenges and drivers to data sharing among stakeholders in the South African construction industry. *Journal of Engineering, Design and Technology, 20*(6), 1698–1715. https://doi.org/10.1108/JEDT-02-2021-0074

Azadeh, A. H., Zadeh, S. A., Saberi, M., Hussain, F. K., & Hussain, O. K. (2014). A trust-based performance measurement modeling using DEA, T-norm and S-norm operators. In *2014 IEEE International Conference on Fuzzy Systems (FUZZ-IEEE)*, 6–11 July. Piscataway: IEEE.

Bobillo, F., & Straccia, U. (2007). A fuzzy description logic with product T-norm. In *2007 IEEE International Fuzzy Systems Conference*, 23–26 July. Piscataway: IEEE.

Borg, G. (1998). *Borg's Perceived Exertion and Pain Scales*. Human Kinetics.

Brandl, C., Mertens, A., & Schlick, C. M. (2017). Effect of sampling interval on the reliability of ergonomic analysis using the Ovako Working Posture Analysing System (OWAS). *International Journal of Industrial Ergonomics, 57*, 68–73. https://doi.org/10.1016/j.ergon.2016.11.013

Caban-Martinez, A. J., Santiago, K. M., Stillman, J., Moore, K. J., Sierra, D., Chalmers, J., Baniak, M., & Jordan, M. M. (2018). Physical exposures, work tasks and osha-10 training among temporary and payroll construction workers. *Journal of Occupational and Environmental Medicine, 60*(4), e159.

Chen, J., Liu, Y., & Li, D. (2015). Enhancing recommender diversity using gaussian cloud transformation. *International Journal of Uncertainty, Fuzziness and Knowledge-Based Systems, 23*(4), 521–544. https://doi.org/10.1142/s0218488515500233

Cheng, T., Teizer, J., Migliaccio, G. C., & Gatti, U. C. (2013). Automated task-level activity analysis through fusion of real time location sensors and worker's thoracic posture data. *Automation in Construction, 29*, 24–39. https://doi.org/10.1016/j.autcon.2012.08.003

Chu, W., Han, S., Luo, X., & Zhu, Z. (2020). Monocular vision-based framework for biomechanical analysis or ergonomic posture assessment in modular construction. *Journal of Computing in Civil Engineering*, *34*(4), 04020018. https://doi.org/10.1061/(ASCE)CP.1943-5487.0000897

Delp, S. L., Anderson, F. C., Arnold, A. S., Loan, P., Habib, A., John, C. T., Guendelman, E., & Thelen, D. G. (2007). OpenSim: Open-source software to create and analyze dynamic simulations of movement. *IEEE Transactions on Biomedical Engineering*, *54*(11), 1940–1950. https://doi.org/10.1109/TBME.2007.901024

DiDomenico, A., McGorry, R. W., Huang, Y.-H., & Blair, M. F. (2010). Perceptions of postural stability after transitioning to standing among construction workers. *Safety Science*, *48*(2), 166–172.

Diraco, G., Leone, A., & Siciliano, P. (2013). Human posture recognition with a time-of-flight 3D sensor for in-home applications. *Expert Systems with Applications*, *40*(2), 744–751. https://doi.org/10.1016/j.eswa.2012.08.007

Dudek, G., & Jenkin, M. (2016). Inertial sensing, GPS and odometry. In B. Siciliano & O. Khatib (Eds.), *Springer Handbook of Robotics* (pp. 737–752). Springer International Publishing. https://doi.org/10.1007/978-3-319-32552-1_29

Elders, L., & Burdorf, A. (2001). Interrelations of risk factors and low back pain in scaffolders. *Occupational and Environmental Medicine*, *58*(9), 597–603.

Feyen, R., Liu, Y., Chaffin, D., Jimmerson, G., & Joseph, B. (2000). Computer-aided ergonomics: A case study of incorporating ergonomics analyses into workplace design. *Applied Ergonomics*, *31*(3), 291–300. https://doi.org/10.1016/s0003-6870(99)00053-8

Fitts, R. H. (1994). Cellular mechanisms of muscle fatigue. *Physiological Reviews*, *74*(1), 49–94. https://doi.org/10.1152/physrev.1994.74.1.49

Frone, M. R., & Tidwell, M.-C. O. (2015). The meaning and measurement of work fatigue: Development and evaluation of the three-dimensional Work Fatigue Inventory (3D-WFI). *Journal of Occupational Health Psychology*, *20*(3), 273–288. https://doi.org/10.1037/a0038700

Gatti, U. C., Schneider, S., & Migliaccio, G. C. (2014). Physiological condition monitoring of construction workers. *Automation in Construction*, *44*, 227–233. https://doi.org/10.1016/j.autcon.2014.04.013

Ghafoori, M., Clevenger, C., Abdallah, M., & Rens, K. (2023). Heart rate modeling and prediction of construction workers based on physical activity using deep learning. *Automation in Construction*, *155*, 105077. https://doi.org/10.1016/j.autcon.2023.105077

Ghasemi, F., & Mahdavi, N. (2020). A new scoring system for the Rapid Entire Body Assessment (REBA) based on fuzzy sets and bayesian networks. *International Journal of Industrial Ergonomics*, *80*, 103058. https://doi.org/10.1016/j.ergon.2020.103058

Golabchi, A., Han, S., & Fayek, A. R. (2016). A fuzzy logic approach to posture-based ergonomic analysis for field observation and assessment of construction manual operations. *Canadian Journal of Civil Engineering*, *43*(4), 294–303. https://doi.org/10.1139/cjce-2015-0143

Golabchi, A., Han, S., Fayek, A. R., & AbouRizk, S. (2017). Stochastic modeling for assessment of human perception and motion sensing errors in ergonomic analysis. *Journal of Computing in Civil Engineering*, *31*(4), 04017010. https://doi.org/10.1061/(ASCE)CP.1943-5487.0000655

Hermens, H. J., Freriks, B., Disselhorst-Klug, C., & Rau, G. (2000). Development of recommendations for SEMG sensors and sensor placement procedures. *Journal of Electromyography and Kinesiology*, *10*(5), 361–374. https://doi.org/10.1016/S1050-6411(00)00027-4

Hignett, S., & McAtamney, L. (2000). Rapid entire body assessment (REBA). *Applied Ergonomics*, *31*(2), 201–205. https://doi.org/10.1016/S0003-6870(99)00039-3

Huo, F., & Hendriks, E. A. (2012). Multiple people tracking and pose estimation with occlusion estimation. *Computer Vision and Image Understanding*, *116*(5), 634–647. https://doi.org/10.1016/j.cviu.2011.12.006

Ibrahim, A., Nnaji, C., Namian, M., Koh, A., & Techera, U. (2023). Investigating the impact of physical fatigue on construction workers' situational awareness. *Safety Science*, *163*, 106103. https://doi.org/10.1016/j.ssci.2023.106103

Ionescu, C., Papava, D., Olaru, V., & Sminchisescu, C. (2014). Human3.6M: Large scale datasets and predictive methods for 3d human sensing in natural environments. *IEEE Transactions on Pattern Analysis and Machine Intelligence*, *36*(7), 1325–1339. https://doi.org/10.1109/tpami.2013.248

Jiang, S., Gao, Q., Liu, H., & Shull, P. B. (2020). A novel, co-located EMG-FMG-sensing wearable armband for hand gesture recognition. *Sensors and Actuators A: Physical*, *301*, 111738. https://doi.org/10.1016/j.sna.2019.111738

Jo, D., & Bilodeau, M. (2021). Rating of perceived exertion (RPE) in studies of fatigue-induced postural control alterations in healthy adults: Scoping review of quantitative evidence. *Gait & Posture*, *90*, 167–178. https://doi.org/10.1016/j.gaitpost.2021.08.015

Karhu, O., Härkönen, R., Sorvali, P., & Vepsäläinen, P. (1981). Observing working postures in industry: Examples of OWAS application. *Applied Ergonomics*, *12*(1), 13–17. https://doi.org/10.1016/0003-6870(81)90088-0

Kee, D. (2021). Comparison of OWAS, RULA and REBA for assessing potential work-related musculoskeletal disorders. *International Journal of Industrial Ergonomics*, *83*, 103140. https://doi.org/10.1016/j.ergon.2021.103140

Kirk, P. M., & Sullman, M. J. M. (2001). Heart rate strain in cable hauler choker setters in New Zealand logging operations. *Applied Ergonomics*, *32*(4), 389–398. https://doi.org/10.1016/S0003-6870(01)00003-5

Kong, Y. K., & Kim, D. M. (2015). The relationship between hand anthropometrics, total grip strength and individual finger force for various handle shapes. *International Journal of Occupational Safety and Ergonomics*, *21*(2), 187–192. https://doi.org/10.1080/10803548.2015.1029726

Kurien, M., Kim, M.-K., Kopsida, M., & Brilakis, I. (2018). Real-time simulation of construction workers using combined human body and hand tracking for robotic construction worker system. *Automation in Construction*, *86*, 125–137. https://doi.org/10.1016/j.autcon.2017.11.005

Lee, W., Lin, K.-Y., Johnson, P. W., & Seto, E. Y. W. (2022). Selection of wearable sensor measurements for monitoring and managing entry-level construction worker fatigue: A logistic regression approach. *Engineering, Construction and Architectural Management*, *29*(8), 2905–2923. https://doi.org/10.1108/ECAM-02-2021-0106

Lee, Y.-C., & Lee, C.-H. (2022). SEE: A proactive strategy-centric and deep learning-based ergonomic risk assessment system for risky posture recognition. *Advanced Engineering Informatics*, *53*, 101717. https://doi.org/10.1016/j.aei.2022.101717

Li, D., & Du, Y. (2017). *Artificial Intelligence with Uncertainty*. CRC Press. https://doi.org/10.1201/9781315366951

Li, G., Chen, B., Zhu, L., He, Q., Fan, H., & Wang, S. (2021). PUGCQ: A large scale dataset for quality assessment of professional user-generated content. In *Proceedings of the 29th ACM International Conference on Multimedia*, Virtual Event, China. https://doi.org/10.1145/3474085.3475183

Li, X., Han, S., Gül, M., & Al-Hussein, M. (2019b). Automated post-3D visualization ergonomic analysis system for rapid workplace design in modular construction. *Automation in Construction*, *98*, 160–174. https://doi.org/10.1016/j.autcon.2018.11.012

Li, X., Han, S., Gül, M., Al-Hussein, M., & El-Rich, M. (2018). 3D visualization-based ergonomic risk assessment and work modification framework and its validation for a lifting task. *Journal of Construction Engineering and Management*, *144*(1), 04017093. https://doi.org/10.1061/(ASCE)CO.1943-7862.0001412

Li, Z., & Li, D. (2022). Action recognition of construction workers under occlusion. *Journal of Building Engineering*, *45*, 103352. https://doi.org/10.1016/j.jobe.2021.103352

Li, Z., Yu, Y., Xia, J., Chen, X., Lu, X., & Li, Q. (2024). Data-driven ergonomic assessment of construction workers. *Automation in Construction*, *165*, 105561. https://doi.org/10.1016/j.autcon.2024.105561

Lingard, H., & Wakefield, R. (2019). *Integrating Work Health and Safety into Construction Project Management*. John Wiley & Sons.

Majumder, N., Hazarika, D., Gelbukh, A., Cambria, E., & Poria, S. (2018). Multimodal sentiment analysis using hierarchical fusion with context modeling. *Knowledge-Based Systems*, *161*, 124–133. https://doi.org/10.1016/j.knosys.2018.07.041

Maruish, M. E. (2004). *The Use of Psychological Testing for Treatment Planning and Outcomes Assessment (Volume 1: General Considerations)*. Routledge.

McAtamney, L., & Nigel Corlett, E. (1993). RULA: A survey method for the investigation of work-related upper limb disorders. *Applied Ergonomics*, *24*(2), 91–99. https://doi.org/10.1016/0003-6870(93)90080-S

Mehmood, I., Li, H., Umer, W., Arsalan, A., Anwer, S., Mirza, M. A., Ma, J., & Antwi-Afari, M. F. (2023). Multimodal integration for data-driven classification of mental fatigue during construction equipment operations: Incorporating electroencephalography, electrodermal activity, and video signals. *Developments in the Built Environment*, *15*, 100198. https://doi.org/10.1016/j.dibe.2023.100198

Merletti, R., Botter, A., Troiano, A., Merlo, E., & Minetto, M. A. (2009). Technology and instrumentation for detection and conditioning of the surface electromyographic signal: State of the art. *Clinical Biomechanics*, *24*(2), 122–134. https://doi.org/10.1016/j.clinbiomech.2008.08.006

Moriguchi, C. S., Carnaz, L., Veiersted, K. B., Hanvold, T. N., Hæg, L. B., Hansson, G.-Å., & Cote Gil Coury, H. J. (2013). Occupational posture exposure among construction electricians. *Applied Ergonomics*, *44*(1), 86–92. https://doi.org/10.1016/j.apergo.2012.05.002

Nath, N. D., Akhavian, R., & Behzadan, A. H. (2017). Ergonomic analysis of construction worker's body postures using wearable mobile sensors. *Applied Ergonomics*, *62*, 107–117. https://doi.org/10.1016/j.apergo.2017.02.007

Occhipinti, E. (1998). OCRA: A concise index for the assessment of exposure to repetitive movements of the upper limbs. *Ergonomics*, *41*(9), 1290–1311. https://doi.org/10.1080/001401398186315

Park, M.-W., & Brilakis, I. (2016). Continuous localization of construction workers via integration of detection and tracking. *Automation in Construction*, *72*, 129–142. https://doi.org/10.1016/j.autcon.2016.08.039

Parr, B., Legg, M., & Alam, F. (2022). Analysis of depth cameras for proximal sensing of grapes. *Sensors*, *22*(11), 4179. https://www.mdpi.com/1424-8220/22/11/4179

Penmetsa, S., Minhuj, F., Singh, A., & Omkar, S. (2014). Autonomous UAV for suspicious action detection using pictorial human pose estimation and classification. *ELCVIA: Electronic Letters on Computer Vision and Image Analysis*, *13*(1), 0018–0032.

Petrlík, M., Krajník, T., & Saska, M. (2021). LIDAR-based stabilization, navigation and localization for UAVs operating in dark indoor environments. In *2021 International Conference on Unmanned Aircraft Systems (ICUAS)*, 15–18 June. Piscataway: IEEE.

Poria, S., Cambria, E., Bajpai, R., & Hussain, A. (2017). A review of affective computing: From unimodal analysis to multimodal fusion. *Information Fusion*, *37*, 98–125. https://doi.org/10.1016/j.inffus.2017.02.003

Roman-Liu, D. (2014). Comparison of concepts in easy-to-use methods for MSD risk assessment. *Applied Ergonomics*, *45*(3), 420–427. https://doi.org/10.1016/j.apergo.2013.05.010

Sengupta, A., Jin, F., Zhang, R., & Cao, S. (2020). mm-Pose: Real-time human skeletal posture estimation using mmwave radars and CNNs. *IEEE Sensors Journal*, *20*(17), 10032–10044. https://doi.org/10.1109/JSEN.2020.2991741

Seo, J., Starbuck, R., Han, S., Lee, S., & Armstrong, T. J. (2015). Motion data-driven biomechanical analysis during construction tasks on sites. *Journal of Computing in Civil Engineering*, *29*(4), B4014005. https://doi.org/10.1061/(ASCE)CP.1943-5487.0000400

Seth, A., Hicks, J. L., Uchida, T. K., Habib, A., Dembia, C. L., Dunne, J. J., Ong, C. F., DeMers, M. S., Rajagopal, A., Millard, M., Hamner, S. R., Arnold, E. M., Yong, J. R., Lakshmikanth, S. K., Sherman, M. A., Ku, J. P., & Delp, S. L. (2018). OpenSim: Simulating musculoskeletal dynamics and neuromuscular control to study human and animal movement. *PLOS Computational Biology*, *14*(7), e1006223. https://doi.org/10.1371/journal.pcbi.1006223.

Shao, S., Li, Z., Zhang, T., Peng, C., Yu, G., Zhang, X., Li, J., & Sun, J. (2019). Objects365: A large-scale, high-quality dataset for object detection. In *Proceedings of the IEEE/CVF International Conference on Computer Vision*, 27 October–02 November. Piscataway: IEEE.

Sherafat, B., Ahn, C. R., Akhavian, R., Behzadan, A. H., Golparvar-Fard, M., Kim, H., Lee, Y.-C., Rashidi, A., & Azar, E. R. (2020). Automated methods for activity recognition of construction workers and equipment: State-of-the-art review. *Journal of Construction Engineering and Management*, *146*(6), 03120002. https://doi.org/10.1061/(ASCE)CO.1943-7862.0001843

Simeonov, P. I., Hsiao, H., Dotson, B. W., & Ammons, D. E. (2003). Control and perception of balance at elevated and sloped surfaces. *Human Factors*, *45*(1), 136–147.

Singh, A. K., Kumbhare, V. A., & Arthi, K. (2022). Real-time human pose detection and recognition using mediapipe. In *2021 International Conference on Soft Computing and Signal Processing (ICSCSP)*, 18–19 June. Singapore: Springer.

Slembrouck, M., Luong, H., Gerlo, J., Schütte, K., Van Cauwelaert, D., De Clercq, D., Vanwanseele, B., Veelaert, P., & Philips, W. (2020). Multiview 3d markerless human pose estimation from openpose skeletons. In *2020 Advanced Concepts for Intelligent Vision Systems (ACIVS)*, 10–14 February. Cham: Springer.

Techera, U. (2017). *Measuring and Managing Construction Worker Fatigue*. Polytechnic University of Catalonia.

Tian, Y., Li, H., Cui, H., & Chen, J. (2022). Construction motion data library: an integrated motion dataset for on-site activity recognition. *Scientific Data*, *9*(1), 726. https://doi.org/10.1038/s41597-022-01841-1

Toro, S. F. D., Santos-Cuadros, S., Olmeda, E., Álvarez-Caldas, C., Díaz, V., & San Román, J. L. (2019). Is the use of a low-cost semg sensor valid to measure muscle fatigue? *Sensors*, *19*(14), 3204. https://www.mdpi.com/1424-8220/19/14/3204

Umer, W., Li, H., Lu, W., Szeto, G. P. Y., & Wong, A. Y. L. (2018). Development of a tool to monitor static balance of construction workers for proactive fall safety management. *Automation in Construction*, *94*, 438–448. https://doi.org/10.1016/j.autcon.2018.07.024

Wang, D., Dai, F., & Ning, X. (2015). Risk assessment of work-related musculoskeletal disorders in construction: State-of-the-art review. *Journal of Construction Engineering and Management*, *141*(6), 04015008. https://doi.org/10.1061/(ASCE)CO.1943-7862.0000979

Wang, J., Han, S., & Li, X. (2021a). 3D fuzzy ergonomic analysis for rapid workplace design and modification in construction. *Automation in Construction*, *123*, 103521. https://doi.org/10.1016/j.autcon.2020.103521

Wang, J., Li, X., Han, S., & Al-Hussein, M. (2023a). 3D standard motion time-based ergonomic risk analysis for workplace design in modular construction. *Automation in Construction*, *147*, 104738. https://doi.org/10.1016/j.autcon.2022.104738

Wang, Y., Huang, Y., Gu, B., Cao, S., & Fang, D. (2023b). Identifying mental fatigue of construction workers using EEG and deep learning. *Automation in Construction*, *151*, 104887. https://doi.org/10.1016/j.autcon.2023.104887

Wang, Y., Yu, S., Ma, N., Wang, J., Hu, Z., Liu, Z., & He, J. (2020). Prediction of product design decision making: An investigation of eye movements and EEG features. *Advanced Engineering Informatics*, *45*, 101095. https://doi.org/10.1016/j.aei.2020.101095

Wang, Z., Chen, J., & Hoi, S. C. H. (2021b). Deep learning for image super-resolution: A survey. *IEEE Transactions on Pattern Analysis and Machine Intelligence*, *43*(10), 3365–3387. https://doi.org/10.1109/TPAMI.2020.2982166

Wu, S., Hou, L., Zhang, G., & Chen, H. (2022). Real-time mixed reality-based visual warning for construction workforce safety. *Automation in Construction*, *139*, 104252. https://doi.org/10.1016/j.autcon.2022.104252

Yan, X., Li, H., Li, A. R., & Zhang, H. (2017). Wearable IMU-based real-time motion warning system for construction workers' musculoskeletal disorders prevention. *Automation in Construction*, *74*, 2–11. https://doi.org/10.1016/j.autcon.2016.11.007

Yan, X., Li, H., Wang, C., Seo, J., Zhang, H., & Wang, H. (2017). Development of ergonomic posture recognition technique based on 2D ordinary camera for construction hazard prevention through view-invariant features in 2D skeleton motion. *Advanced Engineering Informatics*, *34*, 152–163. https://doi.org/10.1016/j.aei.2017.11.001

Yu, Y. (2020). *Automatic Physical Fatigue Assessment for Construction Workers based on Computer Vision and Pressure Insole Sensor*. The Hong Kong Polytechnic University.

Yu, Y., Li, H., Umer, W., Dong, C., Yang, X., Skitmore, M., & Wong, A. Y. L. (2019). Automatic biomechanical workload estimation for construction workers by computer vision and smart insoles. *Journal of Computing in Civil Engineering*, *33*(3), 04019010. https://doi.org/10.1061/(ASCE)CP.1943-5487.0000827

Yu, Y., Li, H., Yang, X., Kong, L., Luo, X., & Wong, A. Y. L. (2019). An automatic and non-invasive physical fatigue assessment method for construction workers. *Automation in Construction*, *103*, 1–12. https://doi.org/10.1016/j.autcon.2019.02.020

Yu, Y., Umer, W., Yang, X., & Antwi-Afari, M. F. (2021). Posture-related data collection methods for construction workers: A review. *Automation in Construction*, *124*, 103538. https://doi.org/10.1016/j.autcon.2020.103538

Yu, Y., Yang, X., Li, H., Luo, X., Guo, H., & Fang, Q. (2019). Joint-level vision-based ergonomic assessment tool for construction workers. *Journal of Construction Engineering and Management*, *145*(5), 04019025. https://doi.org/10.1061/(ASCE)CO.1943-7862.0001647

Zhang, H., Yan, X., & Li, H. (2018). Ergonomic posture recognition using 3D view-invariant features from single ordinary camera. *Automation in Construction*, *94*, 1–10. https://doi.org/10.1016/j.autcon.2018.05.033

Zhang, J., Liu, Q., Hu, Z., Lin, J., & Yu, F. (2017). A multi-server information-sharing environment for cross-party collaboration on a private cloud. *Automation in Construction*, *81*, 180–195. https://doi.org/10.1016/j.autcon.2017.06.021

Zhang, W., Yu, J., Hu, H., Hu, H., & Qin, Z. (2020). Multimodal feature fusion by relational reasoning and attention for visual question answering. *Information Fusion*, *55*, 116–126. https://doi.org/10.1016/j.inffus.2019.08.009

Zhao, M., Li, T., Alsheikh, M. A., Tian, Y., Zhao, H., Torralba, A., & Katabi, D. (2018). Through-wall human pose estimation using radio signals. In *2018 IEEE/CVF Conference on Computer Vision and Pattern Recognition*, 18–23 June. Piscataway: IEEE.

7

SMART OR NOT? PERSPECTIVES ON DIGITAL FABRICATION AND DIGITAL CONSTRUCTION

Qian Chen[1]*, Ming Shan Ng*[2] *and Jürgen Hackl*[3]

[1]Assistant Professor, University of British Columbia Okanagan Campus, Kelowna, Canada. Email: qian.chen@ubc.ca

[2]Professor, Kyoto Institute of Technology, Kyoto, Japan. Email: ng@kit.ac.jp

[3]Assistant Professor, Princeton University, Princeton, USA. Email: hackl@princeton.edu

7.1 Introduction

Construction is one of the largest industries in the world (Barbosa et al., 2017). However, in contrast to other sectors, such as manufacturing, the construction industry has fallen behind, characterized by stagnant productivity, project delays, cost overruns, and low levels of digitization (Weinfass and Kelly, 2020) evidenced by a slow labor-productivity growth averaged at only 1.0% a year over the past two decades. On average, construction laborers spend more than 50% of their workday on paper-based administrative tasks and waiting on equipment, materials, and information. Traditional construction project implementation has been limited in scope to the perspective of a single project stakeholder and project phase (Biörck et al., 2020; Vrijhoef and Koskela, 2000); however, quite a few recent studies have been leading the way in showing that the entire construction supply chain processes are evolving from a single logistics perspective into an interrelated multi-functional perspective, which highlights the interfaces across design, procurement, production or fabrication, logistics, and site installation (Chen et al., 2020a; Lu et al., 2021). Stakeholders usually have disparate and unorganized information exchange and coordination processes when managing projects. This disjointed dynamic frequently inhibits collaboration and information exchange among project teams and increases the likelihood that a project will not meet its schedule and cost expectations. Stakeholders need to develop novel concepts, processes, methods, and technologies to facilitate collaboration and supply chain integration and enable projects to be completed on time, on budget, and on quality.

The Industry 4.0 (Fourth Industrial Revolution) technologies (e.g., smart factories connected by the industrial Internet of Things (IoT) sensors and systems) have a streamlined design and manufacturing processes that are controlled by decentralized control and have a high degree of connectivity. Within such contexts, construction researchers advocated the concept of Construction 4.0 and the digital twin systems (Garcia de Soto and Skibniewski, 2020). Researchers coined digital construction and digital fabrication concepts to refer to data-driven production processes based on digital design information (Bock and Linner, 2015). The two terms are often interchangeably used

DOI: 10.1201/9781003383840-8

and bear close semantic proximity. In this book chapter, "digital fabrication" focuses on advanced manufacturing of customized construction components through economies of scale, and "digital construction" focuses on efficient site installation of construction components to avoid site rework. Both share the ultimate goal of balancing on-site and off-site production of building modules and components and how they are handled, assembled, and integrated.

Examples of digital technologies used for parametric design and material flow management include the integrated use of building information modeling (BIM)-based systems. Prevalent examples include the 4D BIM planning platform and cloud-based BIM systems and integrated BIM with wireless transmission technologies (e.g., radio frequency identification (RFID), near-field communication, Bluetooth low-energy, etc.) (Chen et al., 2022; Dave et al., 2018). Much of this research, however, focuses primarily on detailed technical aspects to improve information exchange efficiency and accuracy. The inherently dynamic nature of processes across multiple construction supply chains and the gap between design, fabrication, and construction have not been adequately addressed in previous work (Chen, 2020; Ng et al., 2022b). The construction industry has been slow to see the benefits of the technologies and is reluctant to digital transformation. Project stakeholders must find an efficient approach to incorporate a dynamic workflow and fully integrate technologies throughout the supply chain process (Chen, 2020). Exactly how the information needs could be understood and shared consistently among stakeholders, from the upstream design stage to the downstream fabrication and construction stage, would have to be investigated to enable supply chain integration and substantially improve project performance.

This book chapter focuses on developing such consistent information flows among construction stakeholders by systematically illustrating an implementation framework of digital fabrication and digital construction. The implementation framework aims to help guide stakeholders to manage design data and material flows efficiently, integrate the construction supply chain activities, and help stakeholders reduce rework and improve project quality. This chapter is structured in the following sections. Section 7.2 presents the literature review on the technological advancement of digital integration in construction. Section 7.3 describes the proposed implementation framework of digital fabrication and digital construction, followed by two respective case studies in Section 7.4 to demonstrate the potential of digital fabrication (the clerestory steel structure project with a bespoke roof structure located on the West Coast of the United States) and digital construction (the prefabricated office building project with a complex design of prefab concrete columns located in Zurich Switzerland). The advantages and areas for future improvement of the implementation framework are discussed in Section 7.5, followed by the conclusions and future work in Section 7.6.

7.2 Literature review

7.2.1 Digital fabrication and prerequisites

One aspect of digital transformation and industrialized construction in the current practice is adopting digital fabrication, which digitizes and automates construction processes to different extents. Existing digital fabrication methods usually include subtractive manufacturing, such as computer numerical control (CNC) production and robotic drilling; additive manufacturing, such as 3D printing and robotic welding; and robotic assemblies, such as robotic nailing and spatial assembly. Designers are a key stakeholder group that must be convinced to adopt emerging digital design and integration methods. There has been no well-established approach to integrating the stakeholders and supply chain activities during the digital fabrication design process. It can be challenging to implement digital fabrication projects if stakeholders are skeptical, uncertain, or unaware of how the process differs from existing design routines (Ng et al., 2022b, 2023).

With a focus on understanding the design for digital fabrication and the prerequisites, scholars note that adopting digital fabrication needs several specific technical requirements of the fabrication technologies. For example, Ng et al. (2022a) specify the requirements for fabrication information in early design phases to assist design development. This enables a continuous feedback loop and a shared virtual environment to ensure integrated communication between design and fabrication teams. A few studies present a digital twin for digital fabrication with sensing technologies to provide real-time fabrication process monitoring to ensure the digital design integrates with the physical products fabricated digitally (Tran et al., 2021).

Some researchers argue that the technological requirements of digital systems are the key cornerstone of implementing digital fabrication projects. For example, visual programming interfaces such as Grasshopper in Rhinoceros® 3D are commonly used tools during design for digital fabrication processes. These tools can help designers and fabricators create a shared data environment and linked data in design management for digital fabrication. In addition to technical processes, the adoption of digital fabrication relies on new organizational arrangements (Ng et al., 2023). Downstream fabrication-related information has to be considered in the upstream design processes. However, this clashes with many of the common existing design approaches and contractual approaches in project delivery models (e.g., the conventional design-bid-build model) in the industry, where design is intentionally separated from the means of construction and fabrication. Thus, new organizational models or structures are proposed to bring downstream stakeholders into the upstream design process. For example, vertical integration and integrated project delivery have been widely studied to enable organizational integration, where digital fabrication experts and design teams participate in the design process. For that, Ng et al. (2022a) propose that new roles such as the digital fabrication manager – who defines strategies to adopt digital fabrication and manage the technologies in design and construction processes – and digital fabrication programmer, who works on programming to control the digital fabrication information integration.

While digital fabrication has been slowly adopted for customized building construction and industrialized construction, digital fabrication entails a rethinking of the technical, technological, and organizational requirements as well as novel management strategies, which are essential to remove stakeholders' skeptical attitudes and enable fabrication to align with project target values and requirements in design and construction processes in practice. Digital fabrication differs from the processes of conventional manual construction, where human interpretation of the design drawings and their skills significantly determine the fabrication process and outcomes. Similar to digital construction information integration within the supply chain, digital fabrication requires the early involvement of key stakeholders, such as the construction teams, in the design process to engage their knowledge and fabrication information for design development. In this chapter, the main difference between digital fabrication and digital construction is the outcome of the two processes, where digital construction does not necessarily require the automation of construction tasks, which include subtractive manufacturing, additive manufacturing, and robotic assembly of building components.

7.2.2 BIM-based digital construction

Digital construction represents a new way of working in construction through digital tools across the management, delivery, and operation of construction projects, which brings about more collaborative and efficient processes in projects. A large body of literature on digital construction integration is focused on the BIM-based shared data environment (or platforms) and BIM-enabled digital twins. Generally, the technological innovations related to Industry 4.0 or digital twins aim to ensure data connection between virtual models with physical infrastructure. It is important to construct a connection loop between the virtual models and physical buildings for the construction supply chain network where different stakeholders from different expertise domains access data and react to changes promptly (Esfahani et al., 2021; Sacks et al., 2020). For example, architects and construction engineers have overlapping tasks to provide building designs that can meet all the required building codes. They would require a way to communicate their ideas quickly to each other throughout the technical drawing processes. Some virtual building design software, such as BIM authoring platforms, provide opportunities for architects and construction engineers to collaborate and exchange information within a shared environment.

In construction, BIM has been the starting point for innovating the traditional way of working (e.g., data silos, multiple 2D drawings, and numerous spreadsheets) and offers consistent information flows through open standards, such as Industry Foundation Class (IFC) (Boje et al., 2020). In some real-world cases, the general contractor or chief architect is primarily responsible for coordinating their BIM data occasionally by exchanging proprietary files, which is also greatly encouraged by other stakeholders. For suppliers, BIM collaboration occurs by merging the BIM reference model with fabrication software via BIM authoring add-ins (Chen et al., 2020b). This allows less data redundancy and inaccuracy for the suppliers to interpret material design data that architects or contractors provide (Bakhshi et al., 2022). To increase the capabilities of BIM, wireless transmission technologies have been integrated with BIM platforms to create digital control models for site assembly and facility operations (Zhao et al., 2019). For example, the BIM design platform was used as a spatial and temporal reference point for all IoT devices deployed in the office buildings (Heaton and Parlikad, 2020) to visualize the status of building elements in real time.

To focus on supply chain requirements and facilitate design to construction processes, the integrated BIM-IoT platforms help suppliers and contractors visualize building information to correctly understand the supply chain processes, thereby reducing supply chain risks in terms of late material deliveries and resource constraints. For example, the integrated BIM-RFID management system can provide data transparency so that suppliers and contractors can track both the site schedules and material status (e.g., delivery time and location) in real time (Chen et al., 2020b). While integrating BIM and wireless transmission technologies helps facilitate the connectivity of data and stakeholders, it remains challenging for stakeholders to uncover the hidden problems in the supply chain processes caused by site demand uncertainties. It was found that design and engineering complexities compound these uncertainties and result in a considerable amount of schedule and design changes. For highly engineered and customized engineering-to-order (ETO) building elements, managing supply chain processes and coordinating material flows is even more difficult. Despite the potential of integrated BIM-IoT technologies, the improvement of supply chain coordination must consider the uncertainties of the dynamic site environment.

Most of the existing technologies are designed for stakeholders to automate information exchange, track the real-time status of resources, and create a shared data environment for collaboration. There is, however, a gap in digital integration implementation frameworks and methodologies to support stakeholders' decision-making processes in dealing with supply chain uncertainties and risks. Stakeholders need both the technologies and associated consistent information flows and workflows that guarantee quick identification of uncertainties and agility to solve the problems caused by uncertainties. This is particularly relevant to building elements that are subject to prompt design changes and continuously changing demand time. This book chapter will introduce the implementation framework of digital fabrication and digital construction to address such a knowledge gap.

7.3 Implementation framework

7.3.1 Complex survey method to derive the implementation framework

Building upon a content review of the existing research, the authors conducted systematic steering workshops and discussion sessions with five construction professionals to compose an online questionnaire with 21 questions categorized under five areas – technical, technological, organization, contractual, and business model types – which affect the implementation of digital fabrication and digital construction projects (Ng et al., 2022a). Key terminologies and concepts related to digital fabrication and digital construction were used to design the questions, including the "BIM," "interface," "virtualization," "supply chain matrix," "simulation," "platforms," "BIM-based platforms," "platform-based integration" as well as emerging roles in practices such as DFAB (digital fabrication) manager. The questionnaire uses a complex survey design approach where each question includes seven to 12 multiple-response answers for selection, which is considered a common and reliable approach to elicit expert opinions (Hsieh and Shannon, 2005). Then, the questionnaire was published and shared online with selected groups of industry stakeholders via emails and multiple social media platforms in May 2020.

Since this survey aims to examine digital fabrication/construction adoption using digital systems to allow design for downstream activity integration, the seven key stakeholder groups, i.e., architects, project managers, project owners, design engineers, BIM specialists, and contractors (general contractors and digital fabrication trade contractors) and consultants, were invited to participate. The questionnaire aligns with the digital construction concepts and stakeholders. Therefore, the survey is compatible to cover insights into digital fabrication and digital construction. The survey possessed 144 participants. An analysis of the participant data revealed several demographic characteristics of the sample. First, participants noted different overlapping roles amongst different groups. For example, some participants work as digital fabrication researchers and platform-related professionals. The sample, however, included few policymakers; it is possible that the technical nature of the questions incentivized more participation by the day-to-day practitioners and less by high-level policymakers or upper management. Moreover, the participants' market focuses cover different continents to ensure the survey results cover the perceptions of industry stakeholders in different regions regarding the implementation of digital fabrication and digital construction projects. Amongst all, 66% of indicated market foci are in Asia, 46% in Europe, and 23% and 22% in North America and the Middle East, respectively. It is important to note that the survey measures how industry participants suggest a consistent view on the importance of variables affecting the implementation, but heterogeneity may remain in various types of real-world projects. To ensure that their perceptions match reality as much as possible, all survey participants were asked to share their work experience in digital fabrication (or digital construction) to different extents. Amongst all, one-third and one-fourth of the participants have more than five years of digital fabrication experience in design and in construction, respectively. An example of the question and the associated distribution of responses is shown in Figure 7.1. The complete questionnaire and the participants' background information are reported online as supplementary materials, which can be accessed via this link.

The response data collected from the survey were further analyzed through Rao–Scott chi-squared tests of independence to investigate the correlations of the variables derived from the questionnaire survey. Amongst the total 144 responses, the responses from 114 participants were selected for the statistical analysis. Since the effective respondents were selected on purpose to serve the goal of this research, the sample of 114 participants is intended to represent the larger

Figure 7.1 An example of the question and the associated distribution of responses on design modeling information for digital fabrication (Ng et al., 2020)

population target of industry stakeholders characterized by sufficient experience in digital fabrication. This can ensure some homogeneity of the sample while acknowledging that some possible variations according to the geographical location are not accounted for in this study. Within the survey scope, the categories of variables in the design process, including (a) fabrication information, (b) design modeling, (c) design review, and (d) design documentation, represent the needs in the design process. The results from the statistical mapping of the variables are described in Section 7.3.2.

7.3.2 Introduction of the implementation framework

A tripartite diagram, shown in Figure 7.2, was derived from the complex survey that encapsulates the implementation framework of digital fabrication and digital construction, which differentiates the variable correlations under six aspects of **who**, **what**, **when**, **where**, **why**, and **how** to implement future construction projects. The tripartite diagram provides an indication of whether a project would be smart in terms of the suitable processes and people to achieve digital fabrication and digital construction outcomes (Ng et al., 2022a). The following paragraphs present the findings from Ng et al. (2022a) that explain the variable correlations under six aspects in detail.

The **who** aspect includes variables under the profession (i.e., roles) category (e.g., DFAB engineer's participation and DFAB program participation). It was found that the variables are significantly correlated to the design specification and configurations of digital fabrication information as the industry needs in design. They are also highly correlated to modeling following the fabrication approach for design modeling, process visualization and simulation for design review, and model ontology and configuration for design documentation. Some findings also indicate unconventional practices possibly adopted by the survey participants. For example, only a digital fabrication contractor among all roles is a highly correlated demonstration of the fabrication intents for design review. Also, only the DFAB engineer's participation in the client team has a high correlation with connections with computer-aided manufacturing/process planning (CAM/CAPP) systems for design modeling.

The **what** aspect includes variables under the programming language category. Many of the variables are correlated to project implication, design specification, and the configurations of digital fabrication information needed in the design. They are also highly correlated with a shared virtual collaborative environment for design modeling, process visualization and simulation for design review, and model ontology and configuration for design documentation. Besides, using

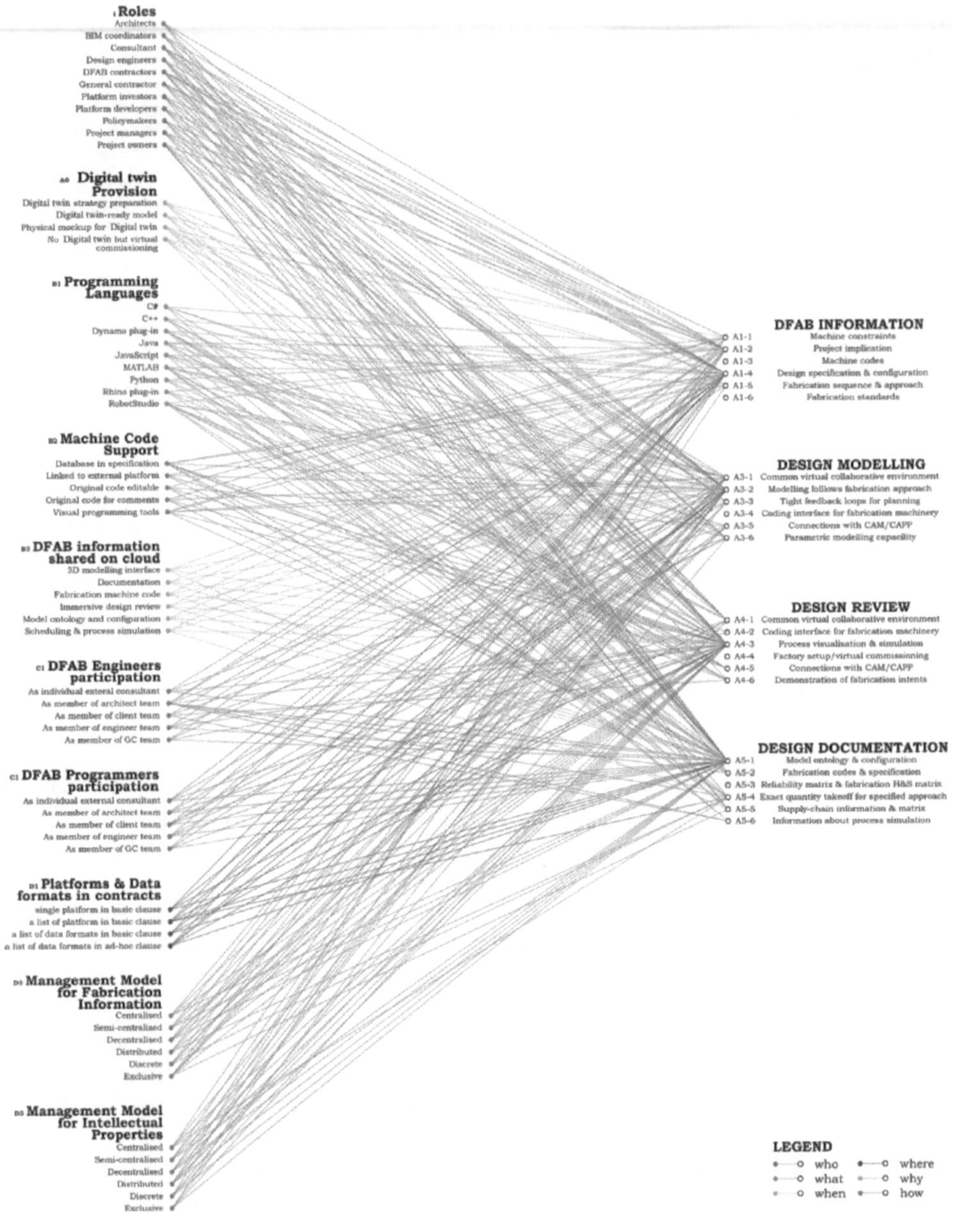

Figure 7.2 The tripartite diagram to encapsulate the implementation framework of digital fabrication and digital construction

different programming languages is correlated to various variables for design documentation. This indicates that stakeholders tend to suggest that the programming and coding choices may impact the documentation efficiency to instruct the fabrication and construction processes.

The **when** aspect includes variables under the digital fabrication information on the cloud category. The correlations will concern when the information sharing should occur, e.g., during the design review or documentation stage. Overall, most of the variables are correlated to design specification and fabrication information needed in design, common virtual collaborative environment modeling that follows fabrication to enrich design modeling, and model ontology and configuration for design documentation. All variables are correlated to process visualization and simulation for design review. The most significant variable of information shared on the cloud – 3D modeling interface – is the only variable, amongst all, that is correlated to supply-chain information and matrix for design documentation.

The **where** aspect includes variables under the platforms and data formats in the contracts category. The correlations concern where the platform(s) and format(s) are included. Overall, many of the variables are correlated to design specification and configurations of digital fabrication information needed in design, to modeling that follows the fabrication approach for design modeling, and to process visualization and simulation for design review. The variables are correlated to model ontology and configuration for design documentation. The single platform in the basic clause variable seems to have different correlations in the design process amongst other variables under the same category.

The **why** aspect includes variables under the digital provision category. The correlations concern why the provisions are needed (e.g., to acquire a digital twin-ready model or physical mock-up). Overall, most of the variables are correlated to design specification and configuration digital fabrication information needed in design; shared virtual collaborative environment; a modeling fabrication approach for design modeling, process visualization, and simulation for design review; and model ontology and configuration for design documentation. Surprisingly, the most significant variable – digital twin strategy preparation for tendering – is the only one that is correlated to project implications fabrication information in the design. This finding suggests that stakeholders prepare digital formats of early design information to be passed throughout the supply chain to downstream fabrication and construction.

The **how** aspect includes variables under the machine code support, management model for fabrication information, and management model for intellectual properties categories. The correlations concern fabrication-aware models (e.g., how machine code is being supported) for adoption. Overall, many of the variables are correlated to design specification and configuration of digital fabrication information needed in design, to project implications of digital fabrication information needed in design, to modeling that follows the fabrication approach for design modeling, and to model ontology and configuration for design documentation. Almost all how variables are correlated to process visualization and simulation for design review. It is worth noting that only the exclusive management model for fabrication information variable, amongst others, is correlated to supply chain information and matrix for design documentation. It indicates that digital fabrication and digital construction stakeholders emphasize design and fabrication information protection when only one set of data owned and managed by the project client is allowed to be shared.

7.4 Examples of implementation

Two relevant case studies are presented to demonstrate the potential of digital fabrication and digital construction, respectively: the clerestory steel structure project with a bespoke roof geometry located on the West Coast of the United States, and the prefabricated office building project with a complex design of prefab concrete columns located in Zurich Switzerland. Both case studies will showcase how the implementation framework (who, what, when, where, why, and how) was realized to bring impact to the project performance, e.g., fabrication-related and interface information such as the integrated schedules should be fed back to the design development for manufacture and assembly so that rework and waste can be substantially reduced.

7.4.1 *Digital fabrication project*

The illustrated digital fabrication project is a steel structural clerestory system in a large-scale commercial project located on the West Coast of the United States. The design involves a complex geometry roof structure with off-the-shelf unitized curtain walling profiles, and the system is supported by the roof structure. The project follows the American Institute of Architects (AIA) standard design stages while adopting Integrated Project Delivery (IPD) as the project delivery model. The construction work began in 2017 and was completed in 2022. The project client required BIM at the beginning of the project for design development and design coordination throughout the entire project. Figure 7.3 shows the integrated digital design process of the project. The BIM-based design process integrated workflow and information flow throughout the project (Ng et al., 2021). The following paragraph explains the variable correlations under six aspects of who, what, when, where, why, and how included in this project through a case study based on Ng et al. (2022a).

Who: the project shows that the early involvement of several key project stakeholders, such as the general contractor and the project manager and trade contractors, could enable integrated information from architects and engineers to coordinate design for digital fabrication from the early design phase through integrated information flows in BIM. **What:** throughout the BIM-based integrated design process, project stakeholders coordinated and exchanged design information and the required digital fabrication information in a shared virtual environment through a cloud BIM platform. **When:** the information sharing on the cloud BIM platform provides real-time information amongst project stakeholders to simultaneously simulate design ideas and outputs. **Where:** the project adopted IPD as the project delivery model. The project contract that includes platform usage and data required, such as data-rich 3D BIM models, had been issued since the design started. **Why:** the project specified digital provision throughout the design process and also during and after the construction process. Detailed BIM models are required to match the existing construction progress to different extents. A detailed as-built BIM model could help facility managers to manage the operations and maintenance of the buildings after construction. **How:** the project adopted a centralized management model for the intellectual properties, including the BIM models, design information,

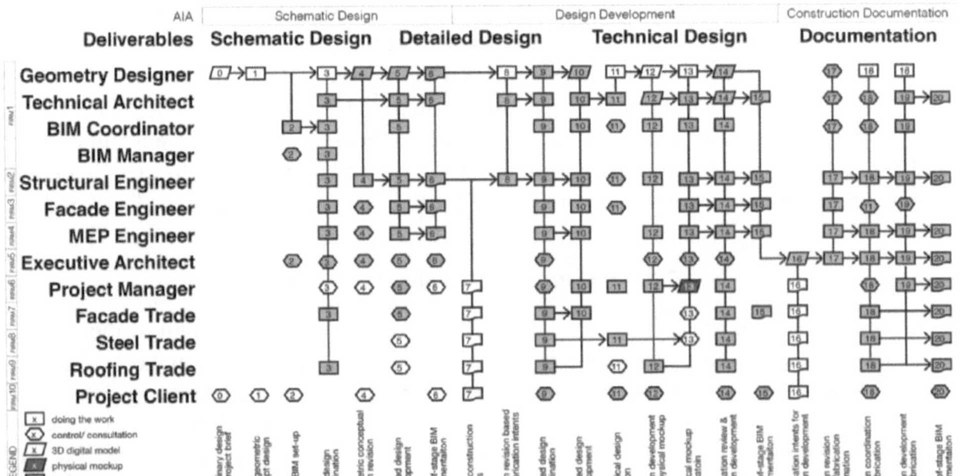

Figure 7.3 Integrated digital design process to facilitate the steel clerestory project

and communication email records. This could facilitate top-down information management owned by the project client for progress and quality monitoring to avoid miscommunication and mistakes in terms of, for example, building model information, after construction in operations.

7.4.2 Digital construction project

The illustrated digital construction project is a 22-floor high-rise office building subject to repetitive manufacturing and installation of prefabricated columns. Though each floor is identical, they contain highly engineered prefabricated structural elements. The building was modeled parametrically using Autodesk Revit® software (Chen et al., 2022). The design data included the material properties of the columns, i.e., the family and type, the length, the volume, and the load-bearing capacity of each column. The prefabricated structural columns were all defined to a LOD300; this was a level sufficient for sharing among contractors and suppliers for making decisions on the required quantities and need time of the materials. Why: to allow efficient digital construction information sharing, an integrated BIM platform prototype was developed for this project that specifically features the BIM information flows for design and schedule changes, as illustrated in Figure 7.4.

What: for the coordination of design changes, contractors and suppliers need to store and exchange object information, construction information, and material delivery information. Object information is the information pertaining to the prefabricated components (e.g., a column, a slab, or a wall to be constructed), including the length, weight, color, and strength. **When and where:** construction information is the information pertaining to construction progress, e.g., the forecasted start and finish time of a task and the requirements for the task to start. Delivery information is the information pertaining to the delivery of the components, e.g. the material order approval date, the material preparation date, the material release date and time, the onsite arrival date and time, the material delivery method, and the deviations from the initial plans (Chen et al., 2022). The information is accessed, collected, and updated by suppliers and contractors with role-based access control. An element ID is assigned automatically to a component and linked to a task ID representing a specific construction task. **Who:** the contractors are able to either query material information by its

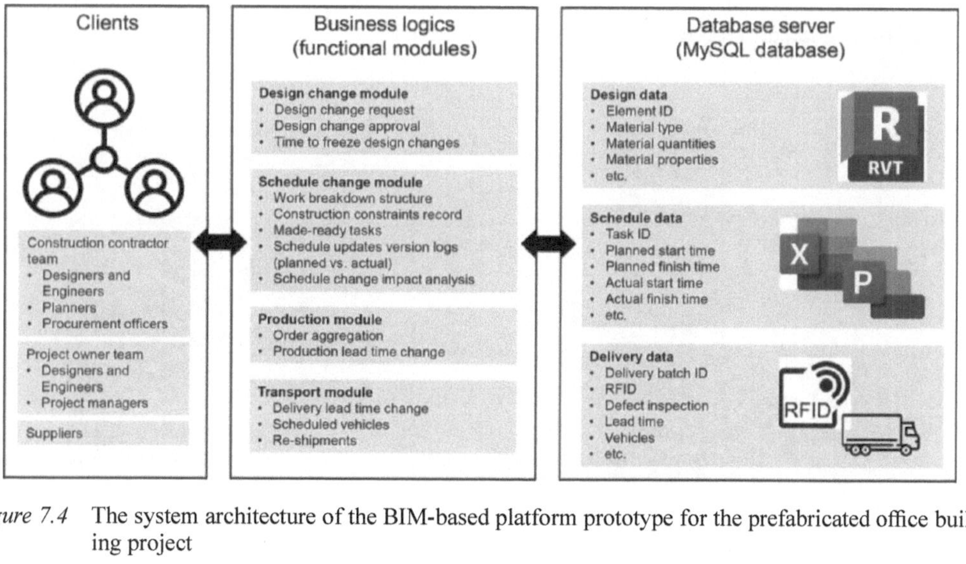

Figure 7.4 The system architecture of the BIM-based platform prototype for the prefabricated office building project

element ID or task ID. The bill of materials (i.e., a list of material requirements) for that design type or that installation task is then automatically generated. When any design change is required, change requests will be initiated either by suppliers or contractors in this module. The design-change module, therefore, includes functions to enable the exchange of change request forms. **How:** in the proposed system, both contractors and suppliers can access the design-change module to register the change request forms. If any conflicts in design are resolved and the changes have been approved by the contractors, they will freeze the design for all the components required for upcoming tasks.

What: for the coordination of schedule changes, contractors have the ability to query milestone dates and edit task information for the specified look-ahead time period by using the platform. It shows how many installation tasks have been completed and what constraints exist for the upcoming construction tasks. **Who:** this allows the contractors to identify what upcoming tasks can be committed to and will be done in the look-ahead time period. The information for each committed task is recorded and updated by contractors, including its task ID, the planned start date, the planned finish date, and whether the constraints can be removed before the tasks start. The commitments of upcoming tasks by each site trade are suggested to be made on Friday on a weekly basis. **How:** considering the input of commitments, the planned start dates can be updated during the look-ahead week if any changes to the originally planned dates are required. Since the planned task start dates are the actual material need dates, they are important information for both contractors and suppliers to communicate and coordinate. The task information from this module is seamlessly and automatically shared with suppliers to enable proactive decision-making on the two milestone dates for production and transport. For the contractors to better foresee the changes and estimate the planned task start dates, a material status function is designed for the schedule-change module. The function allows the contractors to track the status of the required components continuously. A list of conditional statements in programming calculates the status of each component. **Where:** the status of each component can be queried by contractors using the element ID, order ID, and task ID. The query results include different states of each component as it moves through material flow processes; "material ready for transport (holding at the factory)," "material in design coordination," "release for production," "release for transport," and "installed." Each component is given a unique ID for information tracking regarding the associated construction information, as shown in Figure 7.5. In this way, contractors can quickly identify the material status for the whole building or designated look-ahead weeks in the 3D model.

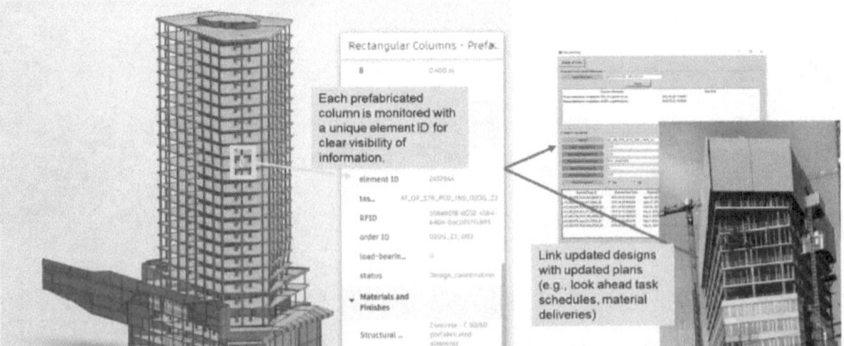

Figure 7.5 An example of prefabricated column given a unique ID for information tracking regarding the associated construction information

7.5 Discussion

Both the digital fabrication and digital construction project implementation would require consistent information sharing and communication across the design, fabrication, and site assembly. BIM provides an opportunity for stakeholders to exchange information coherently and link multiple supply chain stages to enable vertical integration of a construction supply chain as envisioned earlier (Vrijhoef and Koskela, 2000). Therefore, BIM-based digital fabrication and digital construction will promote a data-rich provision in projects that could enhance systemic innovation and foster sustainable development with improved installation quality and higher-performing buildings to different extents.

In the digital fabrication example, mass customization has been demonstrated in line with the "how-what-who" principles of the implementation framework. Mass customization requires continuous input from clients regarding the design needs and is usually subject to multiple design changes and schedule changes to the make-to-order production processes, similar to the digital construction example. However, the success of the example projects potentially shows the benefits of digital fabrication and digital construction to promote economy of scale and provide added values – to satisfy the varieties of individual customer's needs without a corresponding increase in cost and lead time by integrating these varieties and the efficiency of standard mass production (i.e., mass prefabrication of components). The positive learning is consistent with the digital fabrication benefits indicated in the case study of facade fabrication and installation by (Tran et al., 2021).

In the digital construction example, the functionalities of the BIM-based prototype system can be potentially scaled up to coordinate changes in the design and scheduling of various types of construction materials and work packages. Functional developments can be tailored to show the extent of the possible improvement for entire construction projects. By fitting the construction task plans into structured iterative processes through digital data sharing, stakeholders have more capability to capture required change information and resolve associated problems quickly. The evidence of such structured digital control processes echoed with the continuous improvement suggestions shown in (Sacks et al., 2020). This would also be useful for the management of bespoke construction supply chains, as in the digital fabrication project example, where prefabricated elements such as roofing structure, ductwork, plumbing, and mechanical systems typically have to be modified because of last-minute design and schedule changes.

Although the illustrated examples did not showcase the direct sustainability aspects, the best practices derived from them can be adjusted and improved to tie the carbon and energy performance to the variables in the implementation framework. For example, the framework could be revised with a focus on carbon measurement and circular economy strategies such as the digital design for disassembly. In addition, the safety and workforce dimensions need to be added to the framework to ensure the complete success of projects. For example, Daiwa House Group (2022) optimized organizational structure by launching the D-Succeed successive plan in 2020 to restructure the workforce and advance training to promote co-creation for corporate sustainable development. This also aims to enhance productivity and promote responsible procurement in collaboration with other stakeholders to ensure construction safety and quality through digital transformation (Ng, 2022; Ng et al., 2022b).

7.6 Conclusion and future work

This book chapter examines the current industry practice of digital fabrication and digital construction and provides an implementation framework to help stakeholders establish a consistent view of major variables for project success. Whereas BIM platforms have been widely used nowadays by project teams in many countries in the design and construction processes, digital fabrication and digital construction together with the associated digital system needs are still not standardized. The framework provides information for stakeholders' reference when adopting digital fabrication and digital construction in projects, including at which stage in design specific needs are most present. A tripartite diagram was derived from the complex survey that differentiates the variable correlations under six aspects of who, what, when, where, why, and how to implement future construction projects. Two relevant case studies are presented to demonstrate the potential of digital fabrication and digital construction, respectively: the steel frame clerestory project with a bespoke roof structure located on the West Coast of the United States and the prefabricated office building project with a complex design of prefab concrete columns located in Zurich Switzerland. Both case studies indicate the importance of stakeholder integration and BIM-based digital systems to enable information sharing, which ensures processes are smart and informative across the supply chain – design, fabrication, and construction. Future work will be conducted to clarify the technological integration's relationship with process integration, information integration, and organizational integration to enhance the performance of digital fabrication and digital construction projects.

References

Bakhshi, S., Chenaghlou, M. R., Rahimian, F. P., Edwards, D. J., and Dawood, N. (2022). Integrated BIM and DfMA parametric and algorithmic design based collaboration for supporting client engagement within offsite construction. *Automation in Construction*, *133*, 104015.

Barbosa, F., Woetzel, J., and Mischke, J. (2017). *Reinventing construction: A route of higher productivity*. McKinsey Global Institute.

Biörck, J., Blanco, J. L., Mischke, J., Ribeirinho, M. J., Rockhill, D., Sjödin, E., and Strube, G. (2020). *How Construction can Emerge Stronger after Coronavirus*. McKinsey & Company.

Bock, T. and Linner, T. (2015). *Robot oriented design*. Cambridge University Press.

Boje, C., Guerriero, A., Kubicki, S., and Rezgui, Y. (2020). Towards a semantic construction digital twin: Directions for future research. *Automation in Construction*, *114*, 103179.

Chen, Q. (2020). *A systematic approach to enabling digital supply chain coordination in construction projects*. Doctoral Dissertation, ETH Zurich.

Chen, Q., Adey, B. T., Haas, C., and Hall, D. M. (2020b). Using look-ahead plans to improve material flow processes on construction projects when using BIM and RFID technologies. *Construction Innovation*, *20*(3), 471–508.

Chen, Q., Hall, D. M., Adey, B. T., and Haas, C. T. (2020a). Identifying enablers for coordination across construction supply chain processes: A systematic literature review. *Engineering, Construction and Architectural Management*, *28*(4), 1083–1113.

Chen, Q., Adey, B. T., Haas, C. T., and Hall, D. M. (2022). Exploiting digitalization for the coordination of required changes to improve engineer-to-order materials flow management. *Construction Innovation*, *22*(1), 76–100.

Daiwa House Group. (2022). Accessed from: https://www.daiwahouse.com/

Dave, B., Buda, A., Nurminen, A., and Fr¨amling, K. (2018). A framework for integrating BIM and IoT through open standards. *Automation in Construction*, *95*, 35–45.

Esfahani, M. E., Rausch, C., Sharif, M. M., Chen, Q., Haas, C., and Adey, B. T. (2021). Quantitative investigation on the accuracy and precision of Scan-to-BIM under different modelling scenarios. *Automation in Construction*, *126*, 103686.

Garcia de Soto, B. and Skibniewski, M. J. (2020). Future of robotics and automation in construction. In *Construction 4.0* (pp. 289–306). Routledge.

Heaton, J. and Parlikad, A. K. (2020). Asset information model to support the adoption of a digital twin: West Cambridge case study. *IFAC-PapersOnLine*, *53*(3), 366–371.

Hsieh, H. F. and Shannon, S. E. (2005). Three approaches to qualitative content analysis. *Qualitative Health Research*, *15*(9), 1277–1288.

Lu, W., Tan, T., Xu, J., Wang, J., Chen, K., Gao, S., and Xue, F. (2021). Design for Manufacture and Assembly (DfMA) in construction: The old and the new. *Architectural Engineering and Design Management, 17*(1–2), 77–91.

Ng, M. S. (2022). *Design for Digital Fabrication (DfDFAB): Management for sustainable adoption of emerging technologies and innovations*. Doctoral Dissertation, ETH Zurich.

Ng, M. S., Bonanomi, M. M., Hall, D. M., and Hackl, J. (2020). Design for digital fabrication: An industry needs analysis of collaboration platforms and integrated management processes. In *ISARC: Proceedings of the International Symposium on Automation and Robotics in Construction* (Vol. 37, pp. 318–325). IAARC Publications.

Ng, M. S., Chen, Q., Hall, D. M., Hackl, J., and Adey, B. T. (2022a). Designing for digital fabrication: An empirical study of industry needs, perceived benefits, and strategies for adoption. *Journal of Management in Engineering, 38*(5), 04022052.

Ng, M. S., Graser, K., and Hall, D. M. (2021). Digital fabrication, BIM and early contractor involvement in design in construction projects: A comparative case study. *Architectural Engineering and Design Management, 19*(1), 39–55.

Ng, M. S., Hall, D. M., and Hsieh, S. H. (2023). Liability factors and conceptual framework for contracts to manage design for digital fabrication in construction projects. *Journal of Legal Affairs and Dispute Resolution in Engineering and Construction, 15*(1), 04522043.

Ng, M. S., Hall, D., Schmailzl, M., Linner, T., and Bock, T. (2022b). Identifying enablers and relational ontology networks in design for digital fabrication. *Automation in Construction, 144*, 104592.

Sacks, R., Brilakis, I., Pikas, E., Xie, H. S., and Girolami, M. (2020). Construction with digital twin information systems. *Data-Centric Engineering, 1*, e14.

Tran, H., Nguyen, T. N., Christopher, P., Bui, D. K., Khoshelham, K., and Ngo, T. D. (2021). A digital twin approach for geometric quality assessment of as-built prefabricated facades. *Journal of Building Engineering, 41*, 102377.

Vrijhoef, R. and Koskela, L. (2000). The four roles of supply chain management in construction. *European Journal of Purchasing & Supply Management, 6*(3–4), 169–178.

Weinfass, I. and Kelly, M. (2020). Coronavirus delays hit£ 26bn-worth of projects. *Construction News*.

Zhao, L., Liu, Z., and Mbachu, J. (2019). Development of intelligent prefabs using IoT technology to improve the performance of prefabricated construction projects. *Sensors, 19*(19), 4131.

8

METHODICAL REVIEW OF EXOSKELETON USE IN THE CONSTRUCTION INDUSTRY

Abiola Akanmu[1], Anthony Yusuf[2] and Nihar Gonsalves[3]

[1]ASSOCIATE PROFESSOR, MYERS-LAWSON SCHOOL OF CONSTRUCTION, VIRGINIA TECH, BLACKSBURG, VA 24061, U.S. EMAIL: ABIOLA@VT.EDU

[2]PHD CANDIDATE MYERS-LAWSON SCHOOL OF CONSTRUCTION, VIRGINIA TECH, BLACKSBURG, VA 24061, UNITED STATES. EMAIL: ANTHONYY@VT.EDU

[3]PHD CANDIDATE, MYERS-LAWSON SCHOOL OF CONSTRUCTION, VIRGINIA TECH, BLACKSBURG, VA 24061, UNITED STATES. EMAIL: NIHAR.GONSALVES@CLARKCONSTRUCTION.COM

8.1 Introduction

The International Labor Organization (ILO, 2019) noted that the construction industry provides on average 7% of total global employment (220 million people). Out of this global employment figure, the construction industry in the United States employs over 10 million people as of 2020 (Bureau of Labor Statistics, 2020). However, Cho et al. (2018) affirmed that most of these workers are frequently exposed to work-related musculoskeletal disorders (WMSDs). The United States Bureau of Labor Statistics (2019) noted that, as of 2017, the number of cases of WMSDs reported in the construction industry was 9% higher than other industries combined. WMSDs are the third among the top 25 global burdens affecting over 1.7 billion people worldwide (Cieza et al., 2021). Most of the temporary and permanent disabilities reported among construction workers are mainly due to WMSDs (Eaves et al., 2016). According to the United States Bureau of Labor Statistics (BLS, 2020), more than 70% of construction workers reported having WMSDs. This is due to overexertion of body parts triggered by construction work (Wang et al., 2017), of which the risk factors include lifting and lowering, pushing, and static non-neutral postures.

WMSD does not only affect the health of construction workers, but it also affects other socio-economic activities related to workers (Cheng et al., 2010). Kim et al. (2019) reported that a significant number of the construction workers affected by WMSDs lose their jobs annually. In 2016, 30% of all days away from work were attributed to WMSDs. Moreover, Schwatka et al. (2012) recorded that most construction workers retire early due to WMSDs. This has dire implications for the retention of skilled labor in the construction industry. In economic terms, the cost of addressing WMSDs continues to increase (Yelin et al., 2016). It is estimated that the construction industry in the United States loses approximately $54 billion annually due to WMSDs (Marcum and Adams, 2017). In the construction industry, the lower back is one of the most affected body parts (Cho et al., 2018). This is corroborated by the World Health Organization (2022) report that showed that low back pain is the main contributor to the overall burden of WMSDs with

DOI: 10.1201/9781003383840-9

570 million prevalent cases worldwide. A survey by BLS (2018) showed the breakdown of body parts affected by WMSDs as back (38.5%), shoulder (14.9%), leg (11.5%), arm (5.1%), abdomen (5%), and multiple parts (5.4%), and all other body parts accounted for 19.7%.

To address WMSDs, exoskeletons have been touted as a prospective transformative technology to reduce the occurrence of WMSDs in the construction industry (Kim et al., 2019). de Looze et al. (2016) noted that exoskeletons can physically relieve workers while keeping them in control of their task execution. Previous studies have shown how the use of exoskeletons has improved other industry sectors such as manufacturing (Alemi et al., 2020), military (Hong et al., 2019), and healthcare (Baltrusch et al., 2018). Like other industries, construction researchers have been exploring the enormous potential of exoskeletons for a wide range of construction activities (Zhu et al., 2021). To facilitate further research on the use of exoskeletons, the objective of this book chapter is to provide an overview of the current studies on exoskeletons in the construction industry and summarize gaps to be addressed in the future. A systematic literature review was conducted to (a) identify the classification and types of exoskeletons and (b) examine exoskeletons in the construction industry and other industry sectors. The study concluded by examining future research directions on exoskeletons in the construction industry to facilitate the adoption of the device.

8.2 Methodology

A systematic review of relevant extant literature was conducted. A process called Preferred Reporting Items for Systematic Reviews and Meta-Analyses for Protocols 2015 (PRISMA-P 2015) provided by Moher et al. (2015) was adopted to ensure a meticulously planned review. To ensure the quality of the systematic review only academic databases of high repute were considered for materials to be included. Relevant journal articles, book chapters, conference, and workshop papers regardless of geographical location within the last 16 years (2007–2023; as shown in Figure 8.1) in Google Scholar, Scopus, Science Direct, EBSCOHOST, and ASCE library were considered. However, some statistics were obtained from official internet sources (such as the U.S. Bureau

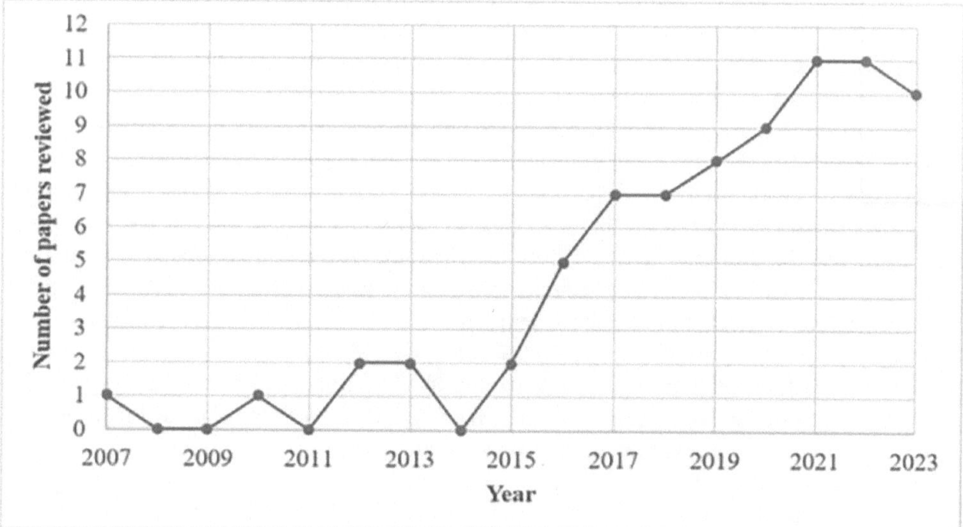

Figure 8.1 Year distribution of papers reviewed

Table 8.1 Breakdown of literature reviewed

S/N	*Paper Type*	*Number*
1	Journal articles	59
2	Conference/symposium proceedings	15
3	Book chapters	2
4	Internet sources	5

of Labor Statistics). The materials used were those written in English language, available online, and peer reviewed. The method suggested by Kitchenham et al. (2009) was used for a literature search. Hence, to ensure an effective search, both main strings and substrings were used. Due to the different configurations of the databases, different search strings were used but the keywords include "wearable robot", "wearable technology", "exoskeleton", "robotics", "construction", and "engineering". The papers retrieved from the databases were screened to determine their relevance to the subject matter of this book chapter. Duplicated papers and those with unrelated titles were removed. Thereafter, the abstracts of the remaining papers were examined for relevance. Finally, the overall foci of the papers were assessed by skimming through; this amounted to a total of 76 papers and five internet sources. The breakdown is shown in Table 8.1.

8.3 Exoskeletons – classifications and types

In addition to the growing body of brand names, exoskeletons have been described with diverse names in literature. These include "wearable robot", "muscle suit", "exosuit", and "assistive device". Commercially available exoskeletons are of various types and can be classified into diverse categories based on three parameters. These include by modus operandi (Antwi-Afari et al., 2021), supported body parts (Gonsalves et al., 2021), and the nature of constituent material (Schwartz et al., 2021). By modus operandi, exoskeletons can be active, passive, hybrid, semi-active/passive, and quasi-active/passive. When exoskeletons are neither completely active nor passive but contain both operational mechanisms, they are regarded as hybrid, semi-, or quasi-active/passive. Active exoskeletons as shown in Figure 8.2 (also known as powered exoskeletons) use electric actuators (or motors) to generate power for providing support (Gonsalves et al., 2021). They are usually more expensive and heavier than passive exoskeletons. However, they provide a greater amount of support. Passive exoskeletons use mechanical actuators (such as springs and dampers) to generate power and provide support to users (Gonsalves et al., 2023b) as shown in Figure 8.2. They are easier to maintain and cheaper than active exoskeletons.

By the body part supported, exoskeletons can be lower body support, (e.g., back support and leg support), upper body support (e.g., shoulder support and arm support) and full body support. As the name implies, these exoskeletons support specific parts of the body (Gonsalves et al., 2021). However, there are exoskeletons that support more than one body part. For instance, CrayX, an active exoskeleton, supports both the back and leg. These exoskeletons are usually classified by the primary support they provide. Based on the nature of constituent materials, exoskeletons can be soft or rigid. Soft exoskeletons are made of flexible materials such as textile fabric and elastomers, while rigid exoskeletons are made of stiff materials such as plastic and metal (Schwartz et al., 2021). However, rigid exoskeletons provide greater support (Chiaradia et al., 2019). Hence, most exoskeletons in the construction/engineering domain are passive and rigid. A summary of exoskeleton classification, types, and examples is presented in Table 8.2. Although there are many commercially available exoskeletons by different manufacturers, those that fall under the same classification category and type are usually of similar designs and operation mechanisms.

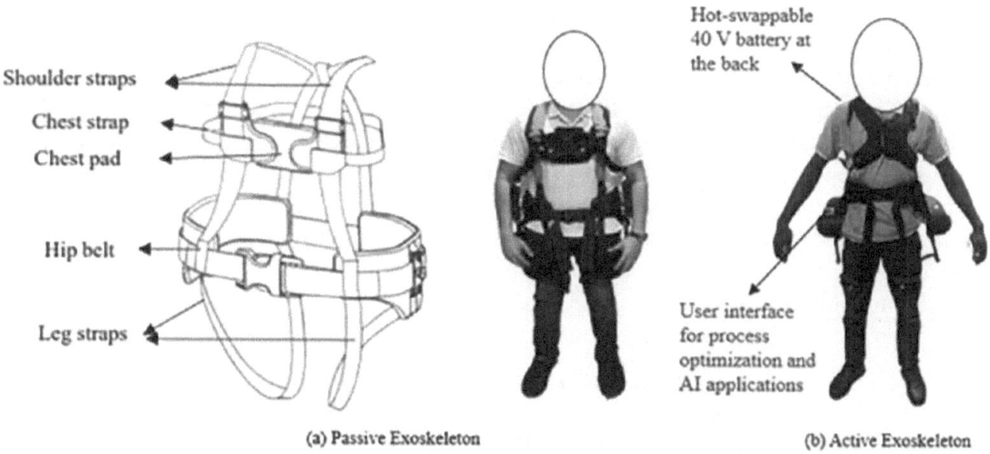

Figure 8.2 Passive and active exoskeletons

Table 8.2 Exoskeleton classification, types, and examples

Classification Category	*Types*	*Examples*
Modus operandi	Active	CrayX, XoTrunk, Apogee, Muscle suit, Ironhand.
	Passive	BackX, Bendezy, FLx ErgoSkeleton, Laevo, Paexo Back, PLAD.
	Semi-active	FEXO knee, XoSoft.
Supported body parts	Lower-body support	Exo-Push, SuperFlex, LegX, Keeogo, SPEXOR, PLAD.
	Upper-body support	FLx ErgoSkeleton, EksoVest, EksoBionics EVO, Ottobock Shoulder, ShoulderX.
	Full-body support	Flex Lift, SuitX MAX.
Nature of constituent material	Soft	CRUX, SUIT, Paexo Soft Back, XoSoft.
	Rigid	BackX, Laevo, Ottobock Shoulder, Paexo Back, CrayX, XoTrunk.

8.4 Theoretical frameworks

Okpala et al. (2021) opined that theoretical underpinning plays a huge role in technology-related research. One of the main reasons for employing theories in technology-related research is to understand how to situate technologies in the workplace – this can be achieved via adoption theories. Generally, these theories help to answer questions relating to the applications of technology and factors that could support or hinder the use of technology (Lee et al., 2013; Gonsalves et al., 2023a). Several researchers have used different theories to explain the acceptance of exoskeletons (Dwivedi et al., 2017; Elprama et al., 2020; Baltrusch et al., 2021; Siedl and Mara, 2022; Gonsalves et al., 2023b). These theories can be categorized as individual level (e.g., theory of reasoned action (TRA), theory of planned behavior (TPB), and technology acceptance model (TAM)), organizational level (e.g., technology, organization, and environment (TOE) theory), and organizational and individual level (e.g., diffusion of innovation (DOI) theory) theories (Hameed et al., 2012). In the construction domain, TOE, human factors principles have been used to explain the adoption of exoskeletons (Gonsalves et al., 2023b; Pan and Pan, 2019, 2020). In other industry sectors diverse theories such as the unified theory of acceptance and use of technology (UTAUT) (Dwivedi et al., 2017; Elprama et al., 2020; Gumasing and Castro, 2023) and TAM (Spada et al., 2017; Gilotta et al., 2019; Siedl and Mara, 2022; Chen et al., 2023) have been used. Through TRA, the construct of behavioral intention has been used to predict exoskeleton adoption (Gilotta et al., 2019; Kim et al., 2019). However, to overcome limitations in TRA (such as how personal decisions are made or how decisions once made influence behavior), the theory of planned behavior (TPB) has been explored in other exoskeleton studies (Giger and Piçarra, 2017; Shore et al., 2018) to explain human and social variables that could influence intention to use.

8.5 Exoskeletons in other industry sectors

8.5.1 Applications

Industry sectors such as manufacturing (Alemi et al., 2020), agriculture (Omoniyi et al., 2020), baggage handling (Baltrusch et al., 2021), logistics (De Bock et al., 2020), health care (Settembre et al., 2020), automotive (Iranzo et al., 2020), ship maintenance (Moyon et al., 2018), and the military (Anam and Al-Jumaily, 2012) have found different use cases for exoskeletons. However, Dahmen and Constantinescu (2020) opined that exoskeleton technology usage in most industries is still in its infancy with very few exoskeleton products in the market.

8.5.1.1 Healthcare

One of the first applications of exoskeletons was in the healthcare sector (Liu et al., 2016). The purpose was to support or treat the upper and lower limbs. The devices help in assisting or treating medically diagnosed conditions such as being physically weak, injured, or physically disabled (Zheng et al., 2022). Studies (Viteckova et al., 2013; Yan et al., 2015) have shown how exoskeletons have been able to allow injured or disabled persons to achieve a wider range of motion such as walking, traversing stairs, sitting and standing up, reaching, and grasping. Besides, exoskeletons are also applicable to workers in the healthcare sector because they are also affected by WMSDs. Previous studies (Baltrusch et al., 2018; Hwang et al., 2021) have shown how healthcare professionals such as caregivers, nurses, residents, and surgeons have used exoskeletons in patient handling and supporting other tasks in their work practices.

8.5.1.2 Manufacturing industry

Most recently, the use of exoskeletons was translated from the healthcare sector and military applications into the manufacturing sector (de Looze et al., 2016; Bogue, 2018). Exoskeletons are being used in aircraft manufacturing (Jorgensen et al., 2022), automobile assembly plants (Iranzo et al., 2020), painting assembly lines (Pinho and Forner-Cordero, 2022), and food manufacturing facilities (Schwerha et al., 2021). Exoskeletons can help in manufacturing tasks such as manual lifting (Alemi et al., 2020), assembly-related tasks (Madinei et al., 2019), painting and welding (Butler, 2016), repetitive standing or acting as a chair when the worker wants to rest (Karvouniari et al., 2018), and during overhead line tracing (Spada et al., 2017). Other manufacturing tasks in industrial settings such as overhead work, stooping/squatting, and carrying the load in factory, logistics, and warehouse worksites are readily carried out with exoskeletons (Lowe et al., 2019).

8.5.1.3 Other industries

There are other industries that have found use for exoskeletons. Anam and Al-Jumaily (2012) stated that a few exoskeletons have been designed for military applications for soldiers to lift or carry heavy loads. In military operations, Zheng et al. (2022) noted that exoskeletons have been used to enhance the physical capabilities of dismounted soldiers. The agricultural industry has also found use for exoskeletons. Fethke et al. (2015) reported that farmers were twice as likely to have back pain compared to workers in the general working population. Davis and Kotowski (2007) argued that most agricultural tasks pose ergonomic risks for WMSDs. Many farm operations are characterized by heavy physical labor, and an aging workforce may be at a higher risk of disabilities and reduced productivity resulting from WMSDs (Omoniyi et al., 2020). To mitigate these, Upasani et al. (2019) suggested the use of exoskeletons to support workers in agricultural settings. Table 8.3 shows an overview of some research efforts on exoskeletons in other industry sectors other than the construction industry.

Table 8.3 An overview of current research efforts on exoskeletons in other industry sectors

Authors	*Exoskeleton Type*	*Tasks*	*Lab/Field Study*	*Outcome Measures*	*Data Collection Means/Instrument*
Alemi et al. (2020)	BackX and Laevo	Repetitive lifting (6.8 kg)	Lab	Muscle activity, energy expenditure (metabolic demand), perceived discomfort, perceived balance, perceived exertion, usability, and perception.	Electromyography (EMG), calorimeter, Borg CR-10 scale, usability scale, questionnaire, and verbal feedback
Omoniyi et al. (2020)	Laevo V2.5	Lifting, bending (2.67 kg), and general agricultural tasks	Field	Safety, comfort, jobs and timing, mobility, ease of use, health, and productivity.	Interview
Baltrusch et al. (2021)	SPEXOR	Baggage handling and assembly line tasks	Field	Self-efficacy	Modified spinal function Sort questionnaire, Focus group, interview
De Bock et al. (2020)	ShoulderX and Skelex	Horizontal lifting, overhead lifting, squatting, stooping, and placing	Field and lab	Muscle activity, heart rate, acceleration, perceived discomfort, and workload.	EMG, ECG, accelerometers, body part discomfort scale, and NASA TLX
Settembre et al. (2020)	CORFOR, Laevo v1, BackX, and CrayX	Lifting, placing, and bending	Field and lab	Acceptance, perceived effort, safety, comfort, efficacy, installation, intention to use, and cardiac activity.	TAM questionnaire, Borg CR-10 scale, interview, and ECG
Iranzo et al. (2020)	Levitate AIRFRAMETM	Overhead task	Field	Muscle activity and range of motion.	EMG and inertial sensors

(*Continued*)

Table 8.3 (Continued)

Authors	*Exoskeleton Type*	*Tasks*	*Lab/Field Study*	*Outcome Measures*	*Data Collection Means/Instrument*
Moyon et al. (2018)	PAD by Skel-Ex	Finishing tasks: sanding, coating, polishing, and painting	Field	Perceived exertion, heart rate, cardiac cost, usability.	Borg scale (CR-10), POLAR RS800CX, and usability questionnaire.
Baltrusch et al. (2018)	Laevo	Lower lifting	Lab	Perceived task difficulty, comfort (general and local), user impression (range of motion, adjustability, efficacy), task completion time.	Visual-Analog Scale Questionnaires and stopwatch
Hwang et al. (2021)	FLx ErgoSkeleton, V22 ErgoSkeleton, and Laevo V2.5	Patient transfer task	Field	Muscle activity, usability, kinematic and kinetic data.	EMG, System Usability Scale (SUS), motion capture system (Flex 13, Motive 2.0 Optitrack)
Jorgensen et al. (2022)	EVO, Skelex 360XFR, and Paexo	Sealing tasks in aircraft manufacturing	Lab	Muscle activity, preference, and feedback.	EMG and survey questions
Pinho and Forner-Cordero. (2022)	ShoulderX (V1)	Screwing and unscrewing a M12 hex head cap screw	Lab	Muscle activity, task completion time, perceived effort, and comfort.	EMG, questionnaire, and timestamp video
Schwartz et al. (2021)	SUIT and SKEL	Load lifting/ lowering task (8kg)	Lab	Range of motion and muscle activity.	Magneto inertial measurement unit and EMG

8.6 Exoskeletons in the construction industry

Researchers in the construction sector have also started exploring the potential of exoskeletons. For example, Zhu et al. (2021) presented a comprehensive review of exoskeletons used in manual material handling. The study explores the potential applications of different exoskeletons in the construction industry, providing valuable insights and implications for enhancing work efficiency and reducing physical strain. Babalola et al. (2023) conducted a systematic review of existing literature, case studies, and empirical data related to exoskeleton applications in the construction sector. The study identified that exoskeletons have the potential to improve occupational safety by reducing the risk of WMSDs among construction workers. Their implementation could lead to decreased physical strain, increased work efficiency, and improved productivity on construction sites. However, the researchers also noted that challenges such as cost and adaptability need to be addressed for broader adoption of exoskeletons in the industry.

Gutierrez et al. (2023) conducted a cross-sectional survey among construction workers, supervisors, and managers from various construction sites. The study reveals that perceptions and readiness to use exoskeletons in the construction industry differ significantly among demographic groups. Younger workers and those with prior experience in technology show higher receptiveness to exoskeleton adoption. Barriers related to cost and discomfort are more pronounced in certain age groups, while perceived benefits like injury prevention and increased productivity are more pronounced in specific job roles. Understanding these differences can guide targeted interventions for successful exoskeleton integration in the construction sector. Some researchers (as shown in Table 8.4) have also evaluated different exoskeleton types such as back support, shoulder support, and leg support for construction tasks in laboratory and field explorations. For example, Golabchi et al. (2023) conducted a laboratory experiment to evaluate the performance of BackX during manual material handling tasks involving

Table 8.4 An overview of current research efforts on exoskeletons in the construction industry

Authors	*Exoskeleton Types*	*Tasks*	*Lab/Field Study*	*Outcome Measures*	*Data Collection Means/ Instruments*
Bennett et al. (2023)	HeroWear Apex, Ekso EVO, and Hilti EXO-001.	Gandola pushing and dumping Wooden block installation and removal	Field	Rage of motion, postures, heart rates, and task completion time, workload, usefulness, comfortability, and satisfaction.	XSens motion capture suit, PolarH10 heart rate sensor, NASA TLX, and usability questionnaire.
Golabchi et al. (2023)	BackX	Dynamic (20lb box) and static (less than 0.5kg) lifting	Lab	Perceived exertion, discomfort, overall fit and comfort, effectiveness, and interference levels.	Borg CR-10 scale, usability questionnaire, and verbal feedback.
Gonsalves et al. (2023a)	BackX	Pipework	Field	Usability, perceived discomfort, benefits, barriers, and design modifications.	Borg CR-10 scale, usability questionnaire, and verbal feedback.
Gonsalves et al. (2023b)	BackX	Concrete crews	Field	Perceived discomfort, usability, user perception.	Borg CR-10 scale, usability questionnaire, semi-structured interview.
Meyer (2023)	V3 ShoulderX, Ottobox Paexo Shoulder, Exobionics EVO	Donning and doffing, maneuvering through constrained spaces, ambulating on a balance beam and around cones, and climbing stairs and a ladder	Lab	Completion time, number of errors, usability, comfort, and preference.	GoPro Cameras, usability questionnaire, and system usability scale (SUS).

(*Continued*)

Table 8.4 (Continued)

Authors	*Exoskeleton Types*	*Tasks*	*Lab/Field Study*	*Outcome Measures*	*Data Collection Means/ Instruments*
Yan et al. (2023)	PER-A and PER-B	Roofing operation	Lab	Posture analysis.	GOM Aramis SRX, Optitrack System
Bances et al. (2022)	Passive shoulder support exoskeleton	Timber prefabrication	Field	Body posture and movements, time spent in non-neutral positions, technology expectations.	Interview, direct observation, and ergonomic analysis.
Bennett et al. (2022)	Hilti EXO-01 and Ekso Evo	Removing wooden blocks from a steel-frame wall	Field	Range of motion, comfort, pain, and perceived effectiveness.	XSens MVN Awinda, video recording, and questionnaire.
Gonsalves et al. (2022a)	BackX	Pipework	Field	Usability, benefit, barrier, design modifications, discomfort, and interference with work.	Usability questionnaire, open-ended survey, and verbal feedback.
Gonsalves et al. (2022b)	Laevo	Rebar work	Lab	Usability: ease of use, ease of learning and comfort, perceived discomfort.	Usability questionnaire and Borg CR-10 scale.
Kopp et al. (2022)	Parcour P4	Timber beam assembly, wooden strip installation	Field	Perceived exertion, muscle activity, heart rate, oxygen consumption, time, and quality.	Borg CR-10 scale, EMG, smartwatch, duration, and accuracy.
Okpala et al. (2022)	Kinetic edge flex lift	Flooring task	Lab	Usability: ease of use, ease of learning, and comfort.	Usability questionnaire.
Ogunseiju et al. (2022)	FLx ErgoSkeleton	Manual material handling	Lab	Perceived discomfort, range of motion, participants' experiences, and completion time.	Borg CR-10 scale, IMU, verbal feedback, and video camera.
Antwi-Afari et al. (2021)	Designed passive trunk exoskeleton	Manual material handling	Lab	Muscle activity, perceived discomfort, perceived musculoskeletal pressure, and usability.	EMG, Borg CR-10 scale, local perceived pressure scale, and SUS.

(*Continued*)

Table 8.4 (Continued)

Authors	*Exoskeleton Types*	*Tasks*	*Lab/Field Study*	*Outcome Measures*	*Data Collection Means/ Instruments*
Chen et al. (2021)	Portable knee exoskeleton.	Roofing tasks	Lab	Muscle activity, kinematics, and knee/foot contact pressures.	EMG, optical motion capture system, IMU, and a portable pressure mat.
de Vries et al. (2021)	Skelex 360	Plastering	Lab	Muscle activity and perceived exertion.	EMG and Borg CR-20 scale
Gonsalves et al. (2021)	BackX	Rebar work	Lab	Muscle activity, perceived discomfort, completion time.	EMG, Borg CR-10 scale, time-stamped camera.
Pillai et al. (2020)	LegX	Panel work, sustained groundwork	Lab	Muscle activity.	EMG.
Otten et al. (2018)	Exoskeleton Lucy	Fastening objects above head level, setting screws above head level, grinding at the wall	Lab	Muscle activity, perceived exertion, user support rating, and perceived force.	EMG, Borg CR-10 scale, and questionnaire.

both static and dynamic movements. The assessment focused on perceived exertion, discomfort, and usability. The results showed a decrease in discomfort in the lower back but an increase in discomfort around the chest. Overall, the participants rated the usability as moderate. Table 8.4 provides an overview of some exoskeleton research works in the construction industry.

8.7 Future research directions

Exoskeleton research is still in its infancy in most industries. There is a need for concerted efforts to increase awareness and adoption among workers and organizations. From this study, the research gaps in the construction domain point to future research directions that could be considered. For instance, current studies on exoskeletons are mostly laboratory based with simulated construction tasks and participants who are not actual construction workers. Also, most current studies either in the laboratory or in the field focused on passive exoskeletons, thus there is paucity of studies on active exoskeletons. Active exoskeletons offer new potential for workers and there is a need to explore the opportunities. Besides, most studies have focused on the back-support exoskeletons. There is a scarcity of research on other kinds of exoskeletons (for example, shoulder, leg, and arm support). Furthermore, there is the need to consider exoskeletons that support multiple parts of the body once adorned. This could translate into research that focuses on adaptive designs that can support body parts based on perceived risk.

Future research should leave the realms of laboratory studies due to the scarcity of field studies. Also, current developments in exoskeletons have focused on technical details rather than the humans or organizations that would adopt them. Hence, researchers need to provide practical demonstrations that take cognizance of long-term and real-life experiences in exoskeleton use. There is also the need to assess measures that have been assessed in other industry sectors e.g., the psychobiological state of construction workers, thermal effect, depth perception of exoskeletons, user anthropometry in vertical and horizontal confined spaces, user and exoskeleton kinematics, and metabolic cost of donning exoskeletons. Additionally, there is the need for means of evaluating exoskeletons e.g., using virtual models for both performance evaluation and real time performance monitoring. This will help contractors select the most suitable type and enhance decisions regarding the extent of exoskeleton use. This could be achieved through the development of algorithms and training sets to enable tracking of both user and exoskeleton kinematics. Last, there is a scarcity of step-by-step implementation plans for construction organizations that may want to integrate exoskeletons into their existing systems.

8.8 Conclusion

The study presented a systematic literature review of exoskeleton use in the construction industry. Exoskeleton use in the construction industry has become imperative in the face of increasing dire consequences of WMSDs on construction workers and the industry. From a review of extant literature, researchers have been able to classify exoskeletons based on the modus operandi, supported body parts, and the nature of exoskeleton constituent materials. To situate the adoption of exoskeletons in the workplace, several theories have been explored by exoskeleton researchers. However, the intricacies associated with exoskeleton adoption and acceptance require the exploration of more theories in different contexts. Despite the benefits of exoskeleton use in the construction industry in reducing the risk of WMSDs among construction workers, the study identified some existing barriers to its adoption. With few studies carried out in the field, it gives an opportunity for future research works to explore practical demonstrations in long-term and real-life scenarios. The study identified other research directions that exoskeleton researchers should focus on in the future.

References

Alemi, M. M., Madinei, S., Kim, S., Srinivasan, D. and Nussbaum, M. A. (2020). Effects of two passive back-support exoskeletons on muscle activity, energy expenditure, and subjective assessments during repetitive lifting. *Human Factor: The Journal of the Human Factors and Ergonomics Society*, *62*(3), 458–474. https://doi.org/10.1177/0018720819897669

Anam, K. and Al-Jumaily, A. A. (2012). Active exoskeleton control systems: State of the art. *Procedia Engineering*, *41*, 988–994. https://doi.org/10.1016/j.proeng.2012.07.273

Antwi-Afari, M. F., Li, H., Anwer, S., Li, D., Yu, Y., Mi, H.-Y. and Wuni, I. Y. (2021). Assessment of a passive exoskeleton system on spinal biomechanics and subjective responses during manual repetitive handling tasks among construction workers. *Safety Science*, *142*, 105382. https://doi.org/10.1016/j.ssci.2021.105382

Babalola, A., Urhal, P., Manu, P., Bartolo, P. J. D. S., Cheung, C., Yunusa-Kaltungo, A., Perera, S., Gao, S., Francis, V. and Paton-Cole, V. (2023). Application of exoskeletons for occupational safety and health management in construction: A systematic review. In: Manu, P., Shang, G., Silva Bartolo, P. J., Francis, V. and Sawhney, A. (eds.), *Handbook of Construction Safety, Health and Well-being in the Industry 4.0 Era* (1st ed., pp. 249–264). Routledge. https://doi.org/10.1201/9781003213796

Baltrusch, S. J., Houdijk, H., Van Dieën, J. H. and de Kruif, J. T. C. (2021). Passive trunk exoskeleton acceptability and effects on self-efficacy in employees with low-back pain: A mixed method approach. *Journal of Occupational Rehabilitation*, *31*(1), 129–141. https://doi.org/10.1007/s10926-020-09891-1

Baltrusch, S. J., van Dieen, J. H., van Bennekom, C. A. M. and Houdijk, H. (2018). The effect of a passive trunk exoskeleton on functional performance in healthy individuals. *Applied Ergonomics*, *72*, 94–106. https://doi.org/10.1016/j.apergo.2018.04.007

Bances, E., Wortmeier, A.-K., Bauernhansl, T., Garcia, B., Kropp, C., Schneider, U. and Siegert, J. (2022). Applicability of exoskeletons in timber prefabrication: Actions for exoskeleton research. *Procedia CIRP*, *107*, 1210–1215. https://doi.org/10.1016/j.procir.2022.05.133

Bennett, S. T., Adamczyk, P. G., Dai, F., Veeramani, D. and Senior, M. W. (2022). Upper extremity exoskeletons in construction, a field-based study. *ICRA 2022 Future of Construction Workshop Papers*, 17–20. https://doi.org/10.22260/ICRA2022/0007

Bennett, S. T., Han, W., Mahmud, D., Adamczyk, P. G., Dai, F., Wehner, M., Veeramani, D. and Zhu, Z. (2023). Usability and biomechanical testing of passive exoskeletons for construction workers: A field-based pilot study. *Buildings*, *13*(3), 822. https://doi.org/10.3390/buildings13030822

Bogue, R. (2018). Exoskeletons – A review of industrial applications. *Industrial Robot*, *45*(5), 585–590. https://doi.org/10.1108/IR-05-2018-0109

Bureau of Labor Statistics (BLS). (2018). *U.S. Department of labor, the economics daily, back injuries prominent in work-related musculoskeletal disorder cases in 2016*. https://www.bls.gov/opub/ted/2018/back-injuries-prominent-in-work-related-musculoskeletal-disorder-cases-in-2016.htm (Accessed July 18, 2023).

Bureau of Labor Statistics (BLS). (2019). *Median days lost*. https://data.bls.gov/pdq/SurveyOutputServlet (Accessed March 25, 2023).

Bureau of Labor Statistics (BLS). (2020). *Injury and illness rate per 10,000 full-time workers*. https://data.bls.gov/pdq/SurveyOutputServlet (Accessed April 28, 2023).

Butler, T. (2016). Exoskeleton technology: Making workers safer and more productive. *Professional Safety*, *61*(9), 32–36.

Chen, K., Lou, V. W. and Cheng, C. Y. M. (2023). Intention to use robotic exoskeletons by older people: A fuzzy-set qualitative comparative analysis approach. *Computers in Human Behavior*, *141*, 107610. https://doi.org/10.1016/j.chb.2022.107610

Chen, S., Stevenson, D. T., Yu, S., Mioskowska, M., Yi, J., Su, H. and Trkov, M. (2021). Wearable knee assistive devices for kneeling tasks in construction. *IEEE/ASME Transactions on Mechatronics*, *26*(4), 1989–1996. https://doi.org/10.1109/TMECH.2021.3081367

Cheng, C.-W., Leu, S.-S., Lin, C.-C. and Fan, C. (2010). Characteristic analysis of occupational accidents at small construction enterprises. *Safety Science*, *48*, 698–707. https://doi.org/10.1016/j.ssci.2010.02.001

Chiaradia, D., Xiloyannis, M., Solazzi, M., Masia, L. and Frisoli, A. (2019). Comparison of a soft exosuit and a rigid exoskeleton in an assistive task. In: Carrozza, M., Micera, S. and Pons, J. (eds.), *Wearable Robotics: Challenges and Trends. WeRob 2018. Biosystems & Biorobotics*, Vol. 22. Springer, Cham. https://doi.org/10.1007/978-3-030-01887-0_80

Cho, Y. K., Kim, K., Ma, S. and Ueda, J. (2018). A robotic wearable exoskeleton for construction worker's safety and health. In: *Construction Research Congress 2018*. ASCE, Reston, VA, 19–28.

Cieza, A., Causey, K., Kamenov, K., Hanson, S. W., Chatterji, S. and Vos, T. (2021). Global estimates of the need for rehabilitation based on the global burden of disease study 2019: A systematic analysis for the global burden of disease study 2019. *The Lancet*, *396*(10267), 2006–2017.

Dahmen, C. and Constantinescu, C. (2020). Methodology of employing exoskeleton technology in manufacturing by considering time-related and ergonomics influences. *Applied Sciences*, *10*, 1591. https://doi.org/10.3390/app10051591

Davis, K. G. and Kotowski, S. E. (2007). Understanding the ergonomic risk for musculoskeletal disorders in the United States agricultural sector. *American Journal of Industrial Medicine*, *50*(7), 501–511. https://doi.org/10.1002/ajim.20479

De Bock, S., Ghillebert, J., Govaerts, R., Elprama, S. A., Marusic, U., Serrien, B., Jacobs, A., Geeroms, J., Meeusen, R. and De Pauw, K. (2020). Passive shoulder exoskeletons: More effective in the lab than in the field? *IEEE Transactions on Neural Systems and Rehabilitation Engineering*, *29*, 173–183. https://doi.org/10.1109/TNSRE.2020.3041906

De Looze, M. P., Bosch, T., Krause, F., Stadler, K. S. and O'Sullivan, L. W. (2016). Exoskeletons for industrial application and their potential effects on physical workload. *Ergonomics*, *59*, 671–681. https://doi.org/10.1080/00140139.2015.1081988

de Vries, A. W., Krause, F. and de Looze, M. P. (2021). The effectivity of a passive arm support exoskeleton in reducing muscle activation and perceived exertion during plastering activities. *Ergonomics*, *64*(6), 712–721. https://doi.org/10.1080/00140139.2020.1868581

Dwivedi, Y. K., Rana, N. P., Jeyaraj, A., Clement, M. and Williams, M. D. (2017). Re-examining the unified theory of acceptance and use of technology (UTAUT): Towards a revised theoretical model. *Information Systems Frontiers*, *21*, 719–734. https://doi.org/10.1007/s10796-017-9774-y

Eaves, D. L., Behmer, L., Jr. and Vogt, S. (2016). EEG and behavioral correlates of different forms of motor imagery during action observation. *Brain and Cognition*, *106*, 90–103. https://doi.org/10.1016/j.bandc.2016.04.013

Elprama, S. A., Vannieuwenhuyze, J. T., De Bock, S., Vanderborght, B., De Pauw, K., Meeusen, R. and Jacobs, A. (2020). Social processes: What determines industrial workers' intention to use exoskeletons? *Human Factors, 62*(3), 337–350. https://doi.org/10.1177/0018720819889534

Fethke, N. B., Merlino, L. A., Gerr, F., Schall, M. C., Jr. and Branch, C. A. (2015). Musculoskeletal pain among Midwest farmers and associations with agricultural activities. *American Journal of Industrial Medicine, 58*(3), 319–330. https://doi.org/10.1002/ajim.22398

Giger, J. C. and Piçarra, N. (2017). Who wants to work with social robots? Using the theory of reasoned action and the theory of planned behavior to predict intention to work with social robots. In: de Jesus, S. N. and Pinto, P. (eds.), *Proceedings of the II International Congress on Interdisciplinarity in Social and Human Sciences University of Algarve, Faro, Portugal 11th–12th May 2017*. CIEO – Research Centre for Spatial and Organizational Dynamics University of Algarve Gambelas Campus, Faro.

Gilotta, S., Spada, S., Ghibaudo, L., Isoardi, M. and Mosso, C. (2019). Acceptability beyond usability: A manufacturing case study. In: *Proceedings of the 20th Congress of the International Ergonomics Association (IEA 2018) Volume VII: Ergonomics in Design, Design for All, Activity Theories for Work Analysis and Design, Affective Design 20*, 922–934. https://doi.org/10.1007/978-3-319-96071-5_95

Golabchi, A., Jasimi Zindashti, N., Miller, L., Rouhani, H. and Tavakoli, M. (2023). Performance and effectiveness of a passive back-support exoskeleton in manual material handling tasks in the construction industry. *Construction Robotics*, 1–12. https://doi.org/10.1007/s41693-023-00097-4

Gonsalves, N., Akanmu, A., Gao, X., Agee, P. and Shojaei, A. (2023a). Industry perception of the suitability of wearable robot for construction work. *Journal of Construction Engineering and Management, 149*(5), 04023017. https://doi.org/10.1061/JCEMD4.COENG-12762

Gonsalves, N., Khalid, M., Akinniyi, A. and Akanmu, A. (2022a). Industry perspectives of the potential of wearable robot for pipe installation work. In: *Proceedings of the International Symposium on Automation and Robotics in Construction*. IAARC Publications, 437–443. https://doi.org/10.22260/ISARC2022/0060

Gonsalves, N., Khalid, M., Akinniyi, A., Ogunseiju, O. and Akanmu, A. (2022b). Subjective evaluation of passive back-support wearable robot for simulated rebar work. In: *Proceedings of the International Symposium on Automation and Robotics in Construction*. IAARC Publications, 430–436.

Gonsalves, N. J., Ogunseiju, O. O., Akanmu, A. A. and Nnaji, C. A. (2021). Assessment of a passive wearable robot for reducing low back disorders during rebar work. *Journal of Information Technology in Construction, 26*, 936–952. https://doi.org/10.36680/j.itcon.2021.050

Gonsalves, N. J., Yusuf, A., Ogunseiju, O. and Akanmu, A. (2023b). Evaluation of concrete workers' interaction with a passive back-support exoskeleton. *Engineering, Construction and Architectural Management*. Advance online publication. https://doi.org/10.1108/ECAM-12-2022-1156

Gumasing, M. J. J. and Castro, F. M. F. (2023). Determining ergonomic appraisal factors affecting the learning motivation and academic performance of students during online classes. *Sustainability, 15*, 1970. https://doi.org/10.3390/su15031970

Gutierrez, N., Ojelade, A., Kim, S., Barr, A., Akanmu, A., Nussbaum, M. A. and Harris-Adamson, C. (2023). *Perceived benefits, barriers, perceptions, and readiness to use exoskeletons in the construction industry: Differences by demographic characteristics*. SSRN: http://dx.doi.org/10.2139/ssrn.4361395

Hameed, M. A., Counsell, S. and Swift, S. (2012). A meta-analysis of relationships between organizational characteristics and IT innovation adoption in organizations. *Information & Management, 49*(5), 218–232. https://doi.org/10.1016/j.im.2012.05.002.

Hong, M. B., Gwang, T. K. and Yeo, H. Y. (2019). ACE-ankle: A novel sensorized RCM (remote-center-of-motion) ankle mechanism for military purpose exoskeleton. *Robotica, 37*(12), 2209–2228. https://doi.org/10.1017/S0263574719000845

Hwang, J., Yerriboina, V. N. K., Ari, H. and Kim, J. H. (2021). Effects of passive back support exoskeletons on physical demands and usability during patient transfer tasks. *Applied Ergonomics, 93*, 103373. https://doi.org/10.1016/j.apergo.2021.103373

International Labor Organization (ILO). (2019). *Developing the construction industry for employment-intensive infrastructure investments*. https://www.ilo.org/wcmsp5/groups/public/---ed_emp/---emp_policy/---invest/documents/publication/wcms_734235.pdf (Accessed June 23, 2023).

Iranzo, S., Piedrabuena, A., Iordanov, D., Martinez-Iranzo, U. and Belda-Lois, J.-M. (2020). Ergonomics assessment of passive upper-limb exoskeletons in an automotive assembly plant. *Applied Ergonomics, 87*, 103120. https://doi.org/10.1016/j.apergo.2020.103120

Jorgensen, M. J., Hakansson, N. A. and Desai, J. (2022). The impact of passive shoulder exoskeletons during simulated aircraft manufacturing sealing tasks. *International Journal of Industrial Ergonomics, 91*, 103337. https://doi.org/10.1016/j.ergon.2022.103337

Karvouniari, A., Michalos, G., Dimitropoulos, N. and Makris, S. (2018). An approach for exoskeleton integration in manufacturing lines using virtual reality techniques. *Procedia CIRP*, *78*, 103–108. https://doi.org/10.1016/j.procir.2018.08.315

Kim, S., Moore, A., Srinivasan, D., Akanmu, A., Barr, A., Harris-Adamson, C., Rempel, D. M. and Nussbaum, M. A. (2019). Potential of exoskeleton technologies to enhance safety, health, and performance in construction: Industry perspectives and future research directions. *IISE Transactions on Occupational Ergonomics and Human Factors*, *7*(3–4), 185–191. https://doi.org/10.1080/24725838.2018.1561557

Kitchenham, B., Brereton, O. P., Budgen, D., Turner, M., Bailey, J. and Linkman, S. (2009). Systematic literature reviews in software engineering – A systematic literature review. *Information and Software Technology*, *51*(1), 7–15. https://doi.org/10.1016/j.infsof.2008.09.009

Kopp, V., Holl, M., Schalk, M., Daub, U., Bances, E., García, B., Schalk, I., Siegert, J. and Schneider, U. (2022). Exoworkathlon: A prospective study approach for the evaluation of industrial exoskeletons. *Wearable Technologies*, *3*, e22. https://doi.org/10.1017/wtc.2022.17

Lee, S., Yu, J. and Jeong, D. (2013). BIM acceptance model in construction organizations. *Journal of Management in Engineering*, *31*(3), 04014048. https://doi.org/10.1061/(ASCE)ME.1943-5479.0000252

Liu, W., Yin, B. and Yan, B. (2016). A survey on the exoskeleton rehabilitation robot for the lower limbs. In: *2nd International Conference on Control, Automation and Robotics (ICCAR)*. IEEE, 90–94. https://doi.org/10.1109/ICCAR.2016.7486705

Lowe, B. D., Billotte, W. G. and Peterson, D. R. (2019). ASTM F48 formation and standards for industrial exoskeletons and exosuits. *IISE Transactions on Occupational Ergonomics and Human Factors*, *7*, 230–236. https://doi.org/10.1080/24725838.2019.1579769

Madinei, S., Kim, S., Alemi, M. M., Srinivasan, D. and Nussbaum, M. A. (2019). Assessment of two passive back-support exoskeletons in a simulated precision manual assembly task. *Proceedings of the Human Factors and Ergonomics Society Annual Meeting*, *63*(1), 1078–1079. https://doi.org/10.1177/1071181319631192

Marcum, J. and Adams, D. (2017). Work-related musculoskeletal disorder surveillance using the Washington state workers' compensation system: Recent declines and patterns by industry, 1999–2013. *American Journal of Industrial Medicine*, *60*(5), 457–471. https://doi.org/10.1002/ajim.22708

Meyer, R. (2023). The effectiveness and efficacy of exoskeletons in construction. *Seminar Paper for School of Public Health University of California*, Berkeley, 1–25. https://www.marshallplan.at/images/All-Papers/MP-2023/Meyer_Rieke_2044.PDF (Accessed June 23, 2023).

Moher, D., Shamseer, L., Clarke, M., Ghersi, D., Liberati, A., Petticrew, M., Shekelle, P. and Stewart, L. A. (2015). Preferred reporting items for systematic review and meta-analysis protocols (PRISMA-P) 2015 statement. *Systematic Reviews*, *4*(1), 1–9. https://doi.org/10.1186/2046-4053-4-1

Moyon, A., Poirson, E. and Petiot, J.-F. (2018). Experimental study of the physical impact of a passive exoskeleton on manual sanding operations. *Procedia CIRP*, *70*, 284–289. https://doi.org/10.1016/j.procir.2018.04.028

Ogunseiju, O., Olayiwola, J., Akanmu, A. and Olatunji, O. A. (2022). Evaluation of postural-assist exoskeleton for manual material handling. *Engineering, Construction and Architectural Management*, *29*(3), 1358–1375. https://doi.org/10.1108/ECAM-07-2020-0491

Okpala, I., Nnaji, C., Awolusi, I. and Akanmu, A. (2021). Developing a success model for assessing the impact of wearable sensing devices in the construction industry. *Journal of Construction Engineering and Management*, *147*(7). https://doi.org/10.1061/(ASCE)CO.1943-7862.0002064

Okpala, I., Nnaji, C., Ogunseiju, O. and Akanmu, A. (2022). Assessing the role of wearable robotics in the construction industry: Potential safety benefits, opportunities, and implementation barriers. In: Jebelli, H., Habibnezhad, M., Shayesteh, S., Asadi, S. and Lee, S. (eds.), *Automation and Robotics in the Architecture, Engineering, and Construction Industry*. Springer, Cham. https://doi.org/10.1007/978-3-030-77163-8_8

Omoniyi, A., Trask, C., Milosavljevic, S. and Thamsuwan, O. (2020). Farmers' perceptions of exoskeleton use on farms: Finding the right tool for the work(er). *International Journal of Industrial Ergonomics*, *80*, 103036. https://doi.org/10.1016/j.ergon.2020.103036

Otten, B. M., Weidner, R. and Argubi-Wollesen, A. (2018). Evaluation of a novel active exoskeleton for tasks at or above head level. *IEEE Robotics and Automation Letters*, *3*(3), 2408–2415. https://doi.org/10.1109/LRA.2018.2812905

Pan, M. and Pan, W. (2019). Determinants of adoption of robotics in precast concrete production for buildings. *Journal of Management in Engineering*, *35*(5). https://doi.org/10.1061/(ASCE)ME.1943-5479.0000706

Pan, M. and Pan, W. (2020). Understanding the determinants of construction robot adoption: Perspective of building contractors. *Journal of Construction Engineering and Management*, *146*(5), 04020040. https://doi.org/10.1061/(ASCE)CO.1943-7862.0001821

Pillai, M. V., Van Engelhoven, L. and Kazerooni, H. (2020). Evaluation of a lower leg support exoskeleton on floor and below hip height panel work. *Human Factors*, *62*(3), 489–500.

Pinho, J. P. and Forner-Cordero, A. (2022). Shoulder muscle activity and perceived comfort of industry workers using a commercial upper limb exoskeleton for simulated tasks. *Applied Ergonomics*, *101*, 103718. https://doi.org/10.1016/j.apergo.2022.103718

Schwartz, M., Theurel, J. and Desbrosses, K. (2021). Effectiveness of soft versus rigid back-support exoskeletons during a lifting task. *International Journal of Environmental Research and Public Health*, *18*(15), 8062. https://doi.org/10.3390/ijerph18158062

Schwatka, N. V., Butler, L. M. and Rosecrance, J. R. (2012). An aging workforce and injury in the construction industry. *Epidemiologic Reviews*, *34*, 156–167. https://doi.org/10.1093/epirev/mxr020

Schwerha, D. J., McNamara, N., Nussbaum, M. A. and Kim, S. (2021). Adoption potential of occupational exoskeletons in diverse enterprises engaged in manufacturing tasks. *International Journal of Industrial Ergonomics*, *82*, 103103. https://doi.org/10.1016/j.ergon.2021.103103

Settembre, N., Maurice, P., Paysant, J., Theurel, J., Claudon, L., Kimmoun, A., Levy, B., Hani, H., Chenuel, B. and Ivaldi, S. (2020). The use of exoskeletons to help with prone positioning in the intensive care unit during COVID-19. *Annals of Physical and Rehabilitation Medicine*, *63*, 379–382. https://doi.org/10.1016/j.rehab.2020.05.004

Shore, L., Power, V., De Eyto, A. and O'Sullivan, L. W. (2018). Technology acceptance and user-centred design of assistive exoskeletons for older adults: A commentary. *Robotics*, *7*(1), 3. https://doi.org/10.3390/robotics7010003

Siedl, S. M. and Mara, M. (2022). What drives acceptance of occupational exoskeletons? Focus group insights from workers in food retail and corporate logistics. *International Journal of Human – Computer Interaction*, 1–10. https://doi.org/10.1080/10447318.2022.2108969

Spada, S., Ghibaudo, L., Gilotta, S., Gastaldi, L. and Cavatorta, M. P. (2017). Investigation into the applicability of a passive upper-limb exoskeleton in automotive industry. In: *Proceedings of the 27th International Conference on Flexible Automation and Intelligent Manufacturing*, Modena, Italy, 1255–1262. https://doi.org/10.1016/j.promfg.2017.07.252

Upasani, S., Franco, R., Niewolny, K. and Srinivasan, D. (2019). The potential for exoskeletons to improve health and safety in agriculture – Perspectives from service providers. *IISE Transactions on Occupational Ergonomics and Human Factors*, *7*(3–4), 222–229. https://doi.org/10.1080/24725838.2019.1575930

Viteckova, S., Kutilek, P. and Jirina, M. (2013). Wearable lower limb robotics: A review. *Biocybernetics and Biomedical Engineering*, *33*(2), 96–105. https://doi.org/10.1016/j.bbe.2013.03.005

Wang, X., Dong, X. S., Choi, S. D. and Dement, J. (2017). Work-related musculoskeletal disorders among construction workers in the United States from 1992 to 2014. *Occupational and Environmental Medicine*, *74*(5), 374–380. https://doi.org/10.1136/oemed-2016-103943

World Health Organization (WHO). (2022). *Musculoskeletal health*. https://www.who.int/news-room/fact-sheets/detail/musculoskeletal-conditions (Accessed June 23, 2023).

Yan, T., Cempini, M., Oddo, C. M. and Vitiello, N. (2015). Review of assistive strategies in powered lower-limb orthoses and exoskeletons. *Robotics and Autonomous Systems*, *64*, 120–136. https://doi.org/10.1016/j.robot.2014.09.032

Yan, X.-T., Browne, J., Swanson, C., Niu, C., Bisland, G., Li, Y. and Johnston, A. (2023). An investigation of exoskeleton robotic systems in assisting construction tasks. *Intelligent Information Management*, *15*(3), 216–241. https://doi.org/10.4236/iim.2023.153011

Yelin, E., Weinstein, S. and King, T. (2016). The burden of musculoskeletal diseases in the United States. *Seminars in Arthritis and Rheumatism*, *46*(3), 259–260. https://doi.org/10.1016/j.semarthrit.2016.07.013

Zheng, L., Hawke, A. L. and Evans, K. (2022). Critical review on applications and roles of exoskeletons in patient handling. *International Journal of Industrial Ergonomics*, *89*. https://doi.org/10.1016/j.ergon.2022.103290

Zhu, Z., Dutta, A. and Dai, F. (2021). Exoskeletons for manual material handling – A review and implication for construction applications. *Automation in Construction*, *122*, 103493. https://doi.org/10.1016/j.autcon.2020.103493

9

ONTOLOGICAL MODELS FOR CONSTRUCTION SCHEDULING AND RESOURCE PLANNING

Zigeng Fang[1], Songbo Hu[2], Junlin Wang[3], Yihai Fang[4] and Qiuchen Lu[5]

[1]Assistant Professor, Department of Civil Engineering, University of Nottingham, Nottingham NG7 2RD, UK. Email: zigeng.fang@nottingham.ac.uk

[2]Associate Lecturer, School of Engineering, Design, and Built Environment, Western Sydney University, Second Ave, Kingswood NSW 2747, Australia, Email: s.hu@westernsydney.edu.au

[3]PhD Student, Department of Civil Engineering, Monash University, Wellington Rd, Clayton VIC 3800, Australia, Email: junlin.wang@monash.edu

[4]Senior Lecturer, Department of Civil Engineering, Monash University, Wellington Rd, Clayton VIC 3800, Australia, Email: yihai.fang@monash.edu

[5]Associate Professor, Bartlett School of Sustainable Construction, University College London, London WC1E 6BT, UK. Email: qiuchen.lu@ucl.ac.uk

9.1 An introduction to ontological models

Projects for Architecture, Engineering, and Construction (AEC) often require the cooperation of multidisciplinary professionals. Each discipline has its way of collecting, storing, and representing information (Haymaker, et al., 2005). Semantic web technology, an open and decentralised alternative to the current centralised data-keeping and sharing (Poinet, et al., 2020), was proposed to promote the use of standardised formats and ontologies to support decentralised data integration and sharing across the web. Before delving into the details of applying ontological models in construction scheduling and resource planning, it is important to know what ontology is. In artificial intelligence, Gruber defined ontology as the "explicit specification of a conceptualisation" (Guber, 1993), which means modelling domain knowledge as concept sets and concept relationships. The aim of domain ontology is never a comprehensive enumeration of all concepts in a domain but the provision of a representation of the essence of knowledge in a domain (El-Diraby, 2013). Ontology has been developed as a semantic model for construction scheduling and resource planning applications, capturing conceptualisations for representing, managing, reusing, and sharing construction knowledge. A typical ontological model involves elements such as concepts, properties (of concepts), instances (of a particular concept), relations (between instances), constraints, and axioms (logical assertions) (Chen & Luo, 2016). By using the ontological model, the AEC industry knowledge can be represented in a formal and machine-manipulable format.

Knowing what ontology is, the next question is: what are the main technologies that enable the implementation of ontological models? The semantic web technology stack provides a set of

DOI: 10.1201/9781003383840-10

standardised technologies to implement the ontological model successfully. These technologies provide industry professionals with tools and standards to create, manage, reuse, retrieve, and reason ontological models. The main layers of standardised semantic web technologies in the entire stack start from the foundation, Resource Description Framework (RDF), providing an ontology language for representing data and relationships as subject-predicate-object triples, which means RDF can achieve data interchange among multi-source heterogeneous data by stating those data in uniform standards. RDF Schema (RDFS) with schema vocabulary is the extension of RDF, which can define hierarchies of classes and properties based on taxonomy for concepts/classes (sets of indBased) on generic logical models; the ontology developer can define the concepts and their properties, specify logical rules and constraints, and correctly establish and maintain a hierarchy. Web Ontology Language (OWL) is the current standard ontology language for authoring ontologies for formal knowledge representation (Chen & Luo, 2016). Semantic Web Rule Language (SWRL) is the language built to further enhance the OWL by allowing semantic and reasoning interoperability between ontologies and rules at the level of instances and their properties. Furthermore, Rule Interchange Format (RIF) provides a standard for rules that are interoperated among different rule systems whose paradigms have few syntactic and semantic similarities. Last but not least, SPARQL is a query language that can query any RDF-based data, using the multi-source heterogeneous data stored in RDF that can be retrieved and manipulated. These technologies have been widely utilised in construction scheduling and resource planning-related ontological models, and the details of how they were used are illustrated in the later sections (9.3 and 9.4).

9.2 Ontological models' major functions in construction

Project management has several core functions (e.g., Contract Administration, Quality Management, Operation Management, Resource Management, etc.) in the construction context. Construction scheduling and resource planning, requiring coordination efforts by general contractors, are the core elements of operation and resource management functions. Both these two elements play a crucial role in determining the success of construction project management.

Construction scheduling involves the time management process that is needed for creating and maintaining a detailed plan and timeline for the sequential execution of construction activities. Although different projects may require different construction scheduling elements, in general, it includes (a) the dissection of the entire project into smaller and manageable elements; (b) the determination of the sequence and duration of activities to perform; (c) the identification of the lead and lag time, as well as the longest sequence of activities to determine the minimum project duration; and (d) the regular updating and monitoring of the project schedule and progress and the preparation of the project contingency plan. The configuration and processes of construction scheduling can be significantly impacted by the project conditions and ongoing changing factors (e.g., site conditions, weather conditions, delay in supply, etc.). The ontological model is suggested to help tackle this challenge by providing product features and design-related work conditions that help make construction planning decisions and predict labour hours (Staub-French, et al., 2003; Lee, et al., 2014). In contrast, construction resource planning involves the effective management and distribution of resources required for construction projects. Resource planning includes both the management of physical resources and non-physical resources. Physical resource planning involves allocating and managing physical items including materials and equipment. In comparison, the non-physical resource includes the managing and planning of knowledge resources (e.g., information and data), human resources (e.g., labour and engineers), and financial resources (e.g., project budget, liabilities, etc.). The traditional construction resource planning process lacks a scalable, integratable, and semantic reasoning mechanism to help validate the planning of the construction activities and resources regarding the multiple constraints due to the use of the singleness mechanism that focuses on satisfying one constraint or objective at a time before moving on to the next one (Tavakolan, et al., 2021). Ontological models were suggested to be capable of providing reasoning abilities regarding construction processes and demanded resources and organising construction processes with higher levels of detail (Zhong, et al., 2015).

After knowing the capability of ontology in representing domain knowledge, it is important to be familiar with the main functions of ontology-related studies in the construction sector, especially for construction scheduling and resource planning. According to Li et al., there are currently three main functions for the utilising of ontology in construction scheduling and resource planning, namely: (a) information extraction, (b) knowledge modelling, and (c) reasoning and conformance check (Li, et al., 2022).

Information extraction generally refers to retrieving the desired types of information from certain sources (e.g., natural language text). Its objective is to process unstructured or semi-structured data to retrieve the instances of a specific class of objects or events and the instances of relationships between them (Teng, et al., 2017). For the AEC context, information extraction is needed to automatically or semi-automatically recognise and extract data/information (e.g., bridge components' conditions and maintenance actions) from BIM-related databases or unstructured textual bridge inspection reports for data analytics (Liu & El-Gohary, 2017). In general, the information extraction for AEC-related ontology applications can be broadly categorised into two categories (according to different information sources): (a) BIM-related sources and (b) other data sources and documents. For BIM-related sources, although the recent BIM applications (e.g., Autodesk Revit, Innovaya, and Solibri Model Checker (SMC), etc.) can utilise the "information/material take-off" tools to locate geometric and material information in BIM (Nepal, et al., 2013), they were suggested to be unable to encode the domain-specific (e.g., construction) knowledge and lack flexibility (Drogemuller, et al., 2004). Flexibility is important and needed for the construction ontology because each construction professional has their viewpoints and rationales to express, and spatial relationships between building components require customisation from case to case. Also, the existing BIM-oriented tools were unable to identify the locations and types of intersections (Nepal, et al., 2013). Therefore, the ontological model is needed for BIM-related information extraction, while, for information extraction from other sources and documents, the processes can be even more challenging than the BIM-related documents. First, the information extraction from the documents (e.g., building inspection reports) is subjected to a wide variety of text characteristics and patterns due to the multi-stakeholders (e.g., different inspectors, engineers, etc.) involved in file composing processes (Liu & El-Gohary, 2017). Second, some technical documents may involve text descriptions with high technical details and complex concept relationships, which make the sentence structure hierarchically complex (Zhou & El-Gohary, 2017). Last, multiple exceptions might occur to fail compliance with the requirement even if one or a set of exception conditions was met. Therefore, it is important to review how and whether these challenges have been overcome.

In the ontology context, knowledge modelling implies the process of representing and organising knowledge in a structured and formalised way using ontologies. Knowledge modelling involves a process of formalising the domain knowledge so that the contained concepts can be explicitly described in the common vocabulary (Anumba, et al., 2008). Knowledge modelling aims to provide a standardised and explicit representation of knowledge to enable better understanding, reasoning, and communication about the domain. The recent interest focus of the construction-related ontology researcher groups and communities (e.g., buildingSMART International (bSI), Linked Data Working Group (LDWG), etc.) for knowledge modelling is to make the data available and link the data in architecture and construction through semantic web format, rather than documents (Curry, et al., 2013). By adopting or abiding by the terminology from other ontologies, the construction ontology communities attempt to create aligned ontologies and work them towards standardisation (Rasmussen, et al., 2019). One of the main challenges of construction ontology knowledge modelling is to agree on a shared standard ontology (or knowledge representation approach) for the buildings and infrastructures (Gursel, et al., 2009). Many obstacles contribute to this challenge. First, the AEC industry is a sector characterised by its fragmentation. It is fragmented not only because of its isolated geographical location or the fully interconnected functional systems but also the fragmented digital data storage and management systems. These sources of fragmentation, combined with the semantic heterogeneity caused by the wide variety and number of professionals involved, make reaching an agreement on a shared standard knowledge base a goal that is far from reach. Second, gathering multiple organisations and stakeholders for a typical construction project also introduces conflicts of interest between different parties and raises further customisation requirements. This means the original single worldview domain conceptualisation has been executed and implemented in a diverse and locally biased manner (Gursel, et al., 2009). Third, the construction project, compared with product manufacturing, is characterised by having multiple phases that differ in major priorities and functional content. In the built environment field, the frequently adopted standard representational approach might hardly find a way to flexibly handle the dynamic information that emerged during the facilities' whole lifecycle. Thus, identifying how and whether the existing project has tackled these obstacles is crucial for the future development of ontology technologies in the AEC industry.

Reasoning and conformance check is another major function of the construction ontology. Reasoning implies the process of deriving new knowledge based on explicitly defined relationships, rules, and axioms. In general, reasoning is used for two major reasons: (a) to examine the context consistency and (b) to use the low-level explicit context to infer the high-level context (Wang, et al., 2004). Other than that, reasoning mechanisms can be used to classify instances (e.g., assigning the instances' class), derive new information (e.g., deducing new facts based on existing construction knowledge), and query answering with natural language (e.g., enabling more precise and comprehensive building element explicitly/implicitly information queries). Utilising the ontological model for reasoning and conformance checks has also suffered some obstacles (Zhang & El-Gohary, 2017). One of the main challenges is the heterogeneity issue between the universal ontologies (e.g., ifcOWL) and the regulatory ontologies (for reasoning and compliance check) when conducting the reasoning and compliance checking process. Three main reasons contribute to the conceptualisation mismatches. First, as universal ontology is intended to cover the full IFC schema, the content scope of universal ontology is different from the regulatory ontologies, where the experts' focus was more on a certain domain (Pauwels, et al., 2017). Second, the terminologies used for regulatory ontology and universal ontology are different. In other words, "window" may be defined differently in regulatory ontology and universal ontology. Last, the universal ontology and the regulatory ontology are different in their description perspectives (Jiang, et al., 2022). Given the aforementioned challenges faced by different functions of construction ontology applications, the selected studies on construction scheduling and resource planning will be reviewed and analysed in the next section.

In the context of this chapter, ontology-related studies in the AEC field are broadly categorised into three domains: creation, integration, and application. For ontology creation, the focus is on the development of domain-specific ontologies that represent the knowledge and concepts within the AEC domain. This type of study aims to create a standardised and structured representation of construction domain information that can be reused within different applications. Ontology integration involves mainly the combining of different existing ontologies or, in some cases, integrating the ontologies with different data sources to improve the coverage of the knowledge representation. A common aim of ontology integration is to create a unified view of the information. Last, for ontology application, the focus is on the practical use of ontologies in real-world AEC cases (e.g., building a practical tool that is ready for use or a platform or system for specific AEC processes). This domain's studies aim to leverage ontologies to improve the AEC processes and overcome some practical construction challenges. Although some studies may cover more than one domain (in the summary Tables 9.1 and 9.2), they will be classified according to their most representative characteristics. In comparison, one study might cover one, two, or even three different ontology application functions.

9.3 The review of ontological models in construction scheduling and resource planning

As mentioned previously, the use of an ontological model in construction scheduling can boost the efficiency and intelligence of construction project management and coordination by providing standardised and formal interlinked project knowledge. By applying ontology's constraint theory, the constraint relationship of components can be used to reflect the construction program's logic (Budayan, et al., 2018). Construction scheduling is a process that requires the cooperation of different professionals. Therefore, ontology can be used to help reveal the implicit or complicated relationship and reach an even greater amount of information that was originally left alone due to different professionals' language complexity (Guévremont & Hammad, 2021). Furthermore, BIM, as a major information source, stores data in a multilayered and diversified manner, which can be helped by the ontological models to optimise the schedule using massive data available. Thus, in this section, relevant studies were selected and reviewed to summarise the current progress and challenges for applying ontological models for construction scheduling.

In general, the quantity of construction scheduling-related ontological model research was much less than that of resource planning, which suggested that the territory of the ontological model's application on construction scheduling has not been widely explored. Out of these three ontology study domains, the ontology integration study had the largest number (three), which was followed by the ontology creation study (two) and ontology application study (one). For the application domain, there were four knowledge modelling studies, two information extraction studies and two reasoning and conformance check studies. For ontology application, Budayan et al. have utilised the delay analysis ontology to identify the causes of activity-level delay and evaluate the frequency and impact of risks according to the user experience (Budayan, et al., 2018). This helped the decision-maker learn the activity-level delay risk for the project scheduling process and further enabled effective delay risk allocation and mitigation strategies. For ontology creation, the focus has been put on providing a new approach or framework to automate the generation of the work packages, with improved accuracy and efficiency (Li, et al., 2022) or formalising a constraint ontology and classification mechanism, with rationale consistency and manual tracing network chains (Koo, et al., 2007). The application of ontological models allowed planners and other decision-makers to provide constraint-loaded schedules with decreased erroneous schedule interpretation and individual sequence logic checks (Koo, et al., 2007). Finally, for ontology integration, some efforts have been put into proposing an integrated ontology or an ontology set for the visual analytics of delay claims (Guévremont & Hammad, 2021). In contrast, more researchers were dedicated to improving construction process management through effective information and data integration (Zheng, et al., 2021; Viljamaa & Peltomaa, 2014). Ontology integration provided the claims analysis enhancement by adding another time dimension to form the 4D simulation that outlines the delay responsibilities, entitlements, and causalities (Guévremont & Hammad, 2021). The integration of information sources (e.g., construction workflow information, intensified infrastructure building process information, etc.) could provide a comprehensive vocabulary related to construction workflow domain knowledge (Zheng, et al., 2021) and ensure better interoperability and model scalability (Viljamaa & Peltomaa, 2014).

Despite the fact that numerous benefits were given by applying ontology in construction scheduling, the review of these studies has also suggested that human involvement can present several challenges during the ontological models' implementation. From one perspective, the development process of a construction scheduling ontology was heavily dependent on expert knowledge and experience. For example, much of the selected research mentioned that they still required skilled personnel, experts, and experienced estimators' input for the construction of the ontology (Budayan, et al., 2018; Zheng, et al., 2021; Viljamaa & Peltomaa, 2014). From another perspective, acquiring all relevant rules, concepts, and relationships from experts with different backgrounds could be challenging and time-consuming due to construction projects' diverse and complex nature. Therefore, researchers normally were required to propose a draft of the ontological model and improve or amend it further with experts' confirmation and on-site consolidation. For instance, Zheng et al. suggested that construct ontology based on the expert workshop and criteria-based evaluation processes are subjective and hardly quantifiable (Zheng, et al., 2021). Viljamaa et al. argued the success of ontology application would require high-technology orientation from the using party with skilled personnel before developing the routines to use the evolved method (Viljama & Peltomaa, 2014). Therefore, it was fair for Budayan et al. to note the urgent need to further address the risk of human bias by using artificial intelligence techniques [24]. Other than the challenge of human involvement, there were also challenges like limitations in scarce use cases (Koo, et al., 2007) and data sources (Zheng, et al., 2021). More detailed information regarding applying ontological models in construction scheduling is summarised in Table 9.1.

9.4 Review of ontological models for resource planning

The previous section (9.3) indicates that, if the knowledge and information related to construction resources are represented in a structured and formalised manner, ontological modelling can benefit construction resource planning with an optimised resource allocation, coordination, and utilisation. Resource planning is about ensuring that the right resources can be available at the right time with the right quantity to complete the project within the budget. Ontological models can provide the decision support and reasoning to optimise resource allocation by identifying potential conflicts and risks based on predefined rules or constraints. For instance, Zhong et al. enhance the construction resource planning with reasoning capability, through their developed ontology framework, for achieving a higher level of detail (Zhong, et al., 2015). Therefore, relevant studies were reviewed and analysed in this section to summarise the current progress and challenges for applying ontological models for construction resource planning.

As stated in the previous section, applying ontology in resource planning has received a fiercer debate in recent years compared with construction scheduling. Out of these three ontology study domains, compared with the case of construction scheduling, where the ontology integration study has received the largest amount of attention, the ontology integration study for resource planning has not been actively investigated. Instead, the ontology application study has the largest amount of research (Drogemuller, et al., 2004), followed by the ontology creation study (Lee, et al., 2014). For the application domain, 12 information extraction studies, nine knowledge modelling studies, and seven reasoning and conformance check studies were found.

In general, all three application functions have received similar attention in the ontology application. For the information extraction function, the ontological models were heavily utilised for the cost estimation and quantity take-off (Lee, 2014; Alzraiee, 2022; Akanbi & Zhang, 2021;

Table 9.1 Summary of literature review for construction scheduling ontological model studies

Reference	*Domain of Study*	*Main Application Function(s)*	*Research Aim*	*Main Benefits/Contributions*	*Main Limitations/ Challenges*	*Year*
(Budayan, et al., 2018)	Application	Reasoning and conformance check	Propose a delay analysis ontology that enables users to identify causes of delay at the activity level and assess the frequency and impact of risks using their previous experience, expertise, and judgement.	1) Decision-maker can easily understand the level of delay risk at the activity level. 2) Indicate the criticality of activities that may change under different risk occurrence scenarios. 3) Enable effective risk response and allocation strategies to mitigate the delay risk. 4) Communicate delay risk information between various project participants.	1) Test in one modified version of the fuzzy network calculation model, which is limited in the tested model types. 2) Requires experienced estimators. 3) Needs to further address the risk of human bias by using artificial intelligence techniques.	2018
(Li, et al., 2022)	Creation	Information extraction, knowledge modelling	Propose a dynamic ontology-based mapping (DOM) approach to automatically generate semantic-enriched work packages.	1) Increase the accuracy of the dynamic work packing process. 2) Improve the efficiency of dynamic work packing with reduced time.	1) Only implement work packaging in the product-oriented MC project.	2022
(Koo, et al., 2007)	Creation	Knowledge modelling or reasoning and conformance check	Propose a way of improving the CPM framework's limitation of representing sequencing rationale using precedence relationships and distinguishing activities only concerning their time-criticality by formalising a constraint ontology and classification mechanism.	1) Provide a simple and consistent way for planners to describe their rationale for activity sequences. 2) Allow planners to create constraint-loaded schedules, thereby reducing misinterpretation of schedule logic and obviating the need to keep track of sequence logic individually. 3) Develop a generic reasoning process that leverages the project-independent representation to identify project-specific instances of component similarity in a given 3D product model. 4) Alleviate planners from having to manually trace network chains and infer the role and status of activities in their minds.	1) Limit by the number of case usage (all three schedules had fewer than 30 activities). 2) Need to use a more efficient transitive algorithm to increase the run-time of the network chain search algorithm.	2007

(*Continued*)

Table 9.1 (Continued)

Reference	*Domain of Study*	*Main Application Function(s)*	*Research Aim*	*Main Benefits/Contributions*	*Main Limitations/ Challenges*	*Year*
(Guévremont & Hammad, 2021)	Integration	Knowledge modelling	Propose a new method for using 4D simulation for visual analytics of delay claims based on an integrated ontology (called Claim4D-Onto).	1) Clarify the causality within the time and space dimensions illustrated in the 4D simulation. 2) Analyse delay responsibilities and entitlements. 3) Enhance the analysis of claims by using the 4D simulation considering the time dimension.	1) Unable to visualise and analyse the specifics of delay claims about effects and causes and assigning responsibility. 2) Lack of the ontologies integration in a more formal and overarching ontology-based approach to grasp essential concepts (e.g., liability, causality, and quantum) in a delay claim.	2021
(Zheng, et al., 2021)	Integration	Knowledge modelling	Propose an ontology set for formalising and integrating construction workflow (CW) information within the digital construction context.	1) Provide a comprehensive vocabulary to represent CW knowledge and information based on domain knowledge. 2) Fill the gaps of existing ontologies that lack information-related entities to support the integration of heterogeneous and multi-context information. 3) Enable further applications based on linked data and Semantic Web technologies for information retrieval. 4) Provide an unambiguous information model to promote construction management systems.	1) Capture only the essential higher-level concepts and properties to represent CWs within the digitalised construction context. 2) Do not build up a taxonomy that would cover the detailed classification hierarchies of the domain entities. 3) Ontology based on the expert workshop and criteria-based evaluation approaches are subjective and difficult to quantify and assess. 4) Lacks more digital data sources to cover all content required by the design ontology.	2021

(*Continued*)

Table 9.1 (Continued)

Reference	*Domain of Study*	*Main Application Function(s)*	*Research Aim*	*Main Benefits/Contributions*	*Main Limitations/ Challenges*	*Year*
(Viljamaa & Peltomaa, 2014)	Integration	Information extraction	Propose to improve process management through more effective information integration, processing, and exploitation, leading to an intensified infrastructure-building process with more effective process control and reaction to process status changes.	1) Provide formalised and easily combined data formats, ensuring better interoperability and enabling automatic inferring and decision-making based on the collected data. 2) Provide a scalable and general-purpose method.	1) Embed problem when adding all needed parties to the system in practice due to scattered infrastructure building process. 2) Require high-technology orientation from the using party with the skilled personnel before the routines to use the method evolve. 3) Do not allow hard real-time utilisation due to heavy data formats and system structure. 4) Require a heavy standardisation process for the development of the real shared ontology. 5) Lack of clear earning principles.	2014

Khosakitchalert, et al., 2019). For instance, Alzralee developed a system for the cost estimation of building project (Alzraiee, 2022); Akanbi and Zhang automated the manual processes of extracting additional design information from construction specifications to generate a cost estimation (Akanbi & Zhang, 2021); Khosakitchalert et al. proposed a method to eliminate the excess quantities and add missing quantities caused by incomplete or incorrect BIM models (Khosakitchalert, et al., 2019). The major goal of these studies was to ensure software interoperability and information consistency when using BIM tools to provide an automated, accurate, and efficient quantity take-off and cost estimation. It was suggested that only a few studies looked into natural language processing (NLP)-based methods to reduce the manual effort needed (Akanbi & Zhang, 2021). For other information extraction studies, Ren et al. aligned the goal of building information modelling (BIM) as the necessary information support with the ontology for depicting the relationships and providing a way of handling a large amount of information flow efficiently (Ren, et al., 2021). In comparison, Ma and Liu developed a BIM-R platform based on ontology and freeware to provide usability for both researchers and developers (Ma & Liu, 2018). Next, for knowledge modelling, there are also some studies focused on cost estimation (Staub-French, et al., 2003; Abanda, et al., 2017). Abanda et al., for instance, building an ontological model based on the New Rules of Measurement (NRM) for tendering stage cost estimation, make the information reusable and sharable without the loss of generality (Abanda, et al., 2017). Staub-French et al. proposed an ontology to calculate construction costs by considering the relating building product model features to construction resources and activities with estimators' rationale (Staub-French, et al., 2003). Although it enabled estimators to have project-specific rationale independently, it also indicated that different estimators might have different preferences for cost impacts' different features. Other researchers, like Yildiz et al., looked into the development of a knowledge-based risk mapping tool for assessing the risk-related variables that lead to cost overrun (Yildiz, et al., 2014). The major benefit of this tool was that it could be utilised for the different project stages (e.g., preparation, contracting, construction, etc.) and has the potential to develop as an organisational risk memory that shares risk-related knowledge among organisation members. Finally, for ontology reasoning and conformance checks, there were also a substantial number of studies focused on code-compliant related quantity take-off and cost estimation (Soman, et al., 2020; Akanbi, et al., 2020; Liu, et al., 2022). Soman et al. constructed a BIM-based semi-automated reasoning system to enable cost estimation of projects based on different design scenarios (Soman, et al., 2020). This approach reduced the time of cost estimation and enabled an agile and accurate tracing of changes in the project cost. Liu et al. proposed a new knowledge-based framework that takes the standard method of measurement (SMM) rules and semantic information in BIM for automatic code-compliance checking (Liu, et al., 2022). Other researchers like Tacakolan et al. built an intelligent information collection and decision support system for short-term collaborative construction and resource planning through semantic reasoning techniques and ontological knowledge representation (Tavakolan, et al., 2021). This approach allowed the imposing of complex multi-variable constraints in an effective manner and a visualisation tool. In contrast, Staub-French and Nepal developed a feature-based framework for representing and reasoning the component similarity with ontological modelling (Staub-French & Nepal, 2007).

For ontology creation, the existing studies have been focused on information extraction and knowledge modelling. There are five information extraction studies and four knowledge modelling studies, in which three of them overlapped with each other. For both information extraction and knowledge modelling functions, although most of the studies still examined the ontology's effectiveness on cost estimation, Ma et al. extended the specification ontology of national specifications to represent regional quota specifications (Ma, et al., 2016); Niknam and Karshenas used an ontology-based cost-estimating application that integrated information provided by different stakeholders (Niknam & Karshenas, 2015); Lee et al. developed an ontological inference to enable the automated search of the most suitable work items (Lee, et al., 2015); Kim et al. looked into the analysis and identifying the quantity discrepancy (QD) for the quantity extraction from different building interior components (Kim, et al., 2019); some other researchers have turned their interests to other areas. Caneparo, for instance, came up with a methodology to differentiate the knowledge about objects, spatial relationships, and constraints from a generative process (Caneparo, 2022).

Although incorporating ontology into construction resource planning provided many benefits, many limitations and challenges still require further attention. The first limitation was that the existing studies were overly focused on cost estimation and quantity take-off, which is reasonable as an accurate ontology-based cost estimation can bring substantial benefits for construction resource planning, especially during the tendering stages (Abanda, et al., 2017). Nonetheless, the problem was it is further correlated with the second limitation, which was that most of the cost estimation and quantity take-off-related studies are gaining data mainly from BIM-related data sources. This limited the interlink between the different project knowledge and raised the need for more exploration of integrating different data sources (Tavakolan, et al., 2021). Third, although there were a few studies that looked into the NLP-based method (Akanbi & Zhang, 2021) to reduce the manual effort, especially during the information extraction, many existing studies' important ontological model-building processes (e.g., extracting of named entities) were still not yet automated with the NLP and machine learning (ML) techniques. Fourth, it was suggested that more work is required to be carried out to ensure the ontology can be reused and shared for other intelligent applications without the loss of generality (Staub – french, et al., 2003). Again, more detailed information regarding applying ontological models in construction resource planning is summarised in Table 9.2.

9.5 Case study: dynamic scheduling for crane operations

Following a detailed examination of ontological models and their relevance in construction scheduling and resource planning, this chapter further presents a case study demonstrating the creation and application of a function-specific ontological model. This model specifically aids in the dynamic scheduling for crane operations, ensuring the crane's availability aligns with construction progress.

To do this, the model integrates information from various sensors, analyses the crane's transient behaviours (e.g., loading and unloading), and correlates as-is crane operations with scheduled lifts. By performing information extraction and integration functions, this ontological model can automatically identify delays in executing scheduled crane operations and inform site managers to update outdated schedules accordingly. Details of this case study can be found in the following sections, exemplifying how ontologies could help construction scheduling and resource planning, particularly in the allocation of critical site infrastructure.

9.5.1 Crane operations: significance in operational efficiency and advancements in planning

The indispensable role of cranes in modern construction is accentuated by the bottleneck scenarios often arising from limited crane access (Hu & Fang, 2023; Soman, et al., 2020). Traditional crane resource planning necessitates site managers to meticulously track completed lifts and arrange real-time lift requests from subcontractors, which becomes increasingly cumbersome with growing lift requests. For efficient management of cranes, several lift scheduling applications like HammerTech (Bookings – Plan Ahead and Stay Ahead, 2023), DataScope (Crane Management Software System, 2023), Veyor (Collaborative Digital Scheduling, 2023), and CraneTime (Cranetime, 2023)

Table 9.2 The result summary of literature review for construction resource planning ontological model studies

Reference	*Domain of Study*	*Main Application Function(s)*	*Research aim*	*Main Benefits/Contributions*	*Main Limitations/Challenges*	*Year*
(Alzraiee, 2022)	Application	Information extraction, knowledge modelling	Propose a system to develop a cost estimate for building projects using a structured query language (SQL).	1) Develop a cost estimate using BIM tools to improve software interoperability. 2) Enable the attachment of work items and subjective factors to BIM elements for an accurate cost estimate. 3) Implement the "switchback" feature that allows the estimator to author BIM in Revit and complete the estimate in Navisworks.	1) Lack of ability to access BIM data for a BIM-based cost estimation system.	2023
(Liu, et al., 2022)	Application	Reasoning and conformance check	Propose a new knowledge model-based framework that incorporates the semantic information and standard method of measurement (SMM) rules in BIM for automatic code-compliant QTO.	1) Identify the semantic errors in BIM models. 2) Obtains code-compliant quantities. 3) Improves the interoperability and open digital workflow for automatic quantity take-off (QTO) and/or cost estimation in construction.	1) Extract information (e.g., named entities) manually based on natural language process (NLP) and machine learning (ML) techniques. 2) Ignore large amounts of semantic information – unmodeled items (e.g., finishing).	2022
(Akanbi & Zhang, 2021)	Application	Information extraction	Propose a method to automate the manual processes of extracting additional design information from construction specifications required to generate a cost estimate.	1) Use the NLP-based method to reduce the amount of manual effort needed. 2) Improve the efficiency and objectivity in specifications reading and processing. 3) Enable the use of extracted design information to match and retrieve material unit prices from a database. 4) Improve the variability and ambiguity issues in item descriptions.	1) Test the developed information and extraction and matching (IEM) algorithm only on the design information instances for wood elements and the construction specifications that followed CSI MasterFormat.	2021

(Continued)

Table 9.2 (Continued)

Reference	*Domain of Study*	*Main Application Function(s)*	*Research aim*	*Main Benefits/Contributions*	*Main Limitations/Challenges*	*Year*
(Tavakolan, et al., 2021)	Application	Reasoning and conformance check	Propose an intelligent information collection and decision support system framework for short-term collaborative construction and resource planning through implementing ontological knowledge representation and semantic reasoning techniques.	1) Enable imposing complex multi-variable constraints simply and effectively. 2) Help facilitate collaboration in short-term planning meetings. 3) Collect planning process information from different sources. 4) Provide a visualisation tool for project planners to enable better decision-making.	1) Suffer spatial limitations and site congestion in project planning. 2) Develop only based on a specific industry section. 3) Do not consider long-term planning. 4) Lack of integration of more project information sources. 5) Store project information on the spreadsheet. 6) Lack of capability to acquire real-time data.	2021
(Ren, et al., 2021)	Application	Information extraction; reasoning and conformance check	Propose a comprehensive approach that aligns with the goals of building information modelling (BIM) as the necessary information support and ontology for the knowledge process.	1) Allow efficient writing of the rules and ontology used to depict relationships and vocabulary. 2) Provide a way of handling a large amount of information flow due to the increased information flow in the semantic and syntactical levels based on IFC to OWL/RDF conversion. 3) Improve people's trust in the computer system. 4) Improve the overall decision-making e fficiency.	1) Are prone to vast variations, and not all the information can be extracted from the BIM and project documents, as multiple sources of heterogeneous data exist. 2) Extract information from a limited range of PPP project models. 3) Lies in the automation process from the existing standardised environment, which is a BIM-based system. 4) Lacks aligning established knowledge development with various data sources to enrich their practical application.	2021

(*Continued*)

Table 9.2 (Continued)

Reference	*Domain of Study*	*Main Application Function(s)*	*Research aim*	*Main Benefits/Contributions*	*Main Limitations/Challenges*	*Year*
(Soman, et al., 2020)	Application	Reasoning and conformance check	Propose a semi-automated BIM-based cost estimation approach that enables practitioners to estimate the cost of projects based on different design scenarios by an accurate and agile system.	1) Enables an agile and accurate BIM-based cost estimation of different scenarios during the design process. 2) Reduces the time of cost estimation in comparison to the traditional manual approaches, especially for complex structures. 3) Helps trace changes in project cost according to any changes in components and materials of the BIM model. 4) Increases cost estimation accuracy while decreasing the probability of human errors.	1) Cannot consider some "ExtraWork" cost items of Fehrest-Baha as a limitation, which influenced its estimation.	2020
(Akanbi, et al., 2020)	Application	Reasoning and conformance check	Propose a new data-driven reverse engineering algorithm development (D-READ) method for developing QTO algorithms based on IFC geometric analysis.	1) Develop a new data-driven method for automated QTO algorithm development using IFC-based BIMs. 2) Can be accumulatively grown into a comprehensive set to cover different objects in different types of construction projects. 3) Extend the current available architectural MVD specifications to one that checks IFC instance files for architectural QTO purposes and leverages it in the new D-READ method.	1) Produce QTO algorithms that can only address geometric representations. 2) Test only on explicitly modelled elements (e.g., walls, slabs) but not the unmodeled elements (e.g., scaffolding).	2020
(Khosakitchalert, et al., 2019)	Application	Information extraction	Propose a method to improve the accuracy of the extracted quantities of compound elements from BIM models that are incomplete or incorrect by using information from BIM-based clash detection to eliminate excess quantities and add missing quantities.	1) Help reduce modelling or editing time. 2) Improve the performance of BIM-based quantity take-off and cost estimation. 3) Eliminate the excess quantities of compound elements that overlap with other elements. 4) Calculate the quantities of each layer of a compound element that has a distinct dimension without editing the BIM model	1) Unable to classify the not well-defined wall or floor types into different groups. 2) Can cover the entire exterior surfaces of a building with only one material due to the calculation method for the exterior finish layer. 3) Cannot separate the material types if multiple wall materials exist in a room.	2019

(*Continued*)

Table 9.2 (Continued)

Reference	*Domain of Study*	*Main Application Function(s)*	*Research aim*	*Main Benefits/Contributions*	*Main Limitations/Challenges*	*Year*
					4) Subject to inaccurate quantity take-off results after transfer of BIM models to other software products.	
(Abanda, et al., 2017)	Application	Knowledge modelling	Propose a way of an ontology based on new rules of measurement (NRM) development for cost estimation during the tendering stages.	1) Facilitate information extraction from BIM models by combining ontologies with a 3D BIM model 2) Develop BIM-based ontology using NRM meaning it can be reused, shared, and used for other intelligent applications without loss of generality.	1) Unable to capture names of concepts as they appear exactly in NRM. 2) Unable to edit names of concepts using Protégé-OWL if they are too long. 3) Unable to avoid the repetition of some concepts in different hierarchies.	2017
(Yildiz, et al., 2014)	Application	Knowledge modelling	Propose a knowledge-based risk mapping tool for systematically assessing risk-related variables that may lead to cost overrun in international markets.	1) Provide an effective way to visualise risk-related variables and risk paths that may emerge in international markets that improve systematic risk identification and classification. 2) Assess the magnitudes of vulnerability sources by decision-makers and the tool automatically quantifies the magnitudes of all other variables. 3) Can utilise the tool in different project stages: bid evaluation and preparation, contracting, construction, post-project commission, and evaluation. 4) Can further be used to develop an organisational risk memory in which risk-related knowledge of previous projects is stored, documented, and shared among organisation members.	1) Cannot prevent an actual cost overrun of 7% in practice. 2) Construct the tool based on data gathered from the Turkish contractor, therefore may not be applicable and reliable for other contractors. 3) Suffer performance fluctuation due to the knowledge sources of the company and the quality of the lessons learned database. 4) Require further improvement from company feedback and feeding of more knowledge sources.	2014

(*Continued*)

Table 9.2 (Continued)

Reference	*Domain of Study*	*Main Application Function(s)*	*Research aim*	*Main Benefits/Contributions*	*Main Limitations/Challenges*	*Year*
					5) Cannot prevent inconsistencies between ratings of different decision-makers. 6) Do not have the capability to consider "risk perceptions" or "fuzziness" or explore risks as "feelings" and "politics".	
(Staub-French & Nepal, 2007)	Application	Reasoning and conformance check	Propose a feature-based framework for representing and reasoning about component similarity that builds on ontological modelling, model-based reasoning, and cluster analysis techniques	1) Formalise an ontology that represents the building component, the component attributes, the direction, the range of acceptable variation for geometric attributes, and the degree of variation required to assess component similarity. 2) Enable practitioners to represent their varied preferences for defining component similarity generically and consistently. 3) Develop a generic reasoning process that leverages the project-independent representation to identify project-specific instances of component similarity in a given 3D product model. 4) Provide a general framework that enables a more complete, consistent, and efficient cost-estimating process.	1) Require account for the subtle similarities between components that are not accounted for by considering the attribute values. 2) Rely solely on the user to identify the relevant properties and degrees of variation that could be improved by allowing some combination of user-driven and system-driven identification of the relevant information. 3) Require further extension to provide feedback to designers so that they can better understand how practitioners view component similarity and optimise the degree of similarity in their designs.	2007
(Staub-French, et al., 2003)	Application	Knowledge modelling or reasoning and conformance check		1) Provide estimators with a structured way to relate building product model features to construction activities and associated resources to represent estimators' rationale in the computer. 2) Enable estimators to represent their rationale independent of a specific project or product model.	1) Focus only on features that result from components, intersections of components, and the similarity of components. 2) Exclude other types of features and certain parameters for defining component similarity.	2003

(Continued)

Table 9.2 (Continued)

Reference	*Domain of Study*	*Main Application Function(s)*	*Research aim*	*Main Benefits/Contributions*	*Main Limitations/Challenges*	*Year*
			Propose an ontology formalised to represent estimators' rationale for relating building product model features to construction activities and associated construction resources to calculate construction costs.	3) Develop a generic reasoning process that leverages the project-independent representation to identify project-specific instances of component similarity in a given 3D product model. 4) Demonstrate that estimators have different preferences for accounting for the cost impacts of different features.	3) Exclude factors exogenous to product design that affect construction costs, such as site characteristics, environmental conditions, and resource skill and availability.	
(Ma & Liu, 2018)	Application	Information extraction, Knowledge modelling	Propose to use a platform to rapidly develop BIM-R applications (BIM-R platform) based on ontology and freeware components.	1) Establish the functional requirements of the BIM-R platform. 2) Provide a mechanism to transform BIM data into ontology data. 3) Provide the architecture of the BIM-R platform. 4) Help reduce the cost and improve the efficiency of the development of BIM-R applications. 5) Provide a platform that can be used by both researchers and developers.	1) Needs many configuration operations, as development is based on IfcOpenShell. 2) Face high time and labour costs for processing and reasoning BIM data.	2018
(Lee, et al., 2014)	Application	Information extraction	Propose an ontological inference process to automate the process of searching for the most appropriate work items, which is limited to tiling in this case study.	1) Help engineers find work items more easily and consistently, given the BIM data. 2) Provide a methodology to fully automate the estimation of cost. 3) Assist cost estimators to use BIM data more easily and help the expansion of BIM-based construction management.	1) Limit the application scope of work condition ontology to tiling work due to work item ontology and semantic rules. 2) Require continually revising and updating work item ontology due to construction type and technological advancements. 3) Require further development of an IFC data parser and an XML-to-RDF transformation engine to fully automate the inference process.	2014

(*Continued*)

Table 9.2 (Continued)

Reference	*Domain of Study*	*Main Application Function(s)*	*Research aim*	*Main Benefits/Contributions*	*Main Limitations/Challenges*	*Year*
(Kim, et al., 2019)	Creation	Information extraction	Propose a way of identifying and analysing the quantity discrepancy (QD) in quantities extracted from different building interior components to improve the accuracy of cost estimation using BIM.	1) Suggests modelling and QT strategies to minimise the discrepancy in the quantity of materials caused by inconsistencies in the model or components. 2) Help identify work types and materials that might be problematic in estimations using BIM.	1) Requires manual take-off to be applied to calculate the quantities of any materials, as quantity estimates were not extracted from the CMO model. 2) Quantities of some materials are overestimated while others are underestimated. 3) Does not necessarily generate greater quantities.	2019
(Ma, et al., 2016)	Creation	Information extraction, Knowledge modelling	Propose to establish an effective approach to formalise such specifications by using ontology.	1) Generate SWRL rules automatically without letting estimators learn the syntax. 2) Extend the specification ontology of national specifications to represent regional quota specifications or construction-firm-specified quotas.	1) Cannot guarantee the consistency of the rule library, as rules are extracted manually and extended by different estimators. 2) Can only use the developed ontology for cost estimation. 3) Need to further identify the missing information from a given BIM model.	2016
(Niknam & Karshenas, 2015)	Creation	Information extraction, knowledge modelling	Propose an estimating application that integrates distributed information provided by project designers, contractors, and material suppliers for preparing cost estimates.	1) Facilitate finding, accessing, and combining the information necessary for construction cost estimating. 2) Makes the mapping machine processable and eliminates the need for estimator involvement in the element property mapping process. 3) Improve estimators' efficiency by saving time for mapping. 4) Can submit material resource specifications to suppliers' Semantic Web Services and retrieve material unit cost data for updating material cost databases.	1) Require several knowledge bases. 2) Require standard ontologies to allow knowledge sharing and create estimating knowledge bases that represent estimating assemblies, work items, and construction resources.	2015

(*Continued*)

Table 9.2 (Continued)

Reference	*Domain of Study*	*Main Application Function(s)*	*Research aim*	*Main Benefits/Contributions*	*Main Limitations/Challenges*	*Year*
(Lee, et al., 2015)	Creation	Information extraction	Propose an ontological inference of work items that enables an automated search of the most appropriate work items and their associated unit costs.	1) Eliminate subjective decision-making via the search for appropriate work items for cost estimation. 2) Use BIM data extracted from IFCXML. 3) Help automate the whole estimation process.	1) Require further ontologies development that consider all components of work and project environments to capture actual work conditions. 2) Require further expanded ontologies of each specific work item to consider all work items.	2015
(Staub – french, et al., 2003)	Creation	Knowledge modelling	Propose the vocabulary formalised to represent the different types of features of building product models that are important to cost estimators of building construction.	1) Provide a general way to describe design conditions using features and represent features independent of a particular project or product model. 2) Allow estimators to customise features according to their preferences. 3) Allow estimators to specify the specific features and properties that affect a component's construction cost.	1) Do not represent all the types of features that affect construction costs. 2) Unable to represent the similarity of location with the feature ontology. 3) Require additional testing to validate the generality of the feature ontology.	2003
(Caneparo, 2022)	Creation	Information extraction, knowledge modelling	Propose a methodology to separate the knowledge about objects, spatial relationships, and constraints from the generative process.	1) Implements multidimensional visual interactive tools to assist designers, planners, and decision-makers in exploring the layouts and the criteria. 2) Increase decision-maker confidence in identifying good and effective solutions. 3) Improve ontology and knowledge reuse and sharing. 4) Assist the early stage planning and design.	1) Conduct the reuse of ontologies and the integration of the existing knowledge manually. 2) Write formal and explicit conceptualisation ontologies in an expert-intensive manner, which is time-consuming and inefficient. 3) Need to amend missing objectives to the generative process. 4) Provide multiple layouts and solutions, which are unsatisfied by designers.	2022

have emerged. These applications streamline crane booking, lift operation sequencing, and asset management, incorporating features such as drag-and-drop lift scheduling and crane maintenance tracking. Despite these advancements, a notable gap remains as they fall short in providing comprehensive schedule fulfilment insights, resulting in outdated crane availability data and unreported idle times. Modern cranes are becoming increasingly intelligent, equipped with integrated sensors to capture data on pose, strains, and imagery, facilitating automated control, mechanical failure alerts, and hazard identification. However, it reveals an unbalanced development, with an emphasis on safety over productivity. For instance, advancements in crane resource optimisation are scarce due to the difficulty in understanding crane behaviour. This challenge can be addressed by adopting semantic web techniques (i.e., ontology) to organise and standardise sensor data and initialise automated and accurate interpretation of lift schedule fulfilment.

9.5.2 Dynamic scheduling: concept and application in crane operations

Dynamic scheduling is a contemporary paradigm in construction management that has burgeoned in popularity due to the proliferation of real-time monitoring systems, the Internet-of-Things (IoTs), and digitised construction management software (Kim, et al., 2021). At its core, dynamic scheduling seeks to transform traditional static scheduling practices into an iterative and responsive process (Fahmy, et al., 2020). Stemming from the manufacturing industry, dynamic scheduling is not a novel concept. Yet, its application in the construction sector is still in its nascent stages. This is primarily due to the unique challenges and variables that construction projects present. While manufacturing settings often have consistent and predictable processes, construction sites are dynamic and involve numerous external factors such as weather conditions, site-specific challenges, and multi-tiered coordination between various stakeholders. Fortunately, as construction projects increasingly adopt prefabrication techniques, the boundary between construction and manufacturing begins to blur (Luo, et al., 2020). For example, crane operations in construction involving prefabricated components bear striking similarities to workstations in a production line: in both scenarios, there are standardised, repeatable operations that assemble parts of a product. This perspective invites the possibility of transferring established scheduling and resource planning techniques from the controlled environment of manufacturing to the dynamic world of construction.

Nevertheless, the unique nature of construction projects still restricts the capacity to formalise and standardise machinery operation tracking, casting doubt on the possibility of devising a robust dynamic scheduling application. Consider, for instance, the fluctuating circumstances surrounding prefabricated building components: these components are often wrapped in protective material, the details of which are typically not updated within information systems. The type, colour, and pattern of these protective wrappings can vary considerably, not just between construction projects but also between individual components within the same project. These variations could disrupt some image-based technologies, such as object recognition using computer vision (CV), leading to data gaps and the potential misapplication of previously gathered knowledge (e.g., invalid object recognition features).

In addition, inherent challenges that beset sensor technologies, such as inaccuracies, harsh environmental conditions, physical occlusions, and unstable internet connections, also exacerbate the difficulties of ensuring robustness in automated progress tracking methodologies (Han & Golparvar-Fard, 2015; Dave, et al., 2016). These obstacles have the potential to further undermine the quality and reliability of collected data and hinder successful automated progress tracking.

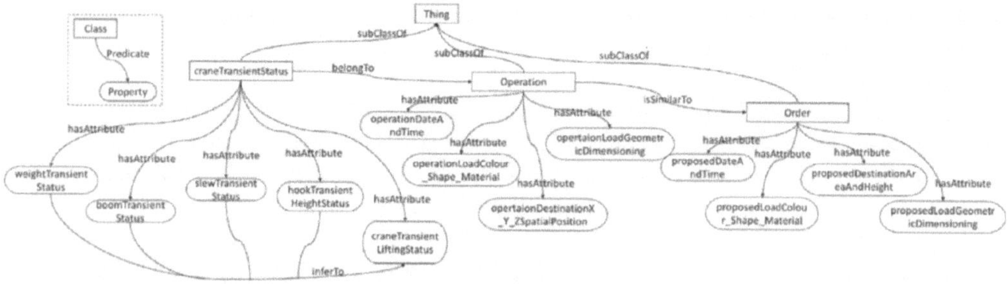

Figure 9.1 The ontology of crane operation monitoring and scheduling (Hu, et al., 2023)

Drawing from prior literature reviews, ontological models are crafted to encapsulate explicit knowledge while also integrating articulated implicit insights. This feature makes them adept at tackling challenges arising from incomplete data and allows for adaptability to project-specific constraints, like the ones mentioned earlier.

9.5.3 *Advancing crane operation management: an ontological model for dynamic scheduling*

To realise the full potential of dynamic scheduling, it is crucial to bridge existing knowledge gaps and devise an ontological model that is flexible enough to handle project-specific factors and robust enough to reason with incomplete data. In this regard, Hu et al. have proposed a pioneering method, the architecture of which consists of two main components: a data acquisition module and a data integration module (Hu, et al., 2023).

The data acquisition module is designed to monitor pressures on crane hooks, identify on-site objects, and track crane poses, employing a trio of sensors: weight sensors, cameras, and inertia measurement unit (IMU) sensors. Initially, each sensor's raw data undergoes individual pre-processing. For instance, weight data is linearly interpolated to estimate missing values, while images captured by cameras are processed using a convolutional neural network (CNN)-based object-detection model. Furthermore, the IMU data is filtered using a Savitzky-Golay filter to eliminate noise. After pre-processing, all sensor data is synchronised and down-sampled (i.e., to 1 Hz) using boxcar averaging.

The data integration module applies ontological models to track crane operation progress, utilising data from IoT devices and lift scheduling systems. This model consists of three key classes: transient crane behaviour, lift operations, and lifting orders (Figure 9.1). By analysing pre-processed data from the three IoT devices, transient crane behaviours can be determined. The Apache Jena framework was employed to construct the functional ontological model supplemented by declarative inference rules. Apache Jena, a distinguished open-source Java framework, is tailored for the development of Semantic Web applications. Within its suite of tools, the Jena generic rule and Reasoner were used to codify inference rules. The rule syntax provided by the Jena generic rule bears similarities to that of SWRL in facilitating ontology augmentation. However, Jena's framework presents enhanced capability and flexibility for articulating intricate and customised logical constructs.

As an example, a Jena generic rule (rule 1) is shown here, clearly expressing the process of using domain knowledge to interpret a situation where a crane is moving without loads in an instant. That is, the crane is in the no-load moving condition when the total weight of the hook is constant and the total weight of the hook is less than or equal to the weight of the empty hook, and the boom is moving during that period.

[rule1: (?x fa:weightTransientStatus?y) equal(?y,"WeightSteady") (?x fa:startTimePointComparedWithEmptyWeight ?z) equal(?z,"equalOrSmall") (?x fa:endTimePointComparedWithEmptyWeight ?a) equal(?a,"equalOrSmall") (?x fa:boomTransientStatus ?b) notEqual(?b,"boomSteady")->(?x fa:craneTransientLiftingStatus "movingWithNoLoad")]

These behaviours are then segmented to identify lift operations, adhering to established rules. For example, transient behaviours have to appear in a unique pattern (i.e., loading, moving with loads, and unloading) to make an operation. The data integration module also organises lift scheduling system information into the ontological model via a user interface (UI) designed to gather lift schedule data. Subsequently, the data integration module correlates actual lift operations with lift orders, applying the Jaccard similarity measure to calculate attributes' value similarities between instances of the two concepts. To be specific, the generalised Jaccard similarity algorithm is used to infer the correlations between the lifting orders and the actual lifting activities. That is, as shown later, the number of attributes considered to be the same in the buffers interval (the intersection of A and B) is divided by the total number of attributes (the concatenation of A and B). This process considers nine attributes, including load weight, colour, material, shape, length, width, and unloading destination coordinates. Combined with buffers that accommodate uncertainties, the method can effectively match actual crane operations with as-planned lifts.

$$J(A,B)=\frac{|A\cap B|}{|A\cup B|}$$

A scaled model of such a structure, along with a mobile crane and simulated construction site, forms a lab testbed. An Intel RealSense D435i camera, a Yost Labs 3-Space™ sensor, and an

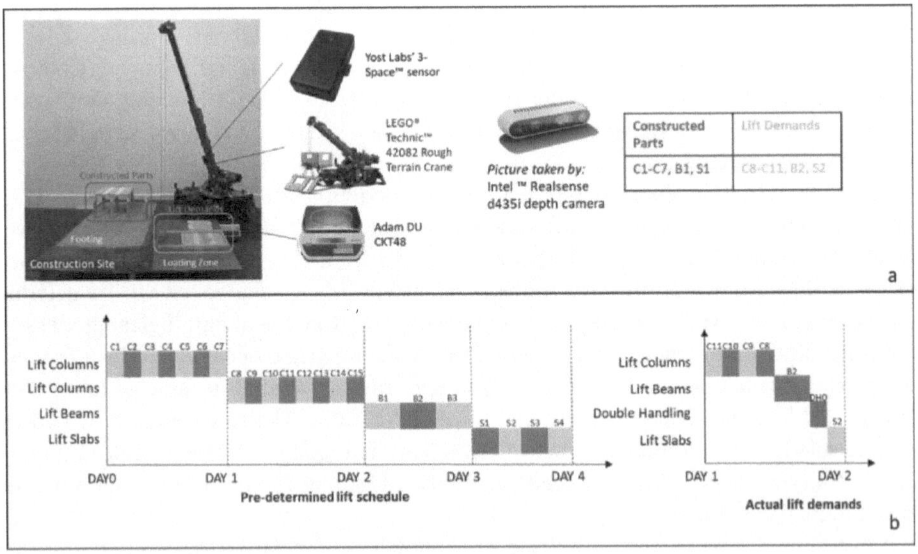

Figure 9.2 The scaled model of mass timber structure (a) and the pre-determined/actual lift schedules (b), adapted from (Hu, et al., 2023)

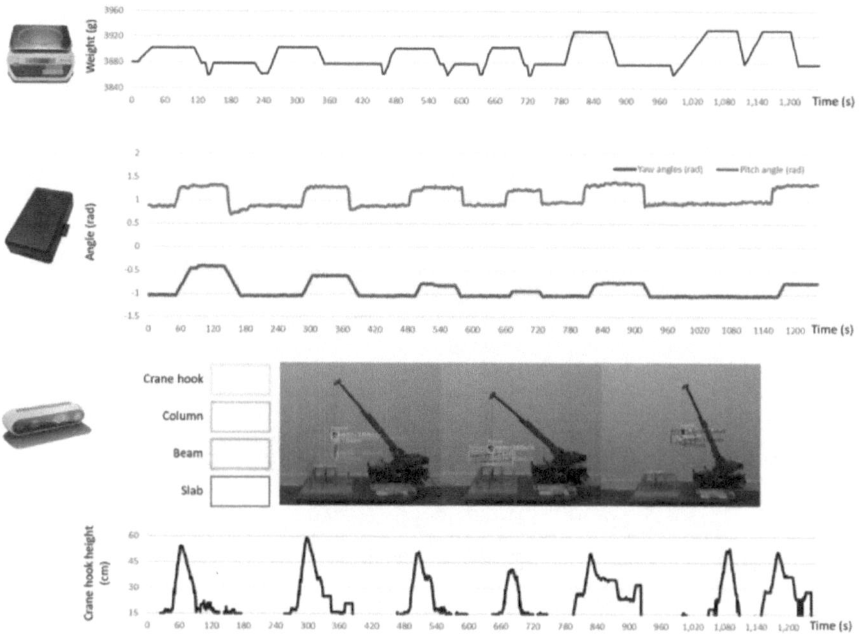

Figure 9.3 Pre-processed sensor data

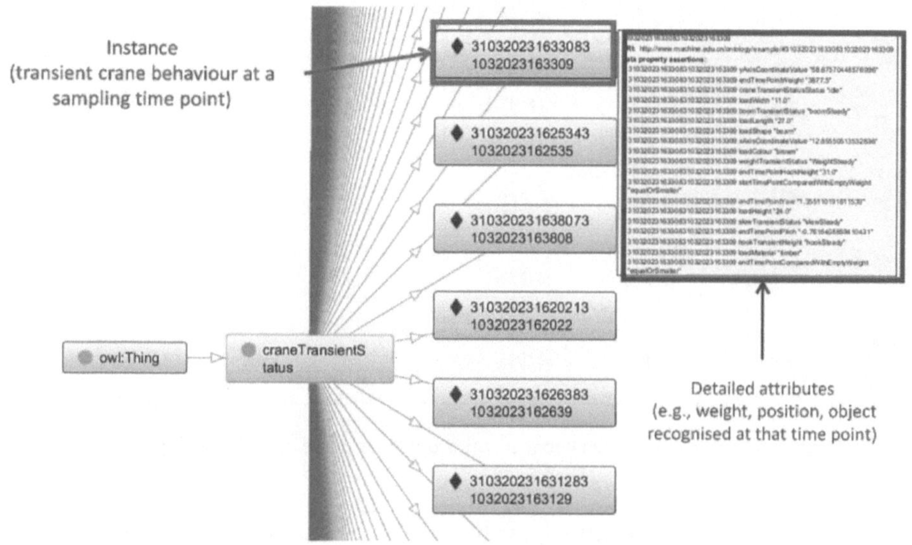

Figure 9.4 Transient crane behaviours

Adam DU CKT48 digital scale were used to capture 3D images, track crane rotation, and measure the crane-load system weight respectively (Figure 9.2a). Pre-programmed lift orders, following a level-by-level installation sequence, were implemented by the crane operator, with an additional operation to simulate double handling (Figure 9.2b). The operator, provided with full discretion

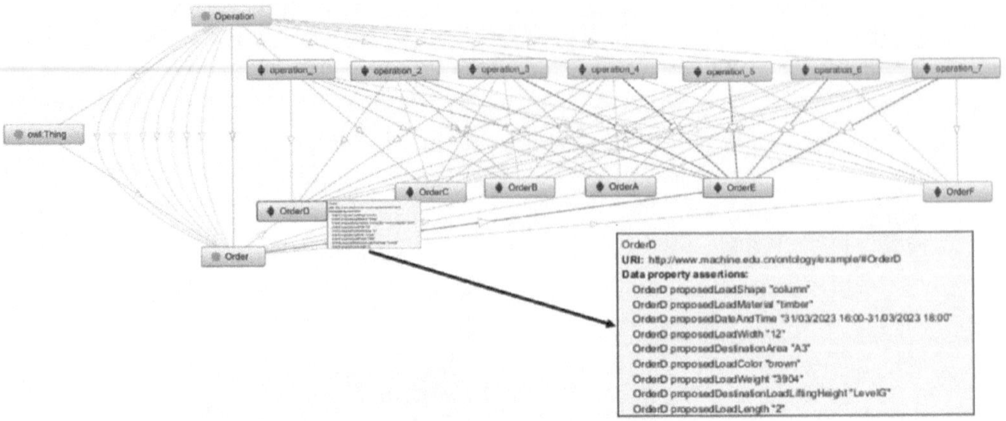

Figure 9.5 Information on the crane operations, lift orders and their relationships

	Op. 1 (C11)	Op. 2 (C10)	Op. 3 (C9)	Op. 4 (C8)	Op. 5 (B2)	Op. 6 (DH0)	Op. 7 (S2)
Order A (C8)	78%	78%	78%	89%	44%	0%	44%
Order B (C9)	78%	78%	89%	78%	44%	0%	44%
Order C (C10)	78%	89%	78%	78%	56%	0%	56%
Order D (C11)	89%	78%	78%	78%	44%	0%	44%
Order E (B2)	33%	33%	44%	22%	100%	11%	78%
Order F (S2)	33%	33%	44%	22%	78%	11%	100%

Figure 9.6 Attributes' value similarities between lift orders and crane operations (highest similarity in red)

in selecting the lift path, transported the load from a fixed location to designated points on the concrete slab footing. The load was then carefully detached by a second participant to maintain accurate scale readings. Throughout the process, which involved installing six components in four minutes, sensors uninterruptedly gathered data without human intervention.

The outcomes of the data pre-processing, depicted in Figure 9.3, highlight the changes in weight, crane rotational movements, object recognition, and crane hook height. These processed data are then transformed into transient crane behaviours via the rules in the ontological model (Figure 9.4). The proposed method segmented this behaviour into seven individual lift operations, accurately correlating six of these with corresponding lift orders, while correctly dismissing double handling activity. The relationships among lift orders and crane operations are shown in Figure 9.5.

SPARQL queries were programmed to extract detailed information about lift operations and corresponding lift orders. According to Figure 9.6, despite some mapping criteria not being met, the method was able to map lift orders to corresponding operations due to the range of attributes' value similarities identified, from 0% to 88%. The proposed method's resilience against errors or missing data from IoT devices was highlighted by its 100% accuracy in the lab test. However, in case of potential errors, querying the data can help comprehend the underlying causes and adjust the rules or sensors accordingly, enhancing the interpretability of the results.

Additionally, the proposed method was accurately able to record crane behaviour. A benchmark was established by observing video footage and manually noting the start and end times of each transient crane behaviour for a specific lift operation. The results of the experiment were then compared to the benchmark, revealing an average error of 0.32 seconds in measuring the start and end times of each crane behaviour (e.g., loading) during the 242.7-second operation.

9.6 Discussion

The case study exemplifies the capacity of ontological models in information extraction, especially in standardising information retrieval from diverse sensors and systems (e.g., lift scheduling software) and amalgamating such data to interpret the planning and utilisation of crucial resources, including crane availability. It showcases the potential of transitioning from previous static resource planning practices to a responsive process that captures real-time data for calibration, a gap identified in the earlier review. Transitioning the test bed from the laboratory setting to a real construction site is expected to further underscore the merits of this method.

Additionally, this case study illustrates how the perspectives of experts or the designers of ontological models significantly impact the data structure. Notably, the weighting of different features used to map as-is crane operations and scheduled lifts is subject to human bias, leading to inaccuracy and unreliability in data interpretation. This aligns with literature findings indicating that many ontological models developed from human knowledge tend to over-rely on expert input, overlooking project-specific traits. For example, as noted, variable and project-specific features could limit experts' ability to transfer knowledge seamlessly from one site to another, rendering the concept of a unified reasoning logic elusive. To address these challenges, data-driven methods, such as machine learning, can be used to determine feature weights in response to different projects' characteristics. The data integration module, as touched upon, is instrumental in aligning actual lift operations with corresponding lift orders, using nine attributes for similarity assessment. However, uniformly weighing these attributes may not always yield accurate results, especially across diverse projects with different conditions. For example, in projects requiring the lifting of numerous similar prefabricated units (sharing characteristics such as colour, material, and shape but meant for different locations at varying times), the accurate matching between operations and orders hinges on more distinct features. Here, attributes such as material, colour, shape, length, and width might not be as crucial as the coordinates of the unloading destination. In such dynamic settings, unsupervised learning becomes invaluable, dynamically adjusting attribute importance based on specific project nuances.

However, data-driven methods often grapple with limited access to extensive datasets. Gathering a substantial volume of sample data can be tedious, leading to resource constraints in learning. In these scenarios, knowledge-driven feature enhancement methods, such as ontology embedding, could serve as viable solutions to counteract the challenges posed by sample paucity.

Despite the limitations, semantic web technologies have changed the construction industry's way of collecting, managing, and using information. The adoption of the ontological model has brought many benefits for effective scheduling and resource planning. For construction scheduling, there are benefits like:

- An easier-to-understand structured activity-level project risk
- A more effective project response, a more efficient and accurate dynamic work process
- A reduced misinterpretation of schedule logic
- A comprehensive vocabulary to represent construction workflow knowledge
- An unambiguous and scalable information model to promote the construction management system.

For resource planning, there are benefits like:

- An improved interoperable and opened digital workflow for automatic quantity take-off
- Improvements in the variability and ambiguity issues in item descriptions
- Enabling the imposing of complex multi-variable constraints simply and effectively
- Enabling the collection of planning process information from different sources
- An improved overall decision-making efficiency and an increased cost estimation accuracy.

Despite these benefits and contributions that the ontological models brought to construction scheduling and resource planning, there are also some limitations for the current studies. For construction scheduling, there are limitations in the area of model development (e.g., lack of capturing of the non-essential detailed-level concepts and properties to represent the construction workflow), data quality (e.g., lack more digital data sources to cover all content required by the design ontology), model utilisation (e.g., incapable in visualising and analysing the specifics of delay claims concerning effects and causes and assigning responsibility), and commercialisation (e.g., lack of clear earning principles). For resource planning, there are limitations like model development (e.g., ignore large amounts of semantic information such as finishing), data quality (e.g., lack of integration of more project information sources), and implementation (e.g., lack aligning established knowledge development with various data sources to enrich their practical application).

9.7 Conclusion

To sum up, this book chapter provides a comprehensive overview of ontological models in scheduling and resource management for construction projects and offers valuable insights to researchers and practitioners on the future development of ontological models in the construction project management field.

Besides the identified benefits and limitations in the previous discussion section, some insights and suggestions are summarised from reviewing the recent studies and analysis of the case study. They are: (a) the existing ontological modelling process heavily relied on human (i.e., experts) input. Therefore, some machine learning techniques should be used to address experts' subjectivity and automate the existing labour-intensive standardisation process; (b) for resource planning studies, especially, attention has been overly focused on quantity take-off studies. The remaining research directions warrant further exploration. This encompasses the automatic validation of resource planning with real-time data, as well as documenting resource usage in coordinating equipment and labour to better predict resource requirements, as exemplified by the case study; (c) software interoperability remains one of the most significant challenges for construction-related ontological model applications, awaiting effective solutions; (d) a better way needs to be found for extracting and integrating information from different sources; and (e) the demand for making the ontological framework more general is still not met. These insights are intended to guide the future research directions of the ontological model-enabled construction project management.

References

Abanda, F. H., Kamsu-Foguem, B., and Tah, J. H. M., BIM – New rules of measurement ontology for construction cost estimation, Engineering Science and Technology, An International Journal. 20 (2017) 443–459.

Akanbi, T., and Zhang J. Design information extraction from construction specifications to support cost estimation, Automation in Construction. 131 (2021) 103835.

Akanbi, T., Zhang, J., and Lee, Y.-C. Data-driven reverse engineering algorithm development method for developing interoperable quantity take-off algorithms using IFC-based BIM, Journal of Computing in Civil Engineering. 34 (2020) 04020036.

Alzraiee, H. Cost estimate system using structured query language in BIM, International Journal of Construction Management. 22 (2022) 2731–2743.

Anumba, C. J., Issa, R. R. A., Pan, J., and Mutis, I. Ontology-based information and knowledge management in construction, Construction Innovation. 8 (2008) 218–239.

Bookings – Plan Ahead and Stay Ahead. HammerTech (2023). https://www.hammertech.com/en-au/solutions/site-operations/bookings (accessed March 11, 2023).

Budayan, C., Dikmen, I., Talat Birgonul, M., and Ghaziani, A. A computerised method for delay risk assessment based on fuzzy set theory using MS ProjectTM, KSCE Journal of Civil Engineering. 22 (2018) 2714–2725.

Caneparo, L. Semantic knowledge in generation of 3D layouts for decision-making, Automation in Construction. 134 (2022) 104012.

Chen, G., and Luo, Y. A BIM and ontology-based intelligent application framework, in: 2016 IEEE Advanced Information Management, Communicates, Electronic and Automation Control Conference (IMCEC), IEEE, 2016: pp. 494–497.

Cranetime. (2023). https://voyagecontrol.com/cranetime (accessed March 11, 2023).

Crane Management Software System. DataScope (2023). https://www.datascopesystems.com/product/crane-management-system/ (accessed March 11, 2023).

Collaborative Digital Scheduling. Veyor (2023). https://www.veyordigital.com/veyor-story (accessed March 11, 2023).

Curry, E., O'Donnell, J., Corry, E., Hasan, S., Keane, M., and O'Riain, S. Linking building data in the cloud: Integrating cross-domain building data using linked data, Advanced Engineering Informatics. 27 (2013) 206–219.

Dave, B., Kubler, S., Främling, K., and Koskela, L. Opportunities for enhanced lean construction management using internet of things standards, Automation in Construction. 61 (2016) 86–97. https://doi.org/10.1016/j.autcon.2015.10.009.

Drogemuller, R., Jupp, J., Rosenman, M. A., and Gero, J. S. Automated code checking, in: CRC for Construction Innovation, Clients Driving Innovation International Conference, 2004.

El-Diraby, T. E. Domain ontology for construction knowledge, Journal of Construction Engineering and Management. 139 (2013) 768–784.

Fahmy, A., Hassan, T., Bassioni, H., and McCaffer, R. Dynamic scheduling model for the construction industry, Built Environment Project and Asset Management. 10 (2020) 313–330. https://doi.org/10.1108/BEPAM-02-2019-0021.

Guber, T. A translational approach to portable ontologies, Knowledge Acquisition. 5 (1993) 199–229.

Guévremont, M., and Hammad, A. Ontology for linking delay claims with 4D simulation to analyse effects-causes and responsibilities, Journal of Legal Affairs and Dispute Resolution in Engineering and Construction. 13 (2021) 04521024.

Gursel, I., Sariyildiz, S., STOuFFS, R., Akin, Ö., Tidafi, T., and Dorta, T. Contextual ontology support as external knowledge representation for building information modeling, Joining Languages, Cultures and Visions: CAADFutures. 2009 (2009) 487–500.

Han, K. K., and Golparvar-Fard, M. Appearance-based material classification for monitoring of operation-level construction progress using 4D BIM and site photologs, Automation in Construction. 53 (2015). https://doi.org/10.1016/j.autcon.2015.02.007.

Haymaker, J., Kam, C., and Fischer, M. A methodology to plan, communicate and control multidisciplinary design processes, Construction Informatics Digital Library. (2005) 408–419.
Hu, S., Wang, J., and Fang, Y. Semantic web-assisted progress monitoring of tower crane operations in construction projects, in: The 30th EG-ICE: International Conference on Intelligent Computing in Engineering, 2023.
Hu, S., Fang, Y., and Moehler, R. Estimating and visualising the exposure to tower crane operation hazards on construction sites, Safety Science. 160 (2023) 106044. https://doi.org/10.1016/j.ssci.2022.106044.
Jiang, L., Shi, J., and Wang, C. Multi-ontology fusion and rule development to facilitate automated code compliance checking using BIM and rule-based reasoning, Advanced Engineering Informatics. 51 (2022) 101449.
Khosakitchalert, C., Yabuki, N., and Fukuda, T. Improving the accuracy of BIM-based quantity take-off for compound elements, Automation in Construction. 106 (2019) 102891.
Kim, S., Chin, S., and Kwon, S. A discrepancy analysis of BIM-based quantity take-off for building interior components, Journal of Management in Engineering. 35 (2019) 05019001.
Kim, T., Kim, Y.-W., and Cho, H. A simulation-based dynamic scheduling model for curtain wall production considering construction planning reliability, Journal of Cleaner Production. 286 (2021). https://doi.org/10.1016/j.jclepro.2020.124922.
Koo, B., Fischer, M., and Kunz, J. Formalization of construction sequencing rationale and classification mechanism to support rapid generation of sequencing alternatives, Journal of Computing in Civil Engineering. 21 (2007) 423–433.
Lee, S., Kim, K., and Yu, J. Ontological inference of work item based on BIM data, KSCE Journal of Civil Engineering. 19 (2015) 538–549.
Lee, S.-K., Kim, K.-R., and Yu, J.-H. BIM and ontology-based approach for building cost estimation, Automation in Construction. 41 (2014) 96–105.
Li, X., Wu, C., Xue, F., Yang, Z., Lou, J., and Lu, W. Ontology-based mapping approach for automatic work packaging in modular construction, Automation in Construction. 134 (2022) 104083.
Liu, H., Cheng, J. C. P., Gan, V. J. L., and Zhou, S. A knowledge model-based BIM framework for automatic code-compliant quantity take-off, Automation in Construction. 133 (2022) 104024.
Liu, K., and El-Gohary, N. Ontology-based semi-supervised conditional random fields for automated information extraction from bridge inspection reports, Automation in Construction. 81 (2017) 313–327.
Luo, L., Jin, X., Shen, G. Q., Wang, Y., Liang, X., Li, X., and Li, C. Z. Supply chain management for prefabricated building projects in Hong Kong, Journal of Management in Engineering. 36 (2020). https://doi.org/10.1061/(asce)me.1943-5479.0000739.
Ma, Z., and Liu, Z. Ontology-and freeware-based platform for rapid development of BIM applications with reasoning support, Automation in Construction. 90 (2018) 1–8.
Ma, Z., Liu, Z., and Wei, Z. Formalised representation of specifications for construction cost estimation by using ontology, Computer-Aided Civil and Infrastructure Engineering. 31 (2016) 4–17.
Nepal, M. P., Staub-French, S., Pottinger, R., and Zhang, J. Ontology-based feature modeling for construction information extraction from a building information model, Journal of Computing in Civil Engineering. 27 (2013) 555–569.
Niknam, M., and Karshenas, S. Integrating distributed sources of information for construction cost estimating using semantic web and semantic web service technologies, Automation in Construction. 57 (2015) 222–238.
Pauwels, P., de Farias, T. M., Zhang, C., Roxin, A., Beetz, J., De Roo, J., and Nicolle, C. A performance benchmark over semantic rule checking approaches in construction industry, Advanced Engineering Informatics. 33 (2017) 68–88.
Poinet, P., Stefanescu, D., and Papadonikolaki, E. Collaborative workflows and version control through open-source and distributed common data environment, in: International Conference on Computing in Civil and Building Engineering, Springer, 2020: pp. 228–247.
Rasmussen, M. H., Lefrançois, M., Pauwels, P., Hviid, C. A., and Karlshøj, J. Managing interrelated project information in AEC Knowledge Graphs, Automation in Construction. 108 (2019) 102956.
Ren, G., Li, H., Liu, S., Goonetillake, J., Khudhair, A., and Arthur, S. Aligning BIM and ontology for information retrieve and reasoning in value for money assessment, Automation in Construction. 124 (2021) 103565.
Staub-French, S., Fischer, M., Kunz, J., Ishii, K., and Paulson, B. A feature ontology to support construction cost estimating, Ai Edam. 17 (2003) 133–154.

Staub-French, S., Fischer, M., Kunz, J., and Paulson, B. An ontology for relating features with activities to calculate costs, Journal of Computing in Civil Engineering. 17 (2003) 243–254.

Staub-French, S., and Nepal, M. P. Reasoning about component similarity in building product models from the construction perspective, Automation in Construction. 17 (2007) 11–21.

Soman, R. K., Molina-Solana, M., and Whyte, J. K. Linked-Data based Constraint-Checking (LDCC) to support look-ahead planning in construction, Automation in Construction. 120 (2020) 103369. https://doi.org/10.1016/j.autcon.2020.103369.

Soman, R. K., Molina-Solana, M., and Whyte, J. K. Linked-Data based Constraint-Checking (LDCC) to support look-ahead planning in construction, Automation in Construction. 120 (2020) 103369.

Tavakolan, M., Mohammadi, S., and Zahraie, B. Construction and resource short-term planning using a BIM-based ontological decision support system, Canadian Journal of Civil Engineering. 48 (2021) 75–88.

Teng, Y., Mao, C., Liu, G., and Wang, X. Analysis of stakeholder relationships in the industry chain of industrialised building in China, Journal of Cleaner Production. 152 (2017) 387–398.

Viljamaa, E., and Peltomaa, I. Intensified construction process control using information integration, Automation in Construction. 39 (2014) 126–133.

Wang, X. H., Zhang, D. Q., Gu, T., and Pung, H. K. Ontology based context modeling and reasoning using OWL, in: IEEE Annual Conference on Pervasive Computing and Communications Workshops, 2004. Proceedings of the Second, IEEE, 2004: pp. 18–22.

Yildiz, A. E., Dikmen, I., Birgonul, M. T., Ercoskun, K., and Alten, S. A knowledge-based risk mapping tool for cost estimation of international construction projects, Automation in Construction. 43 (2014) 144–155.

Zhang, J., and El-Gohary, N. M. Semantic-based logic representation and reasoning for automated regulatory compliance checking, Journal of Computing in Civil Engineering. 31 (2017) 04016037.

Zheng, Y., Törmä, S., and Seppänen, O. A shared ontology suite for digital construction workflow, Automation in Construction. 132 (2021) 103930.

Zhong, B. T., Ding, L. Y., Love, P. E. D., and Luo, H. B. An ontological approach for technical plan definition and verification in construction, Automation in Construction. 55 (2015) 47–57.

Zhou, P., and El-Gohary, N. Ontology-based automated information extraction from building energy conservation codes, Automation in Construction. 74 (2017) 103–117.

10

ITERATIVE DIGITAL TWIN DEVELOPMENT FOR HIGH-TECHNOLOGY HEALTHCARE BUILDINGS

The case of a new ophthalmology diagnostic hub

Grant Mills[1], Duncan Wilson[2], Anne Symons, Irinie Roufaeel[3], Peter Scully, David Shanks, Cephas Bhaskar[4], Ian Eames[5], Hari Jayaram, Sobha Sivaprasad and Paul Foster[6]*

[1]*UCL Bartlett School of Sustainable Construction, London (Corresponding author: g.mills@ucl.ac.uk)

[2]UCL Bartlett Centre for Advanced and Spatial Analysis, London (d.j.wilson@ucl.ac.uk)

[3]UCL Bartlett School of Sustainable Construction, London (anne.symons.14@ucl.ac.uk; irinie.roufaeel.19@ucl.ac.uk)

[4]UCL Bartlett School of Architecture, London (p.scully@ucl.ac.uk; d.r.shanks@ucl.ac.uk; cephas.bhaskar.20@ucl.ac.uk)

[5]UCL Department of Mechanical Engineering, Faculty of Engineering Science, London (i.eames@ucl.ac.uk)

[6]NIHR Biomedical Research Centre at Moorfields Eye Hospital and UCL Institute of Ophthalmology, London (h.jayaram@ucl.ac.uk; s.sivaprasad@ucl.ac.uk; p.foster@ucl.ac.uk)

10.1 Introduction

Smart built environment technologies are developing at a rapid pace, but their combined use requires evaluation in specific settings such as healthcare. In this chapter we apply a number of new technologies and then propose what this might mean for the development of a digital twin for ophthalmology building design. Although this study is focused on an ophthalmology healthcare diagnostics environment, it more generally illustrates how a digital twin will support innovation in a fast-moving and high-technology setting. Our primary objective was to build an interdisciplinary approach to iterative design and digital twin development. As such, we build on the concept of digital twin development and application in hospitals (Karakra et al. 2019; Elayan et al. 2021; Shetty, 2023), through a process of learning through doing. We integrated digital technologies into hospital business operations and developed a collaborative digital platform (Moretti et al. 2023)

DOI: 10.1201/9781003383840-11

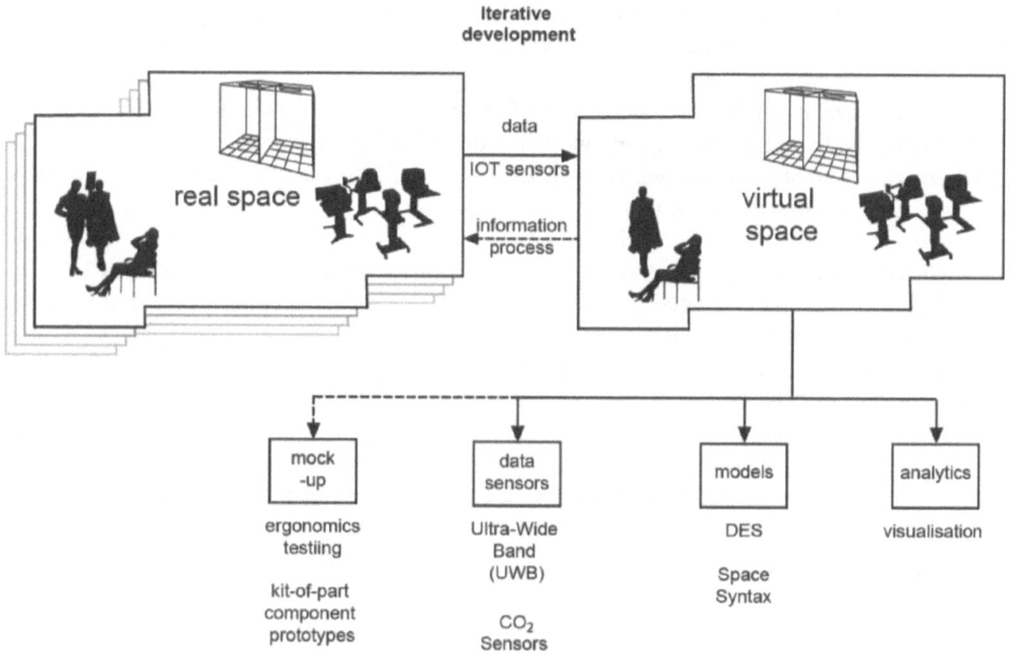

Figure 10.1 Digital twin connection between the real and virtual ophthalmology space utilizing data, IoT sensors, models, analytics, and physical mock-ups

within a specific healthcare context, but our priority was first to innovate in design and to influence a practical and live process of operational delivery.

A focus on healthcare (Zhong et al. 2017) provides an ideal case study setting to understand the adoption of smart and digital approaches (Aceto et al. 2018 and 2020). We demonstrate in this case study how an iterative approach to digital twin development emerged. We describe how Internet of Things (IoT) people tracking, CO_2 sensors, discrete event simulation (DES), augmented and virtual reality, and the use of physical ergonomics mock-ups came together to test complex patient-staff-equipment interactions. This interdisciplinary (Mills et al. 2022) application of advanced technologies (Wilson et al. 2023; Roufaeel et al. 2023) was combined into a real-life outpatient diagnostic clinic for Moorfields Eye Hospital in the UK. We show how the physical process of real-space reconfiguration (Figure 10.1) was connected to a virtual-world space that consisted of various data sensors, models, and analytics.

10.2 Digital twins applications in healthcare and hospital design

The digital twin was conceived as a method to predict behaviour (Tuegel et al. 2011) that mirrored (mocked), health checked, and protected the future of NASA spacecraft (Negri et al. 2017). While some definitions take a fully integrated "digital model", "integrating digital platform", or "top-down" approach (Laamarti et al. 2020; Mukherjee et al. 2023; Moretti et al. 2023), others have proposed the collection of "heterogeneous" (Ali et al. 2023) and "multi-domain" (Peng et al. 2020) data. This is an incremental digital twin development view, whereby iteration creates interconnection (Grieves et al. 2016).

There are good examples of digital twin development in healthcare, for example the connection between medical AI and control rooms or remote surgery (Laaki et al. 2019), but less is known about how digital twins might advance hospital design, operation management, and patient outcomes. Like Liu et al. (2019) we wanted to further enable the testing of a building layout design solution virtually, before scheduling and implementing actual asset changes.

Advances in building information modelling, Internet of Things, and simulation models have been integrated into the digital twin (Gunal, 2012; Mielczarek et al. 2012; Chahal et al. 2013; Feili, 2013; Jones et al. 2019), however, real-time data feedback loops and correction of physical-digital mismatches remain a challenge (Matarneh et al. 2019). These advances have meant that we can evaluate and predict building performance without the cost of physical alteration. For example, Arisha et al. (2016) and Roufaeel et al. (2023) show how healthcare layout design can be achieved through building a connection between the simulation model and human behaviour. Others have focused on the overcrowding problem and bottlenecks (Brailsford et al. 2009; Gonsalves et al. 2009, Bhattacharjee et al. 2014; Morgareidge et al. 2014) or the reduction in hospital-acquired infection (Mielczarek, 2012). These combined with physical prototyping provide a means to rapidly test assumptions (Villegas et al. 2019), test what will work (Dunston et al. 2011; Kumar et al. 2011; Peavey et al. 2012), and predict resource use (Brailsford et al. 2001; Thorwarth, 2009). We now explore how a digital twin for a high-technology healthcare building was developed using a single healthcare case study.

10.3 Methodology

An interdisciplinary process was taken to develop a digital twin that integrated multiple technologies into a single case study environment. The process engaged end-users, clinicians, technicians, and operations managers in four unique building configurations (iterations) which were devised and tested, with circa 1,000 patients visiting a site for a series of eye diagnostics tests (Mills et al. 2022). Circa 30 staff were involved in the interdisciplinary co-design process to encourage an innovative approach to equipment configuration, layout design, and emerging scientific evidence base. The ophthalmology clinic was located in a North London shopping centre retail unit. It provided technician-led outpatient appointments for glaucoma, medical retina, and cataract care where the patient travels through a combination of several standardized tests.

Figure 10.2 shows some of the kit-of-part components that were developed and the various sensors and modelling technologies that were implemented. The establishment of a single integrated digital twin has significant merit, although we started with each discipline bringing their expertise and evidence of development methodology to the iterative design, manufacture, installation, operation, and innovation process. As such the digital twin was established from various models that were not always digitally integrated but always collectively discussed to understand where there might be opportunities for greater integration.

Different layout iterations were tested between October 2022 and March 2023.

10.4 Use of IoT tracking of people and CO_2 sensors

Internet of Things (IoT) and CO_2 sensors were applied to the ophthalmology clinic to understand the influence of patient throughput on the minimization of infection risk throughout waves of COVID-19. This use of IoT technology (Wilson et al. 2023) sought to test the automated tracking of people to better model patient flow and airborne hazards to inform building layout design and reduce patient journey time. Two primary technologies of automated data capture were used. They included patient tracking through ultra-wide band (UWB) technology provided by Ubisense and environmental

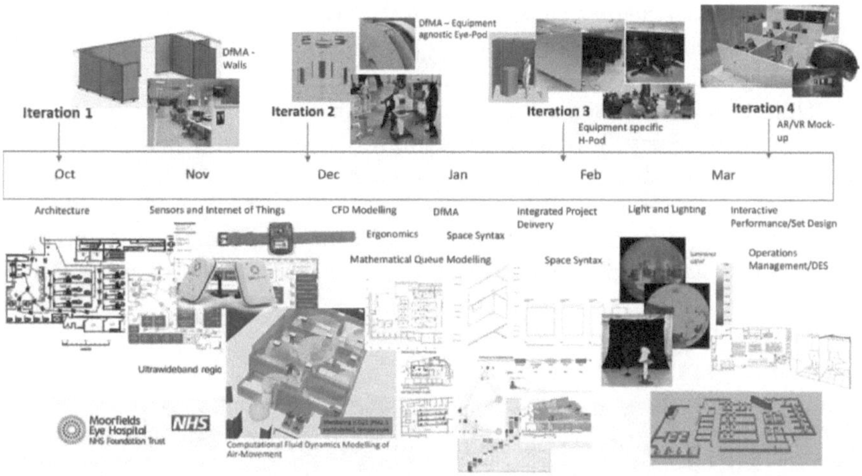

Figure 10.2 Application of data, IoT sensors, models, and analytics over the process of delivering a digital twin for ophthalmology building design

Figure 10.3 Photograph of clinic showing typical layout of equipment, wall design, and sensor (Wilson et al. 2023). It shows the clinic's fourth iteration. Including reconfigurable low-height partitions [middle] and light-controlled pods [right] for use by technicians in diagnostic tests and full-height partitions [left] for private consult and pre-/post-op assessment (performed by a nurse, ophthalmologist, or optometrist)

measurements from air samplers located throughout the clinic. This enabled the assessment of particulate loading and sizes, temperature, humidity, and CO_2 concentration over a working day (e.g. people as a source of heat, humidity, and airborne pathogens against the environmental layout, equipment, and exchange surfaces, the entrance, other floors, and air conditioning units).

Seventeen Ubisense UWB sensors located around the ceiling of the clinic support the real-time location of 100+ credit-card-sized tags allocated for tracking. An example UWB sensor can be seen in Figure 10.3. Each tag was worn by a patient participating in the research and it tracked their movement throughout their visit to an accuracy of 15 cm. Data from these technical platforms were validated against ethnographic observations following an individual patient through a complete journey and qualitative feedback from clinicians, technicians, and the design team responsible for the interior layout of the clinic (Wilson et al. 2023).

Basic statistics showed patient flow variations (days of week and the four different layouts). The highest CO_2 concentrations (> 1000 ppm), located at the bottom of the stairs, were not correlated to a high local concentration of people but due to an anomalous air flow path: hot air rose to the second floor, where it mixed and was cooled by air-conditioning units and then flowed downstairs. By way of an example, Figure 10.4 shows a heat map of patient and staff movements

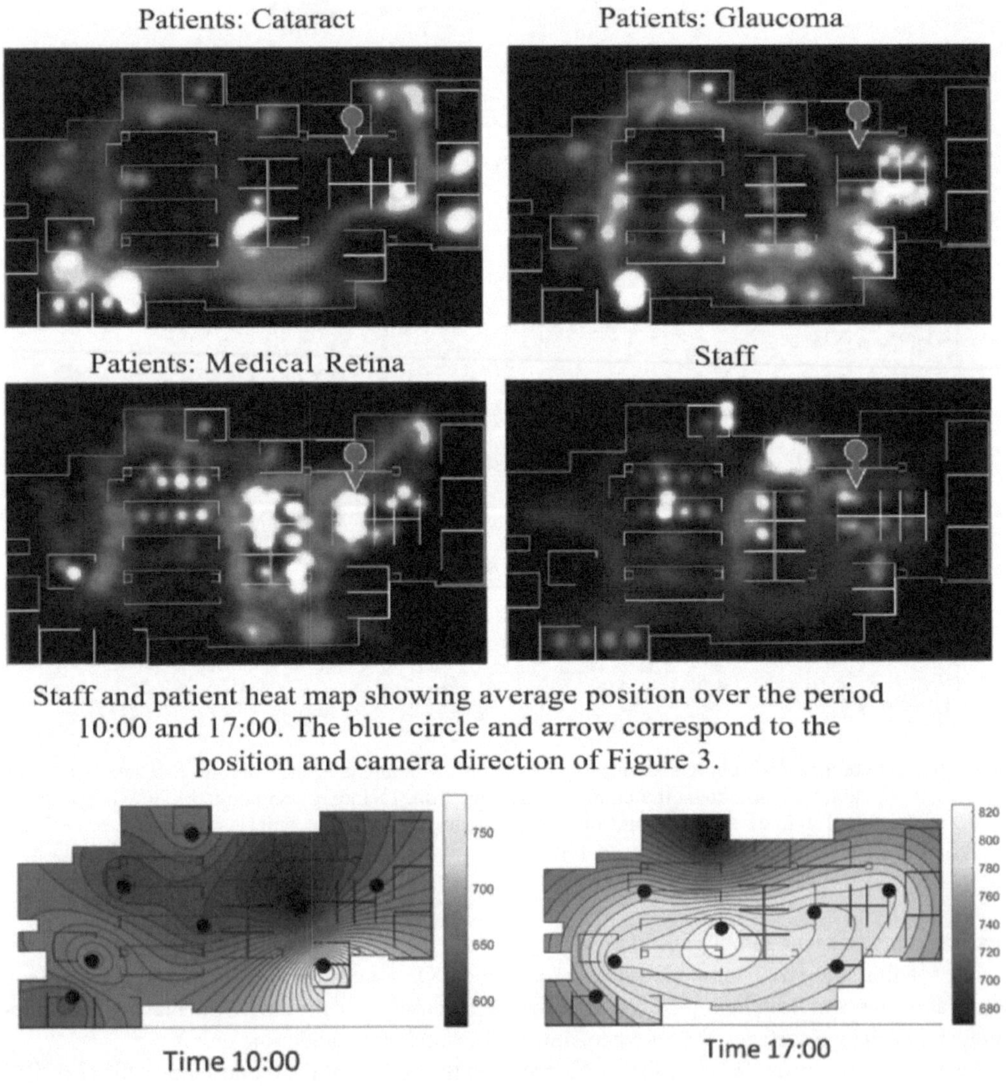

Figure 10.4 Heat maps show a snapshot of the effective source of CO_2 from patients and staff [top] and spatial distribution of CO_2 [bottom], taken from Wilson et al. (2023)

(Iteration 4) and the spatial distribution of CO_2 (Iteration 4). The time of 15:00 corresponded to the maximum CO_2 concentration in the space (1,050 ppm) and incorporates glaucoma, medical retina, and cataract patients, visitors, and staff (Wilson et al. 2023).

The use of IoT tracking of people and CO_2 was useful in the development of the physical and digital design. It provided a useful data set that could be aggregated into a digital twin, although there was complexity in the assimilation of large volumes of IoT data from multiple technical systems. The challenge was the various data formats and visual representations.

10.5 Use of discrete event simulation to understand staff and patient flow

Modelling was used to demonstrate the effects of different layouts. Figure 10.5 shows an indicative DES model used to illustrate the clinical flow and facilitate a discussion on efficiency and smooth running with no bottlenecks and the consideration of working hours. These digital DES models provided an important data set that could be assimilated into a digital twin that attempts to mimic and so test the designs for each iteration. Iteration 1 had a looped design with four pathway activities separated for medical retina (MR) and glaucoma (GL). Iteration 2 had equipment pools that fully amalgamated patient pathways into either five or six pathway steps. Iteration 3 had a five-activity hybrid design with dedicated and shared clusters and pods. Iteration 4 added cataract diagnostics and included nurse and ophthalmologist/optometrist interactions, pre-/post-operation assessment, and consultation that added complexity to the ophthalmology clinic.

Separating patient pathway circulation and staggering the arrival of patients by ten minutes reduced patient contact and created a swift and even flow (Roufaeel et al. 2023). Iteration 3 was delivered through an interdisciplinary approach that led to clear/open circulation routes (as preferred by staff and patients), improved patient wayfinding/tech sightlines, and utilized pooled/banked equipment (to improve flow). This iteration was judged to be the most successful.

Taking a sample of the ultra-wideband real-time location system data (from a single day) showed the time spent by patients in each pod/station and the total time spent for each patient across the full pathway of machines (Roufaeel et al. 2023). There appeared to be a difference between the DES simulated data and ultra-wideband data after grouping equipment/tests into vision, dilation, and MR for the medical retina patients and AF, HFA, and GL for the glaucoma patients, which

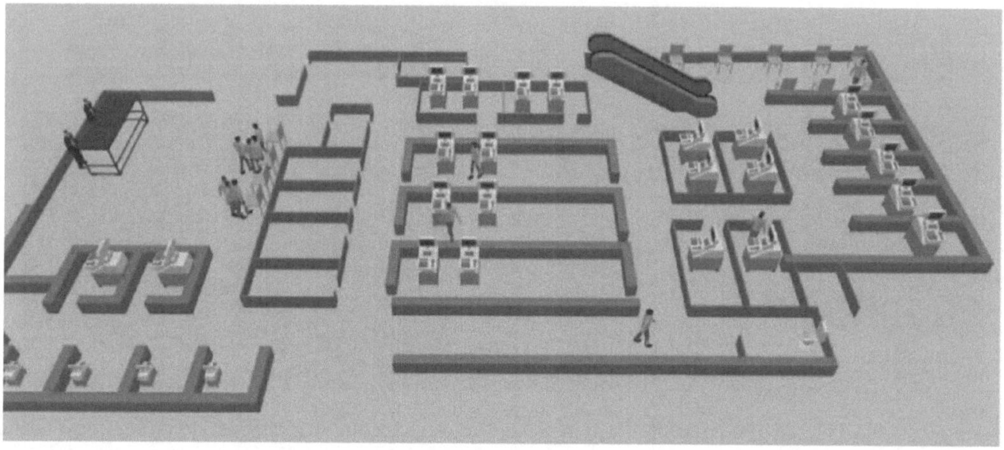

Figure 10.5 Discrete event simulation (DES) 3D simulation run for Iteration 1 using Anylogic software

enabled model adjustment. The analysis showed that the HFA machines for the Medical Retina patients' lane required extra time in order to recalibrate the machine after conducting each test (Roufaeel et al. 2023). In terms of the glaucoma patients' lane, journeys to the toilet were often necessary and so this behaviour in elderly patients often increased the time taken to complete the pathway and disrupted the swift and even linear flow (resulting in bottlenecks and waits). The use of discrete event simulation as a technique provided a useful means of prediction and could be aggregated into a digital twin and there appeared to be a number of possible data connections.

10.6 Use of virtual/augmented reality mock-up and consultation

The XR Studio at the Bartlett School of Architecture worked with extended reality (AR/VR) and 3D scanning to develop the plan and equipment visualizations. Figure 10.6 shows the use of three tools (Sketchfab – a browser-based tool, an augmented reality mobile device, and virtual reality) as approaches to support the implementation of the digital twin.

Figure 10.6 3D The XR Studio at the Bartlett School of Architecture demonstration of Iteration 4 using Sketchfab [top], an augmented reality mobile device [middle], and virtual reality – Edify.ac [bottom]

These digital mock-ups of the layout design showed that AR and VR were useful during design development, component selection and manufacture, and in the process of engaging key stakeholders in the design decision-making process. The development of the models allowed a continuous process of 3D documentation that could be utilized on a range of hardware platforms (including phones, tablets, and headsets).

10.7 Use of a physical mock-up to test the staff-patient-equipment interactions

Various disciplines[1] were included in the development of design information gathering. All disciplines had different scientific preferences and capabilities and they used mock-ups and developed design information in different ways. All came together in Iteration 3. A good example of this was a single day that was spent on site to gather information and generate new insights. The healthcare planner observed the patient journey, the lighting designer mocked up a dark room with controllable lights, and the DES modeller and healthcare planner investigated equipment ergonomics, while the design for manufacture specialist role-played as a patient and marked up the floor into zones and considered the design of specific modular components. This was possible in a real-world space.

Figure 10.7 shows part of the physical process of trial and error, ergonomic drawing, and design exploration. The information gathered was translated into various design models. Some used physical drawings, while others used advanced digital design capabilities. There was no singular BIM model in Iteration 3, although working together disciplines were able to respond to many questions – for example how and when will patients access amenities? How will patients move and interact with staff and technology? Where will patients and staff wait? And how are technologies connected to each other and is there a sequence that must be followed?

The iteration in physical mock-up seemed to support digital twin development, allowed learning, and provided evidence.

10.8 Development of physical and digital modular prototypes

A key design feature of the outpatient clinic was reconfigurability, to allow for the testing of different layouts (e.g. equipment number and sequence). In Iteration 3 advanced digital models and physical prototypes were developed by B-Made at the Bartlett School of Architecture.

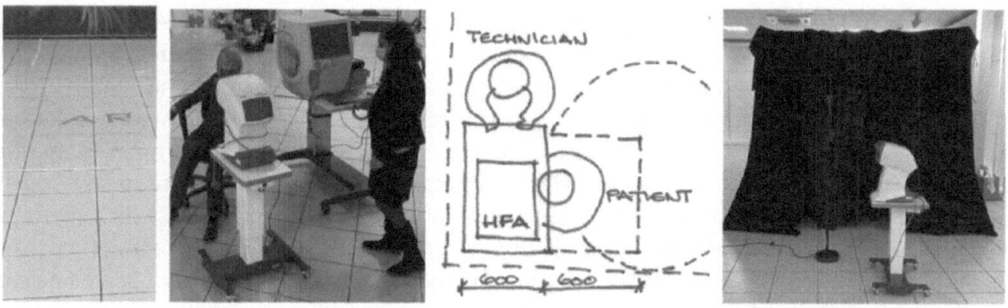

Figure 10.7 Setting out of the physical space needed around equipment [left] testing the interaction and configuration of the patient, technician and equipment, and the hand-drawn ergonomic layouts that positioned the equipment effectively in the physical space [middle]. The mock-up of a dark room to test the ideal lighting levels for the ophthalmology diagnostic equipment [right]

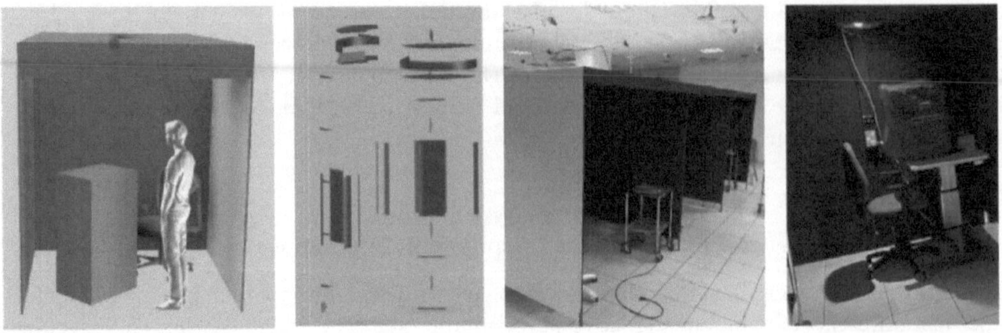

Figure 10.8 Pod and lighting design. The use of digital design mock-ups [left] developed by Bartlett Architecture B-Made and then translated by them into a manufactured and installed component [middle and right]

Procurement deadlines and design supplier and contractor preferences constrained Iteration 1, while in Iteration 2 a physical installation that differed from what was 2D designed influenced prototype pod installation. In Iteration 3, however, a prototype Humphrey Field Analyser module (that we called a HFA-pod) was developed which integrated ergonomic principles, with lighting and a digital mock-up (Figure 10.8). The wider environment of the centre was designed for low lighting, although the manufactured pod created increased lighting flexibility for both patients and staff.

During the process of physical mock-up various new functions emerged. For example, modular partitions used in consulting and examination (pre- and post-operation assessment) were adapted with hemp acoustic panels to maintain privacy and control acoustics. Both digital and physical prototyping enhanced the design, although the adoption of a digital twin perhaps created unique procurement challenges.

10.9 Discussion

In this chapter we have applied new technologies to develop a digital twin for ophthalmology building design. Although this study was focused on an ophthalmology healthcare diagnostics environment, it more generally illustrates the use of a digital twin to support innovation in a fast-moving and high-technology setting. We show the application of a diversity of IoT sensors (movement and CO_2 sensors) and models (building models, space syntax, discrete event simulation, and augmented/virtual reality) that were combined in the process of design. We developed both a digital and physical collaborative process of digital twin development. In fact, it is this which is the chapter's contribution. The novelty is the rapid and iterative process of digital twin development, which involves collaboration towards an innovative solution. Data idiosyncrasies were integrated through iteration. We tested, trialled, and mocked up in a process of experimentation using a repetitive and iterative design: manufacture, install, operate, and then change the process. This created an interconnection among data which partially mimicked real-life behaviour (Wagner et al. 2017; Arthur et al. 2020). For example, we used patient movement sensors to keep the behaviour of physical assets dynamically updated across four life cycles – through design, manufacture, installation, and operation, each time making improvements as we went. Although others have looked to standardize digital twins, we looked for incremental improvements in both the physical space and its digital twin, during both routine and high-stress situations.

The physical and virtual models enabled us to exchange viewpoints among design, engineering, and clinical/technical disciplines. While we attempted to integrate various complex data sources, we found ourselves developing simplifications. For example, in order to cope with the complexity of the real-time movement sensors (by minute, hour, day, week, month, and iteration) and make this usable in a discrete event simulation (DES), a single typical day was taken in each iteration. In line with Mukherjee et al. (2023) we found some commonalities between disciplines, but there was also a significant number of evidence variations that impacted the digital twin. We extend the four digital twin manifestations proposed by Grieves and Vickers (2017) – 1) prototype, 2) interaction across a lifecycle, 3) interdisciplinary evaluation, and 4) an aggregate environment. We have shown how iterative digital twin development can utilize IoT people tracking, CO_2 sensors, and discrete event simulation alongside virtual/augmented reality, physical mock-up, and modular prototyping. The aggregation of data sources using an interdisciplinary process is complex, but this is supported by an iterative and action-based digital twin development process.

We have attempted to mirror real-world and virtual-world space (VanDerHorn et al. 2021) and demonstrate how multiple sources and types of information require aggregation into a digital twin. We show how a co-design process can be developed with various disciplines to build capability and ultimately deliver, through iteration, a digital twin for ophthalmology building design. The next step is to develop a common data model/hierarchy and knowledge base (e.g. real-time sensors) to continuously improve ophthalmology healthcare building.

10.10 Conclusion

A new digital twin development process is proposed that incorporates simulation modelling and advanced IoT sensor systems. This chapter has shown how to iteratively design complex healthcare built environments using smart technologies. This living and breathing process of learning was useful in establishing a unique ophthalmology digital twin platform, combining evidence, data sets, and expertise that can be openly shared. We believe that similar iterative digital-physical twin development collaborations will support innovation in other high-technology building types. This work implies that innovation in healthcare building design can be facilitated by an iterative digital twin development approach.

Acknowledgments

Research submitted on behalf of The HERCULES Consortium. This work is funded by the NIHR Biomedical Research Centre at Moorfields Eye Hospital, UCL Institute of Ophthalmology and UCL Bartlett. The views expressed in this publication are independent and are not necessarily those of the National Institute for Health Research. The authors thank the staff and patients at Moorfields Brent Cross diagnostic hub who participated in this research. Project senior responsible officer was Mr Jonathan Wilson. Collaboration and in-kind support gratefully received from Paul Webster, Ubisense Ltd.

Note

1 Expertise in architecture, healthcare planning, lighting design, ergonomics, technology use, acoustics, Ophthalmology, project and value management, design for manufacture and assembly and modelling for example.

References

Aceto, G., Persico, V., and Pescapé, A. (2018). The role of information and communication technologies in healthcare: Taxonomies, perspectives, and challenges. Journal of Network and Computer Applications, 107, 125–154.

Aceto, G., Persico, V., and Pescapé, A. (2020). Industry 4.0 and health: Internet of things, big data, and cloud computing for healthcare 4.0. Journal of Industrial Information Integration, 18, 100129 (1–13).

Ali, S., Armand, T. P. T., Athar, A., Hussain, A., Ali, M., Yaseen, M., Joo, M.-I., Kim, H.-C., et al. (2023). Metaverse in healthcare integrated with explainable AI and blockchain: Enabling immersiveness, ensuring trust, and providing patient data security. Sensors, 23(2):565.

Arisha, A., and Rashwan, W. (2016, December 11–14). Modeling of healthcare systems: Past, current and future trends. In: T. M. K. Roeder, P. I. Frazier, R. Szechtman, E. Zhou, T. Huschka, & S. E. Chick (Eds.), Proceedings of the 2016 winter simulation conference (pp. 3213–3224). IEEE Press, Arlington Virginia.

Arthur, R., French, M., Ganguli, J. S., Kinard, D. A., Kraft, E.M., Marks, I., Matlik, J.F., Pinon, O. J., Sangid, M.D., Bruhn, E.F., Seal, D., Tucker, E., and Vickers, J. (2020). American Institute of Aeronautics and Astronautics (AIAA) Digital Engineering Integration Committee. "Digital twin: Definition & value, an AIAA and AIA position paper. Available from: digital-twin-institute-position-paper-(december-2020).pdf (aiaa.org). [December 2020].

Bhattacharjee, P., and Ray, P. K. (2014). Modelling and optimisation of healthcare systems: An overview. International Journal of Logistics Systems and Management, 19(3), 356–371. https://doi.org/10.1504/IJLSM.2014.065502.

Brailsford, S. and Hilton, N. (2001) A comparison of discrete event simulation and system dynamics for modelling health care systems. In: J. Riley (Ed.), Planning for the future: Health service quality and emergency accessibility. Operational research applied to health services (ORAHS) (01/01/01). Glasgow Caledonian University.

Brailsford, S. C., Harper, P. R., and Pitt, M. (2009). An analysis of the academic literature on simulation and modelling in health care. Journal of Simulation, 3(3), 130–140. https://doi.org/10.1057/jos.2009.10.

Chahal, K., Eldabi, T., and Young, T. (2013). A conceptual framework for hybrid system dynamics and discrete event simulation for healthcare. Journal of Enterprise Information Management, 26(1), 50–74. https://doi.org/10.1108/17410391311289541.

Dunston, P.S., Arns, L.L., Mcglothlin, J.D., Lasker, G.C., and Kushner, A.G. (2011). An immersive virtual reality mock-up for design review of hospital patient rooms. In X. Wang and J.JH. Tsai (Eds.), Collaborative design in virtual environments. Intelligent Systems, Control and Automation: Science and Engineering (Vol. 48). Springer, Dordrecht.

Elayan, H., Aloqaily, M., and Guizani, M. (2021). Digital twin for intelligent context-aware IoT healthcare systems. IEEE Internet of Things J., 8(23):16749–16757.

Feili, H. R. (2013). Improving the health care systems performance by simulation optimization. Journal of Mathematics and Computer Science, 07(01), 73–79. https://doi.org/10.22436/jmcs.07.01.08.

Gonsalves, T., and Itoh, K. (2009). Service optimization with patient satisfaction in healthcare systems. Journal of Simulation, 3(3), 150–162. https://doi.org/10.1057/jos.2009.2.

Grieves, M., and Vickers, J., 2017. Digital twin: Mitigating unpredictable, undesirable emergent behavior in complex systems. In Transdisciplinary perspectives on complex systems (pp. 85–113). Springer, Cham.

Gunal, M. M. (2012). A guide for building hospital simulation models. Health Systems, 1(1), 17–25. https://doi.org/10.1057/hs.2012.8.

Jones, D.E., Snider, C., Kent, L. and Hicks, B. (2019) Early-stage digital twins for early stage engineering design. In Proceedings of the design society: International conference on engineering design. Cambridge University Press, July, Vol. 1, No. 1, (pp. 2557–2566).

Karakra, A., Fontanili, F., Lamine, E., and Lamothe, J. (2019). HospiT'Win: A predictive simulation-based digital twin for patients pathways in hospital. In IEEE, EMBS, ICBHI, Chicago, I. (pp 1–4).

Kumar, S., Hedrick, M., Wiacek, C. and Messner, J. (2011), Developing an experienced-based design review application for healthcare facilities using a 3d game engine, Journal of Information Technology in Construction (ITcon), Vol. 16, pg. 85-104.

Laaki, H., Miche, Y. and Tammi, K., 2019. Prototyping a digital twin for real time remote control over mobile networks: Application of remote surgery. IEEE Access, 7, pp. 20325–20336.

Laamarti, F., Badawi, H.F., Ding, Y., Arafsha, F., Hafidh, B. and El Saddik, A (2020) "An ISO/IEEE 11073 standardized digital twin framework for health and well-being in smart cities". IEEE Access, 2020, 8. pp. 105950–105961.

Liu, Y., Zhang, L., Yang, Y., Zhou, L., Ren, L., Wang, F., Liu, R., Pang, Z. and Deen, J. (2019). A novel cloud-based framework for the elderly healthcare services using digital twin. *IEEE Access*, *7*, 49088–49101. https://doi.org/10.1109/ACCESS.2019.2909828.

Matarneh, S.T., Danso-Amoako, M., Al-Bizri, S., Gaterell, M. and Matarneh, R., 2019. Building information modeling for facilities management: A literature review and future research directions. Journal of Building Engineering, 24, p.100755.

Mielczarek, B. and Uziałko-Mydlikowska, J. (2012). Application of computer simulation modeling in the health care sector: A survey. Simulation, 88(2), 197–216. https://doi.org/10.1177/0037549710387802.

Mills, G. et al. (2022). Driving innovation in healthcare design: The case of an interdisciplinary co-design process for a new rapid virtual diagnostics hub for eyes. ARCH22 Rotterdam: Delft https://doi.org/10.24404/62344823caad020dcc847405.

Moretti, Nicola; Mukherjee, Anandarup; Chan, Yin-Chi; Yilmaz, Gokcen; Merino, Jorge; Sasidharan, Manu; Rosun, Zahrah; Carr, Colin; McFarlane, Duncan; Parlikad and Ajith Kumar (2023) Low-cost digitalisation opportunities in healthcare: A histopathology department case study. In: (Proceedings) 1st workshop on low-cost digital solutions for industrial automation. University of Cambridge: Cambridge, UK. (In press).

Morgareidge, D., Cai, H. and Jia, J. (2014). Performance-driven design with the support of digital tools: Applying discrete event simulation and space syntax on the design of the emergency department. Frontiers of Architectural Research, 3(3), 250–264. https://doi.org/10.1016/j.foar.2014.04.006

Mukherjee, A, N Moretti, YC Chan, G Yilmaz, M Sasidharan, J Merino, Z Rosun, C Carr, D McFarlane, and AK Parlikad (2023). "Healthcare process and variable mapping for identifying digitalisation opportunities and building a hospital digital twin." European Council for Computing in Construction. doi:10.17863/CAM.97072.

Negri, E., Fumagalli, L. and Macchi, M. (2017) A Review of the Roles of Digital Twin in CPS-based Production Systems. Procedia Manufacturing, 11, 939–948. doi: 10.1016/j.promfg.2017.07.198.

Peavey, E.K., Zoss, J. and Watkins, N. (2012) Simulation and mock-up research methods to enhance design decision making, HERD. 5(3), 133–144.

Peng, Y., Zhang, M., Yu, F., Xu, J. and Gao, S. (2020). Digital twin hospital buildings: An exemplary case study through continuous lifecycle integration. Adv. Civ. Eng., 2020:8846667.

Roufaeel, Irinie, Grant Mills, Anne Symons, Duncan Wilson, Paul Webster and Alan Penn (2023), "Simulation modelling as a decision support tool for efficient healthcare layout design". The 30th EG-ICE: International conference on intelligent computing in engineering, 4–7 July, London: University College London.

Shetty, A., Kulkarni, G. S., Sn, R. B. and Paarakh, P. M. (2023). A review on: Metaverse in health care and pharma. J. Community Pharm. Pract., 3(1):1–11.

Thorwarth, M. and Arish, A. (2009) Application of discrete-event simulation in health care: A review. Dublin Institute of Technology, Technical Report No. 3.

Tuegel, E. J., Ingraffea, A. R., Eason, T. G. and Spottswood, S. M. (2011) Reengineering Aircraft Structural Life Prediction Using a Digital Twin. International Journal of Aerospace Engineering, 2011, 1–14. doi: 10.1155/2011/154798.

VanDerHorn, E. and Mahadevan, S., 2021. Digital twin: Generalization, characterization and implementation. Decision Support Systems, 145, p.113524.

Villegas-Ch W, Palacios-Pacheco X, Luján-Mora S. Application of a smart city model to a traditional university campus with a big data architecture: A sustainable smart campus. Sustainability. 2019; 11(10):2857.

Wagner, C., Grothoff, J., Epple, U., Drath, R., Malakuti, S., Grüner, S., Hoffmeister, M. and Zimermann, P. (2017) The role of the Industry 4.0 asset administration shell and the digital twin during the life cycle of a plant. In 2017 22nd IEEE international conference on emerging technologies and factory automation (ETFA) September. IEEE, (pp. 1–8).

Wilson, D., S. Gray, I. Eames and P. Foster (2023) "Understanding patient flow using IoT in a pop up eye clinic", The 30th EG-ICE: International conference on intelligent computing in engineering, 4–7 July, London: University College London.

Zhong, X., Lee, H.K. and Li, J. (2017) From production systems to health care delivery systems: a retrospective look on similarities, difficulties and opportunities. International Journal of Production Research, 55(14), pp. 4212–4227.

11

A THEORETICAL FRAMEWORK FOR INTEGRATING DIGITAL TWINS IN BUILDING LIFECYCLE MANAGEMENT

Karla Saldana Ochoa[1], Yasin Delavar[1], Deepak Balakrishnan[1], Chady Elias[2], Ravi S. Srinivasan[2] and Chimay J. Anumba[2]

[1]University of Florida, College of Design, Construction and Planning, School of Architecture, 1480 Inner Rd, Gainesville, FL 32611, USA

[2]University of Florida, College of Design, Construction and Planning, M.E Rinker, Sr. School of Construction Management, 304 Rinker, Gainesville, FL 32611, USA

Email address: ksaldanaochoa@ufl.edu (Karla Saldana Ochoa), y_delavar@yahoo.com (Yasin Delavar), sravi@ufl.edu (Ravi S. Srinivasan), anumba@ufl.edu (Chimay J. Anumba) (Karla Saldana Ochoa), (Ravi S. Srinivasan)

11.1 Introduction

11.1.1 An overview of digital twins (DT) in building lifecycle

The lifecycle of a building involves several stages, including design, construction, operation, and maintenance, each necessitating efficient information exchange and strategies for interoperability (Vanlande et al., 2008). This entails sharing design details among architects and engineers, communicating progress between stakeholders during construction, obtaining feedback from facilities and occupants during operation, and performing continuous maintenance during the building lifecycle. The concept of building information modeling (BIM), which emerged in the early 90s, facilitates this process by providing a sharable and consistent database containing both geometric and semantic information about building elements (van Nederveen and Tolman, 1992). BIM enables seamless collaboration among architecture, engineering, construction, and operations (AECO) professionals throughout the building's lifecycle, significantly impacting its management and enhancing efficiency across design (Oh et al., 2011), construction (Azhar et al., 2008; Singh et al., 2018; Deng et al., 2019; Deng et al., 2021), and operation and maintenance phases (Bahar et al., 2013; Soust-Verdaguer et al., 2017).

BIM alone typically offers only static data regarding the built environment and cannot automatically update real-time information in models without additional data sources (Tang et al., 2019). The emergence of the Internet of Things (IoT), characterized by the interconnection of sensing devices facilitating information exchange across diverse platforms (Gubbi et al., 2013),

DOI: 10.1201/9781003383840-12

has enabled the integration of real-time sensing data with static information from BIM models (Tang et al., 2019). BIM models can now visualize and analyze real-time data, leveraging smart devices. This integration allows for real-time monitoring of construction processes during the construction phase as well as indoor environmental conditions during the operation phase. Moreover, integrating BIM and IoT technologies can enhance operational and maintenance efficiency (Lee et al., 2016; Dave et al., 2018; Li et al., 2018; Natephraa and Motamedib, 2019). This combination of BIM and IoT forms the foundation for the emergence of digital twin (DT) technologies (Tagliabue et al., 2018), which originated from the aerospace industry (Shafto et al., 2010) and subsequently extended into industrial manufacturing (Negri et al., 2017; Kritzinger et al., 2018; Zhuang et al., 2018; Tchana et al., 2019; Zhang et al., 2019). However, DTs for AECO applications are still in their early stages, necessitating a systematic review of the transition from BIM to DTs and assessing existing state-of-the-art technologies to establish a future research agenda.

11.1.2 IoT data from building life cycle

BIM has become widely adopted in the construction industry, particularly in commercial sectors, and it is employed across all stages of a project's lifecycle (Valinejadshoubi et al., 2021). While its primary focus is on the design, engineering, and construction phases, functioning as a collaborative platform for stakeholders such as owners, architects, engineers, contractors, subcontractors, and suppliers, it converges various aspects into a unified virtual model. This convergence facilitates seamless communication, enhances project transparency, and optimizes workflows for all involved parties throughout the construction process.

However, the current use of BIM is concentrated on static data documentation. Its application in areas such as operation and maintenance has yet to reach its full potential. In contrast, DTs offer real-time data acquisition, simulations, analysis, and controls during the design and construction phases, providing a methodology to integrate IoT and sensor-actuator systems and overcoming these limitations. Integrating IoT and sensor-actuator systems enables real-time data capture and visualization, building analysis, facility management, and optimization of building management through AI and machine learning during operation and maintenance (Saldana Ochoa et al., 2021). The implementation of DT involves creating a virtual replica of a physical entity, as depicted in Figure 11.1. This digital representation utilizes real-time simulations to mimic its physical counterpart's actions and functionalities accurately. Critical components of DTs encompass a physical entity, a virtual entity, and bidirectional connectivity between the two (Yitmen et al., 2021; Fuller et al., 2020).

Integrating sensors to gather data from the physical building and simulation data with the DT building cycle represents a transformative shift, offering significant advantages across the design, engineering, construction, operation, and maintenance phases. A DT coupled with an array of devices such as IoT devices, sensors, and data loggers can acquire data on the indoor environmental conditions of the physical asset (Ozturk et al., 2021; Xie et al., 2023). For example, integrating DTs with EnergyPlus™, a building energy simulation algorithm developed by the U.S. Department of Energy, will allow real-time data gathering to perform real-time simulations for the asset. DTs also provide the ability to predict failures based on real-time data collection using AI/ML algorithms (Srinivasan et al., 2020).

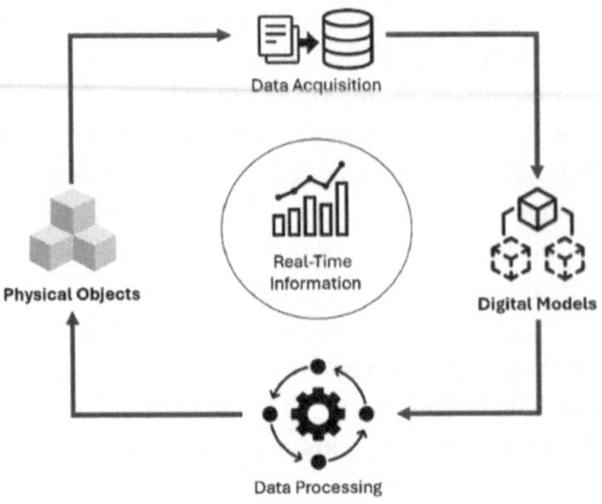

Figure 11.1 A conceptual model of digital twin architecture (based on Omrany et al., 2023)

11.2 BIM in design and construction practice

11.2.1 BIM levels of detail

As mentioned in the previous section, DTs are virtual representations of physical objects or systems that have the potential to revolutionize the construction industry by facilitating more efficient, cost-effective, and sustainable building design, construction, and maintenance. However, the development and implementation of DTs in architecture, engineering, and construction are still in their early stages (Menassa, 2021). There is a need to establish industry standards and best practices to ensure their effective and widespread use. Currently, the conventional tool for virtual/digital representation is BIM. BIM can contain a vast array of data related to a building, including its design, construction, and operation. Depending on the level of detail (LOD) (Pan and Zhang, 2021), BIM can be classified as:

- LOD 100. The model element may be graphically represented in the model with a symbol or other generic representation. However, it needs to satisfy the requirements for LOD 200. Information related to the model element (e.g., cost per square foot, tonnage of AC, etc.) can be derived from other model elements (Menassa, 2021)
- LOD 200. The model element is generically and graphically represented within the model with approximate quantity, size, shape, location, and orientation.
- LOD 300. The model element, as designed, is graphically represented within the model such that its quantity, size, shape, location, and orientation can be measured. (Pan and Zhang, 2021)
- LOD 350. The model element, as designed, is graphically represented within the model so that its quantity, size, shape, location, orientation, and interfaces with adjacent or dependent model elements can be measured
- LOD 400. The model element is graphically represented within the model with detail sufficient for fabrication, assembly, and installation
- LOD 500. The model element is a graphic representation of an existing or as-is condition developed through observation, field verification, or interpolation. The level of accuracy shall be noted or attached to the model element.

A DT can be created without reaching the highest BIM level, as it depends on the project's scale and requirements. DTs encompass categories such as existence or descriptive twin (like BIM

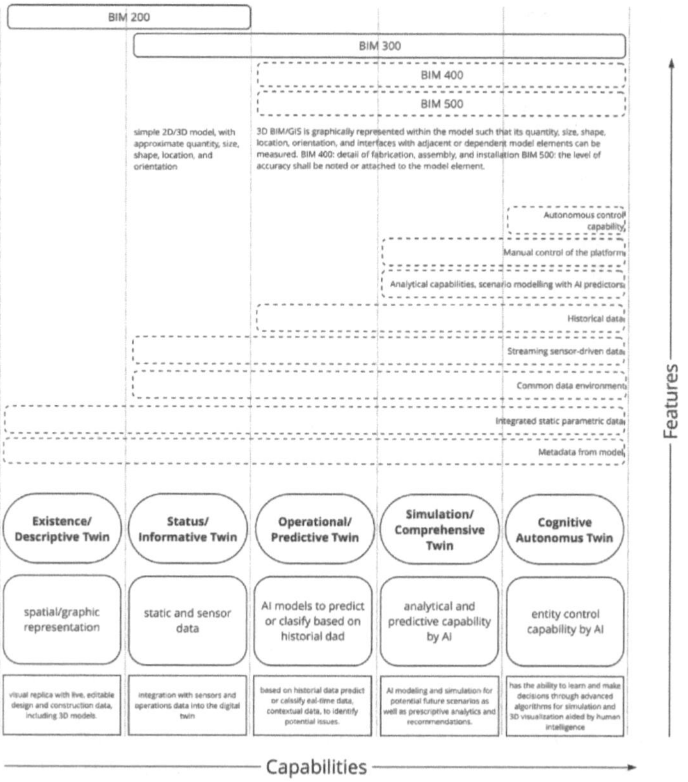

Figure 11.2 DT continuum, from BIM to DT

LOD 200), status or informative twin (providing real-time data), operational or predictive twin (bi-directional data connection), simulation or comprehensive twin (utilizing AI algorithms and historical data for predictions) (Guo et al., 2022), and cognitive or autonomous twin (capable of decision-making guided by humans). The DT continuum aligns with different BIM levels, progressing from essential spatial representation to intricate simulations. Figure 11.2 shows how BIM data is used to create DT models (Architecture Engineering Construction, n.d.).

11.3 Data formats for digital twins

While a universally adopted standard for developing a DT for construction projects is still emerging, several key data formats play a crucial role in the process. The Industry Foundation Classes (IFC) (ISO 16739-1:2018) format is the foundation, providing a structured data model for building components and their properties. BIM collaborating format (BCF) facilitates communication by linking real-world observations and updates to specific elements within the DT. 3D model formats like Revit (.rvt) or SketchUp (.sketchup) further enrich the DT by providing a visual representation. Sensor data formats (CSV, JSON, or proprietary) stream real-time information such as temperature or equipment status (Help, n.d.).

Beyond these core formats, Universal Scene Description (.usd) is gaining traction as a powerful format for enhancing the visual fidelity of DTs. USD is an open-source interchange format specifically designed for complex 3D scenes. It excels at representing a scene's geometry, materials, lighting, and animation data in an efficient and interoperable way between different software applications. This interoperability is particularly valuable in construction, where project stakeholders often use various design and rendering tools (Pixar Universal Scene Description (USD), n.d).

Here's how USD benefits DTs:

- **Enhanced visual quality**: USD allows for the import of high-fidelity 3D models from various sources, resulting in a more visually stimulating and immersive DT experience.
- **Improved collaboration**: the interoperable nature of USD makes it easier for different teams using different software to collaborate on the visual aspects of the DT.
- **Efficient data management**: USD's efficient data structure allows for managing complex scenes without sacrificing performance, which is crucial for real-time rendering within DTs.

While IFC, BCF, and 3D model formats remain the cornerstones of data exchange for DTs, USD offers a compelling option for projects that demand exceptional visual quality and seamless collaboration across diverse software environments. Understanding these data formats enables the user to navigate the DT development process and choose the right tools to construct a robust and informative digital representation of a building project.

11.4 User identification

While the overall concept of DTs follows similar concepts for most applications, the practical implementation varies depending on factors such as the level of detail and use cases for DTs. Additionally, setting up a DT on each platform requires a distinct procedure encompassing different levels of complexity. This implies that a careful decision and thorough study of user identification and case studies will be essential before implementation. Consequently, a comprehensive comparison matrix can serve as a roadmap for selecting the most appropriate platform based on DTs' requirements and potential use cases. Given the focus on the construction field encompassing both academia and industry sectors, the users of DTs within this domain can be broadly classified into three main groups as shown in Table 11.1: User identification is crucial in shaping the constraints and determining the project's scope. A user-defined perspective aids in pinpointing critical aspects of DTs relevant to specific user groups. For instance, integrating BIM data and corresponding sensors and actuators into DTs is a significant consideration. This integration involves exploring specific criteria to ensure a seamless and effective merging of these technologies. Today's widespread adoption of AI algorithms across diverse domains introduces a novel dimension to the DT landscape. This is particularly relevant

Table 11.1 User description

Development group:
The development group consists of architects, engineers, and construction managers involved in the design and construction phases of projects. Their role involves defining and providing essential building-related data while establishing mechanisms for the continuous updating of data within the DT
Management group:
Following the completion of construction, the control group, comprising facility managers, scrutinizes and maintains the data provided by the development group and from any embedded IOT sensors. Their focus lies in ensuring optimal operational performance and efficient maintenance scheduling for the facility.
Innovators:
Researchers and innovators within the realm of DTs explore new frontiers, experiment with emerging technologies, and contribute to the ever-evolving landscape of this technology through dedicated research and development initiatives.

for innovators and research groups exploring the potential of these advanced technologies within the realm of DT technology. Integrating AI algorithms into DTs requires careful exploration, as it introduces new criteria and considerations distinct from traditional aspects (Saldana Ochoa and Comes, 2021).

Each aspect – user-defined perspectives, BIM integration, and the incorporation of AI – comes with its criteria (refer to Section 11.4.1). These criteria are pivotal in selecting a suitable platform for DTs to meet specific requirements. The following section develops detailed criteria in a comparison matrix, taking into account the user's needs.

11.5 Evaluation criteria to analyze DT platforms

11.5.1 Visualization: development group (architects, engineers)

As the digital representation of physical assets and processes becomes increasingly integral to operations, the ability to render and interact with complex 3D models, dynamic data, and diverse levels of detail emerges as a cornerstone requirement. Therefore, it is critical to identify key criteria and the corresponding methods or tools for evaluating them. In this subsection, a selection of criteria to consider will be discussed (see Table 11.2).

Table 11.2 Evaluation criteria – development group

Criteria	*Description*
Rendering and performance	When evaluating DT platforms for the construction sector, rendering capabilities and performance are critical factors. The ability of platforms to efficiently handle intricate 3D models, dynamic data, and various levels of detail directly impacts user experience. Metrics such as real-time adaptability and processing speed for complex computations provide insights into platform effectiveness. Furthermore, responsiveness to user interactions and efficient utilization of system resources play vital roles in preventing lags or delays, ensuring seamless operation throughout the project lifecycle (Cabral et al., 2015).
Real time collaboration	Collaboration features are indispensable for design, construction, and maintenance teams across different project phases. Platforms facilitating real-time interaction and stakeholder coordination enhance communication and streamline decision-making processes. Compatibility and scalability: platforms should support various file formats and seamlessly integrate with other design and 3D modeling tools commonly used in the construction industry. Moreover, accessibility to computer-aided design (CAD) and BIM assets and libraries of materials and textures enhance the platform's utility. Scalability ensures that the platform can accommodate evolving project requirements and scale up to effectively meet the demands of large-scale construction projects (Sobotkiewicz and Milkowski, 2021).
Performance optimization	Platforms that offer performance optimization tools cater to the need for efficient resource utilization and enhanced rendering efficiency. Features such as LOD adjustments and occlusion culling improve overall platform performance, ensuring smooth operation even with complex data sets. DT platforms can deliver a superior user experience by implementing these optimization techniques, enabling stakeholders to focus on project objectives without technical constraints (Bethel et al., 2012).
Customized views	Offering perspectives relevant to stakeholders' needs enhances communication and decision-making within the stakeholders in the building life cycle. By allowing stakeholders to focus on the aspects pertinent to their domain expertise, customized views streamline collaboration and contribute to project success.

11.5.2 User interface (UI) and user experience (UX): development group (architects and engineers), control group (facility managers), innovators group (researchers and academia)

Selecting the right platform for your DT deployments pivots on more than just technical specifications. User experience (UX) and visualization capabilities are crucial to success. To facilitate this decision-making process, a comprehensive evaluation framework has been developed, encompassing various criteria crucial for assessing different platforms. This evaluation framework outlines comprehensive criteria to assess different platforms on critical aspects such as intuitiveness, collaboration features, visual programming support, interactivity, navigation, and community support (see Table 11.3). Weaving together metrics, evaluation methods, and specific considerations, this framework will empower users to make informed decisions prioritizing user empowerment and efficient DT creation and interaction.

11.5.3 Sensor-actuator connections: management group (facility managers) and innovators group (researchers and academia)

The sensor connection will be evaluated based on three main categories: model representation, data streaming and integration criteria, and interoperability and collaboration criteria (Kalantari et al., 2022; Khallaf et al., 2022).

Table 11.3 Evaluation criteria – user groups

Criteria	*Description*
Intuitiveness	Measure user task completion time and error rates for joint actions to gauge the learning curve and ease of use. Evaluate user satisfaction through subjective feedback and assess how well interface design leverages familiar patterns, offers clear labeling and help functions, and promotes discoverability with a logical information hierarchy (Crum, 2020).
Collaboration features	Look for real-time editing, shared workspaces, and collaborative tools that enhance teamwork. Verify alignment with industry-standard formats and assess platform-specific features like commenting annotation and access control (Bergstrom and Schall, 2014).
Visual programming language support	Evaluate the variety and complexity of visual programming functionalities. Test ease of use for non-programmers and their ability to build efficient workflows. Assess the availability of drag-and-drop functionalities, pre-built components, and the ability to extend workflows with scripting or low to no-code approaches.
Interactivity and navigation	Measure time spent navigating and manipulating elements within the DT. Consider user feedback on the fluidity, responsiveness, and intuitiveness of controls. Evaluate support for multiple views, camera angles, drill-down capabilities, real-time data visualization, and interactive simulations (Spolsky, 2008; Bergstrom and Schall, 2014).
Security	Analyze industry compliance certifications (e.g., ISO/IEC 27001:2022) and platform-specific security features. Evaluate user authentication, data encryption, and access control mechanisms. Assess support for multi-factor authentication, role-based access control, audit logging, intrusion detection, and integration with existing security infrastructure.

11.6 Comparison matrix

The increasing utilization of DTs in the construction sector, along with advancements in graphics processing unit (GPU)-related hardware, has led to a significant rise in their adoption. This trend has been accompanied by a continuous introduction of software platforms, each proficient in aspects of DT technology. Major companies such as Autodesk, NVIDIA, and Epic Games have developed platforms such as Tandem (2024), Omniverse (2023), and Unreal Engine (2023) respectively, which are highly relevant and capable in DT development for the building lifecycle of design, construction, operation, and maintenance. Consequently, these platforms were selected for creating a comparison matrix to assist stakeholders in building lifecycle management by choosing the most suitable platform for their DT applications. The criteria for comparison were established based on key aspects crucial for evaluating DT platforms, including visualization capabilities, user interfaces, sensor connectivity, and other relevant features essential for effective DT implementation. These criteria were identified through an analysis of industry standards, best practices, and expert knowledge in the field.

The comparative analysis was established through a systematic evaluation process. Each platform was assessed against the predetermined criteria, and the findings were organized into a comparison matrix (Table 11.4). This matrix provided a structured framework for comparing the strengths and room for improvement of each platform across different categories, allowing stakeholders to make informed decisions. The evaluation process involved input from end users, including two

Table 11.4 Evaluation criteria – management group

Criteria	*Description*
Level of detail (geometry, materials, systems, operational characteristics)	Thoroughly assess the depth of representation, considering geometry intricacies, material specifics, system intricacies, and operational characteristics in the sensor connection model.
Data resolution and sensor coverage	Evaluate the model's capability to achieve high data resolution and ensure comprehensive sensor coverage, allowing for a detailed and encompassing representation.
Validation methods (comparison with physical measurements, user feedback)	Establish rigorous validation procedures, incorporating comparisons with physical measurements and gathering user feedback to ensure the accuracy and reliability of the sensor-connected data.
Real time and historical data streaming	Evaluate the system's ability to seamlessly stream both real-time and historical data, ensuring a dynamic representation of the environment over time.
Data acquisition (sensors, IoT networks)	Assess the efficiency of data acquisition mechanisms, considering the integration of various sensors and IoT networks for comprehensive data collection.
Data processing	Examine the data processing capabilities, focusing on the system's efficiency in managing and processing the diverse data acquired through sensors and IoT networks.
Internal simulation capabilities	Evaluate the model's internal simulation capabilities, assessing its ability to simulate different scenarios and environmental changes.
Connection to external simulation applications	Determine the model's compatibility and seamless integration with external simulation applications, allowing for enhanced simulation capabilities and broader insights.
Integration with other systems and platforms	Assess the model's capacity to integrate seamlessly with other systems and platforms, fostering collaboration and data exchange across diverse environments.

architects, four engineers, and seven researchers, involved in the field of DT. Their engagement in the evaluation process provided valuable insights and perspectives, ensuring a comprehensive assessment of the DT platforms (see Table 11.5).

11.7 Discussion

The comparison matrix of DT platforms encompasses three key dimensions, catering to distinct user groups and their specific needs. In the visualization criteria, we focused on the development group, which includes architects and engineers; the comparison focuses on rendering capabilities, real-time collaboration, compatibility, scalability, performance optimization, and customized views. The criteria for the user interface encompass intuitiveness, collaboration features, visual programming language support, interactivity, navigation, and security. Last, in the context of sensor-actuator connections, particularly for the management group (facility managers) and innovators group (researchers and academia), the evaluation criteria are model representation, data streaming, integration, interoperability, and collaboration. These criteria cover aspects such as the level of detail, data resolution, validation methods, real-time and historical data streaming, data acquisition, data processing, internal simulation capabilities, connection to external simulation applications, and interoperability with other systems and platforms. The comprehensive framework provides a structured approach for assessing DT platforms across diverse functionalities, ensuring alignment with user requirements and optimal performance throughout the project lifecycle.

The findings from evaluating three DT platforms – Autodesk Tandem, Epic Games Unreal Engine, and NVIDIA Omniverse – reveal distinctive features and performance when tested in the context of the building life cycle. Autodesk Tandem demonstrates robust support for various 3D formats, including Revit and IFC, with additional compatibility for Alembic and USD. However, its convertibility and visual programming language support are currently unavailable, posing challenges for beginner users. On the other hand, Epic Games and Unreal Engine have tools for convertibility, visual programming language support, and navigation, making them highly accessible for users with various expertise levels. It offers good compatibility with mixed reality (MR), virtual reality (VR), and augmented reality (AR) tools and supports real-time ray tracing. NVIDIA Omniverse stands out in AI compatibility and real-time ray tracing. Notably, all three platforms exhibit excellent support for 3D asset libraries, material libraries, and extensions, with varying community support for sensor connections. Epic Games Unreal Engine and NVIDIA Omniverse offer better simulation features and real-time analytics support than Autodesk Tandem. However Tandem excels in bi-directional data flow capabilities. Moreover, customizing views is excellent across all platforms. Each platform presents unique strengths and weaknesses, emphasizing the importance of aligning platform choice with specific project requirements and user needs.

11.8 Conclusions

This chapter emphasizes the early stage of DTs in this domain, calling for a systematic review of the transition from BIM to DTs. Advances in technology have transformed construction practices, integrating digital tools for decision-making in facilities and asset management. However, employing BIM for operation and management is still not fully exploited, especially when reflecting dynamic building conditions, highlighting data interoperability and automation as crucial in the AECO domain. The user-defined perspective is critical for delineating constraints and defining the scope of DT platforms' comparative matrix.

Table 11.5 A matrix comparative analysis of the platforms used to create DT: Tandem, Unreal Engine, and Omniverse

Category	*Criteria*	*Reason*	*Autodesk Tandem*	*Epic Games Unreal Engine*	*NVIDIA Omniverse*
Visualization and UI/UX	**3D format support**	DT are 3D environments;, therefore 3D file format is a requirement.	Available, supports Revit (.rvt), AutoCAD (.dwg), Industry Foundation Classes (IFC), and Navisworks (.nwd) data formats.	Available, supports formats such as Alembic and Universal Scene Description (USD).	Available, supports USD.
	Convertibility	**Convertibility** evaluates how accurate the conversion of **lighting** and **materials** from different platforms into the DT platform are.	**Pro:** • Simple export, no adjustment needed. **Con:** • Material properties cannot be changed in Tandem.	**Pro:** • Large community-driven marketplace supported by NVIDIA. **Con:** • Complexity for beginner, same as Omniverse.	**Pro:** • Supported by NVIDIA's material definition language (MDL). **Con:** • Limited material libraries compared to Unreal Engine.
	Visual programming language support	Visual scripting system allows non-programmers to create complex interactions with the available APIs.	Available • Provided with Tandem Connect.	Available • Provided with Blueprint System.	Available • Provided with Action Graph.
	Navigation	How easy is navigating in the virtual 3D scene?	Available.	Available.	Available Following the same system as Unreal, it is hard to get familiar at first with, but it gradually becomes handy.
	MR/VR/AR compatibility	How compatible are the platforms with different MR/AR/VR tools?	Currently unavailable.	Available.	Available.
	Rendering	Support of real-time ray tracing.	Currently unavailable.	Available.	Available.
	3D asset libraries, material libraries, extensions		Available.	Available.	Available.

(*Continued*)

Table 11.5 (Continued)

Category	*Criteria*	*Reason*	*Autodesk Tandem*	*Epic Games Unreal Engine*	*NVIDIA Omniverse*
Sensor connection	**Community support**		Available.	Available.	Available.
	Customizing views		Available.	Available.	Available.
	Bidirectional data flow between DT and sensors	1) Support for bidirectional communication with physical devices 2) Communication protocols for seamless interaction with the physical system.	**Pro:** 1) Support for Azure IoT hub connection, which can handle bidirectional communication. **Con:** 1) No direct support for bidirectional communication.	**Pro:** 1) Support for Azure IoT hub and other IoT hubs connection, which can handle the bidirectional communication. 2) Direct support for bidirectional communication (APIs).	**Pros:** 1) External D.B.B connection to sensor data 2) External API integration for dataflow through Omniverse Nucleus (Connect). **Cons:** 1) Confinement to Omniverse setup 2) API and External D.B.B setup is cumbersome.
	Simulation	1) Inherent simulation features for predictive/energy analysis 2) Capability to connect with other simulation engines.	**Pro:** 1) Data can be visualized (time series plots and heat maps) **Con:** 1) No inherent simulation engine.	**Pros:** 1) Integration capability with other simulation engines 2) External API integration for connection with other simulation engines. **Cons:** 1) No inherent simulation features 2) Deprecated content for API connections.	**Pros:** 1) Integration capability with other simulation engines 2) External API integration for connection with other simulation engines. **Cons:** 1) No inherent simulation features 2) API and External D.B.B setup is cumbersome.

(*Continued*)

Table 11.5 (Continued)

Category	*Criteria*	*Reason*	*Autodesk Tandem*	*Epic Games Unreal Engine*	*NVIDIA Omniverse*
	Realtime analytics*	1) Real time data visualization and monitoring tools 2) Analytics capabilities for extracting meaningful insights from data.	**Pros:** 1) In-built tools for data visualization (time series charts and heat maps) 2) Realtime data visualization from connected sensors. **Con:** 1) Predefined analytics and plots	**Pro:** 1) External DB/API integration for IoT data 2) Realtime data visualization from connected sensors. **Con:** 1) No in-built tools for data visualization.	**Pros:** 1) External DB/API integration for IoT data 2) Realtime data visualization from connected sensors. **Con:** 1) No in-built tools for data visualization.
	User interface and user experience (UI/UX)*	1) Intuitive interface for users to interact with the DT 2) Customization options for different user roles.	**Pros:** 1) In-built U.I.I. 2) Predefined user roles **Con:** 1)Predefined U.I.I. not customizable.	**Pro:** 1) Customizable U.I.I. **Cons:** 1) U.I.I. has to be developed from scratch 2) No predefined user roles.	**Pro:** 1) Customizable U.I.I. **Cons:** 1) U.I.I. has to be developed from scratch 2) No predefined user roles.
Facility management and operations	**BIM data transfer** Relationships	1) Extract relational information about elements 2) Filter system components.	**Pros:** 1) Systems are interconnected 2) Systems can be filtered with built-in features. **Con:** 1) Dependent on BIM modeling accuracy.	**Pro**: 1) BIM data can be imported. **Con**: 1) Systems are not connected by default.	**Pro**: 1) BIM data can be imported. **Con**: 1) Systems are not connected by default.
	Custom views	Create views based on systems/ components to be monitored.	1) Built-in system views. 2) Built-in features to create and save custom views.	Customizable.	Customizable.

(Continued)

Table 11.5 (Continued)

Category	*Criteria*	*Reason*	*Autodesk Tandem*	*Epic Games Unreal Engine*	*NVIDIA Omniverse*
	User alert	Produce event notifications to external users.	Currently unavailable.	Programmable.	Programmable.
	Fault detection and diagnostics using AI/ML	1) Detect anomalies. 2) Provide detailed insights about suboptimal equipment performance.	Currently unavailable.	Programmable.	Programmable.
	Document management	Store asset-related documents (O&M manual, cut sheets, operational reports, maintenance logs).	Available.	Programmable.	Programmable.

Effective communication and collaboration between these user groups are essential for DTs' successful implementation and evolution. The development group's insights should feed into the Management Group's operational decisions, while feedback from facility managers will inform potential refinements in the DT model. Innovators, bridging the gap between academia and practical applications, contribute to the ongoing evolution of DTs. Moving forward, addressing challenges related to data interoperability, security, and user interface design will be crucial for enhancing the collaborative potential of these user groups and unlocking the full capabilities of DTs in the built environment.

This chapter presents a comparison matrix of Autodesk Tandem, Epic Games Unreal Engine, and NVIDIA Omniverse, highlighting their respective strengths and weaknesses within the context of our specific application in the building lifecycle. While Autodesk Tandem demonstrates robust 3D format support, Epic Games Unreal Engine excels in convertibility and user-friendly features. NVIDIA Omniverse stands out in AI compatibility and real-time ray tracing, but it also has limitations and inconvertibility issues.

Finally, integrating BIM data with sensors and actuators and incorporating AI algorithms are critical aspects requiring further exploration, especially for innovators and researchers. The next steps of this study will involve integrating criteria for AI models to be executed within the DT platforms. This includes assessing how well each platform accommodates and supports the integration of AI algorithms, emphasizing factors such as compatibility with machine learning frameworks and ease of implementation for AI models to facilitate enhanced decision-making and analytics within the DT environment. The evaluation should consider the platforms' support for real-time interaction with AI models, seamless integration with APIs, and their ability to process and analyze data generated by AI algorithms. By incorporating these criteria, the analysis will provide insights into the platforms' effectiveness in harnessing the power of AI to optimize and augment the functionalities of DTs across diverse applications and in the construction industry.

References

Architecture Engineering construction. (n.d.). *Architecture-engineering-construction*. https://www.autodesk.com/solutions/digital-twin/

Azhar, S., A. Nadeem, J. Y. Mok, & B. H. Leung. (2008). Building Information Modeling (BIM): A new paradigm for visual interactive modeling and simulation for construction projects. In *Proceedings of the First International Conference on Construction in Developing Countries*.

Bahar, Y. N., C. Pere, J. Landrieu, & C. Nicolle. (2013). A thermal simulation tool for building and its interoperability through the Building Information Modeling (BIM) platform. *Buildings*, *3*(2), 380–398. https://doi.org/10.3390/buildings3020380

Bergstrom, J. R., & A. Schall. (Eds.). (2014). *Eye tracking in user experience design*. Elsevier. https://api.semanticscholar.org/CorpusID:40789776

Bethel, E. W., H. Childs, & C. Hansen. (Eds.). (2012). *High performance visualization: Enabling extreme-scale scientific insight*. CRC Press. https://doi.org/10.1201/b12985

Cabral, I., A. P. Souto, H. Carvalho, & J. Cunha. (2015). Exploring geometric morphology in shape memory textiles: Design of dynamic light filters. *Textile Research Journal*, *85*(18), 1919–1933. https://doi.org/10.1177/0040517515578

Crum, L. (2020). Laws of UX: Using psychology to design better products & services, by Jon Yablonski Sebastopol. *CA: O'Reilly Media*, *152*, 9781492055310 (US 44.99/CAD 59.99). https://doi.org/10.1080/17547075.2020.1822074

Dave, B., A. Buda, A. Nurminen, & K. Främling. (2018). A framework for integrating BIM and IoT through open standard. *Automation in Construction*, *95*, 35–45. https://doi.org/10.1016/j.autcon.2018.07.022

Deng, M., V. J. L. Gan, Y. Tan, A. Joneja, & J. C. P. Cheng. (2019). Automatic generation of fabrication drawings for façade mullions and transoms through BIM model. *Advanced Engineering Informatics*, *42*, 100964. https://doi.org/10.1016/j.aei.2019.100964

Deng, M., Y. Tan, J. Singh, A. Joneja, & J. C. P. Cheng. (2021). A BIM-based framework for automated generation of fabrication drawings for façade panel. *Computers in Industry, 126*, 103395. https://doi.org/10.1016/j.compind.2021.103395

Fuller, Z. Fan, C. Day, & C. Barlow. (2020). Digital twin: Enabling technologies challenges and open resear "h". *IEEE Access, 8*, 108952–108971. https://doi.org/10.1109/ACCESS.2020.2998358

Gubbi, J., R. Buyya, S. Marusic, & M. Palaniswami. (2013). Internet of Things (IoT): A vision, architectural elements, and future direction. *Future Generation Computer Systems, 29*(7), 1645–1660. https://doi.org/10.1016/j.future.2013.01.010

Guo, Z., K. Saldana Ochoa, & P. D'Acunto. (2022, September). Enhancing structural form-finding through a text-based AI engine coupled with computational graphic statics. In *Proceedings of IASS Annual Symposia* (Vol. 2022, No. 8, pp. 1–11). International Association for Shell and Spatial Structures (IASS).

Help. (n.d.). https://help.autodesk.com/view/RVT/2022/ENU/?guid=GUID-0D546BEA-6F88-4D4E-BDC1-26274C4E98AC 2022

Kalantari, S., S. Pourjabar, S., T. B. Xu, & J. Kan. (2022). Developing and user-testing a "digital twins" prototyping tool for architectural design. *Automation in Construction, 135*, 104140. https://doi.org/10.1016/j.autcon.2022.104140

Khallaf, R., L. Khallaf, C. J. Anumba, & O. C. Madubuike. (2022). Review of digital twins for constructed facilities. *Buildings, 12*(11), 2029. https://doi.org/10.3390/buildings12112029

Kritzinger, W., M. Karner, G. Traar, J. Henjes, & W. Sihn. (2018). Digital twin in manufacturing: A categorical literature review and classification. *IFAC-PapersOnLine 51*(11), 1016–1022.

Lee, D., G. Cha, & S. Park. (2016). A study on data visualization of embedded sensors for building energy monitoring using B.I.I. *International Journal of Precision Engineering and Manufacturing, 17*(6), 807–814. http://dx.doi.org/10.1007/s12541-016-0099-4

Li, C. Z., F. Xue, X. Li, J. Hong, & G. Q. Shen. (2018). An Internet of Things-enabled BIM platform for on-site assembly services in prefabricated construction. *Automation in Construction, 89*, 146–161. https://doi.org/10.1016/j.autcon.2018.01.001

Menassa, C. C. (2021). From BIM to digital twins: A systematic review of the evolution of intelligent building representations in the AEC-FM industry. *Journal of Information Technology in Construction (ITcon), 26*(5), 58–83. http://dx.doi.org/10.36680/j.itcon.2021.005

Natephraa, W., & A. Motamedib. (2019). Live data visualization of IoT sensors using Augmented Reality (AR) and BIM. ISARC. In *Proceedings of the International Symposium on Automation and Robotics in Construction.* IAARC Publications. https://doi.org/10.22260/ISARC2019/0084

Negri, E., L. Fumagalli, & M. Macchi. (2017). A review of the roles of digital twin in CPS-based production system. *Procedia Manufacturing, 11*, 939–948. https://doi.org/10.1016/j.promfg.2017.07.198

Oh, S., Y.-J. Kim, C.-S. Park, & I.-H. Kim. (2011). Process-driven BIM-based optimal design using integration of Energy Plus, genetic algorithm, and pareto optimality. In *Proceedings of the IBPSA building simulation 2011 conference*, Sydney, Australia. https://doi.org/10.26868/25222708.2011.1354

Omniverse. (2023) *NVIDIA*. https://www.nvidia.com/en-us/omniverse/ (access 2023).

Omrany, H., K. M. Al-Obaidi, A. Husain, & A. Ghaffarianhoseini. (2023). Digital twins in the construction industry: A comprehensive review of current implementations, enabling technologies, and future directions. *Sustainability, 15*(14), 10908. https://doi.org/10.3390/su151410908

Ozturk, G. B. (2021). Digital twin research in the AECO-FM industry. *Journal of Building Engineering, 40*, 102730.

Pan, Y., & L. Zhang. (2021). A BIM-data mining integrated digital twin framework for advanced project management. *Automation in Construction, 124*, 103564. https://doi.org/10.1016/j.autcon.2021.103564

Pixar Universal Scene Description (USD). (n.d.). *NVIDIA developer*. https://developer.nvidia.com/usd

Saldana Ochoa, K., & T. Comes. (2021). A Machine learning approach for rapid disaster response based on multi-modal data. The case of housing & shelter needs. *arXiv preprint*. https://doi.org/10.48550/arXiv.2108.00887

Saldana Ochoa, K., P. O. Ohlbrock, P. D'Acunto, & V. Moosavi. (2021). Beyond typologies, beyond optimization: Exploring novel structural forms at the interface of human and machine intelligence. *International Journal of Architectural Computing, 19*(3), 466–490. https://doi.org/10.1177/1478077120943062

Shafto, M., M. Conroy, R. Doyle, E. Glaessgen, C. Kemp, J. LeMoigne, & L. Wang. (2010). Draft modeling, simulation, information technology & processing roadmap. *Technology Area, 11*.

Singh, J., M. Deng, & C. P. Cheng. (2018). Implementation of mass customization for MEP layout design to reduce manufacturing cost in one-off projects. In *IGLC 2018-proceedings of the 26th annual conference*

of the international group for lean construction: Evolving lean construction towards mature production management across cultures and frontiers. http://dx.doi.org/10.24928/2018/0519
Sobotkiewicz, M., & A. Milkowski. (2021, January 11). Real-time collaboration for the construction industry. *KanBo*. https://kanboapp.com/en/blog/real-time-collaboration-for-the-construction-industry/
Soust-Verdaguer, B., C. Llatas, & A. García-Martínez. (2017). Critical review of BIM-based LCA method to building. *Energy and Buildings*, *136*, 110–120. https://doi.org/10.1016/j.enbuild.2016.12.009
Spolsky, A. J. (2008). *User interface design for programmers*. Apress.
Srinivasan, R. S., B. Manoharan, & R. R. AIssa. (2020). Urban building energy CPS (UBE-CPS): Realtime demand response using digital twin. In *CPS in the Built Environment*. Springer. http://dx.doi.org/10.1007/978-3-030-41560-0_17
Tagliabue, L. C., S. Maltese, F. Re Cecconi, A. L. C. Ciribini, & E. De Angelis. (2018). *BIM-based interoperable workflow for energy improvement of school buildings over the life cycle*. http://dx.doi.org/10.22260/ISARC2018/0058
Tandem. (2024). Homepage – Autodesk tandem. (2024, February 14). *Autodesk tandem*. https://intandem.autodesk.com/ (access 2023).
Tang, S., D. R. Shelden, C. M. Eastman, P. Pishdad-Bozorgi, & X. Gao. (2019). A review of building information modeling (BIM) and the internet of things (IoT) devices integration: Present status and future trends. *Automation in Construction*, *101*, 127–139. https://doi.org/10.1016/j.autcon.2019.01.020
Tchana, Y., G. Ducellier, & S. Remy. (2019). Designing a unique digital twin for linear infrastructures lifecycle management. *Procedia CIRP*, *84*, 545–549. https://doi.org/10.1016/j.procir.2019.04.176
Unreal Engine 5.3 Documentation. (n.d.). *Epic developer community*. https://dev.epicgames.com/documentation/en-us/unreal-engine/unreal-engine-5-3-documentation. (access 2023)
Valinejadshoubi, M., O. Moselhi, A. Bagchi, & A. Salem. (2021). Development of an IoT and BIM-based automated alert system for thermal comfort monitoring in buildings. *Sustainable Cities and Society*, *66*, 102602. https://doi.org/10.1016/j.scs.2020.102602
van Nederveen, G. A., & F. P. Tolman. (1992). Modelling multiple views on building. *Automation in Construction*, *1*(3), 215–224. https://doi.org/10.1016/0926-5805(92)90014-B
Vanlande, R., C. Nicolle, & C. Cruz. (2008). IFC and building lifecycle management. *Automation in Construction*, *18*(1), 70–78. https://doi.org/10.1016/j.autcon.2008.05.001
Xie, H., M. Xin, C. Lu, & J. Xu. (2023). Knowledge map and forecast of digital twin in the construction industry: State-of-the-art review using scientometrics analysis. *Journal of Cleaner Production*, *383*, 135231. https://doi.org/10.1016/j.jclepro.2022.135231
Yitmen, I., S. Alizadehsalehi, İ. Akıner, & M. E. Akıner. (2021, May). An adapted model of cognitive digital twins for building lifecycle management. *Applied Sciences*, *11*(9), 4276. https://doi.org/10.3390/app11094276
Zhang, H., L. Ma, J. Sun, H. Lin, & M. Thürer. (2019). Digital twin in services and industrial product service systems: Review and analysis. *Procedia CIRP*, *83*, 57–60. https://doi.org/10.1016/j.procir.2019.02.131
Zhuang, C., J. Liu, & H. Xiong. (2018). Digital twin-based smart production management and control framework for the complex product assembly shop-floor. *International Journal of Advanced Manufacturing Technology*, *196*, 1149. https://link.springer.com/article/10.1007%2Fs00170-018-1617-6

12

DIGITAL TWIN-BASED PROTOTYPE SYSTEM FOR HEALTHCARE FACILITIES MANAGEMENT

Obinna C. Madubuike[1] *and Chimay J. Anumba*[2]

[1]M.E. Rinker Sr. School of Construction Management, University of Florida, Gainesville, Florida, USA

[2]College of Design, Construction and Planning, University of Florida, Gainesville, Florida, USA

Email address: madubuikefrancis43@gmail.com (Obinna C. Madubuike), anumba@ufl.edu (Chimay J. Anumba) (Obinna C. Madubuike)

12.1 Introduction

Technology continually develops over time. The technology comprises bits and atoms that continually develop and ensure work coordination that enables much work to be done with little effort (Negroponte, 1995). The growth in technology has also led to the advancement in virtual modeling and data acquisition technology, which has led to the development of the concept of digital twins. digital twin (DT) technology involves the development of a living digital model of the physical asset, which continually adapts to changes in the physical environment or operations and delivers the best result (Parris et al., 2016). DT can improve based on the accumulated data from the physical environment achieved through the effective simulation of data obtained using embedded sensors.

A DT comprises three major components – the physical component, the digital (virtual) component, and data linking the physical to the digital component as shown in Figure 12.1. A DT is created by first defining the purpose and then developing a virtual model of the physical object by using and analyzing the data obtained from the physical object and creating a synchronized link between the virtual model and the physical object for easy monitoring (GAVS, 2017). Additionally, DT can improve a physical entity's efficiency, lifespan, and operational cost by providing opportunities for a predictive and proactive maintenance cycle (Lui et al., 2018). This is made possible because DT employs tools and technologies to program the stored data of the physical object and provide up-to-date information on the physical object (Palit and Datta, 2017). DT uses synchronized real-time information coordination between the physical and virtual models to achieve its aims.

However, DT is yet at a nascent stage, particularly in the field of construction with more focus on building management. Adequate studies are yet to be carried out to show the DT

DOI: 10.1201/9781003383840-13

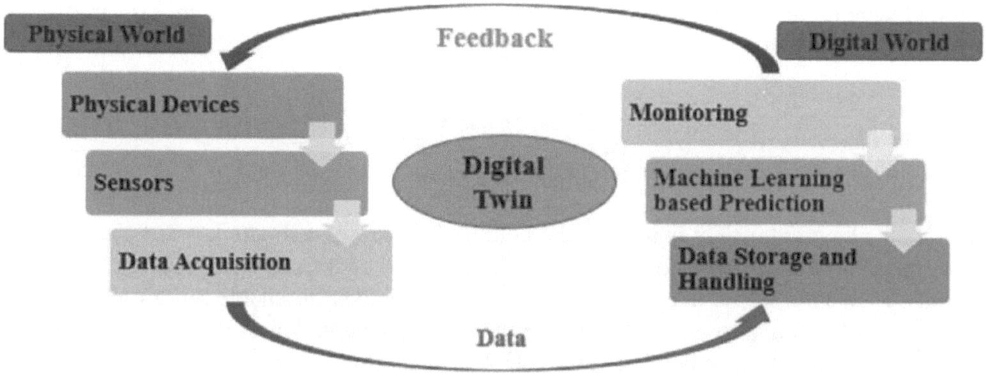

Figure 12.1 DT framework comprises the physical world, digital world, and data linkage

applications in the construction industry. However, DT has shown diverse and acknowledged applications in other industrial sectors such as aerospace, healthcare, manufacturing, and others (Madubuike et al., 2022). This chapter describes a study on the applicability of a DT-based system for healthcare facilities management (HFM). It provides the background to DT, including its history, definitions, and key features. The chapter also discusses the application of DT in both the construction industry and other industries and highlights the study's focus on the application of DT in HFM. This chapter also includes an overview of healthcare facilities, DT applications in HFM, a case study to guide the development of DT deployment scenarios, prototyping of the adopted DT scenarios, evaluation and validation of the developed DT prototype, and conclusions.

12.2 Methodology

The study described here had four main objectives, which include (a) undertaking a review of HFM to identify operational deficiencies and limitations of conventional technologies and approaches, (b) investigating the concept of digital twin and its applications in various industry sectors including construction, (c) conducting empirical studies and experimentation to explore the applicability of DT technology to HFM, and (d) developing and testing a prototype DT application for HFM. To address the objectives, the study adopted a mixed-method approach, which combined both qualitative and quantitative research tools. The tools employed including their justifications for adoption are provided in the following sections:

12.2.1 Extensive literature review

The study adopted an extensive literature approach in gathering some of the selected literature. The main purpose of the literature review was to assist in developing the needed understanding of the whole available research on the topic and to identify the strengths and weaknesses of the study (De Los Reyes and Kazdin, 2008). This study adopted an 'extensive review technique' to critically analyze different studies within the field of DT and use findings to address the main research problem of this study. The extensive literature review provided the basis to review various literature in the area of DT including the concepts, areas where DT has been applied, enabling technologies of previous applications in other fields, and possible cues for implementing DT in any new fields.

12.2.2 Case study

A case study is one of the most common approaches applied in the field of information systems because of its multiple perspectives rooted in specific contexts, and it provides multiple data collection methods. A case study is used in the field of qualitative methodology and accounts for a substantial proportion of the research presented in different books and articles (Rebolj, 2013). In this study, a case study that adopted interviews was undertaken to enable more findings on the problems with HFM and to provide a facility for the implementation of the proposed DT-based for HFM. It identified the following steps established to guide its implementation – select the healthcare facilities, identify underlying features and facility management standards put in place, identify problem areas where the DT concept is applicable, and select a candidate room to implement a DT solution.

The case study adopted was a large healthcare facility belonging to one of the prestigious public universities in Florida. The choice of the healthcare facility was based on accessibility to the required data and the facility possessing the features essential to achieving the aim of the research. The healthcare facility is about 139,000 sq. ft. and has five operating rooms, a waiting room, a family consult room, 19 pre-operations, and post-anesthesia care units. The choice-selected rooms for this research were the MRI and operating rooms. The research limited its scope to these rooms due to the available sensors and difficulty in accessing the other rooms for research purposes since the other spaces were in use at the time of the study. The case study employed interviews that had three different participants discussing the state of the healthcare facilities' management procedures.

12.2.3 Interviews

An interview is a research methodology used to obtain information from a selected population or individual. An interview is a qualitative tool that entails making inquiries and obtaining results from selected participants in a study. The interview approach is applied in certain conditions. Easwaramoorthy and Zarinpoush (2006) identified that the interview method should be employed when the research requires in-depth information on people's opinions, thoughts, experiences, and feelings. Interviews could be carried out in person or through phone calls.

12.2.4 Scenario planning

Scenarios are coherent and consistent depictions of different theoretical futures that reflect different viewpoints on past, present, and future developments, which can serve as a foundation for action (van Notten, 2005). The scenarios developed in this research used diagrams and narrations to explain the potential of DT application for healthcare facilities management (HFM) as provided in the 'Scenario Development' section. This study adopted the exploratory scenario development methodology as it provides bases to use the information obtained from the case study and interview to explore possible DT solutions applicable to HFM (see Figure 12.2).

The developed scenarios were validated by the focus group using the questionnaire method of data collection. The questionnaire was structured such that it aided the participants in understanding the developed scenarios and the posed questions for the scenarios. The questionnaire combined

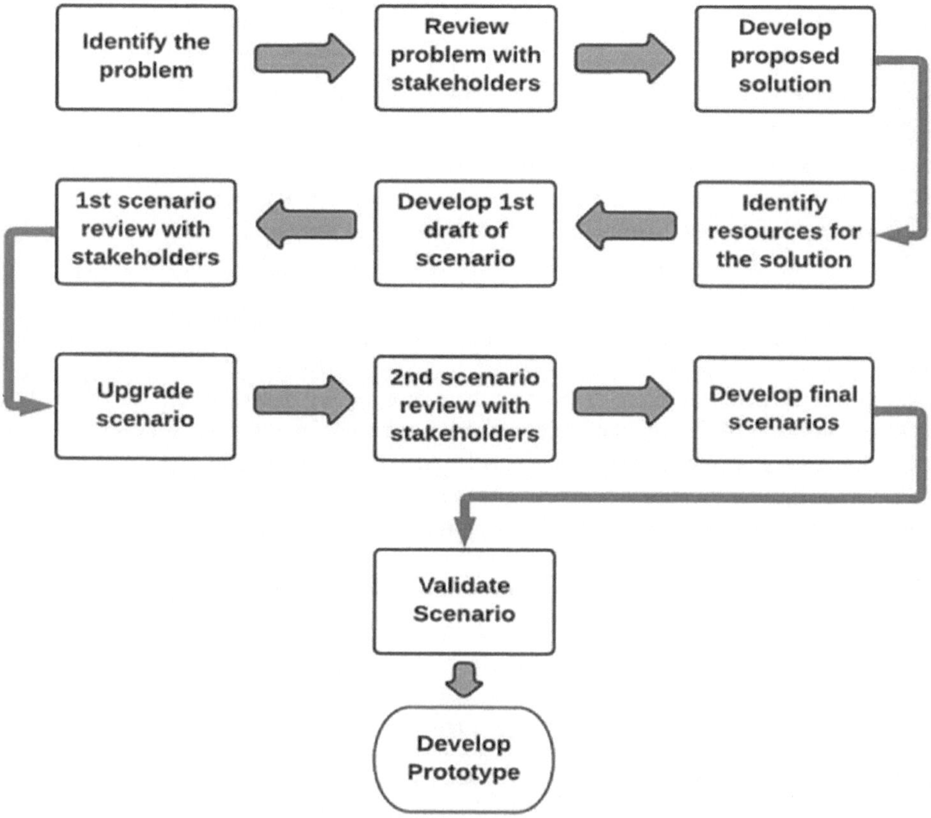

Figure 12.2 Scenario planning process

both the Likert scale and open-ended questions. The Likert scale used different measuring scales as it affects the scenario questions. The responses obtained from the questionnaire were analyzed using statistical tools to select what scenario(s) to adopt for prototyping.

12.2.5 Prototype development

Prototypes are created to study the feasibility of an intended system to be developed. Prototypes are used where a certain amount of artificial intelligence is needed in the decision process of a system. This research adopts the rapid prototyping method to address the related objective. The rapid prototyping method was adopted to enable quick development of the DT-base system for HFM. The rapid prototyping method enables the development of new technologies from the three-dimensional computer-aided design of any object.

12.2.6 Focus group

To validate the developed prototype, a system evaluation technique was adopted. A focus group was adopted for system evaluation. System evaluation entails assessing the operational functionalities of the developed prototype and providing feedback to validate the system. The focus group

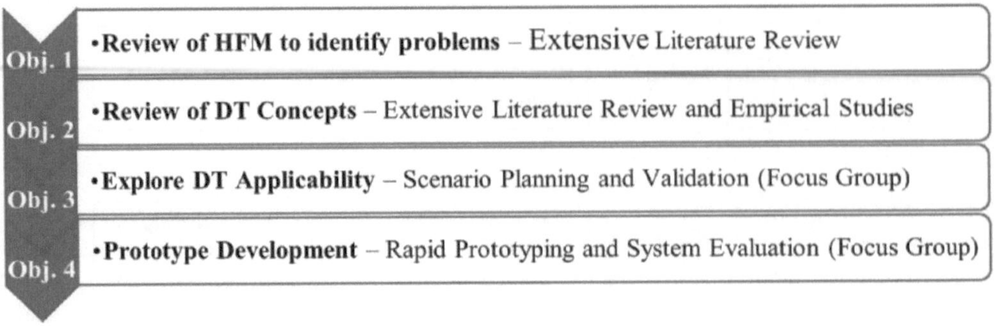

Figure 12.3 Summary of the adopted research methodology

was used for the system evaluation to validate the prototype DT-based system for HFM. A focus group is the most appropriate method adopted when testing a new model to develop new ideas formed within a social context (Breen, 2006). Adequate care was taken to select participants with the right knowledge and technical expertise in building automation systems (BAS) and wireless sensing. The validation through the focus group was achieved using statistical tools such as frequency, graphs, mean, and standard deviations to ascertain the reliability of the stakeholders' responses. Figure 12.3 provides a summary of the research methodology adopted for this study.

12.3 Digital twins – key concepts

12.3.1 History

The earliest mention of DT was barely two decades ago, and it was applied in the field of product lifecycle management (PLM) by Micheal Grieves (Grieves, 2005). PLM integrates different business activities that create, modify, and use data to support all product lifecycle phases from design and production through use and maintenance to recycling and disposal (Schroeder et al., 2016). However, within the DT context, the earliest mention of the term 'twin' was in the field of aerospace, traced back to NASA Technology Roadmaps – the 1960 NASA Apollo project (Shafto et al., 2010; Negri et al., 2017; Boschert et al., 2018). The NASA Apollo project entailed creating another space vehicle on Earth, which was the physical twin to mirror the space vehicle in space while performing the mission (Boschert et al., 2018).

DT has shown applicability in various fields and industries including the construction industry. All DT applications follow the same line of thought – creating a digital model linked to the physical entity using smart devices and communication networks to update the digital model in real time (Madubuike et al., 2022). Grieves and Vickers (2017) opined that DT must include all information about the physical system/asset obtained through diligent inspection of the real world to ensure that DT optimally represents the physical system. Figure 12.4 shows the history and milestone progress made in the development of the concept of DT.

12.3.2 Definitions of digital twins

The first step in any DT development is to define the purpose of the digital twin and what it is meant to accomplish. Authors have coined several definitions to explain what DT technology means. The first DT definition was by Michael Grieves in his PLM course where he defined DT as an information mirroring model (Grieves, 2005). Subsequently, other authors have provided

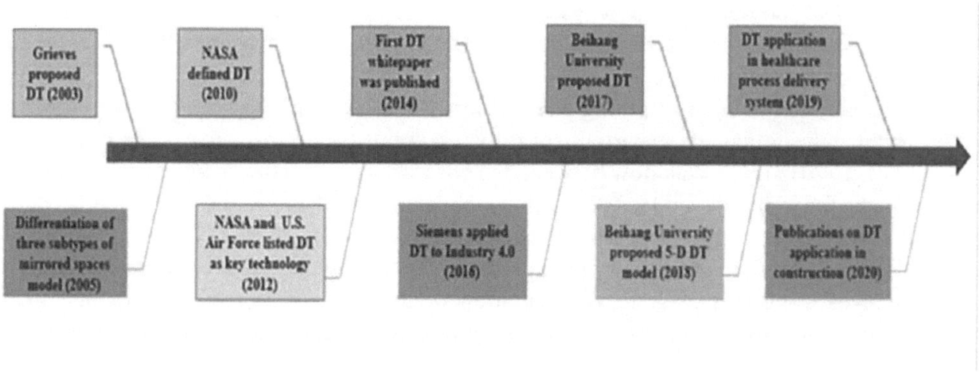

Figure 12.4 History and milestone progress with DT concept development
Source: Madubuike et al., 2022.

more detailed definitions for DT with context in their fields of applications. Rosen et al. (2015) defined DT as the combination of both physical and virtual spaces, mirroring each other to evaluate the lifecycle operations of the physical components. This definition focused on highlighting the major components of DT and how they link to each other. Shortly thereafter, Boschert et al. (2018) coined a more concise DT definition as a technology encompassing all the system's useful physical and functional data.

Grieves and Vickers (2017) expanded on the previous DT definition in the context of PLM as a set of virtual information constructs that provide a true representation of a potential or actual physical manufactured product from its atomic level to the macro geometrical level. Consequently, this study defines DT as the virtual representation of a physical asset with the aid of DT-enabling technologies such as sensors, communication networks, and 3D models to obtain real-time updates and effect bi-directional coordination such that the virtual model truly represents the physical asset (Madubuike et al., 2022). This definition identifies the major components of DT including the enabling technologies and the purpose of the DT.

12.3.3 Key features (DT components and data dynamics)

DT is one of the fast-growing technologies applicable in different fields and industries given its types of features and components. It operates on the principle of the interaction between the physical and virtual worlds, providing the virtual world the opportunity to monitor the performance of the physical world (GAVS, 2017). DT is identified as one of the most promising state-of-the-art technologies with the possibility of creating a virtual twin of a physical asset (Martinez et al., 2018). DT is made up of the physical asset, virtual model, sensors, communication networks, and simulation engine. Each component plays an intricate role in enabling DT technology to carry out its functions.

Glaessgen and Stargel (2012) identified that DT comprises three major parts: the physical system, the virtual system, and the communication link between them. Castaldini (2019), on the other hand, opined that DT has four key components, which include data (such as external data, blueprint data, and building systems), reasoning (such as AI/ML models or nonlinear rules), key performance indicators (such as efficiency, safety indicators, emissions, and net operating income), and context (such as system/device behaviour, occupant behaviour, and workflows). The DT's major parts and key components encompass the dynamics required to define the behavioral pattern of the DT. It is important to note that the dynamics of DT are the core aspects of any DT technology as it defines the performance and outcome of the DT. The dynamics of DT depend on the data obtained from the physical asset through the sensors and then simulating the data with the digital model such that the digital model partially or fully represents the performance of the physical asset (Boschert et al., 2018). The data flow between the physical asset and the virtual model is the brainbox behind the functionality of DT. The continuously obtained data from the physical asset enables the DT to continuously adapt to operational changes of the physical asset based on the information and online data collection, providing future predictions that may impact the physical asset.

DT involves a systems-of-systems integration of the various sources of data that include embedded sensors, wireless sensor networks, digitized building life-cycle data and systems, and integration with other cloud services and data providers (Haag and Anderl, 2018). Sensors play a critical role in any DT endeavors as they help collect the required data from the physical asset, which is then synchronized with the virtual model in real time. The sensors collect data in real time and update the virtual model continuously (Lui et al., 2018). The data obtained from the physical asset is synchronized to the virtual model using communication networks, which is an essential part of the DT (Madubuike et al., 2022). Conclusively, a study by Madni et al. (2019) extensively identified several features and components of DT including the physical asset, virtual model, communication networks, and the data represented on the virtual model about the physical asset.

12.3.4 Steps to implementing digital twin technology

DT works on the principle of the Internet of Things (IoT) combining sensors, cloud computing, and big data analytics for its operation (Palit and Datta, 2017; Lui et al., 2018). Its implementation depends on the level of information needed to create and continuously feed the digital model (Deloitte, 2017). The amount of information uploaded in the virtual model poses a major challenge in the application of DT. Based on the IoT platform, digital twin implementation follows two categories (OFA, 2017): simplistic device models (SDM), which uses JavaScript Object Notation (JSON) document for storage, and it contains two sets of attributes: a set of observed or reported values, which involve the use of sensors attached to the physical object to collect data/values and a set of desired values, which refers to the values to be set on each device (sensors, actuators) by the controlling application (DT). The other IoT platform entails Industrial Twin (IT), which uses information and real-time data from the physical object to model the virtual component.

A study by Deloitte (2017) identified six probable approaches to implementing digital twin technology which include: imagination of opportunities, identification of suitable process, piloting the digital twin, industrializing the process, scaling the twin, and monitoring and measuring.

12.4 Digital twin applications in other industries

The versatility of DT has seen its application in the various industrial sectors including its benefits and prospects for future opportunities. The adaptability of DT makes it possible to implement it in the various industrial sectors. The manufacturing industry has taken more advantage of yielding the powerful and compelling DT application (Deloitte, 2017). It provides a virtual replica, which provides the full instance of what is happening on the factory floor in real time and aids in the monitoring and control of the processes (Madubuike et al., 2022). The study identified some of the industries where DT has been applied such as aerospace, automotive, energy, healthcare, telecommunications, and manufacturing. Table 12.1 provides a summary of the diverse but not exhaustive DT applications in the various industries.

Table 12.1 DT applications in various industries

Industry	*Use Case*
Aerospace	An aircraft engine can be viewed in flight including its temperature and stresses on its parts. The DT model can be used to identify safety risks and act accordingly.
Automotive	Behavioral and operational data of a connected vehicle help in analyzing the overall vehicle performance to deliver a truly personalized/customized service for the customers.
Energy	An example is the use of DT to improve the efficiency of offshore wind turbines through conditioned-based maintenance (CBM), increasing efficiency by up to 20% and generating about $100 million over the wind farm's lifespan.
Healthcare	Prototyping a DT for real-time remote control over mobile networks: application of remote surgery and healthcare supply chain system.
Telecom	Improving onsite interaction with a product (tower, switches) using augmented reality (AR), service data, simulation analysis, and results from big data analytics.
Manufacturing	Simulating products using a DT, before actual manufacturing, can help companies eliminate future processing errors and improve their fabrication throughputs such as for semiconductor production.

Source: Madubuike et al., 2022.

12.5 Digital twin applications in the construction industry

There has been a rise in the design and construction of intelligent and smart buildings which should provide the basis for DT adoption in the construction industry. However, the application of DT in construction is still low compared to other industries. The adoption rate of digitization in the construction industry is 1.4% while it is 4.6% in the information and communications industry Bughin et al. (2016). Brilakis et al. (2019) also opined that DT has the potential to improve the design, construction, and lifecycle performance of construction projects but has yet to fully gain ground in the built environment. DT prospects in the construction industry go beyond buildings with embedded automated systems to the synchronization of systems such that there is tight coordination between the physical buildings and cyber worlds (Madubuike et al., 2022). The buildings are embedded with automated sensors, actuators, and other smart technologies to operate on the principle of real-time updates and bi-directional coordination for performance monitoring (Palit and Datta, 2017).

It is opined that DT provides a dynamic model that is better created at the inception stage (Grieves and Vickers, 2017) of a construction project. This enables the virtual model to be seamlessly integrated with the physical asset receiving all relevant data, which will enable the virtual model to easily monitor and control the physical asset. Therefore, implementing DT at the nascent stage of a project ensures tight synchronization and effective performance of the physical asset (Madubuike et al., 2022). DT functions on accumulated data, which depends on the sensors and data analytics employed in developing it. Keen (2019) identified possible DT features peculiar to construction such as building information modeling (BIM), schedules, contracts, construction documents (submittals, change orders, RFIs, etc.), operational data collected by sensors, and data analytics through artificial intelligence (AI) and machine learning (ML) technologies. The seamless integration of these features and other DT features can enable smart buildings with DT capabilities.

Some smart buildings and structures have been integrated with DT to improve their performance. Singapore's Fraser Tower is integrated with DT to provide connected and conducive working environments (BIM News, 2020). This project was developed by Bentley Systems and Schneider Electric, using a mix of 179 Bluetooth beacons and 900 sensors for lighting, temperature, and air

Figure 12.5 A geometric digital model of a bridge

Source: Sacks et al., 2018.

quality monitoring. Another DT application in the construction industry is seen in monitoring the state of a bridge (Sacks et al., 2018). The DT implemented here was referred to as a geometric DT, a geometric digital model enhanced with semantic information as a starting point for developing a comprehensive DT (Brilakis et al., 2019). The geometric DT is the first step in creating a DT and refers to an automatic process of generating 3D models from point clouds captured by laser scanners (Sacks et al., 2018). Figure 12.5 shows the geometric digital model of a bridge developed using laser scanners. Some other studies looked at developing DT for HFM by developing DT scenarios for prototyping and integrating DT-enabling technologies as identified in the studies (Madubuike and Anumba, 2021, 2022; Madubuike et al., 2023).

12.6 Overview of HFM issues

Problems in HFM are complex and can include but are not limited to inadequate training, poor communication, poor equipment design, systems or equipment failures, poor management systems, and work processes (Vincent, 2006). Additionally, poor maintenance of healthcare facilities can lead to rapid deterioration and unprecedented and adverse failures of its systems and equipment. This depends, to a substantial extent, on the maintenance system put in place to monitor, manage, and control these systems (Shohet et al., 2003). The importance of healthcare facilities and the complex nature of the systems and equipment calls for smart FM with real-time, 3D visualization, and bi-directional coordination features. Unfortunately, this has not been the case, as healthcare facilities still lack adequate maintenance of their systems and equipment. Maintenance procedures implemented still adopt the traditional approaches of corrective maintenance or the use of computer-aided software, which still does not operate in real time. To further identify critical issues affecting HFM amenable to DT, this study adopted a case study approach that employed interviews.

This study adopted a combination of face-to-face and semi-structured interview approaches to obtain data to address the aim of the study. These approaches were adopted as they made it possible to obtain direct information from the facilities managers on the issues with the current FM approach. The first interview focused on the mechanical equipment and identified that the healthcare facility employs a building automation system (BAS) to monitor the systems and facility. However, the system does not integrate a BIM (3D) model of the building, thereby limiting visualization and effective monitoring. The use of 2D images often leads to inadequate capture, analysis, and communication of design intent and rationale and in turn causes poor integration, coordination, and clarity of any communicated information (Anumba et al., 1997). Additionally, the BAS does not effectively monitor all the building systems and medical equipment.

The second interview focused on the maintenance culture of the healthcare facility management team. The interview revealed that the facility management team irrespective of the BAS still uses a manual method to schedule and carry out maintenance. A corrective maintenance approach is that an equipment or system has to break down before maintenance is carried out. The third interview focused on the maintenance culture adopted for medical equipment. It was identified that there are no systems put in place to monitor medical equipment, particularly mobile medical equipment in case of tracking or monitoring performance. Conclusively, healthcare facilities involve complex systems and equipment that require a facility management system that works in real time, provides 3D visualization, and allows for the possibility of bi-directional coordination.

12.7 Digital twin for HFM

Healthcare facilities are important and need to be continually operational. They are constructed to provide treatment and ensure the health and safety of their users (Guenther and Vittori, 2008). Additionally, the operations of healthcare facilities are continuous and as such require building systems and equipment to function continuously and effectively (Madubuike and Anumba, 2023). Thus, they require building systems and medical equipment to function effectively. The healthcare facility, its operations, and its services are expected to be at optimum performance. Currently, aging healthcare facilities fail continually, and facility maintenance teams address these problems with the mentality of 'if it breaks, we fix it' (Facility Executive, 2018). Implementing a proactive and automated type of maintenance would be beneficial to healthcare facilities. To do this, FM executives requested 'Smart Facilities Management Practices' that meet the complex needs of the healthcare facility.

There are numerous computer-aided facility management (CAFM) software such as Fast Track Maintenance (FTMaintenance), e-Maintenance Computerized Maintenance Management System (eMaint CMMS), Facilities Management Express (FMX), Upkeep, Select, and Hippo CMMS. They are either cloud-based or web-based facility management systems compatible with Mac, Windows, and Linux platforms and also useful to small, medium, and large business sizes (Software Advice, 2020). Although the various FM software is efficient, the maintenance updates are typically based on manual input of the maintenance schedules provided by the vendors and not truly on the actual performance status of the equipment or systems. The integration of sensor-based notifications as proposed by this study is lacking. Furthermore, they are not developed to integrate BIM models of the buildings, and they are not tailored specifically to healthcare facilities which are more complex and thus require very effective FM systems that integrate sensors and functions in real time.

DT applications in healthcare go beyond FM to DT applications for remote surgery and the healthcare supply chain (Zhang et al., 2018; Liu et al., 2020). However, this study focused on DT applications for HFM. The FM practices employed in healthcare facilities lack the features of real-time tracking of assets, 3D visualization, and bi-directional information flow (Madubuike and Anumba, 2022). Real-time updates on equipment performance status assist healthcare professionals in planning and performing their services efficiently. Real-time information updates have been identified to have statistical impacts on the efficiencies of systems and services (Gooze et al., 2013). Additionally, enabling bi-directional coordination would help the easy and timely flow of information between the healthcare professionals and the maintenance team or between the maintenance team and systems within the facility.

Understanding the critical nature of the healthcare services with more focus on the timeliness and effectiveness of both the services and the equipment, providing real-time updates and bi-directional coordination is important. For large healthcare facilities, asset tracking is very important, particularly for broken equipment. The bi-directional information flow enables the physical component to be controlled through the virtual model thereby providing a quick response. Although healthcare facilities managers employ the use of computer-aided facilities management (CAFM) software, the various software is either not integrated with BIM or supports real-time updates and bi-directional information flow. BIM comprehends the physical structure of a building and its operability and when integrated into FM will ensure control and reduction in operating cost, labor-intensive reduction, efficiency enhancement, and addition to outsourcing of resources. To develop DT for HFM, the study first developed a system architecture to describe the framework of the DT-based HFM system.

12.8 DT-based HFM system architecture

A system architecture is a framework guiding the design, creation, and performance of any software system. The system architecture developed comprises six layers as shown in Figure 12.6 and explained in the following subsections. This system architecture took its basis from a different study on cyber-physical systems (Akanmu et al., 2013) and was also validated by the focus group together with the scenarios.

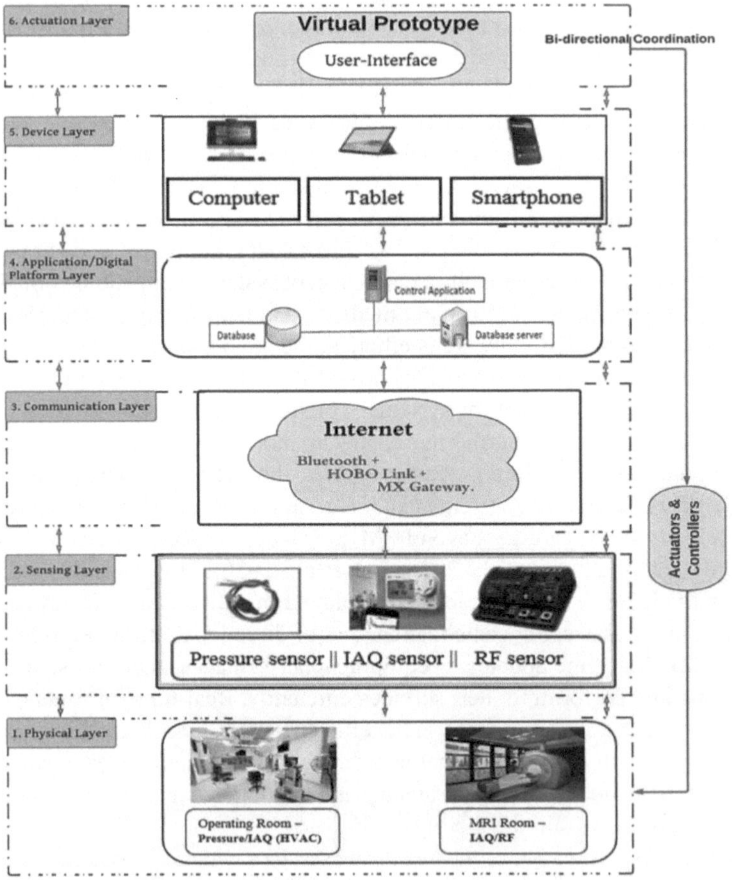

Figure 12.6 Systems architecture for DT-based HFM

Source: Madubuike and Anumba, 2023.

12.8.1 Physical layer

The physical layer contains the physical assets and systems that need to be monitored and controlled. The digital or virtual model created is the replica of the physical components in the physical layer. This study monitored the performance of the HVAC system and MRI machine by monitoring IAQ.

12.8.2 Sensing layer

This layer houses all the various sensors, transmitters, and gateways employed to develop the prototype DT-based system for HFM. The sensors are installed in the MRI and operating rooms to obtain data that is displayed on the 3D Revit model and used to monitor the physical assets.

12.8.3 Communication layer

The communication layer works with the application layer to enable the exchange of data/information between the physical components and the virtual models. The communication layer is also responsible for the performance of the DT-based system.

12.8.4 Application/digital platform layer

The application/digital platform layer contains the varying databases such as the local databases, database server, and control applications that store data, analyze data, and constantly update information obtained from both the physical components (through the communication layer) and the actuation layer (Akanmu et al., 2013). The results from the actuation layer are visualized through the device layer.

12.8.5 Device layer

The device layer contains all the mobile devices such as computers, iPads, Tablet PCs, and smartphones. It contains all the client devices important in the visualization of data and the virtual model. This study uses a computer (laptop) to provide the end-user interface where the model and data are monitored and controlled.

12.8.6 Actuation layer

This layer houses the virtual prototype, which enables the end-user to visualize the sensed information in real-time through the user interface. The actuation layer is visualized through the mobile device providing human-readable information and visuals resulting from the application layer and also the means to control the physical asset through the virtual model.

12.9 DT enabling technologies

The effective application of DT depends on identifying the enabling technologies that would facilitate the seamless integration between the virtual model and the physical object. This research opted to develop a 'DT-based system' for HFM and adopted the following major enabling technologies (see Figure 12.7 for adopted DT enabling technologies and processes for the prototype DT-based system):

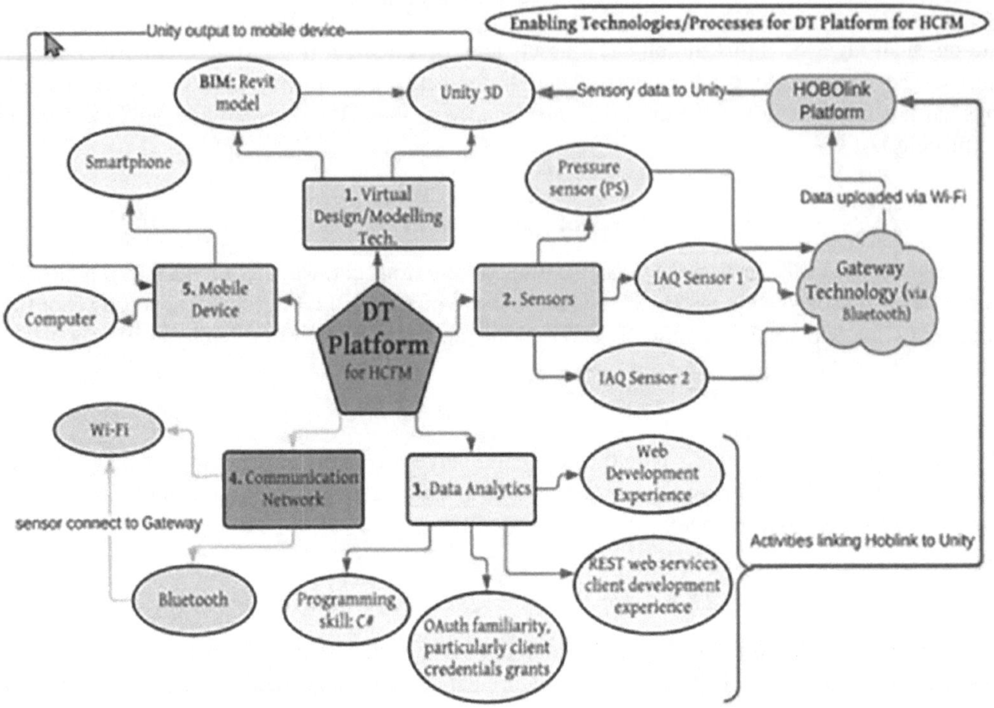

Figure 12.7 Adopted DT-enabling technologies and processes for the DT-based system for HFM

12.9.1 Virtual design/modeling technologies

These are important as they help create the digital model and simulation of the model to develop the DT-based system. To create the digital model, BIM-Revit was used to develop the 3D model of the healthcare facility, which was later uploaded to the simulation engine using 3DS Max. 3DS Max was used to alter the material texture and quality to meet the requirements of the simulation engine (Madubuike and Anumba, 2023). BIM is essential in creating the digital model as it contains physical and functional properties of key components essential in the DT application for building FM (Monteiro and Martins, 2013; Bradley et al., 2016; Cecchini, 2019). The simulation engine provides a virtual environment where the raw data is obtained by the sensors and transformed and reported on the digital model.

12.9.2 Sensors

Sensors are important components of any DT application, and they are used to obtain information from systems and equipment for monitoring and controlling throughout their life cycle (Akinci and Anumba, 2008). Three wireless sensors were adopted for this study including an indoor air quality (IAQ) sensor (MX1104), which measures temperature, relative humidity (RH), and light; another IAQ sensor (MX1102A) measuring carbon dioxide (CO_2), temperature, and RH; and a pressure transmitter (PXULX05S) for pressure reading. The MX1104 sensor and pressure transmitter were connected in the operating room (OR) to monitor temperature, RH, and pressure while the MX1102A sensor was connected in the MRI room to monitor CO_2, temperature, and RH. The data obtained by the sensors were reported on the simulation engine through an MX Gateway, which allows data from the sensors to be uploaded using Wi-Fi to third-party platforms. The sensors in themselves cannot report data directly to the simulation engine but require the MX Gateway.

12.9.3 Data analytic techniques

Data analytic techniques are a very important aspect of DT applications as they define how the DT application would function. They are the data analytics and storage technologies that include processing and visualization (Akanmu et al., 2021). The data analytic techniques transform the raw data obtained by the sensors into useful information that enables decision-making on the digital model. The data retrieved by the sensors are in raw form and need to be defined, structured, compressed, and transformed for DT purposes. The data analytic techniques for this study employed web development experience, REST web services client development experience, Open Authorization (OAuth) familiarity, and programming skills in C# language.

12.9.4 Communication network

There exist several types of communication networks such as wireless local area networks (WLAN), wireless personal area networks (WPAN), wide area networks (WAN), and worldwide interoperability for microwave access (WiMAX) (Akanmu et al., 2013). The communication networks adopted for the DT-based prototype for HFM were Bluetooth and Wi-Fi. The sensors uploaded data to the MX Gateway using Bluetooth while the MX Gateway uploaded the data to the Hobolink platform using a Wi-Fi connection. The Hobolink is a web portal that receives the data from the MX Gateway and stores the data in Amazon Web Services (AWS). The data is sent from the Hobolink platform to the simulation engine using the OAuth command.

12.9.5 Mobile devices

Mobile devices provide the UI to visualize the DT-based system – to visualize the virtual model. Mobile devices play important roles in DT applications as they enable visualization of the sensed data from the sensing layer and provide the user interface to enter information by the end-users (Akanmu et al., 2013).

12.10 Scenario development

Scenario development allows researchers to explore an existing concept to predict future outcomes. Scenario development explores how a particular system or area may be developed in the future (Fancourt, 2016). It uses a quantitative approach and complex modeling procedures to predict the future of a system. The study developed four different scenarios based on the findings from the literature and case study. The developed scenarios were validated by a focus group selected based on their experience with FM procedures. The four scenarios include monitoring and controlling the performance of the heating, ventilation, and air-conditioning (HVAC) system (Scenario 1); monitoring the performance of the pulsed light (PL) (Scenario 2); 3D monitoring of IAQ to monitor equipment performance (Scenario 3); and monitoring radio frequency interference (RFI) concerning the performance of the magnetic resonance imaging (MRI) machine (Scenario 4).

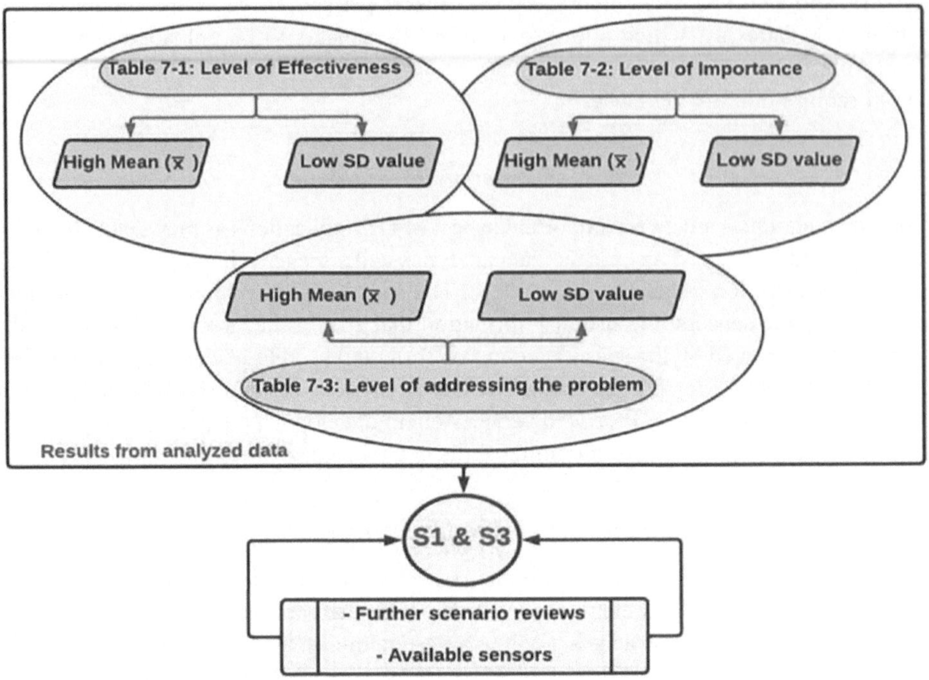

Figure 12.8 Criteria for the adoption of the scenarios

Following the validation of the four developed scenarios, two of the scenarios (Scenarios 1 and 3) were adopted. The validation of the scenarios was done to assess the feasibility of implementing the scenarios and decide which of the scenarios to adopt. The adopted scenarios 1 and 3 were guided by the following criteria: the readily available sensors and the scenarios' responses having higher mean values with the lowest standard deviations. Standard deviation tests the reliability of values by measuring the extent to which the data sets are scattered (Whitley and Ball, 2002), and the smaller values indicate that the data points cluster closer to the mean value showing consistency in the results (Frost, 2022). Figure 12.8 provides a summary of the selection criteria for the adopted scenarios. The details of all the scenarios and selection criteria including the validation process have been identified in a different paper (Madubuike et al., 2023). The two scenarios that were selected for prototyping have been described in the following sections.

12.10.1 Scenario 1: monitoring and controlling the HVAC system

To monitor the performance of the HVAC system, the scenario focused on indoor air pressure as the measurement criterion. This is important as the pressure within a healthcare facility depends on the performance of the HVAC system (Rosone, 2018) and also affects activities within the operating room. The following sequence of actions illustrates the procedures of Scenario 1 as shown in Figure 12.9:

1) The connected pressure sensor in the OR detects pressure abnormalities and reports it to the DT system
2) Using data analytic techniques, the DT system analyzes the received sensory data to identify the cause of the problem and display the data result on the 3D model

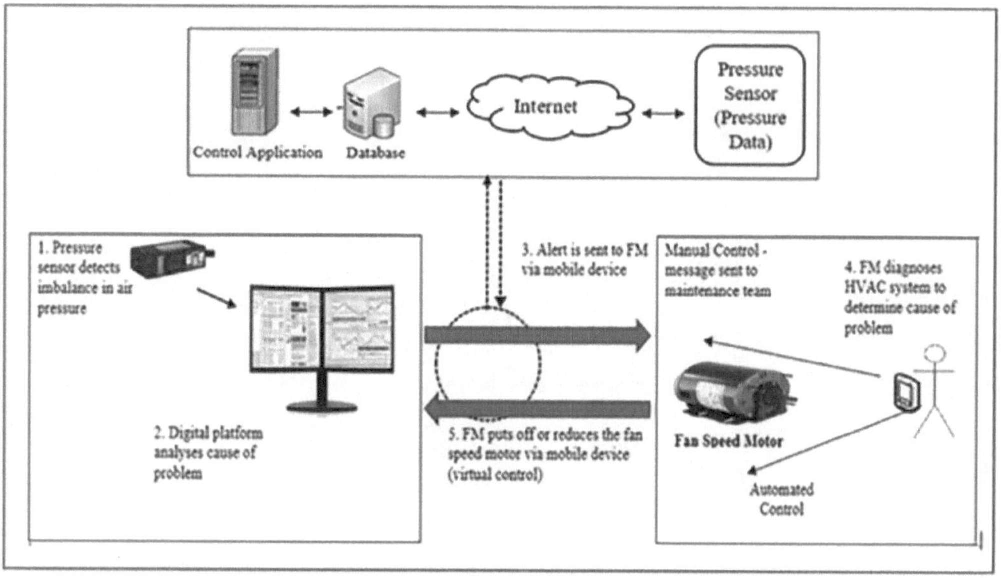

Figure 12.9 Monitoring and controlling the HVAC system – pressure control
Source: Madubuike et al., 2023.

3) A notification is sent to the facility manager on the mobile device stating the problem being encountered
4) The facility manager further explores the probable cause of the problem by taking a further look into the DT system to make decisions
5) This sequence enables bi-directional coordination where the facility manager can control the thermostat of the HVAC through the virtual model.

12.10.2 Scenario 3: 3D monitoring of IAQ to monitor equipment performance

This scenario explains how DT is used to monitor IAQ including temperature, RH, and carbon dioxide. The scenario was developed to monitor IAQ within the MRI room to monitor the performance of the MRI machine. The performance of the MRI machine is affected by the temperature and RH levels within the room (Gillin, 2021). IAQ is a multidisciplinary phenomenon consisting of different trends by which chemical, biological, and physical contaminants become part of the total indoor environmental composition (Tham, 2016). These contaminants become part of the total indoor environmental composition that can affect the performance of some medical equipment. This scenario identified sequences that explain the performance of the scenarios as shown in Figure 12.10:

1) The IAQ sensor already connected to the MRI room detects abnormal air quality
2) The IAQ sensor collects the data and sends it to the DT platform
3) The DT platform analyzes the information and sends it to the 3D model for easy visualization
4) The DT-based system sends a notification to the facility manager and displays it
5) To regulate the IAQ, the facility manager can effect control of the HVAC system through the virtual model

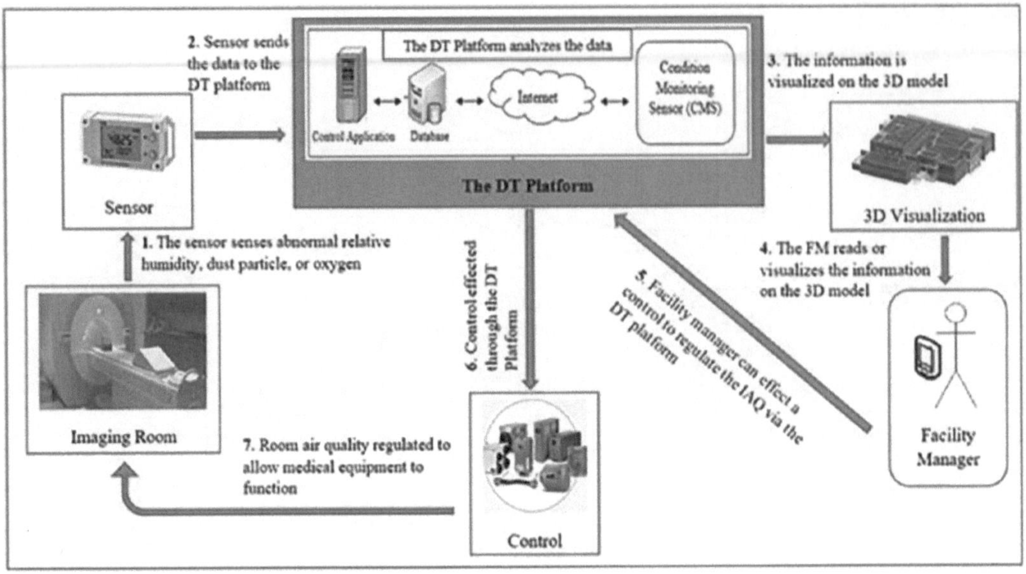

Figure 12.10 3D monitoring of IAQ to monitor equipment performance
Source: Madubuike et al., 2023.

6) The facility manager virtually operates the thermostat of the HVAC system
7) The IAQ inside the room is regulated to an acceptable standard that does not affect the performance of the medical equipment.

12.11 Prototype development approach

Prototyping is a critical activity in system development that enables the development of a new product. Prototyping is considered one of the most essential but the least adopted techniques in design development (Camburn et al., 2013; Menold et al., 2017). A prototype is defined as an early or experimental model involving a series of technical coordination to test or evaluate a concept or process (Lauff et al., 2018). A prototype provides a developer the opportunity to test or experiment with an idea to prove the feasibility and scalability of a concept.

The rapid prototyping method was adopted for system development. The successful application of the rapid prototyping method for system development is traced back to history in the manufacturing industry (Salonitis, 2014). The rapid prototyping method adopted the simulation approach that entails integrating the mock-up system that allows for the prototype system to be developed and tested and results from feedback sent to the developer. The rapid prototype method used in this research followed the principles developed by Chua et al. (2020) and adopted six processes/steps for implementation:

- Develop or obtain the digital model of the healthcare facility
- Integrate the digital model into the simulation engine
- Create the controls, functions, and material qualities in the simulation engine

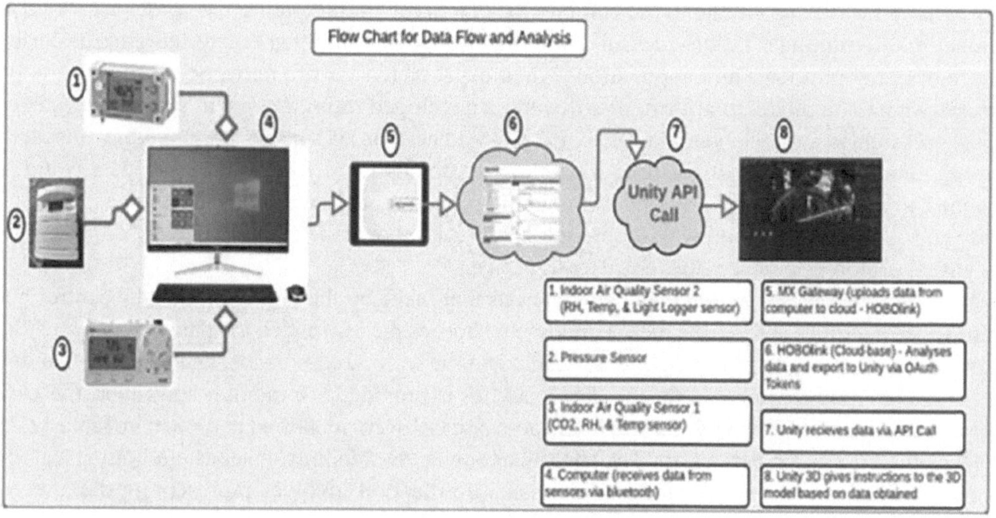

Figure 12.11 Integration of the hardware and software of the DT-based prototype system

- Install the sensors in positions and link sensory data to the simulation engine
- Write functions (program) in the simulation engine for command control
- Test, clean up, and finish the model.

Following the development of the digital (3D) model of the healthcare facility, the integration of the digital model into the simulation engine followed suit. It was difficult to directly integrate BIM into the simulation engine without first passing the 3D Revit model through 3DS Max. 3DS Max was used to convert the Revit model to an FBX file first before integrating it with the simulation engine. A more detailed report on this is being prepared for a different publication. Furthermore, the prototyping process involved the integration of some hardware (sensors, Gateways, and mobile devices) and software (Hobolink and simulation engine) required for the DT-based prototype development.

The integration of the DT prototyping system adopted two IAQ sensors that measure temperature, RH, and CO_2 and a pressure transmitter that measures pressure as previously identified. The sensors obtained the data and transmitted it to the MX Gateway, which uploaded the data to the Hobolink platform. The data in the Hobolink platform is integrated into the simulation engine using API calls to link the data in real-time. Figure 12.11 shows the integration of the hardware and software involved in the development of the DT-based prototype system.

12.12 Operations of the DT prototype

The developed DT-based system was based on the selected scenarios previously identified to monitor IAQ. The DT-based system for this research integrated virtual design/modeling technologies, wireless sensors, data analytics, communication networks, and mobile devices as its enabling technologies. It focused on real-time updates, BIM integration (3D visualization), and bi-directional coordination. The bi-directional coordination aspect was addressed using a schematic diagram to explain its implementation. Additionally, the DT virtual prototype combined real-time updates and emulation to demonstrate its functionality.

Emulation was adopted due to the complexities involved in integrating the whole sensory data in real time. Hemminger (2005) identified the importance of emulation due to the inferior performance of many protocols and applications when exposed to real-life networks. Emulation is the process whereby a model, platform, or software is developed to imitate a real system such that it closely resembles and behaves like the real system. The type of sensors employed in this study proved some limitations that required a series of protocols to be followed before data could be obtained in real time. Accessing and commanding the protocols were beyond the scope of this study and, as such, emulation was adopted. Figure 12.12 shows the actuation layer, which houses the virtual prototype and provides the user interface.

The user interface shows the included materials that make up the digital model, the control buttons to access various spaces, the main window that houses the 3D model, the data update panel that shows the real-time data of the IAQ, color code, and the scripts, commands, and setpoints window where the program responsible for the data analytics is provided. To monitor and report the IAQ, the various setpoints of the IAQ for the MRI room and OR were identified as shown in Table 12.2.

Once the DT system detects any IAQ below or above the setpoints, it sends out both visual and sound notifications on the 3D model. This visual notification involves representing the affected room(s) with different associated colors based on the color codes as shown in Table 12.3.

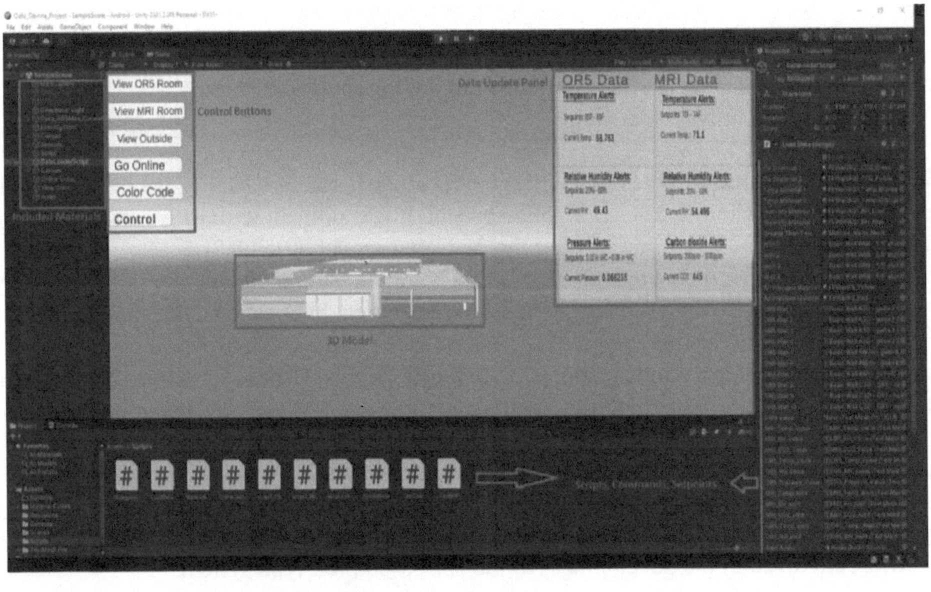

Figure 12.12 The simulation engine window – user interface

Table 12.2 The setpoints for the IAQ

Air components	*Operating room*		*MRI room*	
	High	*Low*	*High*	*Low*
Pressure	> 0.08 in WC	< 0.02 in WC	N/A	N/A
CO_2	N/A	N/A	> 1,000 ppm	< 200 ppm
Temperature	> 740° F	< 70° F	> 80° F	< 65° F
RH	> 60%	< 20%	> 60%	< 20%

The operations of the DT-based prototype system for HFM were able to show alerts using the appropriate color. Figures 12.13–19 show the operations of the DT-based prototype with notifications on when the IAQ exceeds or goes below the thresholds. When the IAQ is within the threshold limits, no notification is sent, and the walls of the room retain their original colors. Figure 12.20 shows the visual notification that all walls are black meaning that all the IAQ components exceeded their threshold simultaneously. Figure 12.21 shows the Hobolink platform window where the sensory data stored in the cloud using the MX Gateway can be accessed and linked to the simulation engine.

Table 12.3 Color code table representing the different IAQ components

Air Components	*High*	*Acceptable*	*Low*	*Multiple Alerts*
Pressure	Red	Retain the material's original colors	Yellow	Black
CO_2	Purple		Lime	
Temperature	Orange		Bronze	
RH	Brown		Blue	

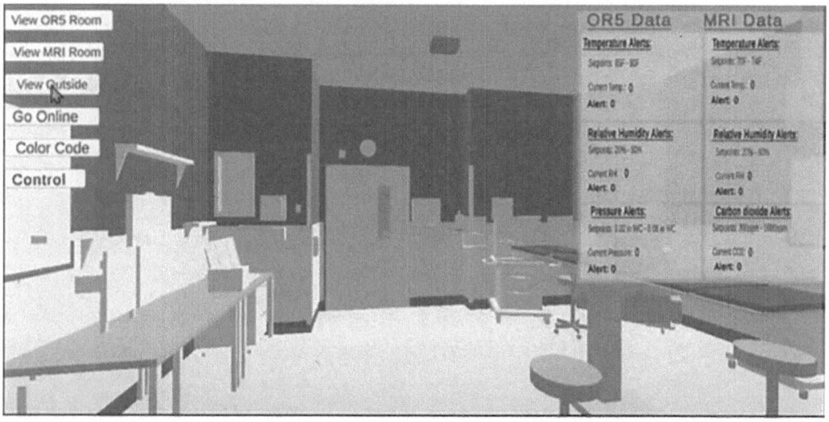

Figure 12.13 The DT-based prototype shows no alerts for the operating room (original color retained)

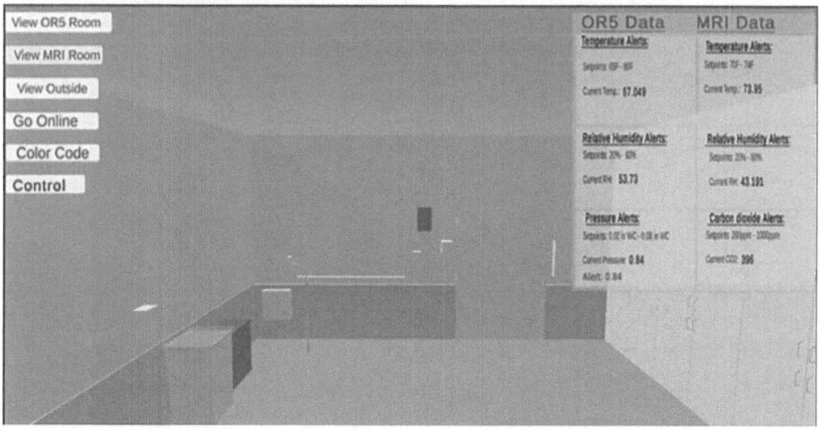

Figure 12.14 DT-based prototype showing no alerts for the MRI room (original color retained)

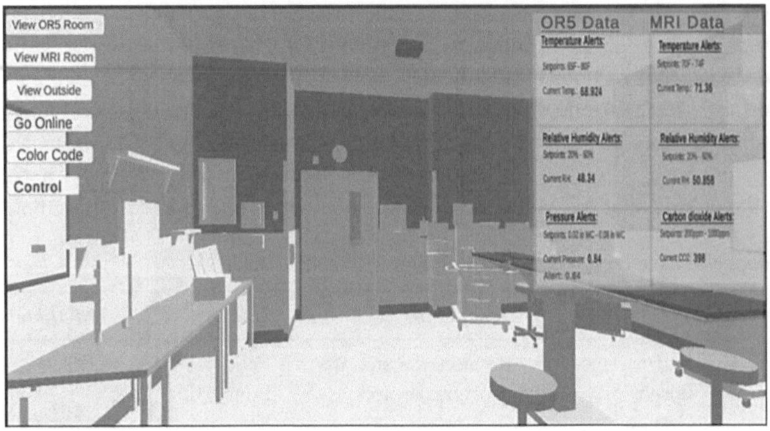

Figure 12.15 The DT-based prototype shows high pressure using the red color

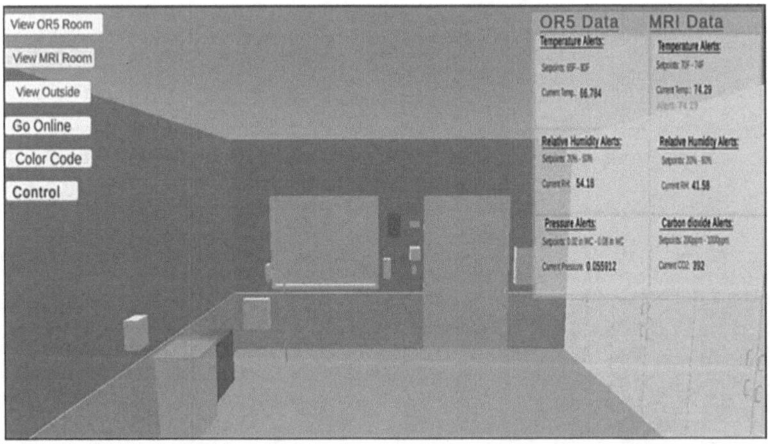

Figure 12.16 The DT-based prototype shows high temperature using the orange color

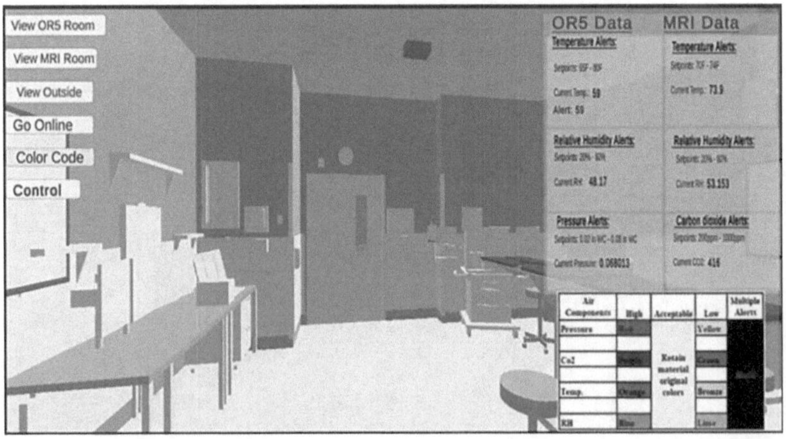

Figure 12.17 The DT-based prototype shows low temperature (bronze) and the color code image

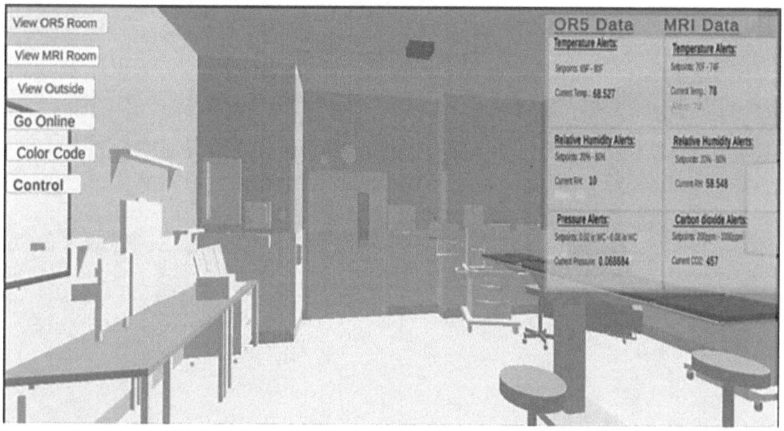

Figure 12.18 The DT-based prototype shows low RH using lime color

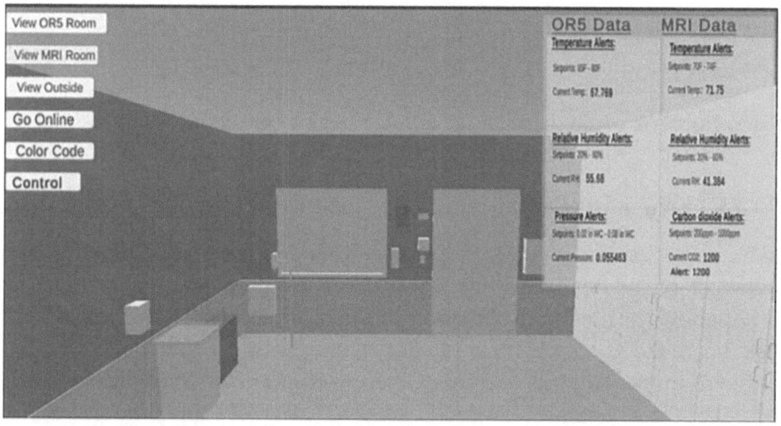

Figure 12.19 The DT-based prototype shows high CO_2 using the purple color

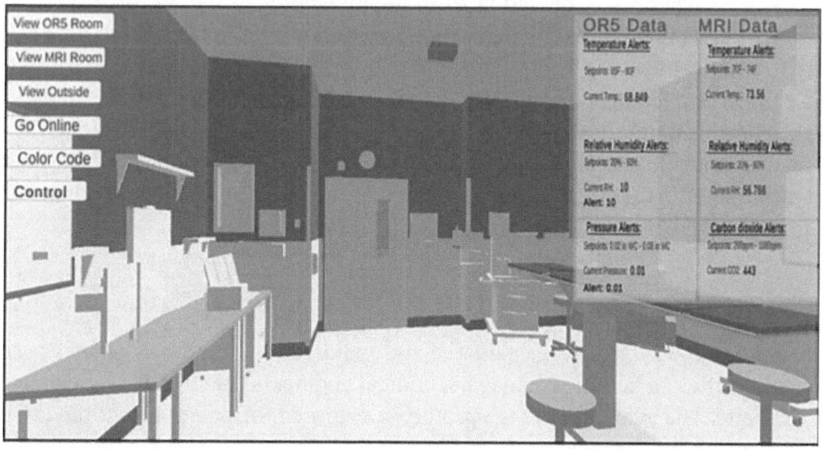

Figure 12.20 The DT-based prototype shows black-colored walls for multiple alerts

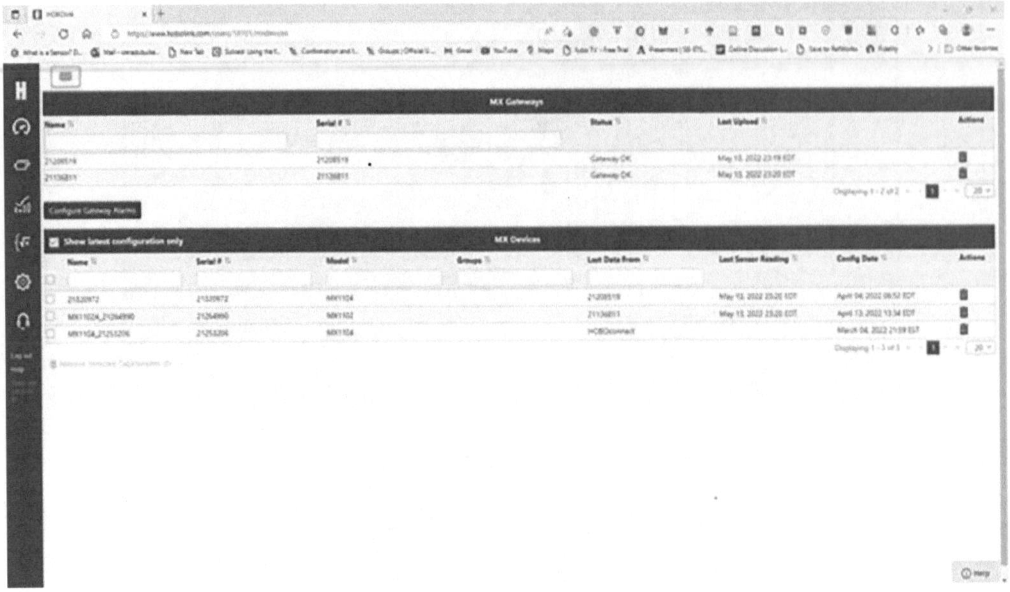

Figure 12.21 The Hobolink portal shows the details and status of the sensors and Gateways

12.13 Schematic representation of the bi-directional coordination

Bi-directional coordination was one of the features intended for the DT-based system for HFM. However, the nature of the healthcare facility that was used for the implementation of the DT prototype did not allow for the implementation of the bi-directional coordination feature. Understandably a healthcare facility is a critical facility dealing with the safety and well-being of lives. It was impossible to implement this feature as there were limitations by the hospital in allowing third-party control of their systems. However, a schematic model explaining the ideal implementation process was developed.

The model is a simple demonstration of the implementation of bi-directional coordination once the DT-based system detects any deviation from the thresholds of the air qualities. For instance, when the sensors detect that the IAQ values are not within the setpoints (thresholds) and send the sensory data to the prototype DT-based system. The prototype DT-based system automatically instructs the controller to control the HVAC system. The controller instructs the actuator, which in turn inputs the control signal for the HVAC system's control, which is also linked to the DT-based system. A manual override can also be integrated into the system to enable facilities managers to manually control the HVAC system when necessary. Figure 12.22 presents the schematic diagram explaining the operation of the bi-directional coordination feature.

12.14 Evaluation of the DT-based prototype system

The DT-based prototype system was evaluated and validated by the focus group consisting of a 16-member team including facilities managers, clinical engineers, BIM/CPS/DT academic experts, and industry experts. The evaluation of the prototype explored the operations of the DT-based system for HFM based on the feedback from the end users. As postulated by Delbert and Neil (2002),

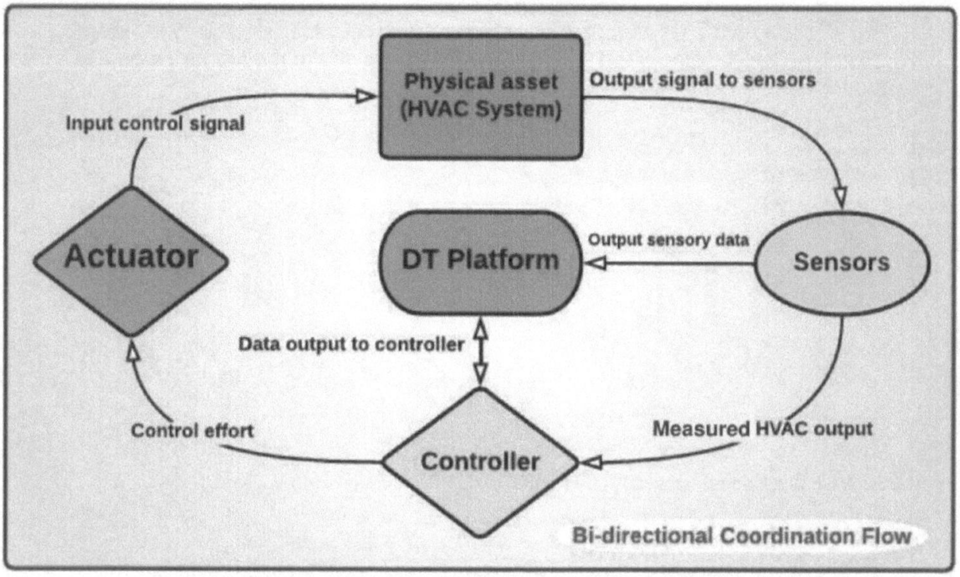

Figure 12.22 Schematic representation of the bi-directional coordination

research evaluation is essential in assessing a concept or technique before scalability. The feedback was necessary to validate the DT virtual prototype and determine ways of improving upon it. The evaluation and validation of the DT-based prototype focused on the general information of the participants, the usability of the DT-based prototype system, real-time and bi-directional coordination capabilities, improved visualization capabilities, and the overall experience and feedback having the following categories: ease of use, interactivity, informative level, and exciting level. Each of the survey questions had an attributed Likert scale for measurement with a detailed description provided in a different publication.

The evaluation showed that the majority of the participants in the validation process had between 0 and ten years of experience while others had 11–20 years. The participants responded that the user interface of the DT-based prototype was 'extremely easy' to use. For real-time updates and bi-directional coordination capabilities, the participants opined that it was easy to access both real-time and historical data and that the system has the potential to enable bi-directional coordination as shown in Figure 12.23. The effectiveness of the 3D visualization was assessed, and it was determined that the visuals were 'extremely effective and that the appearance of the model was 'extremely satisfactory'. For the overall experience and feedback, the DT-based system was deemed easy, interactive, very informative, and very exciting to use as shown in Figure 12.24.

Additional feedback from the participants showed an interest in integrating bi-directional coordination, machine learning, and artificial intelligence amongst other capabilities including improving the system to critically monitor IAQ, particularly in the aspect of pressure, which is very important in emergency and operating rooms. A more detailed report on the evaluation and the implementation of the feedback from the focus group was prepared for a different publication.

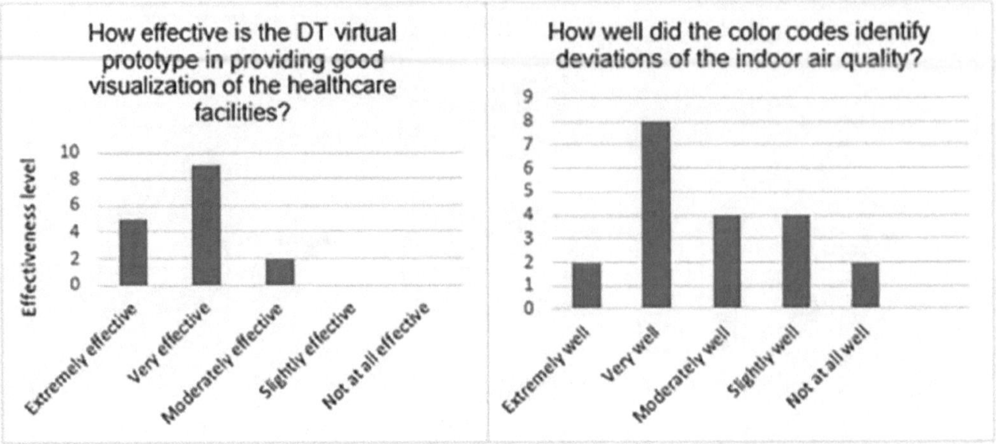

Figure 12.23 Visualization effectiveness and color code effectiveness

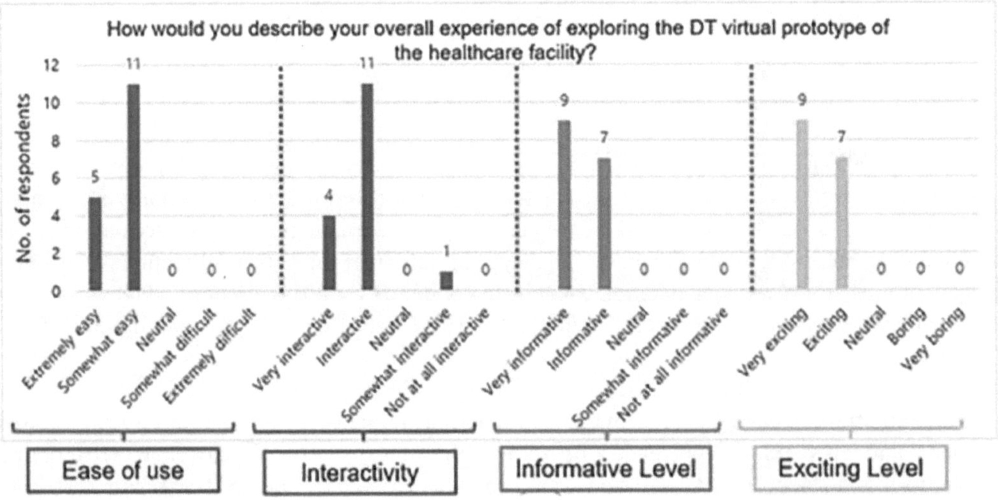

Figure 12.24 Overall user experience

12.15 Discussion and conclusions

This chapter has presented research that investigated the extent to which DT can be applied to improve HFM. It presented the development of a DT virtual prototype for HFM and described seven major procedures for the implementation of the DT prototype. These include (a) concept design and development that identifies major DT concepts, enabling technologies, and implementation processes from other studies; (b) system architecture development that represents the major components for virtual prototyping; (c) scenario planning that clearly outlines steps in implementing DT for the identified HFM problems; (d) scenario validations to select the scenarios that would be implemented; (e) prototyping to develop the DT-based system; (f) validation of the prototype to get end users feedback; and (g) development of the DT-based system implementation guidelines.

The research developed four objectives addressed with different research methods. Thematic analysis and exploratory studies were the major tools employed to address the objectives. Draper et al. (2011) addressed the importance of exploratory study in helping to obtain a deeper knowledge of a topic. The exploratory study aided in identifying some healthcare facilities' management problems amenable to DT solutions and also in the development and evaluation of the scenarios used for the prototype implementation. In validating the scenarios and prototype that were the products of the thematic analysis and exploratory study, the research employed quantitative techniques to evaluate the responses of the focus group. The importance of quantitative tools in analyzing the responses of research respondents has been addressed in different studies (Delbert and Neil, 2002; Whitley and Ball, 2002; Frost, 2022).

One of the important aspects of this research was to develop a DT for the HFM prototype – which would help monitor and report the performance of healthcare facilities in real time. The DT virtual prototype provided a visualization tool to understand the layout of the facility and the function of the prototype. Before the prototype development, scenarios were developed and validated by the facilities managers and DT experts in the industry. Scenario development was employed because it was discovered that one of the best ways to communicate ideas to reviewers would be with the use of scenarios. The scenarios provided practical ways to convey the proposed DT solution idea with images in solving the identified healthcare facilities' management issues. Scenarios provide consistent depictions of different theoretical expectations serving as the foundation for actions and making the ideas being depicted easy to understand (van Notten, 2005).

The developed DT-based prototype combined various enabling technologies termed 'DT Enabling Technologies' comprising virtual/modeling technologies, wireless sensors, data analytic techniques, communication networks, and mobile devices. The DT-based prototype for HFM showed potential for real-time updates and bi-directional coordination between physical building components and their digital replicas. Additionally, the prototype was able to provide improvement in 3D visualization for FM systems. Furthermore, the DT-based prototype system showed potential for improvement through the integration of smart elements such as artificial intelligence and machine learning to improve its performance, which will be considered for future research.

The validation of the prototype revealed that the prototype DT-based system provided a basis for BIM integration with FM systems and real-time updating of data. Based on the interactions and feedback from the validation participants, the prototype DT-based system provided an effective 3D FM system to monitor the performance of healthcare facilities. However, the DT-based system did not fully implement real-time updates but integrated emulation to achieve the aim of the prototype. This study identified that, to implement a fully real-time DT-based system, the DT-enabling technologies, particularly the sensors, should be designed solely for the DT purpose.

This DT-based prototype for HFM, although developed within the scope of a healthcare facility, monitored certain indoor air qualities such as temperature, pressure, carbon dioxide, and relative humidity. These IAQs can affect the performance of some medical equipment and operating room conditions. Thus, it was considered important to use this reason to define the purpose of the DT-based prototype system to improve healthcare facilities management procedures. The developed DT-based prototype, although for HFM, can be applied to other facilities impacted by similar IAQs. Additionally, the prototype developed in this research can serve as the basis to build on for other similar studies. Scalability has been attributed as a major reason behind the prosperity of many IT-related studies (Palmié et al., 2023). Future studies on DT for facilities management will build on this research to develop a DT-based prototype for FM with more included FM features to manage.

Conclusively, the research contributed to knowledge by developing both theoretical and conceptual frameworks and guidelines relevant to the implementation of DT for related concepts or other DT applications. The theoretical framework focused on the theories developed to help in the implementation of the prototype. This research extended existing theories as proposed by Chatterjee and Hambrick (2007). This research extended the DT theories in other fields to develop a DT-based system for HFM. The theories addressed in this research cover HFM processes and their shortcomings, theories addressing DT applications in the various industries leading to identifying the enabling technologies, and proposed solutions via scenario planning and prototyping. The conceptual framework focused on the system architecture that helped in developing the DT-based system prototype. Finally, the research was able to prove that DT can be developed for healthcare and other FM purposes using the prototype DT-based system for HFM.

12.16 Limitations of the study

This research had its limitations, particularly in the areas of the chosen case study and implementing the bi-directional coordination. To streamline the focus of the research, the case study focused on a single healthcare facility for ease of accessibility and the available choice of facility for implementation purposes. The study limited its scope to monitoring pressure and indoor air qualities (temperature, CO_2, and RH) as these were the problem areas identified by the focus group. Furthermore, studies have revealed that the mentioned indoor air qualities impact the performance of healthcare facilities and some equipment such as the MRI (Tham, 2016; Rosone, 2018; Gillin, 2021).

The study achieved the prototyping by adopting both emulation and real-time updates. Emulation is the process whereby a model, platform, or software is developed to imitate a real system such that it closely resembles and behaves like the real system (Hemminger, 2005). The sensory data were obtained in real-time and raw form through the Hobolink – a web portal for the sensors to collect and store data. The selected and available sensors adopted for the study made it difficult to link the sensory data to the DT-based prototype in real time, necessitating the adoption of emulation. Finally, there was difficulty in implementing the bi-directional coordination aspect of this study given the delicate nature of the healthcare facility and the near-zero possibility of gaining control of the systems of the healthcare facility. Understandably, the critical nature of a healthcare facility makes it impossible to access control of their systems by third parties. This study opines that, for a DT concept to be fully and effectively developed, all systems (including physical and virtual) and enabling technologies should be created solely for DT development.

References

Akanmu, A., Anumba, C., and Messner, J. (2013). Scenarios for cyber-physical systems integration in construction. *Journal of Information Technology in Construction, 18*, 240–260. www.itcon.org

Akanmu, A. A., Anumba, C. J., and Ogunseiju, O. O. (2021). Towards next-generation cyber-physical systems and digital twins for construction. *Journal of Information Technology in Construction, 26*, 505–525. https://doi.org/10.36680/j.itcon.2021.027

Akinci, B., and Anumba, C. J. (2008). Editorial – Sensors in construction and infrastructure management. *ITcon, Special Issue on Sensors in Construction and Infrastructure Management, 13*, 69–70.

Anumba, C. J., Baron, G., and Evbuomwan, N. F. O. (1997). Communications issues in concurrent lifecycle design and construction. *BT Technology Journal, 15*(1), 209–216.

BIM News. (2020, March 13). Digital twin creates the blueprint for future smart buildings. *BIM Today*. https://www.pbctoday.co.uk/news/bim-news/digital-twin-smart-building/73253/

Boschert, S., Heinich, C., and Rosen, R. (2018). Next generation digital twin. *Proceedings of TMCE*, Las Palmas de Grad Canaria, Spain.

Bradley, A., Li, H., Lark, R., and Dunn, S. (2016). BIM for infrastructure: An overall review and constructor perspective. *Automation in Construction, 71*(4), 139–152. https://doi.org/10.1016/j.autcon.2016.08.019.

Breen, R. L. (2006). A practical guide to focus-group research. *Journal of Geography in Higher Education, 30*(3). https://doi.org/10.1080/03098260600927575

Brilakis, I., Pan, Y., Borrman, A., Mayer, H. G., Rhein, F., Von, C., Pettinato, E., and Wagner, S. (2019). Built environment digital twinning. *Report of the International Workshop on Built Environment Digital Twinning Presented by TUM Institute for Advanced Study and Siemens AG.* https://api.semanticscholar.org/CorpusID:229249538

Bughin, J., Hazan, E., Labaye, E., Manyika, J., Dahlström, P., Ramaswamy, S., and de Billy, C. C. (2016). Digital Europe: Pushing the frontier, capturing the benefits. https://www.mckinsey.com/~/media/McKinsey/Business%20Fun-ctions/McKinsey%20Digital/Our%20Insights/Digital%20Europe%20Pushing%20the%20frontier%20capturing%20the%20benefits/Digital-Europe-Full-report-June-2016.ashx

Camburn, B. A., Dunlap, B. U., Kuhr, R., Viswanathan, V. K., Linsey, J. S., Jensen, D. D., Crawford, R. H., Otto, K., and Wood, K. L. (2013). Methods for prototyping strategies in conceptual phases of design: Framework and experimental assessment. *ASME Paper No. DETC2013–13072.* https://api.semanticscholar.org/CorpusID:53536091

Castaldini, F. (2019). How digital twin was central to smart buildings. *Facility Executive.* https://facilityexecutive.com/2019/02/how-digital-twin-technology-is-central-to-smart-buildings/

Cecchini, C. (2019). From data to 3D digital archive: A GIS-BIM spatial database for the historical center of Pavia (Italy). *Journal of Information Technology in Construction, 24* (Nov), 459–471.

Chatterjee, A., and Hambrick, D. (2007). It's all about me: Narcissistic chief executive officers and their effects on company strategy and performance. *Administrative Science Quarterly*, 52, 351–386

Chua, C. K., Leong, K. F., and An, J. (2020). Introduction to rapid prototyping of biomaterials. In: *Rapid Prototyping of Biomaterials, Principles, and Applications.* 2nd ed.; 1-15. Woodhead Publishing Series in Biomaterials.

De Los Reyes, A., and Kazdin, A. E. (2008). When the evidence says, "Yes, no, and maybe so": Attending to and interpreting inconsistent findings among evidence-based interventions. *Current Directions in Psychological Science, 17*(1), 47–51.

Delbert, C. M., and Neil, J. S. (2002). *Handbook of Research Design and Social Measurement.* 6th ed. Sage Publications, London.

Deloitte. (2017). Industry 4.0 and digital twin – Manufacturing meets its match. In: *A Deloitte Series on Industry 4.0.* Deloitte University Press.

Draper, J., Dawson, M., and Casey, M. (2011). An exploratory study of the importance of sustainable practices in the meeting and convention site selection process. *Journal of Convention and Event Tourism, 3*(12), 153–178.

Easwaramoorthy, M., and Zarinpoush, F. (2006). Interviewing for research. *Canada Volunteerism Initiative, Imagine Canada.*

Facility Executive. (2018). Five trends in healthcare affecting facility management. https://facilityexecutive.com/2018/06/five-trends-healthcare-affecting-facility-management/

Fancourt, M. (2016). Scenario development: A review of approaches. *United Nations Environment Program.* https://www.unep-wcmc.org/system/comfy/cms/files/files/000/000/803/original/Scenario_Development_2016_WEB.pdf

Frost, J. (2022, April 2). Standard deviation: Interpretations and calculations. https://statisticsbyjim.com/basics/standard-deviation/

GAVS Technologies. (2017). Digital twin concept. https://gavstech.com/insights-trends/whitepaper/.

Gillin, S. (2021). Cybersecurity risk to healthcare from building services. *2021 Cambridge–McKinsey Risk Prize.* Center for Risk Study. www.jbs.cam.ac.uk/wp-content/uploads/2021/06/2021-06-30-risk-prize-entry-gillin.pdf

Glaessgen, E. H., and Stargel, D. (2012). The digital twin paradigm for future NASA and US air force vehicles. In: *53rd structural dynamics material conference – Special session: Digital twin*, Honolulu, HI, US, 1–14.

Gooze, A., Watkins, K. E., and Borning, A. (2013). Benefits of real-time transit information and impacts of data accuracy on rider experience. *Transportation Research Record: Journal of the Transportation Research Board, 2351*, 95–103.

Grieves, M. (2005). Product lifecycle management: The new paradigm for enterprises. *International Journal of Product Development, 2*(1/2), 71–84.

Grieves, M., and Vickers, J. (2017). Digital twin: Mitigating unpredictable, undesirable emergent behavior in complex systems. In: Kahlen, F. J., Flumerfelt, S., Alves, A. (eds) *Transdisciplinary Perspectives on Complex Systems*. Springer, Cham. https://doi.org/10.1007/978-3-319-38756-7_4

Guenther, R., and Vittori, G. (2008). *Sustainable Healthcare Architecture*. John Wiley & Sons, New Jersey, NJ.

Haag, S., and Anderl, R. (2018). Digital twin – Proof of concept. *Journal of Manufacturing Letters*, *15*, 64–66.

Hemminger, S. (2005). Network emulation with NetEm. *Open-Source Development Lab*. https://citeseerx.ist.psu.edu/viewdoc/download?doi=10.1.1.67.1687&rep=rep1&type=pdf

Keen, M. (2019). Construction technology defined. What is digital twin? *Autodesk Construction Cloud*. https://constructionblog.autodesk.com/digital-twin/

Lauff, C. A., Kotys-Schwartz, D., and Rentschler, M. E. (2018). What is a prototype? What are the roles of prototypes in companies? *Journal of Mechanical Design*, *140*, 061102/1–12. https://doi.org/10.1115/1.4039340

Liu, W., Zhang, W., Dutta, B., Wu, Z., and Goh, M. (2020). Digital twinning for productivity improvement opportunities with robotic process automation: Case of Greenfield hospital. *International Journal of Mechanical Engineering and Robotics*, *9*(2), 258–263.

Lui, Z., Meyendorf, N., and Mrad, N. (2018). The role of data fusion in predictive maintenance using digital twin. In: *44th Annual Review of Progress in Quantitative Nondestructive Evaluation, Volume 37* (AIP Conf. Proc. 1949, 020023-1–020023-6). https://doi.org/10.1063/1.5031520

Madubuike, O. C., and Anumba, C. J. (2022). Digital twin application in healthcare facilities management. *Computing in Civil Engineering, 2021*. https://doi.org/10.1061/9780784483893.046

Madubuike, O. C., and Anumba, C. J. (2023). Digital twin-based healthcare facilities management (2022). *Journal of Computing in Civil Engineering*, *37*(2), 1–11. https://doi.org/10.1061/JCCEE5.CPENG-4842

Madni, A. M., Madni, C. C., and Lucero, S. (2019). Leveraging digital twin technology in model-based systems engineering. *Systems*, *7*, 7.

Madubuike, O. C., Anumba, C. J., and Agapaki, E. (2023). Scenarios for digital twin deployment in healthcare facilities management. *Journal of Facilities Management*. Vol. ahead-of-print. https://doi.org/10.1108/JFM-10-2022-0107

Madubuike, O. C., Anumba, C. J., and Khallaf, R. (2022). A review of digital twin applications in construction. *Journal of Information Technology in Construction (ITcon)*, *27*, 145–172. https://doi.org/10.1108/10.36680/j.itcon.2022.008

Martinez, V., Ouyang, A. N., Neely, A., Burstall, C., and Bisessar, D. (2018). Service business model innovations: The digital twin technology. In: *Working Paper; Cambridge Service Alliance*. University of Cambridge.

Menold, J., Jablokow, K., and Simpson, T. (2017). Prototype for X (PFX): A holistic framework for structuring prototyping methods to support engineering design. *Journal of Design Studies*, *50*(1), 70–112.

Monteiro, A., and Martins, J. P. (2013). A survey on modeling guidelines for quantity takeoff-oriented BIM-based design. *Automation in Construction*, *35*(May), 238–253. https://doi.org/10.1016/j.autcon.2013.05.005

Negri, E., Fumagalli, L., and Macchi, M. (2017). A review of the roles of digital twin in CPS-based production systems. *Procedia Manufacturing*, *11*, 939–948.

Negroponte, P. N. (1995). *Being Digital*. Vintage Books, Random House Incorporated, New York, NY.

Oracle Fusion Applications. (2017). Digital twins for IoT applications: A comprehensive approach to implementing IoT digital twins. *Oracle White Paper*. www.oracle.com

Palit, S., and Datta, A. (2017). Emergence of digital twins. *Journal of Innovation Management*, *5*, 14–34.

Palmié, M., Parida, V., Mader, A., and Wincent, J. (2023). Clarifying the scaling concept: A review, definition, and measure of scaling performance and an elaborate agenda for future research. *Journal of Business Research*, *158*, 1–15.

Parris, C. J., Laflen, J. B., Grabb, M. L., and Kalitan, D. M. (2016). The future for industrial services: The digital twin. In: *External Document*. Infosys Limited.

Rebolj, A. B. (2013). The case study is a type of qualitative research. *Journal of Contemporary Educational Studies*, *1*, 28–43.

Rosen, R., von Wichert, G., Lo, G., and Bettenhausen, K. D. (2015). *About the importance of autonomy and digital twins for the future of manufacturing*. IFAC.

Rosone, M. (2018). HVAC troubleshooting: negative and positive air pressure systems. *Arista*. https://aristair.com/blog/hvac-troubleshooting-surprising-symptoms-of-hvac-problems/

Sacks, R., Kedar, A., Borrmann, A., Ma, L., Brilakis, I., Hüthwohl, P., and Barutcu, B. E. (2018). See Bridge as next-generation bridge inspection: Overview, information delivery manual, and model view definition. *Automation in Construction, 90*, 134–145.

Salonitis, K. (2014). Stereolithography: Comprehensive material processing. *Material Science and Material Engineering, 10*, 19–67.

Schroeder, G. N., Steinmetz, C., Pereira, C. E., and Espindola, D. B. (2016). Digital twin data modeling with automationML and a communication methodology for data exchange. *IFAC-PapersOnLine, 49–30*, 12–17.

Shafto, M., Conroy, M., Doyle, R., Glaessgen, E., Kemp, C., LeMoigne, J., and Wang, L. (2010). *Modeling, Simulation, Information Technology & Processing – Technology Area 11*. National Aeronautics and Space Administration, Washington, DC. www.nasa.gov

Shohet, I. M., Lavy-Leibovich, S., and Bar-On, D. (2003). Integrated maintenance monitoring of hospital buildings. *Journal of Construction Management and Economics, 21*, 219–228.

Software Advice (2020, October 30). Facilities management software. https://www.softwareadvice.com/cafm/p/all/

Tham, K. W. (2016). Indoor air quality and its effects on humans – A review of challenges and developments in the last 30 years. *Energy and Buildings, 130*, 637–650.

van Notten, Ph. W. F. (2005). *Writing on the wall: Scenario development in times of discontinuity*. Dissertation. www.dissertation.com

Vincent, C. (2006). *Patient Safety*. Edinburgh, Elsevier Ltd.

Whitley, E., and Ball, J. (2002). Statistics review 4: Sample size calculations. *Journal of Critical Care, 6*(4), 335–341. https://doi.org/10.1186/cc1521

Zhang, Q., Liu, J., and Zhao, G. (2018). *Towards 5G Enabled Tactile Robotic Telesurgery*. Cornell University, Ithaca, NY.

13

ACTIVITY-BASED MODELS FOR SMART AND SUSTAINABLE URBAN ENVIRONMENT

Qian-Cheng Wang[1] *and Li Wan*[2]

[1]PhD candidate, Department of Land Economy, University of Cambridge CB3 9EU United Kingdom, qw250@cam.ac.uk.

[2]Associate Professor, Department of Land Economy, University of Cambridge CB3 9EU United Kingdom, Lw423@cam.ac.uk.

13.1 Introduction

Understanding the operations of cities and informing policymaking through simulating urban change processes has been a long-lasting endeavour of generations of urban modellers (Batty, 2009; Wegener, 1994). Digital twins, as an emerging technology originated in engineering and manufacturing, features a simulation process that amalgamates information from various, real-time sources to construct a virtual representation of physical products or systems (Vanderhorn & Mahadevan, 2021; Wan et al., 2023). Within the realm of urban environments, digital twins can be employed to model complex interdependent urban infrastructure systems as well as enhance collaborations among stakeholders and citizen engagement (Deng et al., 2021; Ferré-Bigorra et al., 2022).

One particular urban modelling challenge is to capture the transformative yet nuanced changes of activity patterns at the individual level and understand the associated wider impacts at the city level such that pre-emptive measures could be considered. These patterns mirror the interaction between demands of urban residents and the resource supply of the urban environment. On the one hand, shifts in activity pattern trends may have intricate long-term effects on urban development (Xiao et al., 2020). Propelled by technological advancement and demographic changes, activity patterns of residents are evolving at an accelerated pace. For instance, flexible working and online shopping have been democratised. These novel activity patterns alter demand for products, facilities, and services, which may in turn reshape the spatial structure of cities (Althoff et al., 2022; Van Nieuwerburgh, 2023). By analysing resident activity patterns, urban researchers can more accurately comprehend and predict the changing trends of crucial indicators such as urban transport demand (Xiao et al., 2020), infectious disease transmission processes, resource allocation (Keirstead & Sivakumar, 2012), employment trends (Wöhner, 2022), and population relocation (Althoff et al., 2022). On the other hand, resident activity patterns can also connect changes in the built environment to urban residents. In smart city decision-making and governance, planners can also forecast how resource allocation, facility planning, land use patterns, and development strategies in the city will impact commuting time, leisure methods, and activity diversity of urban

DOI: 10.1201/9781003383840-14

residents, thereby affecting their public health (Frank & Engelke, 2001), well-being and place attachment (Wang et al., 2023), energy consumption (Ghauche et al., 2010), and carbon emissions (De Lauretis et al., 2017).

In the realm of urban digital twin technology, activity-based models offer an approach to understanding transport, land use, and related social and physical systems by utilising individual actions in space and time as the foundation (Liao et al., 2013; Timmermans & Arentze, 2011). Activity-based approaches possess several advantages over traditional approaches, including the ability to capture complex constraints and linkages that mobility determines more effectively. Additionally, activity-based models allow for a broader range of policy variables, including non-transport solutions to mobility issues (Rasouli & Timmermans, 2014). Activity-based approaches can contribute to digital twins in urban built environments by providing a realistic representation of the demand for infrastructure, services, and products by urban residents based on their activities and travel behaviours. By incorporating activity-based approaches into digital twins, urban planners and decision-makers can gain a more detailed understanding of how people interact with their environment and use this information to make better decisions regarding urban planning and operations.

This chapter aims to introduce activity-based models and their role within the context of smart cities. Section 2 presents the roles of activity patterns of residents in the urban environment. Section 3 introduces activity-based approaches in modelling urban environments. It begins by discussing the development of activity-based models and important concepts based on time geography. Subsequently, this section examines how activity-based approaches can address specific issues in urban digital twins. Section 4 introduces several main categories of activity-based models, including single facet models, constraint-based models, economics models, and rule-based models. This section discusses the features and benefits of different kinds of activity-based models and reviews how they contribute to smart and sustainable urban decision-making. This section also presents recent cases of activity-based models in smart city governance scenarios. Section 5 draws the conclusion.

13.2 Activity patterns in the urban built environment

Within the field of urban studies, digital twins have transcended their novelty from a technical standpoint. Before the extensive discussion on city digital twins became prevalent, urban modelling and analytics had experienced over six decades of development (Batty et al., 2012; Hansen, 1959), thereby laying a solid theoretical and practical groundwork (Wan et al., 2023). Extant studies have focused on land use planning strategies, the built environment, economic sectors, and their interrelations (He et al., 2024). Nonetheless, traditional urban models often overlook the role of activity patterns in urban development (Rasouli & Timmermans, 2014). This section aims to explore the role of activity patterns in urban environments through a selection of typical cases. These discussions encompass both the influence of the critical elements in the urban environment on activity patterns and the reciprocal impact of activity patterns on the urban environment. The discussions are not exhaustive or definitive.

13.2.1 The influence of the urban environment on activity patterns

The supply of infrastructure services and amenities can significantly shape the activity patterns of urban dwellers. In particular, strategic urban land use and transport planning can encourage more sustainable activity patterns. For example, the Netherlands, a country with more bikes than people, demonstrates how quality and integrated active travel infrastructure has been a key enabler for promoting sustainable travel behaviour (Cabral Dias & Gomes Ribeiro, 2021; Ploeger & Oldenziel, 2022). The country is home to over 35,000 kilometres of dedicated and integrated cycling network and trips involving bicycles make up more than a quarter of all transport demand.

In contrast, the built environment can also impose certain restrictions on people's activity choices which may have detrimental effects on quality of life and sustainability. For example, inadequate public green spaces in dense urban environment can lead to limited outdoor leisure activities and protracted indoor activity time, particularly for those residing in apartment complexes (Feng et al., 2021; Mytton et al., 2012; Sugiyama et al., 2014; Wang et al., 2023). The availability and accessibility of such spaces are crucial in promoting physical activity and fostering social interactions, thereby enhancing the overall quality of life (Mytton et al., 2012).

The natural environment plays an important role in shaping the activity patterns of residents. Natural environment encompasses all living organisms as well as non-man-made physical features such as air, landforms, climate, soil, and bodies of water within an urban context. A quintessential example of this influence is the impact of climatic conditions on activities of residents (Edwards et al., 2015; Fan et al., 2023). During the sweltering summer months, residents may recalibrate their daily routines to evade the peak heat periods of the day. This could entail rescheduling outdoor activities to the cooler segments of the day, such as the early morning or late evening hours (Edwards et al., 2015; Shaaban et al., 2018). Conversely, during the frigid winter months, residents may opt to spend a greater proportion of their time in indoor environments compared to the summer months (Chen et al., 2015). The optionality of flexible/remote working is likely to further complicate such seasonal variations in the timing and location of daily activities.

The cultural environment and the interpersonal interactions among residents can also shape their activity patterns (Bader et al., 2018; Park et al., 2009). The cultural environment of a city encompasses a collective set of beliefs, practices, customs, behaviours, and values prevalent among its residents (Opoku, 2015). These elements are moulded by myriad factors including historical context, religious beliefs, political climate, available resources, economic status, and the education system. For instance, household type and structure present significant influence on time-use pattern and trip chains (Lee et al., 2007). Also, concerns about social security may compel parents to dedicate more time to the transport of their children to and from school (Barker, 2003). This not only reflects the societal emphasis on safety but also underscores the role of parental responsibilities in shaping daily routines. Similarly, a robust work ethic might motivate employees to devote more hours to their professional commitments (Kim et al., 2018; Park et al., 2009; Sirianni & Negrey, 2000). This could result in an increase in the average working hours within a specific locale, thereby influencing the overall lifestyle and activity patterns of the residents (Bader et al., 2018).

13.2.2 The effects of activity patterns on urban environment

The activity patterns can exert critical influence on industrial development and the built environment. Industries and the built environment provide the necessary infrastructure, goods, and services to cater to the demands stemming from the living, working, and leisure activities of residents. Over the past decade, online shopping has gained significant momentum. The new shopping patterns have created fresh business opportunities and jobs while also causing far-reaching effects on the urban built environment (Ni, 2022). One notable issue is the increasing demand for "last-mile" delivery, where dispersed short-distance freight demands can lead to increased traffic flow, unexpected congestion, and carbon emissions (Jaller & Pahwa, 2020). Furthermore, online shopping also affects other elements in the built environment and industries. For instance, it has led to increased mixed land use, reduced retail space, and more dispersed warehousing (Koch et al., 2020; Muhanna & Wolf, 2002).

A recent case is the influence of emerging working patterns on the urban environment (Li et al., 2022; Yu et al., 2019). The nature of work, characterised by various working patterns such as night shifts, rotational shifts, flexible office hours, and the "nine-nine-six" work system (Xiao et al., 2020; Yu et al., 2019), significantly influences workers' daily and weekly activity patterns (Olson, 1983). For instance, employees who have the option of flexible/remote working can repurpose their commuting time for leisure or household activities. Such flexibility underscores the dynamic interactions between workplace-based and home-based activities. The rise of remote work has significantly influenced the urban built environment, particularly in suburban areas. As remote and hybrid work patterns are more widely adopted, employees are no longer bound to living in close proximity to their workplaces (Wohlers & Hertel, 2017). This shift may lead to a surge in demand for suburban homes, as certain population groups seek larger living spaces and a quieter lifestyle away from the hustle and bustle of city centres. The change drives a more prosperous housing market in suburban than city centres (Van Nieuwerburgh, 2023). This trend is not only reshaping the housing market but also is affecting commercial real estate, with a decline in demand for office spaces in urban areas and an increase in demand for home offices and co-working spaces in suburban locations (Wohlers & Hertel, 2017). The shift towards suburban living is also driving changes in transport demand (van der Loop et al., 2019) and infrastructure development, with increased demand for amenities like shopping centres, schools, and healthcare facilities in these areas (Yu et al., 2019). Understanding the nuanced changes of activity patterns across different population groups and locations would inform the planning of future infrastructure and enhance spatio-social equality.

Also, the activity patterns of urban residents can influence the natural environment, primarily through the consumption of resources (e.g., energy, water, and food) and the generation of waste (e.g., organic waste, air pollution, and GHG) (De Lauretis et al., 2017; Hatzopoulou et al., 2011). The lived experience of lockdowns, homeworking, and social distancing during the COVID-19 pandemic resulted in a change of electricity and hot water consumption in the domestic sector. Contrast to the peak load usually observed in evenings before the pandemic, energy consumption became more evenly distributed throughout the day (Rouleau & Gosselin, 2021). Some studies have leveraged this window of lifestyle changes by exploring novel interventions to nudge for environmentally friendly behaviours and sustainable activity patterns (Liu et al., 2021; Xu et al., 2023).

Activity pattern is also associated with an individual's physical and mental health (Brajša-Žganec et al., 2011; Trainor et al., 2010). For example, online leisure activities, such as computer gaming, can have a profound impact on the subjective well-being of residents (Pandya & Lodha, 2021; Valkenburg, 2022). These activities can provide a form of escapism, offering temporary relief from reality. However, several studies suggest that the screening time may negatively correlate with the quality of life and overall well-being of individuals. Outdoor activities play a crucial role in promoting physical and mental health. These activities enhance cardiovascular fitness and muscle strength, contributing to overall well-being. Furthermore, exposure to outdoor spaces, such as nature and green/blue spaces, can significantly enhance the subjective well-being of individuals (Huang et al., 2022; Wang et al., 2023).

Given the significant effects of activity patterns on both the natural and built environment and well-being, the next section will review activity-based approaches for modelling the urban environment.

13.3 Activity-based approaches in modelling urban environment

13.3.1 Development of the activity-based approach

Prior to the 1960s, urban modelling often employed concepts derived from physics, such as the gravity model (Hansen, 1959; Lowry, 1964) and the entropy maximising theory (Murchland, 1966; Wilson, 1969, 1970), to simulate interactions in an abstract manner. While these aggregate zonal models provided a foundation for understanding the built environment, they lacked explanations for decision-making processes and behavioural patterns of agents. Since the 1960s, there have been calls for urban environment models to shift their focus from changes in the industries and built environment to activity demands of residents (Chapin, 1968, 1971; Jones et al., 1983).

Hägerstrand (1970) and Chapin (1974) first proposed the concept of time geography that provides an established framework for measuring and simulating how humans allocate time, as a scarce resource, among spatial activities through a bottom-up perspective. The basic assumption of time geography is that an individual's daily activity plan is subject to a limited time budget and the mandatory participation in certain activities (Chapin, 1974; Hagerstrand, 1970). Therefore, individuals' activity pattern is not only driven by their desires but also governed by a set of time and space constraints (Asgari et al., 2019; Ellegård, 1999; Hägerstrand, 1989; Hägerstrand et al., 2009). With the advent of time geography, which will be reviewed in detail in the next section, activity-based approaches to modelling the built environment and transport emerged and have gained traction since the 1970s.

The activity-based approach was initially developed for use in the field of urban transport as a replacement for traditional four-step models (also called the trip-based model). These activity-based models are predicated on the principle that travel demand arises from individuals' needs and desires to engage in activities. Some activities may occur at home, while others may take place outside the home, necessitating travel between the two locations. Early activity-based models often employed single discrete choice models, such as the multinomial logit (MNL) model, to simulate travel destination/mode choices across different types of residents. With advances in discrete choice modelling, more sophisticated models, such as nested logit and probit models, have been adopted to avoid unrealistic assumptions inherent in the MNL model. At this stage, most activity-based studies still rely on aggregate inter-zonal traffic statistics as input data. Similar to traditional trip-based models, these models primarily focus on single-destination and single-purpose travel activities, without considering the independences among activity demands and hence trips.

One catalyst for the rise of the activity-based approach in transport modelling has been the emergence of dual-earner families, which has altered traditional family dynamics. With unleashed productivity of women, time devoted to household activities such as house cleaning and grocery shopping has become increasingly constrained. Therefore, interactions between family members have assumed greater importance, with spouses coordinating task responsibilities such as picking up children and tending the garden. The rise of new lifestyles has also increased the proportion of multi-destination and multi-purpose travel activities (i.e. trip chains), such as grocery shopping or picking up children on the way home from work. Additionally, the widespread use of private cars has enhanced the mobility of urban residents, enabling them to travel greater distances with a more flexible schedule. These changes can affect traffic flow and travel timing but are difficult to capture at aggregate four-step transport models.

Moreover, there has been a growing interest in micro-simulation models for activity-based approaches. Microsimulation models feature high spatio-temporal resolutions at individual, trip, vehicle, or family level, hence representing the agent-based modelling (ABM) framework.

A notable advantage of ABM is the ability to capture complex activity sequences, interpersonal dynamics, and spatiotemporal constraints on individual activity pattern decisions, including activity scheduling, location selection, and travel chains (e.g., Brandon & Price, 2020).

With the increasing availability of data, activity-based approaches are being applied to a broader range of urban simulations. Activity-based models focus on the space, resource, and facility needs that arise directly or indirectly from residents' activities. For instance, by analysing the spatiotemporal characteristics of activity patterns of residents, researchers can predict the spatial distribution of energy and water resource demands and pollutant emissions for individual economic sectors or urban regions across different time periods.

13.3.2 Time geography

From the perspective of time geography, everyone is equal in front of time: people have the same amount of time every day (i.e., 24 hours), and they physically exist in one and only one spatial location. The theory assumes that the passage of time has a constant speed. Since everyone must consume all available time, spatial existence of an individual can be measured by clock time, which contains three important parts: *past*, *now*, and *future*. The *past* reflects what has happened, which cannot be changed but can have an impact on the *future*. The concept of *future* is the coming time, which contains many possible choices of activities, while only one will become reality when *future* becomes *past*. The concept of *now* is the moment when the *future* continuously becomes the *past*, and all individuals exist in this moment. *Now* constitutes a point in time at which an individual can take action and make changes. Clock time and indivisible individuals form the basis of time geography symbols (Ellegård, 2018).

In order to better explain the influence of space-time constraints on individual daily activities, time geography adopts a novel notation system, which accurately describes both the temporal and spatial processes that are difficult to explain by ordinary text-based description (Ellegård, 1999; Pred, 1984). The visual notation system is devised using the following concepts: individual path, bundle of paths, space-time prism, project, and constraint (Hagerstrand, 1970; Hägerstrand, 1985). An individual path is a record of an individual's movement sequence in time and space (Hägerstrand et al., 2009). The individual path cannot present the individual itself; instead it focuses on the individual's unique actions and trajectories in the past (Ellegård, 2018; Hägerstrand, 1985). In Figure 13.1 (a), the solid line shows an individual path in a two-dimensional (i.e., time and

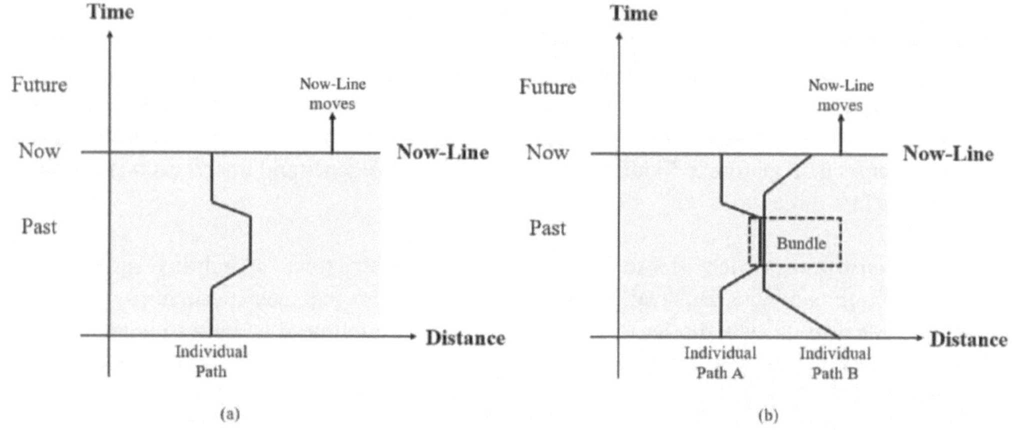

Figure 13.1 Simplified two-dimensional individual paths[1]

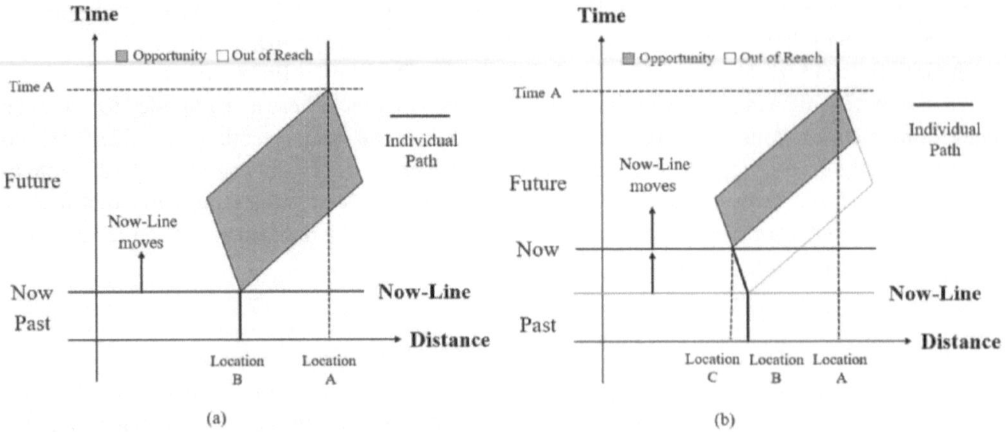

Figure 13.2 Simplified two-dimensional time-space prism

distance) coordinate system. When the research object involves multiple individuals, their paths might appear simultaneously (see Figure 13.1 (b)). This is called a bundle of paths (Ellegård, 2018). Bundles are often used to illustrate collective activities.

The future contains more opportunities than the individual can realise, and the individual must choose only one path among the opportunities. The temporal and spatial scope of an individual's opportunity in the future is called a space-time prism (Hagerstrand, 1970; Hägerstrand, 1989). The space-time prism is determined by the individual's current geographic location, the maximum movable speed, and the time and space coordinates of the next determined behaviour. Figure 13.2 (a) shows a simplified space-time prism: the individual is now at location B and is scheduled to appear at location A at time A. The origin and terminal vertices point out the earliest start time and the latest end time for travel and potential out-of-home activities, which also reflects both perceived and fixed temporal thresholds (Soo, 2009). The shaded part represents all the locations where the individual has opportunities to appear between now and time A. As shown in Figure 13.2 (b), as the Now-Line advances upward, the individual path extends to position C. The time-space prism also shrinks with the extension of the individual's path, which means that there are fewer locations where the individual has opportunities to appear.

In time geography, projects reflect the activities carried out by individuals or groups to achieve a set goal. A project often contains multiple activities that contribute to achieve the set goal (Ellegård, 2018; Hagerstrand, 1970; Hägerstrand, 1985). However, many constraints exist in the process of completing these activities. According to Ellegård (2018), the concept of constraint is to specify the factors that confine individuals' activity choices. Wegener and Fuerst (2004) classified the constraints into three types:

- *Capacity constraints*, which refer to personal non-spatial restrictions on mobility, such as monetary budget, time budget, and availability of transport modes (e.g., car ownership)
- *Coupling constraints*, which refer to "restrictions on the coupling of activities by location and time schedules of facilities and other individuals" (e.g., opening time of shops, availability of families and colleagues)
- *Institutional constraints*, which refer to "restrictions of access to facilities by public or private regulations such as opening hours, entrance fees or prices".

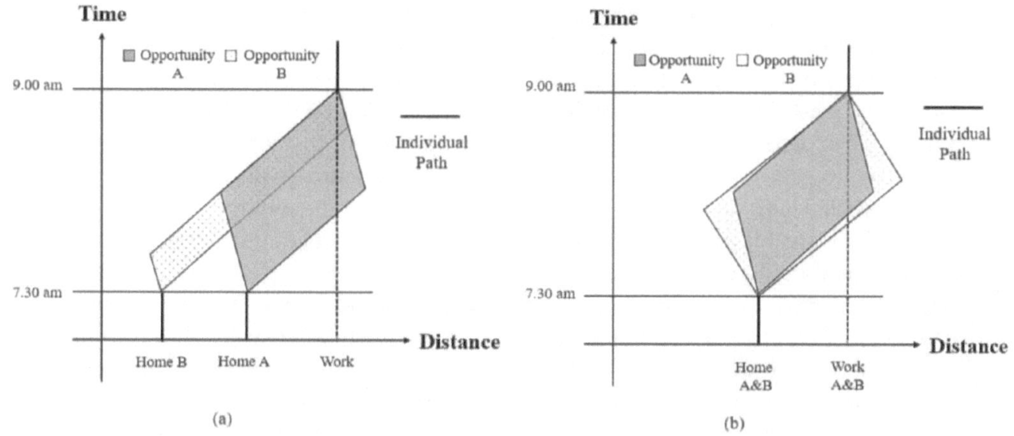

Figure 13.3 The impacts of location choice (a) and travel choice (b)

Time geography has been widely used in activity-based urban development studies that use time-space prisms to evaluate the range of possibilities of conducting individual or organisational projects (Kwan, 1998) or to model activity chains in time and space under various constraints (Asgari et al., 2019; Bradley et al., 2010; Liao et al., 2013; Yoon et al., 2014; Yoon & Goulias, 2010). These space-time prisms can thus be regarded as accessibility measures, i.e., they measure potential areas of opportunity that can be realised under predefined constraints. The main factors that determine the space-time prism include: (a) location choice and (b) travel choice. The location choice determines the vertex of the space-time prism. For example, the two most common locations of a full-time commuter's daily activity frame are the home and work locations (Asgari et al., 2019; Yoon & Goulias, 2010). Both home and work locations are usually considered as long-term choices and are rarely affected by short-term changes, e.g., travel conditions. The travel choice determines the travel speed, which in turn affects the boundaries of the space-time prism.

Figure 13.3 shows the influence of location and travel choices on the space-time prism. Figure 13.3(a) shows that the individual A whose home is closer to the work location obviously has a larger space-time prism than the individual B who lives farther away from work, assuming the same travel speed between their respective home and work locations. Figure 13.3(b) shows that individuals A and B, who have the same distance between the home and work location but use different travel modes, have different space-time prisms. Individual B who uses a faster transport mode has a larger space-time prism.

13.3.3 City digital twins and activity-based modelling

Activity-based modelling has much to contribute to substantiate the emerging technology trend of city digital twins (CDTs). As a new generation of digital technology for urban management and planning, CDTs are enabled by the fast advancement of ubiquitous sensing and cloud and distributive computing. A high-frequency, multi-source data feed is a determining feature of CDTs, which also provide an unprecedented opportunity to further incorporate activity-based modelling techniques in CDTs. Specifically, activity-based modelling can contribute to CDTs on the following aspects.

First, activity-based approaches are capable of capturing the city-level impact of individual-level behavioural and lifestyle changes (proxied by observed or simulated changes of daily/weekly activity patterns). For instance, while flexible/remote working would reduce the need for commuting for many, peak-time traffic congestion has returned, if not yet exceeded, the pre-pandemic level in many cities across the globe. The shift (not reduction) of travel demand across different times of the day and different days of the week requires more granular data and a more nuanced analytical approach. Understanding the interdependence between commuting and non-commuting demand at the individual level is thus crucial for designing targeted policy interventions for tackling car dependence amidst a global decline of public transport ridership (Wan et al., 2021).

Second, the activity-based modelling approach represents an inherent inter-sectoral perspective, which contrasts with the prevalence and limitations of single-sectoral models. The modelling of an individual's activity choices and the derived travel demand naturally connect buildings, infrastructure, public services, and amenities. For example, the wide adoption of homeworking can shift energy demand from commercial to residential buildings. Similarly, the anticipated large-scale uptake of electric vehicles (EVs) brings new challenges for energy and building management, as price-elastic EV owners can change their EV charging location(s) across the city (and adjust their working pattern accordingly), leading to uncertainties in the timing and location of peak load for both the energy grid and building usage.

Third, the activity-based approach enables a well-being-oriented policy evaluation framework beyond the conventional cost-benefit analysis. Empirical studies have confirmed the close, causal relationship between time use (activity) and individual well-being (Chen & Wan, 2023). Unlike aggregate sectoral models, activity-based approaches can reflect the impact of micro-level factors on the economy, the built and natural environment, and the physical and psychological well-being of people. The explicit emphasis on well-being reflects a new societal aspiration for a better quality of life, as opposed to a narrow focus on economic gains.

13.4 Overview of activity-based models

The evolution of activity-based approaches since the 1980s has significantly enriched urban studies by introducing diverse theoretical and disciplinary perspectives. In terms of their theoretical underpinnings, activity-based urban models can be categorised into four distinct types: single facet models, constraint-based models, econometric models, and rule-based models. The ensuing discussion will delve into the unique characteristics of these model types, exemplify with typical models, and illustrate their practical applications.

13.4.1 Single-facet models

Many models focus on a single dimension of activity patterns of urban residents, such as activity frequency, activity timing, activity duration, or location choice. While these models might not incorporate a tour-based travel structure, the flexible use of novel data sources enables these new city-scale models to enhance the modelling accuracy or capture activity patterns from novel perspectives. In particular, the increased availability of activity/time-use diary data and other city-scale panel data has facilitated efforts of urban researchers to examine specific aspects of activity patterns. Single-facet models often introduce new model structures or incorporate important factors that were rarely considered in previous models. The findings of these studies can be integrated with existing models or provide new directions for optimising established activity-based models.

A growing body of literature in urban environment research has examined various aspects of time allocation and activity engagement. For example, some papers employ the structural equation model (SEM) to address complicated causal relationships among socio-demographics, activity participation, time allocation, and travel pattern (Cheng et al., 2019; Chung et al., 2009; Manoj & Verma, 2017; Sharmeen et al., 2014). Sharmeen et al. (2014) employed data collected in the Netherlands to elicit activity and travel needs and their dependencies on life cycle and social network dynamics. Also, Ma and Goulias (1999) presented a case employing Poisson regression models to predict the frequency of subsistence, maintenance, and out-of-home leisure activities.

Some more sophisticated models were also developed for this field (Habib & Miller, 2006, 2008). For example, Habib and Miller (2006) employed an agent-based approach to detail a comprehensive modelling system for shopping activities, consisting of three core components: a model for weekly shopping frequency, a model for daily shopping frequency, and a model for time allocation during individual shopping episodes. The model incorporates household-level influences on an individual's shopping behaviour through fixed effects and a random intercept, and the latent response is modelled using an ordinal response model. In addition, some cross-cultural comparison studies on time allocation and activity frequency were conducted. For instance, there is a study on the daily allocation of time to travel and out-of-home activities across eight European cities in three countries: France, Switzerland, and Belgium (Raux et al., 2011). The Cox proportional hazard model was used to investigate the effects of socio-demographic, spatial context, transport availability, and city-specific variables. The results indicated that socio-demographic characteristics and city-specific effects played a major role in time budgets for travel and out-of-home activities.

Many activity-based transport models focus primarily on the impact of outdoor activity patterns rather than at-home activities. A potential reason is that at-home activities are usually not directly related to travel pattern. Some researchers argued that the interchangeability of home and outdoor activities should be considered in built environment models (Shabanpour et al., 2018; Zhang et al., 2005). Zhang et al. (2005) addressed the lack of household-level models for activity-travel demand in multi-person households with a model grounded in a multi-linear group utility function. This study considered both task allocation to different members and their time allocation to different activities. Shabanpor et al. (2018) estimated a joint discrete-continuous model, where activity type and duration were modelled using a multinomial logit and a log-linear regression model, respectively. The joint model structure was established using a copula approach that captured the unobserved shared factors affecting the two activity attributes: activity type and time allocation. Shabanpor et al. (2018)'s model could be integrated with ADAPTS, a widely used activity-based model, to enhance its predictive power for household activity. This approach provides a more nuanced understanding of the interplay between at-home and out-of-home activities and their impact on transport patterns.

In addition to activity time allocation and activity frequency, some single-facet studies have also paid attention to action location choice (Arentze, 2015; Arentze et al., 2013; Horni et al., 2009; A. Huang & Levinson, 2011). Several studies have investigated the decision-making process of individuals when determining whether to engage in activities within or outside of their homes. These substitution patterns have significant implications for travel demand, as activities conducted within the home do not require travel, whereas those conducted outside of the home necessitate travel. For instance, one study utilised mixed logit models to examine the factors associated with this decision at the individual activity level and discovered that the characteristics of an activity play a crucial role in determining the likelihood of engaging in out-of-home activities (Akar et al., 2011). Also, there are some studies that adopted novel data to explore activity location choices.

Based on metro travellers in Shanghai, China, Wang et al. (2017) and Zhang et al. (2024) extracted trip information from smart transport card records to model location choice of after-work activities and to capture the interpersonal heterogeneity.

There are several single-facet studies to extend the existing comprehensive models. For instance, Horni et al. (2009) developed a singly constrained model based on Hägerstrand's (1970) time-geographic approach, which incorporates competition for slots on the activity infrastructure. This model explicitly integrates time-dependent capacity restraints for each activity location into the agent's location choice process, coupling the actual load with these restraints. As anticipated, this constrained model reduces the occurrence of implausibly overcrowded activity locations, thereby enhancing the behavioural realism of the location choice module of MATSim. Also, to further develop the ALBATROSS model, Arentze et al. (2010) constructed a location choice model within multi-day activity-based frameworks, utilising the random utility maximisation (RUM) framework. This model takes into account not only the broader activity schedule and interactions within households but also introduces an innovative parameter estimation method that does not rely on choice-set observations. Then, the same team investigated the location and travel choices of out-of-home leisure activities to develop a model capable of predicting individuals' social preferences in joint activity choice (Arentze, 2015). This model provides a framework for exploring and forecasting social biases in shared location choices.

This section presents an overview of typical single-facet models that focus on various aspects of activity patterns, including activity time allocation, activity frequency, task allocation, and activity location choice. In addition to these dimensions, numerous studies have also examined other important decisions related to residents' activities, such as travel party, departure time, ITC use, trip chaining, and the number of stops. These examples illustrate the potential contributions of single-facet models to the development of more comprehensive activity-based models. The following sections provide examples of comprehensive activity-based models that incorporate a broader range of activity pattern decisions.

13.4.2 Constraint-based models

In addition to typical single-facet models, the research community has also developed more comprehensive activity-based models. These models often encompass multiple dimensions of activity patterns simultaneously and capture the interdependencies between these dimensions, for instance, the dependence of activity location and time on travel mode (McNally & Rindt, 2007).

Constraint-based models were among the first to fast develop following the concept of time geography. These models are usually based on the feasibility assessment of specific activity patterns within particular time-space environments, also called behavioural paradigms or activity programs. The time-space environment is defined in terms of locations, attributes, available modes of transport network, and travel times between locations for each transport mode. Capacity constraints, coupling constraints, and institutional constraints can all affect the time-space environment. For example, the opening hours of specific venues and the operating hours of vehicles are often employed as typical constraints in these models. Combinatorial algorithms are typically employed to generate all possible activity sequences in order to assess the feasibility of a given activity plan. Arentze and Timmermans (2000) summarised a three-step process for evaluating the feasibility of each sequence in constraint-based models: (a) checking whether the interval between the end time of the previous activity and the start time of the next activity is sufficient to perform the activity plus the associated travel time, (b) testing whether the activity can start after the earliest possible start time and finish before the latest possible finish time, and (c) verifying that conditions regarding the sequence of activities are not violated.

One classic and widely used constraint-based model is the Combinatorial Algorithm for Rescheduling Lists and Activities (CARLA), developed by (Jones et al., 1983). CARLA is a computer model that explicitly represents the temporal constraints and linkages that influence the activity and derived travel patterns of an individual within a certain time period. The model employs a four-rule framework to generate activities and add them to individual schedules, including (a) logical rules that refer to the presumption of one unique activity at a time at one location, (b) environmental rules that refer to authority constraints such as access time restrictions to different places and travel times between locations, (c) inter-personal rules that refer to coupling constraints, and (d) personal rules that reflect personal preferences and individual heterogeneity. Additionally, compared with other early-developed constraint-based models, such as PESASP (Lenntorp, 1976) and BSP (Huigen, 1986), CARLA employs two principles to optimise the activity pattern generation process: first, generating alternative activities in an ordered manner to avoid obviously infeasible activity and travel patterns and second, scheduling the activities with heuristic rules to reduce the total examined alternatives. However, Huigen (1986)'s BSP advanced the constraint-based models by considering the situation of adopting different travel modes to conduct different trips in the same activity chain.

Building upon the PESASP model proposed by Lenntorp (1976) and Dijst and Vidakovic (2000) developed a more advanced constraint-based model: model of action spaces in time intervals and clusters (MASTIC). The goal of the MASTIC is to capture the lifestyle of different groups of people in urban environment. The model is based on major ideas in time geography (Hagerstrand, 1970) and also incorporates the views on the heterogeneity of individual activity patterns (Kutter, 1981; Kwan, 1998, 1999). The core concept of the model is the potential action space, defined as the area containing all the action places that an individual can visit within a certain period of time. This area is located around one or two bases of personal relevance, referring to the fixed places where an individual spends a significant and regular amount of time, such as household and workplace. Individuals typically leave these bases for only a limited amount of time, known as time windows, as they must return to their base(s) to engage in compulsory activities such as paid work, domestic work, and sleeping. The events between departure from and arrival at the bases might lead to visits to accessible non-base locations and derived trip(s). Therefore, a journey is constrained by the time window from the temporal dimension and by the potential action space from the spatial dimension (van Eck et al., 2005). The potential action space is determined by four constraints: the distance between bases, the available time window, the travel time ratio, and the speed of the mode of transport. Considering the activity pattern of urban residents, van Eck et al. (2005) summarised three types of potential action space: circular, linear, and elliptic space. Figure 13.4 illustrates the differences among these three types.

13.4.3 Econometric models

Econometric models, also known as utility-maximising models, are one of the two primary categories of activity-based models that represent the decision-making process of individuals when determining how to allocate their time, engage in activities, and select their travel patterns. These models use statistical methods to estimate the relationships between variables and represent individual activity decisions as choices among a set of alternatives. The econometric models typically capture the activity patterns of individuals reasonably well, with results often stemming from the use of observed distributions in activity generation modules.

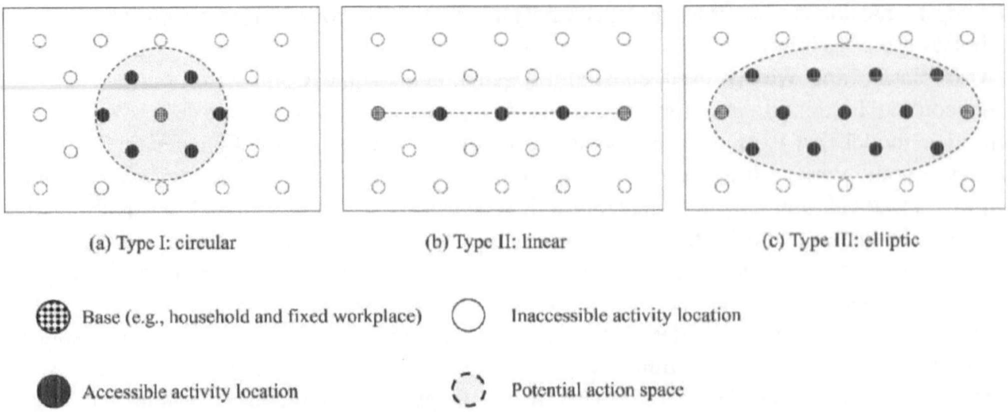

Figure 13.4 Three types of potential action space

Most econometric models employ discrete choice models to capture individual activity pattern decisions, assuming that individuals make decisions based on the principle of utility maximisation. This means that individuals choose the option that provides them with the greatest level of satisfaction or utility. Utility-maximising models use a set of equations to specify a mathematical model, which can either be translated into a model such as Markov chains or differential equations or treated using tools that directly work on the equation-set in place of a translated model.

Simulation of travel/activity responses to complex household interactive logistic decisions (STARCHILD) (Kitamura, 1988) is a well-known representative utility maximising model. STARCHILD is a policy-sensitive approach to modelling travel behaviour based on activity pattern analysis (Recker et al., 1986a, 1986b). The model views travel as an input to a more basic process involving activity decisions and assumes that travel decisions are driven by the collection of activities that form an agenda for participation. The model considers travel behaviours as part of the temporal constraint in the process of scheduling activities. Hafezi et al. (2021) summarised that STARCHILD is based on a three-step process to generate activity patterns of urban residents. In the first step, the model identifies all possible participation activities, subject to all constraints. In the second step, these choices are clustered into three to ten groups. In the third step, representative activity patterns are established for each group, and the model employs a MNL model to estimate the utility of different activity patterns and identify the final activity choices.

The comprehensive utility-based system of activity-travel scheduling options modelling (CUSTOM) (Habib, 2018) is another typical example of an econometric model that employs a random utility-maximising econometric approach to model various dimensions of activity patterns. The model uses a simultaneous modelling approach to schedule time budget-constrained activities, considering continuous time specifications for both work and non-work activity episode duration and start time choices. CUSTOM employs the random utility maximisation (RUM) assumption to jointly model activity sequences, activity durations, start times, and non-work activity location choices. However, the CUSTOM does not consider scheduling priority among the competing activities.

DaySim (Bowman et al., 2006), an integrated microsimulation of personal activities and travel, is another example of an econometric model that employs a highly disaggregate treatment of the purpose, activity, travel timing, and location dimensions of the modelled outcomes. DaySim follows the day activity schedule approach developed by (Bowman & Ben-Akiva, 2001) and operates at four integrated levels: longer-term person and household choices, single day-long activity pattern choices, tour-level choices, and trip-level choices. The structure of DaySim is presented in Figure 13.5. In the Sacramento application, the model was validated for the base year 2000 and applied to forecast years 2005 and 2035.

The advent of new data sources, such as time use data and transit card data, coupled with the rapid advancement of discrete choice models, has paved the way for the development of increasingly sophisticated utility-maximising models. Over the past decade, these utility-maximising models have found numerous significant applications in urban research. For example, a recent econometric model has addressed the methodological challenge of modelling multi-day activity-travel demand by formulating a novel multivariate multiple discrete-continuous probit (MDCP) model system (Astroza et al., 2018). This comprehensive framework integrates two MDCP model components, one pertaining to weekday time allocation and the other to weekend activity-time allocation. A practical case study based on UK time use survey data demonstrated that the integrated MDCP model system effectively captures the relationships in activity-time allocation between weekdays and weekend days. Another notable application focuses on secondary activities in transport

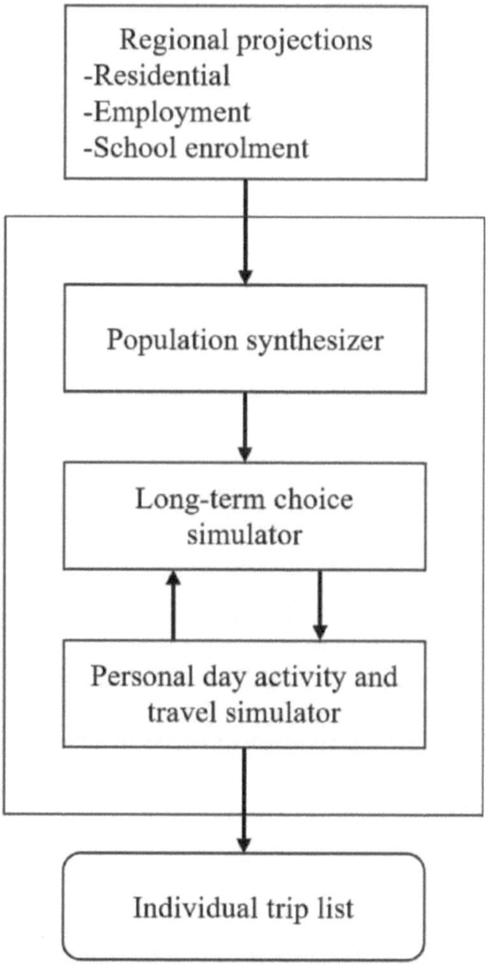

Figure 13.5 The system structure of DaySIM

scenarios (Calastri et al., 2022). Calastri et al. (2022) employed a utility-maximising modelling framework based on multiple discrete continuous extreme values (MDCEV) to predict residents' online activity patterns during train travel. The MDCEV framework allows models to consider both discrete and continuous choices. Using empirical data from UK railway services, Calastri and colleagues incorporated residents' travel purposes, travel time (peak or off-peak), and ticket type into their model. Their model serves as a valuable reference for understanding well-being of railway travellers.

13.4.4 Rule-based model

Rule-based models, also known as computational process models, use decision rules to derive feasible solutions. These models represent individual activity decisions as heuristics or rules. They usually capture the activity patterns of individuals reasonably well, and their results may stem from the use of observed distributions in activity generation modules.

While utility maximising models have bridged several critical research gaps in urban environment modelling, some researchers argued that the individual does not have to follow the optimal choices. To address this research gap, rule-based models have been developed to conceptualise choices as outcomes of heuristics. In contrast to econometric models, which postulate that the process of activity generation and scheduling can be modelled as discrete choices, rule-based models do not rely on utility maximisation. Instead, they use a set of rules that indirectly specifies a mathematical model. The rule set can either be translated into a model such as Markov chains or differential equations or be treated using tools that directly work on the rule set in place of a translated model.

ALBATROSS (Arentze & Timmermans, 2000) is an emblem of rule-based model. ALBATROSS constitutes a learning-based simulation system that centres on transport, grounded in the realm of choice heuristics theories. Figure 13.6 illustrates the ALBATROSS model architecture. It delves into the intricate processes of decision-making undertaken by individuals, with the overarching objective of comprehensively understanding and predicting their actions within a predefined timeframe. This approach encompasses a multifaceted analysis, taking into account the characteristics of activities, their respective locations, timing, duration, occasional companions, and the chosen transport mode. ALBATROSS postulates that the observed transport patterns emerge from a complex decision-making procedure, through which individuals endeavour to achieve specific objectives while navigating the spatial-temporal and institutional limitations imposed by their environment.

The decision-making process asserts that long-term decisions at the family level play a significant role in shaping the composition and structure of the activity calendar. Choices related to marital status, the number of children, and similar factors exert a strong influence on this process. The challenges faced by families wield a considerable impact as well. Following the generation of activities, individuals proceed to schedule their engagements. This encompasses determining the location, mode of transport, companions, and the specific activities to be included in the schedule, as well as establishing connections among travel, activity timing, and location. The process of actual activity scheduling can be succinctly described as individuals striving to achieve specific goals within the framework of various constraints. ALBATROSS categorises these constraints into the following types: situational constraints, institutional constraints, household-related constraints, spatial constraints, temporal constraints, and spatial-temporal constraints. The selection of activities by an individual is referred to as their choice behaviour. ALBATROSS posits that this choice behaviour is influenced by rules

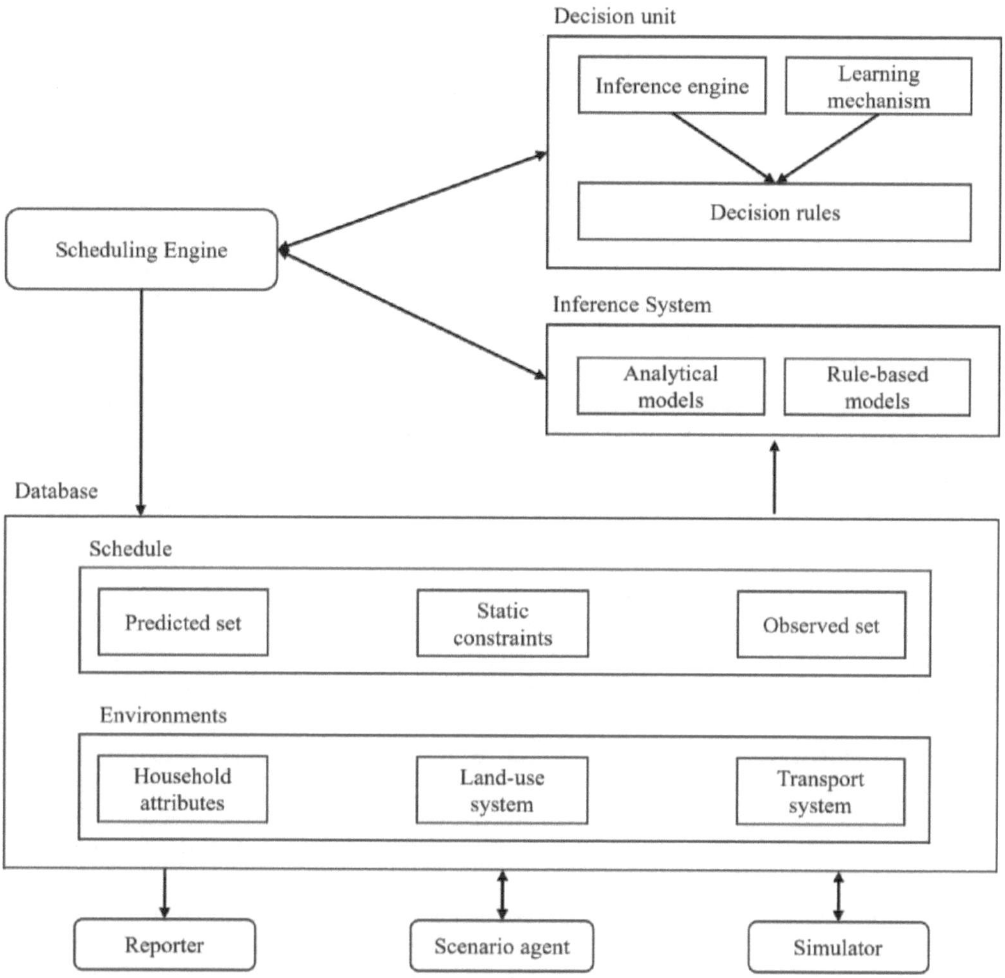

Figure 13.6 The system architecture of a rule-based model

that evolve and adapt gradually over time through individual interactions with the environment (reinforcement learning) or through the exchange of information with others (social learning). These choice behaviours exhibit variations among individuals due to their unique inclinations. Essentially, the learning theory underlying Albatross suggests that the rules governing choice behaviour are heuristic, context-dependent, and inherently adaptable in nature.

The model is composed of three integral components: (a) a model that illustrates the sequential decision-making process; (b) models designed to dynamically compute constraints on available choice options; and (c) a set of decision trees, serving as representations of individuals' decision-making behaviour at each step within the process model. ALBATROSS operates based on the premise of a predefined sequence of choice facets, established through a prioritised ranking of activities by their types and attributes. Drawing upon considerations such as travel companions, activity types, activity instances, activity sequencing, schedules, durations, and relevant details, Albatross generates activity definitions. Its scheduling process model mirrors the way individuals shape their preferences and sequentially arrange them during the scheduling of activities. Subsequently, the model identifies choice options by considering dynamic constraints and devises a method to generate decision trees from data, along with establishing a rule for deducing decisions from the inferred trees.

Another critical case of rule-based model is the forecasting evolutionary activity-travel of households and their environmental repercussions (FEATHERS). FEATHERS is an ALBATROSS-extended activity-based modelling system developed by the Transportation Research Institute of Hasselt University (Bellemans et al., 2010). The FEATHERS facilitates the implementation of activity-based models for transport demand forecasting (Bao et al., 2018). It suggests a four-stage development trajectory for a smooth transition from the traditional four-step models towards static activity-based models in the short term and dynamic activity-based models in the long term. FEATHERS has been applied in many urban management scenarios (Bao et al., 2015) and in several regions and countries (Do Lee et al., 2012). For example, in Belgium, FEATHERS has been applied to investigate the potential impact of light rail initiatives on travel demand at a local network in Flanders (Bao et al., 2018).

There are several practical cases employing rule-based models in urban decision-making process. For example, Kim et al. (2022) employed an ALBATROSS-based model to support spatial distribution of EV chargers. The burgeoning advancements in sustainable transport technology have led to electric vehicles swiftly gaining a larger foothold in the private transport market (Brown et al., 2010; Morrissey et al., 2016). A pivotal aspect of this transition is the strategic planning of optimal locations and capacities for public charging stations catering to plug-in electric vehicles (PEVs) within urban landscapes (Ahmad et al., 2022). The effective planning of PEV charging infrastructure emerges as an indispensable step towards realising the goal of sustainable urban mobility (Kim et al., 2022; Morrissey et al., 2016). Kim and colleagues proposed an ALBATROSS-based innovative framework for determining the optimal location and size of public charging stations for plug-in electric vehicles (PEVs) in urban areas (Kim et al., 2022). This framework takes into account the uncertainty and variability of charging demand. The framework is composed of two primary components.

The first component involves predicting out-of-home charging demand using an activity-based travel demand model. This part of the study utilises a dynamic decision tree model to forecast the multi-day activity-travel patterns of PEV users, subsequently deriving their out-of-home charging demand. Kim et al. (2022) employed a chi-squared automatic interaction detector (CHAID)-based method to extract probabilistic decision rules for these predictions. The second component primarily comprises a public charging station location-allocation model that utilises a scenario-based stochastic programming (SP) approach. A two-stage stochastic mixed-integer programming (TSMIP) model is used to determine the optimal number and locations of charging stations from a set of potential locations.

Kim and colleagues (2022) employed their method in a case in Eindhoven, the Netherlands, effectively predicting different activity periods and the frequency of PEV user charging times. This case study illustrates that the proposed approach can capture the dynamic and heterogeneous aspects of charging behaviour and provide optimal solutions under conditions of uncertainty. Well-executed planning of charging facilities can alleviate the 'range anxiety' experienced by PEV users, thereby enhancing the attractiveness of sustainable transport alternatives. Also, a meticulously planned charging infrastructure not only minimises the squandering of land resources and facilities but also aids in managing grid load. This can prevent power outages and ensure grid stability.

13.5 Conclusion

This chapter provides an overview of activity-based modelling approaches and their theoretical underpinnings. In light of major societal challenges, notably the climate change and the observed behavioural and lifestyle changes induced by the COVID-19 pandemic, the activity-based approach has renewed significance and can offer much-needed, inter-sectoral insights for promoting smart and sustainable cities. The analytical advantages of activity-based modelling are discussed in relation to the emerging technology trend of city digital twins, where the availability of real-time, multi-source data enables new frontiers for activity-based modelling research. The chapter reviews representative modelling techniques, with a focus on their potential contribution to smart and sustainable urban development. It is argued that activity-based perspective provides a suitable framework for understanding the city-level, inter-sectoral impacts derived from individual-level behavioural and lifestyle changes.

References

Ahmad, F., Iqbal, A., Ashraf, I., Marzband, M., & khan, I. (2022). Optimal location of electric vehicle charging station and its impact on distribution network: A review. *Energy Reports*, *8*, 2314–2333. https://doi.org/10.1016/J.EGYR.2022.01.180

Akar, G., Clifton, K. J., & Doherty, S. T. (2011). Discretionary activity location choice: In-home or out-of-home? *Transportation*, *38*, 101–122.

Althoff, L., Eckert, F., Ganapati, S., & Walsh, C. (2022). The geography of remote work. *Regional Science and Urban Economics*, *93*, 103770. https://doi.org/10.1016/J.REGSCIURBECO.2022.103770

Arentze, T. (2015). Individuals' social preferences in joint activity location choice: A negotiation model and empirical evidence. *Journal of Transport Geography*, *48*, 76–84.

Arentze, T. A., Borgers, A. W., Ma, L., & Timmermans, H. J. (2010). An agent-based heuristic method for generating land-use plans in urban planning. *Environment and Planning B: Planning and Design*, *37*(3), 463–482.

Arentze, T., Ettema, D., & Timmermans, H. J. P. (2013). Location choice in the context of multi-day activity-travel patterns: Model development and empirical results. *Transportmetrica A: Transport Science*, *9*(2), 107–123.

Arentze, T., & Timmermans, H. (2000). Albatross: A learning based transportation oriented simulation system. Citeseer.

Asgari, H., Jin, X., & Rojas IV, M. B. (2019). Time geography of daily activities: A closer look into telecommute impacts. *Travel Behaviour and Society*, *16*, 99–107.

Astroza, S., Bhat, P. C., Bhat, C. R., Pendyala, R. M., & Garikapati, V. M. (2018). Understanding activity engagement across weekdays and weekend days: A multivariate multiple discrete-continuous modeling approach. *Journal of Choice Modelling*, *28*, 56–70.

Bader, A. K., Froese, F. J., & Kraeh, A. (2018). Clash of cultures? German expatriates' work-life boundary adjustment in South Korea. *European Management Review*, *15*(3), 357–374. https://doi.org/10.1111/EMRE.12102

Bao, Q., Kochan, B., Bellemans, T., Shen, Y., Creemers, L., Janssens, D., & Wets, G. (2015). Travel demand forecasting using activity-based modeling framework feathers: An extension. *International Journal of Intelligent Systems*, *30*(8), 948–962.

Bao, Q., Kochan, B., Shen, Y., Creemers, L., Bellemans, T., Janssens, D., & Wets, G. (2018). Applying feathers for travel demand analysis: Model considerations. *Applied Sciences*, *8*(2), 211.

Barker, J. (2003). Passengers or political actors? Children's participation in transport policy and the micro political geographies of the family. *Space and Polity*, *7*(2), 135–151.

Batty, M. (2009). Urban modeling. In International *encyclopedia of human geo*graphy. Elsevier.

Batty, M., Axhausen, K. W., Giannotti, F., Pozdnoukhov, A., Bazzani, A., Wachowicz, M., Ouzounis, G., & Portugali, Y. (2012). Smart cities of the future. *The European Physical Journal Special Topics 2012 214:1*, *214*(1), 481–518. https://doi.org/10.1140/EPJST/E2012-01703-3

Bellemans, T., Kochan, B., Janssens, D., Wets, G., Arentze, T., & Timmermans, H. (2010). Implementation framework and development trajectory of feathers activity-based simulation platform. *Transportation Research Record*, *2175*(1), 111–119.

Bowman, J. L., & Ben-Akiva, M. E. (2001). Activity-based disaggregate travel demand model system with activity schedules. *Transportation Research Part a: Policy and Practice*, *35*(1), 1–28.

Bowman, J. L., Bradley, M. A., & Gibb, J. (2006). *The Sacramento activity-based travel demand model: Estimation and validation results*. European Transport Conference.

Bradley, M., Bowman, J. L., & Griesenbeck, B. (2010). SACSIM: An applied activity-based model system with fine-level spatial and temporal resolution. *Journal of Choice Modelling*, *3*(1), 5–31.

Brajša-Žganec, A., Merkaš, M., & Šverko, I. (2011). Quality of life and leisure activities: How do leisure activities contribute to subjective well-being? *Social Indicators Research*, *102*, 81–91.

Brandon, N., & Price, P. S. (2020). Calibrating an agent-based model of longitudinal human activity patterns using the Consolidated Human Activity Database. *Journal of Exposure Science & Environmental Epidemiology*, *30*(1), 194–204.

Brown, S., Pyke, D., & Steenhof, P. (2010). Electric vehicles: The role and importance of standards in an emerging market. *Energy Policy*, *38*(7), 3797–3806. https://doi.org/10.1016/J.ENPOL.2010.02.059

Cabral Dias, G. J., & Gomes Ribeiro, P. J. (2021). Cycle highways: A new concept of infrastructure. *European Planning Studies*, *29*(6), 1003–1020.

Calastri, C., Pawlak, J., & Batley, R. (2022). Participation in online activities while travelling: An application of the MDCEV model in the context of rail travel. *Transportation*, *49*(1), 61–87. https://doi.org/10.1007/S11116-021-10166-8/FIGURES/6

Chapin, F. S. (1968). Activity systems and urban structure: A working schema. *Journal of the American Institute of Planners*, *34*(1), 11–18.

Chapin, F. S. (1971). Free time activities and quality of urban life. *Journal of the American Institute of Planners*, *37*(6), 411–417.

Chapin, F. S. (1974). Human activity patterns in the city: Things people do in time and in space (Vol. 13). Wiley-Interscience.

Chen, J., & Wan, L. (2023). Causality between the built environment and subjective wellbeing: Applying difference-in-differences and synthetic control methods to longitudinal data From England. *SSRN Electronic Journal*. https://doi.org/10.2139/SSRN.4430978

Cheng, L., Chen, X., Yang, S., Wu, J., & Yang, M. (2019). Structural equation models to analyze activity participation, trip generation, and mode choice of low-income commuters. *Transportation Letters*, *11*(6), 341–349.

Chen, L., Wen, Y., Zhang, L., & Xiang, W. N. (2015). Studies of thermal comfort and space use in an urban park square in cool and cold seasons in Shanghai. *Building and Environment*, *94*, 644–653. https://doi.org/10.1016/J.BUILDENV.2015.10.020

Chung, J.-H., Kim, T., Baik, H., & Choi, Y.-S. (2009). A structural equation model of activity participation and travel behavior using longitudinal data. *Transportation Planning and Technology*, *32*(2), 163–185.

De Lauretis, S., Ghersi, F., & Cayla, J. M. (2017). Energy consumption and activity patterns: An analysis extended to total time and energy use for French households. *Applied Energy*, *206*, 634–648. https://doi.org/10.1016/J.APENERGY.2017.08.180

Deng, T., Zhang, K., & Shen, Z. J. (Max). (2021). A systematic review of a digital twin city: A new pattern of urban governance toward smart cities. *Journal of Management Science and Engineering*, *6*(2), 125–134. https://doi.org/10.1016/J.JMSE.2021.03.003

Dijst, M., & Vidakovic, V. (2000). Travel time ratio: The key factor of spatial reach. *Transportation*, *27*, 179–199.

Do Lee, W., Cho, S., Bellemans, T., Janssens, D., Wets, G., Choi, K., & Joh, C.-H. (2012). Seoul activity-based model: An Application of feathers solutions to Seoul metropolitan area. *Procedia Computer Science*, *10*, 840–845.

Edwards, N. M., Myer, G. D., Kalkwarf, H. J., Woo, J. G., Khoury, P. R., Hewett, T. E., & Daniels, S. R. (2015). Outdoor temperature, precipitation, and wind speed affect physical activity levels in children: A longitudinal cohort study. *Journal of Physical Activity and Health*, *12*(8), 1074–1081. https://doi.org/10.1123/JPAH.2014-0125

Ellegård, K. (1999). A time-geographical approach to the study of everyday life of individuals-a challenge of complexity. *GeoJournal*, *48*(3), 167–175.

Ellegård, K. (2018). Time geography in the global context: An anthology. Routledge.

Fan, Y., Wang, J., Obradovich, N., & Zheng, S. (2023). Intraday adaptation to extreme temperatures in outdoor activity. *Scientific Reports 2023 13:1*, *13*(1), 1–11. https://doi.org/10.1038/s41598-022-26928-y

Feng, X., Toms, R., & Astell-Burt, T. (2021). Association between green space, outdoor leisure time and physical activity. *Urban Forestry & Urban Greening*, *66*, 127349. https://doi.org/10.1016/J.UFUG.2021.127349

Ferré-Bigorra, J., Casals, M., & Gangolells, M. (2022). The adoption of urban digital twins. *Cities*, *131*, 103905. https://doi.org/10.1016/J.CITIES.2022.103905

Frank, L. D., & Engelke, P. O. (2001). The built environment and human activity patterns: Exploring the impacts of urban form on public health. *http://Dx.Doi.Org/10.1177/08854120122093339*, *16*(2), 202–218. https://doi.org/10.1177/08854120122093339

Ghauche, A., Ben-Akiva, M., & Turner, E. K. (2010). *Integrated transportation and energy activity-based model*. https://dspace.mit.edu/handle/1721.1/58185

Habib, K. M. N. (2018). A comprehensive utility-based system of activity-travel scheduling options modelling (CUSTOM) for worker's daily activity scheduling processes. *Transportmetrica A: Transport Science*, *14*(4), 292–315.

Habib, K. M. N., & Miller, E. J. (2006). Modeling individuals' frequency and time allocation behavior for shopping activities considering household-level random effects. *Transportation Research Record*, *1985*(1), 78–87.

Habib, K. M. N., & Miller, E. J. (2008). Modelling daily activity program generation considering within-day and day-to-day dynamics in activity-travel behaviour. *Transportation*, *35*, 467–484.

Hafezi, M. H., Daisy, N. S., Millward, H., & Liu, L. (2021). Ensemble learning activity scheduler for activity based travel demand models. *Transportation Research Part C: Emerging Technologies*, *123*, 102972.

Hagerstrand, T. (1970). What about people in regional. *Regional Science Association*, *24*, 7–21.

Hägerstrand, T. (1985). Time-geography: Focus on the corporeality of man, society and environment. *The Science and Praxis of Complexity*, *3*, 193–216.

Hägerstrand, T. (1989). Reflections on "what about people in regional science?" *Papers of the Regional Science Association*, *66*(1), 1–6.

Hägerstrand, T., Ellegård, K., Svedin, U., & Lenntorp, B. (2009). *Tillvaroväven*. Formas.

Hansen, W. G. (1959). How accessibility shapes land use. *Journal of the American Institute of Planners*, *25*(2), 73–76.

Hatzopoulou, M., Hao, J. Y., & Miller, E. J. (2011). Simulating the impacts of household travel on greenhouse gas emissions, urban air quality, and population exposure. *Transportation*, *38*(6), 871–887. https://doi.org/10.1007/S11116-011-9362-9/FIGURES/9

He, P., Wang, Q.-C., & Shen, G. Q. (2024). The carbon emission implications of intensive urban land use in emerging regions: Insights from Chinese cities. *Urban Science*, *8*(3), 75. https://doi.org/10.3390/urbansci8030075

Horni, A., Scott, D. M., Balmer, M., & Axhausen, K. W. (2009). Location choice modeling for shopping and leisure activities with MATSim: Combining microsimulation and time geography. *Transportation Research Record*, *2135*(1), 87–95.

Huang, A., & Levinson, D. (2011). Why retailers cluster: An agent model of location choice on supply chains. *Environment and Planning B: Planning and Design*, *38*(1), 82–94.

Huang, H., Li, Y., Zhao, Y., & Zhai, W. (2022). Analysis of the impact of urban summer high temperatures and outdoor activity duration on residents' emotional health: Taking hostility as an example. *Frontiers in Public Health*, *10*, 955077. https://doi.org/10.3389/FPUBH.2022.955077/BIBTEX

Huigen, P. P. (1986). *Binnen of buiten bereik?: Een sociaal-geografisch onderzoek in Zuidwest-Friesland* (Vol. 7). Koninklijk nederlands aardrijkskundig genootschap.

Jaller, M., & Pahwa, A. (2020). Evaluating the environmental impacts of online shopping: A behavioral and transportation approach. *Transportation Research Part D: Transport and Environment*, *80*, 102223.

Jones, P. M., Dix, M. C., Clarke, M. I., & Heggie, I. G. (1983). *Understanding travel behaviour*. Gower Publishing.

Keirstead, J., & Sivakumar, A. (2012). Using activity-based modeling to simulate urban resource demands at high spatial and temporal resolutions. *Journal of Industrial Ecology*, *16*(6), 889–900. https://doi.org/10.1111/J.1530-9290.2012.00486.X

Kim, S., McLean, G. N., & Park, S. (2018). The cultural context of long working hours: Workplace experiences in Korea. *New Horizons in Adult Education and Human Resource Development*, *30*(2), 36–51. https://doi.org/10.1002/NHA3.20214

Kim, S., Rasouli, S., Timmermans, H. J. P., & Yang, D. (2022). A scenario-based stochastic programming approach for the public charging station location problem. *Transportmetrica B: Transport Dynamics*, *10*(1), 340–367. https://doi.org/10.1080/21680566.2021.1997672

Kitamura, R. (1988). An evaluation of activity-based travel analysis. *Transportation*, *15*, 9–34. https://doi.org/10.1007/BF00167973.

Koch, J., Frommeyer, B., & Schewe, G. (2020). Online shopping motives during the COVID-19 pandemic – Lessons from the crisis. *Sustainability*, *12*(24), 10247.

Kutter, E. (1981). Some remarks on activity pattern analysis in transportation planning. In *New horizons in travel-behavior research*. The Fourth International Conference on Behavioral Travel Modeling.

Kwan, M. (1998). Space-time and integral measures of individual accessibility: A comparative analysis using a point-based framework. *Geographical Analysis*, *30*(3), 191–216.

Kwan, M. (1999). Gender and individual access to urban opportunities: A study using space – Time measures. *The Professional Geographer*, *51*(2), 210–227.

Lee, Y., Hickman, M., & Washington, S. (2007). Household type and structure, time-use pattern, and trip-chaining behavior. *Transportation Research Part A: Policy and Practice*, *41*(10), 1004–1020. https://doi.org/10.1016/J.TRA.2007.06.007

Lenntorp, B. (1976). Paths in space-time environments. A time-geographic study of movement possibilities of individuals. *Lund Studies in Geography. Série B, Human Geography*, *44*.

Li, T., Jing, X., Wei, O., Yinlong, L., Jinxuan, L., Yongfu, L., Li, W., Ying, J., Weipan, X., Yaotian, M., & Yifan, D. (2022). Mobility restrictions and their implications on the rental housing market during the COVID-19 pandemic in China's large cities. *Cities*, *126*, 103712. https://doi.org/10.1016/J.CITIES.2022.103712

Liao, F., Arentze, T., & Timmermans, H. (2013). Incorporating space – Time constraints and activity-travel time profiles in a multi-state supernetwork approach to individual activity-travel scheduling. *Transportation Research Part B: Methodological*, *55*, 41–58.

Liu, X., Wang, Q.-C., Jian, I. Y., Chi, H.-L., Yang, D., & Chan, E. H.-W. (2021). Are you an energy saver at home? The personality insights of household energy conservation behaviors based on theory of planned behavior. *Resources, Conservation and Recycling*, *174*, 105823.

Lowry, I. S. (1964). *A Model of Metropolis* (Issue 第 4035 号). Rand Corporation.

Ma, J., & Goulias, K. G. (1999). Application of poisson regression models to activity frequency analysis and prediction. *Transportation Research Record*, *1676*(1), 86–94.

Manoj, M., & Verma, A. (2017). A structural equation model based analysis of non-workers' activity-travel behaviour from a city of a developing country. *Transportation*, *44*, 241–269.

McNally, M. G., & Rindt, C. R. (2007). The activity-based approach. In *Handbook of transport modelling* (Vol. 1, pp. 55–73). Emerald Group Publishing Limited.

Morrissey, P., Weldon, P., & O'Mahony, M. (2016). Future standard and fast charging infrastructure planning: An analysis of electric vehicle charging behaviour. *Energy Policy*, *89*, 257–270. https://doi.org/10.1016/J.ENPOL.2015.12.001

Muhanna, W., & Wolf, J. (2002). The impact of e-commerce on the real estate industry: Baen and Guttery revisited. *Journal of Real Estate Portfolio Management*, *8*(2), 141–152.

Murchland, J. D. (1966). *Some remarks on the gravity model of traffic distribution, and an equivalent maximization formulation*. London School of Economics and Political Science, Transport Network Theory Unit.

Mytton, O. T., Townsend, N., Rutter, H., & Foster, C. (2012). Green space and physical activity: An observational study using health survey for England data. *Health & Place*, *18*(5), 1034–1041. https://doi.org/10.1016/J.HEALTHPLACE.2012.06.003

Ni, W. (2022). Online and offline integration development of Yiwu cross-border e-commerce in digital economy era. *Procedia Computer Science*, *202*, 307–312.

Olson, M. H. (1983). Remote office work. *Communications of the ACM*, *26*(3), 182–187. https://doi.org/10.1145/358061.358068

Opoku, A. (2015). *The role of culture in a sustainable built environment*. 37–52. https://doi.org/10.1007/978-3-319-14002-5_3

Pandya, A., & Lodha, P. (2021). Social connectedness, excessive screen time during COVID-19 and mental health: A review of current evidence. *Frontiers in Human Dynamics*, *3*, 684137. https://doi.org/10.3389/FHUMD.2021.684137

Park, S. G., Min, K. B., Chang, S. J., Kim, H. C., & Min, J. Y. (2009). Job stress and depressive symptoms among Korean employees: The effects of culture on work. *International Archives of Occupational and Environmental Health*, *82*(3), 397–405. https://doi.org/10.1007/S00420-008-0347-8/TABLES/3

Ploeger, J., & Oldenziel, R. (2022). Bicycle-oriented development: How the Dutch railroad shaped urban planning and discovered cyclists along the way, 1960-1990. *Journal of Urban History*, 00961442221133080.

Pred, A. (1984). Place as historically contingent process: Structuration and the time-geography of becoming places. *Annals of the Association of American Geographers*, *74*(2), 279–297.

Rasouli, S., & Timmermans, H. (2014). Activity-based models of travel demand: Promises, progress and prospects. *International Journal of Urban Sciences*, *18*(1), 31–60.

Raux, C., Ma, T.-Y., Joly, I., Kaufmann, V., Cornelis, E., & Ovtracht, N. (2011). Travel and activity time allocation: An empirical comparison between eight cities in Europe. *Transport Policy*, *18*(2), 401–412.

Recker, W. W., McNally, M. G., & Root, G. S. (1986a). A model of complex travel behavior: Part I—Theoretical development. *Transportation Research Part A: General*, *20*(4), 307–318. https://doi.org/10.1016/0191-2607(86)90089-0.

Recker, W. W., McNally, M. G., & Root, G. S. (1986b). A model of complex travel behavior: Part II—An operational model. *Transportation Research Part A: General*, *20*(4), 319–330. https://doi.org/10.1016/0191-2607(86)90090-7.

Rouleau, J., & Gosselin, L. (2021). Impacts of the COVID-19 lockdown on energy consumption in a Canadian social housing building. *Applied Energy*, *287*, 116565.

Shaaban, K., Muley, D., & Elnashar, D. (2018). Evaluating the effect of seasonal variations on walking behaviour in a hot weather country using logistic regression. *International Journal of Urban Sciences*, *22*(3), 382–391. https://doi.org/10.1080/12265934.2017.1403363

Shabanpour, R., Golshani, N., Fasihozaman Langerudi, M., & Mohammadian, A. (2018). Planning in-home activities in the ADAPTS activity-based model: A joint model of activity type and duration. *International Journal of Urban Sciences*, *22*(2), 236–254.

Sharmeen, F., Arentze, T., & Timmermans, H. (2014). An analysis of the dynamics of activity and travel needs in response to social network evolution and life-cycle events: A structural equation model. *Transportation Research Part A: Policy and Practice*, *59*, 159–171.

Sirianni, C., & Negrey, C. (2000). Working time as gendered time. *Feminist Economics*, *6*(1), 59–76. https://doi.org/10.1080/135457000337679

Soo, J. (2009). Space-time prism vertices: Exploring gender differences and multiple-peak distributions in arrival and departure times. *European Journal of Transport and Infrastructure Research*, *9*(4).

Sugiyama, T., Paquet, C., Howard, N. J., Coffee, N. T., Taylor, A. W., Adams, R. J., & Daniel, M. (2014). Public open spaces and walking for recreation: Moderation by attributes of pedestrian environments. *Preventive Medicine*, *62*, 25–29. https://doi.org/10.1016/J.YPMED.2014.01.030

Timmermans, H., & Arentze, T. A. (2011). Transport models and urban planning practice: Experiences with albatross. *Transport Reviews*, *31*(2), 199–207. https://doi.org/10.1080/01441647.2010.518292

Trainor, S., Delfabbro, P., Anderson, S., & Winefield, A. (2010). Leisure activities and adolescent psychological well-being. *Journal of Adolescence*, *33*(1), 173–186.

Valkenburg, P. M. (2022). Social media use and well-being: What we know and what we need to know. *Current Opinion in Psychology*, *45*, 101294. https://doi.org/10.1016/J.COPSYC.2021.12.006

van der Loop, H., Willigers, J., & Haaijer, R. (2019). Empirical estimation of effects of flexible working on mobility and congestion in the Netherlands 2000 to 2016. *Transportation Research Record*, *2673*(6), 557–565. https://doi.org/10.1177/0361198119845889/ASSET/IMAGES/LARGE/10.1177_0361198119845889-FIG1.JPEG

Vanderhorn, E., & Mahadevan, S. (2021). Digital twin: Generalization, characterization and implementation. *Decision Support Systems*, *145*, 113524. https://doi.org/10.1016/j.dss.2021.113524

van Eck, J. R., Burghouwt, G., & Dijst, M. (2005). Lifestyles, spatial configurations and quality of life in daily travel: An explorative simulation study. *Journal of Transport Geography*, *13*(2), 123–134.

Van Nieuwerburgh, S. (2023). The remote work revolution: Impact on real estate values and the urban environment: 2023 AREUEA Presidential Address. *Real Estate Economics*, *51*(1), 7–48. https://doi.org/10.1111/1540-6229.12422

Wan, L., Nochta, T., Tang, J., Schooling, J., Oti-Sarpong, K., & Melo, V. A. B. (2023). Conceptualising city digital twins (CDTs). *Digital twins for smart cities*, 13–20. https://doi.org/10.1680/DTSC.66007.013

Wan, L., Tang, J., Wang, L., & Schooling, J. (2021). Understanding non-commuting travel demand of car commuters – Insights from ANPR trip chain data in Cambridge. *Transport Policy*, *106*(March), 76–87. https://doi.org/10.1016/j.tranpol.2021.03.021

Wang, Q.-C., Liu, X., Jian, I. Y., Zhang, E.-J., Hou, Y.-T., Siu, K. W. M., & Li, Y.-B. (2023). Community resilience in city emergency: Exploring the roles of environmental perception, social justice and community attachment in subjective well-being of vulnerable residents. *Sustainable Cities and Society*, *97*, 104745.

Wang, Y., de Almeida Correia, G. H., de Romph, E., & Timmermans, H. J. P. (2017). Using metro smart card data to model location choice of after-work activities: An application to Shanghai. *Journal of Transport Geography*, *63*, 40–47.

Wegener, M. (1994). Operational urban models state of the art. *Journal of the American Planning Association*, *60*(1), 17–29. https://doi.org/10.1080/01944369408975547

Wegener, M., & Fuerst, F. (2004). Land-use transport interaction: State of the art. Available at SSRN 1434678. https://dx.doi.org/10.2139/ssrn.1434678

Wilson, A. G. (1969). The use of entropy maximising models, in the theory of trip distribution, mode split and route split. *Journal of Transport Economics and Policy*, 108–126.

Wilson, A. G. (1970). The use of the concept of entropy in system modelling. *Journal of the Operational Research Society*, *21*(2), 247–265.

Wohlers, C., & Hertel, G. (2017). Choosing where to work at work – Towards a theoretical model of benefits and risks of activity-based flexible offices. *Ergonomics*, *60*(4), 467–486.

Wöhner, F. (2022). Work flexibly, travel less? The impact of telework and flextime on mobility behavior in Switzerland. *Journal of Transport Geography*, *102*, 103390. https://doi.org/10.1016/J.JTRANGEO.2022.103390

Xiao, C., Silva, E. A., & Zhang, C. (2020). Nine-nine-six work system and people's movement patterns: Using big data sets to analyse overtime working in Shanghai. *Land Use Policy*, *90*, 104340. https://doi.org/10.1016/J.LANDUSEPOL.2019.104340

Xu, Q., Li, S., Shen, L., Chang, R., Wang, Q.-C., Liu, X., & Chen, Y. (2023). Pricing strategy for Household Energy-Saving Option (HESO): A novel option-based intervention for promoting household energy efficiency. *Environmental Impact Assessment Review*, *98*, 106969.

Yoon, S. Y., & Goulias, K. (2010). Impact of time-space prism accessibility on time use behavior and its propagation through intra-household interaction. *Transportation Letters*, *2*(4), 245–260.

Yoon, S. Y., Ravulaparthy, S. K., & Goulias, K. G. (2014). Dynamic diurnal social taxonomy of urban environments using data from a geocoded time use activity-travel diary and point-based business establishment inventory. *Transportation Research Part A: Policy and Practice*, *68*, 3–17.

Yu, R., Burke, M., & Raad, N. (2019). Exploring impact of future flexible working model evolution on urban environment, economy and planning. *Journal of Urban Management*, *8*(3), 447–457.

Zhang, J., Timmermans, H. J. P., & Borgers, A. (2005). A model of household task allocation and time use. *Transportation Research Part B: Methodological*, *39*(1), 81–95.

Zhang, Y., Manley, E., Martens, K., & Batty, M. (2024). A metro smart card data-based analysis of group travel behaviour in Shanghai, China. *Journal of Transport Geography*, *114*, 103764.

14

SUSTAINABLE URBAN BUILT ENVIRONMENT

Principles, tools, and applications

Phillip Jones[1] and Jianxiang Huang[2]

[1]Welsh School of Architecture, Cardiff University, King Edward VII Avenue, Cardiff, CF10 3NB, UK. E-mail: jonesp@cardiff.ac.uk

[2]Department of Urban Planning and Design, 8/F Knowles Building, The University of Hong Kong, Pokfulam Road, Hong Kong SAR, China. E-mail: jxhuang@hku.hk

14.1 Introduction

To sustain a high-density city is a multi-faceted task (Figure 14.1), leading to low environmental impacts, encouraging a sustainable lifestyle, and providing places where people want to live (Jones, 2021). The built environment includes new building developments, the existing building stock and the supporting infrastructures of energy, waste, water/sewage, transport, green/blue structures, and IT (Figure 14.1a). There are impacts at different scales, experienced by people within a building, in a community, and on the planet (Figure 14.1b). A sustainable approach must also consider the triple bottom line of environment, economy, and society (Figure 14.1c). Urban sustainability is about the continual process of operating the city and not just about end products, which in themselves contribute to this continual urban process. The aims of reducing demand, efficient supply, and integration across all aspects of energy, water, waste, material use and transport, must go hand in hand with meeting the needs of people, creating spaces where people can be healthy, comfortable, and productive; these spaces can be internal, external, and transitional.

Cities are dynamic, with physical, social, and technological structures that are constantly changing. We a need to ensure that sustainability is at the centre of this change, continually improving the environment, achieving greater social equality, and providing economic opportunities. It is not just about avoiding problems down the line but aiming to create a sense of balance, positivity, and delight for all of society.

The world is becoming increasingly urban, with the population of cities rapidly growing. Currently, around 60% of the population of the planet live in cities and this is set to rise to 75% of the population by 2050 (UN, 2018). To put this into perspective, some 1.8 million people will be added to our cities every week from now until 2050. In the developing world, cities are growing through migration from rural areas and expansion into peri-urban areas, often encroaching onto arable land. In the developed world, more people are being attracted to live in cities to engage with city life and culture. Cities already account for some 80% of energy use and 75% carbon dioxide

DOI: 10.1201/9781003383840-15

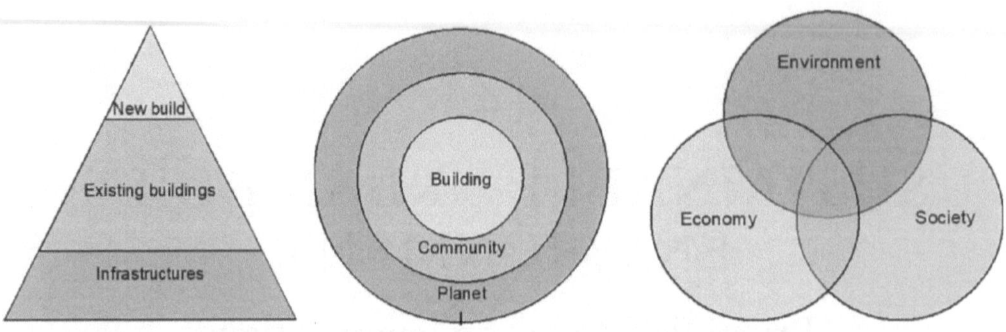

a) Built Environment b) Impacts c) Triple Bottom Line

Figure 14.1a, 1b, 1c Conceptual diagrams for a sustainable urban built environment

emissions, yet they only occupy 2 to 3% of the earth's land surface. Consequently, they become hotspots of poor air quality, pollution, and urban heat. At their best, cities provide a rich mix of culture and diversity, lifestyles, skills, and a home for financial, intellectual, social, and moral capital. However, at their worst they are places of uncontrolled sprawl, congested traffic, social depravation, ecological overload, high energy use, excessive waste, and the spread of infectious disease.

14.2 Top-down bottom-up

Future city thinking encompasses a range of terms including eco, green, low-carbon, compact, and smart. These terms are basically loose definitions of the same thing, with the focus on quality of life, efficient use of resources, and minimal environmental impact. However, they often tend to be top-down political aspirations without much thought on how they can be realised in practice (Jones, 2017). They are often vaguely defined and the sustainability principles at the outset of a project often disappear or are diluted when the economics of development take over. Clearly economic considerations are part of sustainability, and the economic, as well as social, aspects need to be fully understood from the outset, if all sustainability aspirations are to be successfully fulfilled.

The built environment is a main driver of the economy, especially in developing countries. This should respond to societal needs, linking up top-down strategies and bottom-up activities (Figure 14.2) with a shift in emphasis to bottom-up, with the addition of 'middle-out agents' to help activate projects on the ground. Top-down strategies may be best implemented through regional government, which has a greater level of governance and knowledge of the locality. Bottom-up project-based activities can deliver people's needs using a whole systems approach, which can result in multiple benefits. For example, a zero-energy building reduces CO_2 emissions, which is a top-down target, while also reducing energy costs, providing affordable warmth, improving health, reducing pollution, and creating jobs and economic opportunities, which are more related to bottom-up needs.

14.3 Zero-carbon performance and energy supply

The quest for carbon neutrality has pressured cities to reduce their carbon footprints. New buildings and urban developments should strive towards achieving a zero-carbon energy-positive performance, through a systems approach linking reduced energy demand, renewable energy supply, and energy storage. However, it may not be practically possible for every individual building to

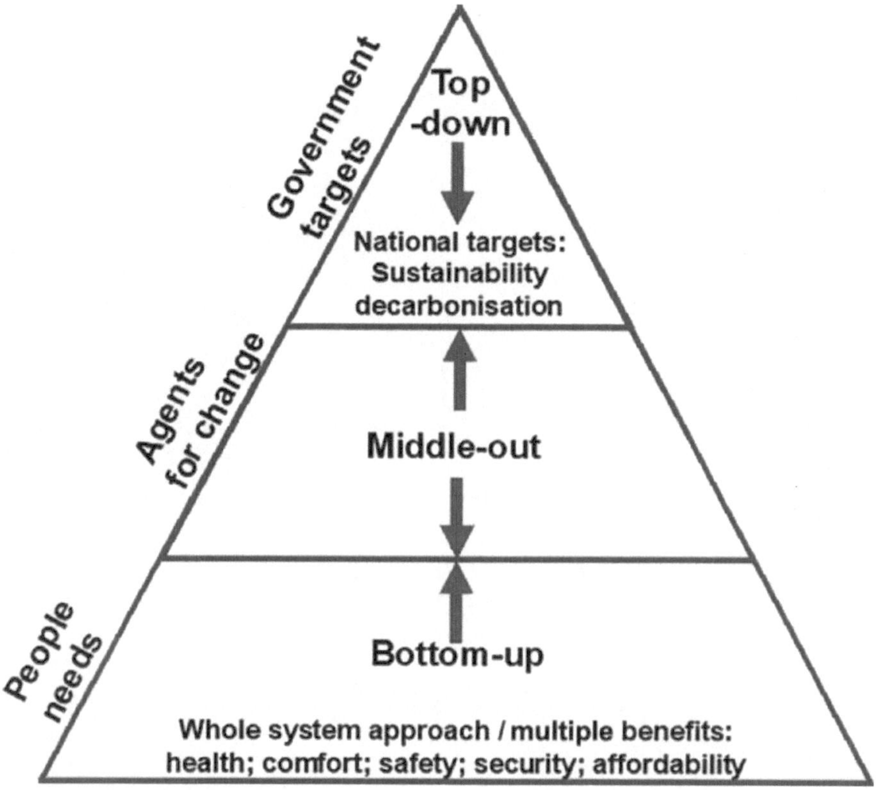

Figure 14.2 Linking up top-down strategies and bottom-up activities for the built environment

achieve a zero-carbon performance with building integrated renewables. For example, high-rise dense urban areas can strive to be low-energy in terms of their own individual performance but will generally need external (renewable) grid energy to achieve complete zero-carbon performance. The same applies to retrofitting existing buildings; where in Europe, around 80% of buildings that will exist in 2050 have already been built and will need some form of retrofit. The balance between generating energy from building integrated generation and importing from a local or centralised grid needs to be optimised on a case-by-case solution. Our future built environment will likely be made up from buildings that have a mix of near-zero-energy, zero-energy, and energy-positive performances, linked to a decarbonised energy grid, all leading towards an overall zero carbon target.

Zero-carbon in the past has generally been associated with the operating energy of a building and has not included embodied carbon, which may increase as a building tends towards zero-energy and low-carbon performance, due to the increase in insulation materials and energy generation and storage technologies. A low embodied carbon building development may use more natural materials, such as timber structures and constructions, plant-based insulation, and internal finishes, all of which are generally non-toxic, and it may even sequestrate carbon throughout its life, for example, using hemp blocks and lime for construction.

The operation of our buildings, with their associated infrastructures for energy supply, water and sewage, waste and transport, determines the efficiency and effectiveness of a city's performance in meeting people's needs. There is a growing interest in our future cities being more resilient

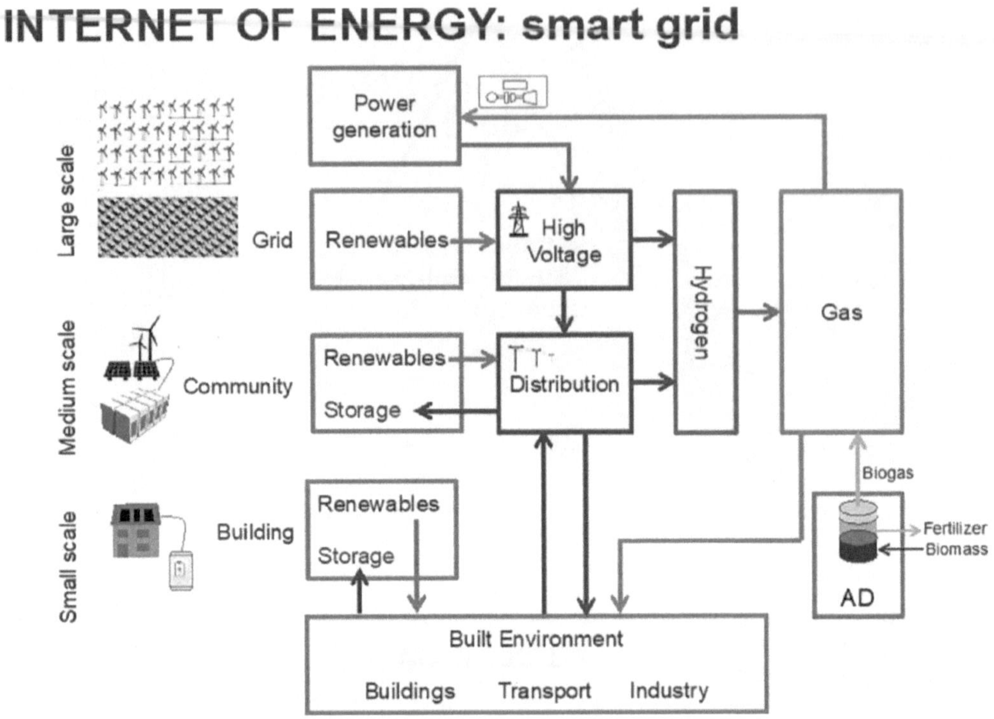

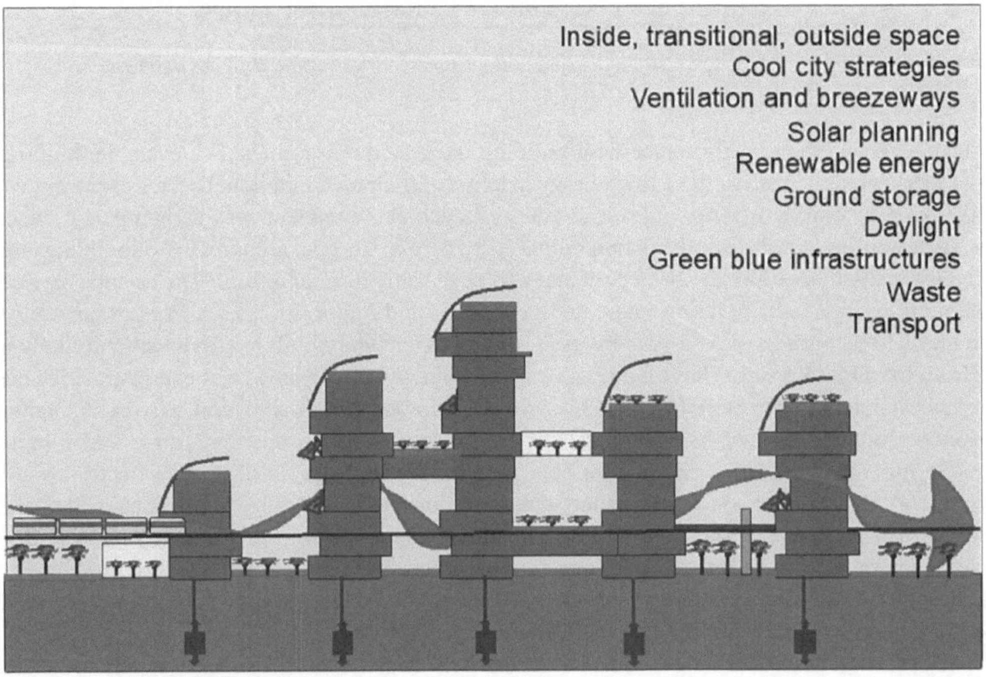

Figure 14.3 (a) Future energy systems will be a combination of building integrated, localised distributed, and central grid. (b) City as a building

and self-sufficient, integrating the built environment with nature-based systems. For example, the 'sponge city' concept aims to collect excess rainfall and integrate flood control into urban infrastructure planning, linking with green and blue structures (Zou et al., 2021). Rainwater harvesting can recharge depleted aquifers and provide water for our homes and businesses. Localised distributed energy clusters, focusing on industrial activities and sewage treatment, can combine wind and solar renewables with anaerobic digestion, generating electricity and heat. Methane from AD plants and hydrogen produced from excess renewable energy can be injected into existing gas mains to decarbonise our gas supply. These combine with the concept of 'buildings as power stations' where a building generates the energy it needs from renewable sources. The future energy supply system is changing to a mix of local and central, with building integrated renewable energy, distributed localised energy clusters, and a centralised grid energy supply, which in the future will itself be decarbonised. Energy storage will play a major role for both buildings and transportation, which may be combined to optimise energy use. Smart control will place the end-user central to decision-making in this 'internet of energy' (Figure 14.3(a)).

14.4 City as a building

This internet of energy can then be related to the concept of the 'city as a building' with its own joined up closed-loop systems-thinking approach across all aspects of the built environment (Figure 14.4(a)). An optimum ratio of a high-density built environment coupled with the allocation of blue/green space is needed to maximise the benefits of urban living within an efficient and

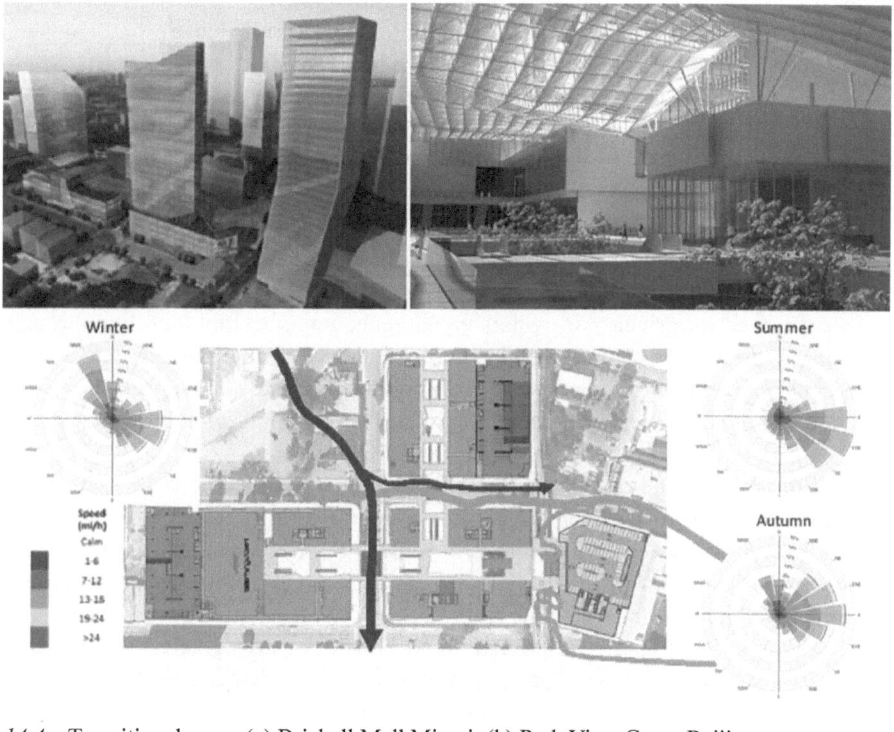

Figure 14.4 Transitional space (a) Brickall Mall Miami; (b) Park View Green Beijing

Source: Jones & Hou, 2021.

Figure 14.4 (Continued)

healthy surrounding. There is the need to integrate all spatial types, that is, internal, external, and transitional spaces. External space can provide amenity, especially in high-density cities, where internal space may be constrained, with shaded areas and breezeways being a major contribution to outdoor comfort (Huang, Hao, et al., 2022). Transitional spaces, such as malls (Figure 14.4(b)), should have more natural environments for comfort and energy efficiency, again providing shade and breezeways, thus eliminating the need for air conditioning.

There is a need for greater clarity in our approach to sustainable urban design with simulation tools such as those for predicting energy use and environmental conditions. These will allow a focus on demand reduction and the provision of efficient and effective supply. Energy will be the main focus, as will the reduction of fossil fuel use for buildings, transport, and other infrastructures. An example of a reduced-demand approach is transportation, which involves first reducing the demand for transportation and then providing efficient and effective mobility. So, the first step is to reduce the need to travel through improving accessibility to leisure and work. This embraces not only the objective of reducing the need to travel but also a further exploration of the full potential of IT. This proceeds the provision of an integrated and efficient transport system that can eliminate the need for private car ownership and encourage healthy mobility, through walking and cycling. This concept of first reducing demand and then providing efficient and effective supply should be applied to all infrastructure, building planning and design projects. Connections within and across infrastructures will enable the reuse and recycling of energy, materials, water, and waste.

Cities need to breathe! The rationale behind Cerda's plan for Barcelona in conformity to the hygienist theory is to ensure sufficient clean air for everyone to breathe, in comparison to the disease-laden, overcrowded old town (Aibar & Bijker, 1997). Hence, air quality remains a major factor in modernist urban planning and the success of future cities. There needs to be more control over what we 'dump' into the air – whether traffic fumes or exhaust air from buildings – and how we access fresh air for breathing. These relate to two of the world's greatest killers, respiratory diseases such as SARS and COVID-19, amongst other infectious diseases. The priority should be to design urban environments with an appropriate balance of green space and building density, with 'breezeways' to take away airborne pathogens and excess heat and to reduce all sources of pollution exhausted into the outdoor air.

14.5 The need for design tools

To facilitate the design of a high-density sustainable urban environment, measurement surveys and simulation tools are needed to improve our understanding and enable the design of both new developments and retrofitting existing projects. The main design issues include building energy use, outdoor comfort, pollution, and infectious disease.

The development of urban scale modelling tools validated with measurement data has generally been an extrapolation from building scale modelling, with external spaces simulated in addition to internal spaces (Huang et al., 2024) with the emphasis on external comfort alongside building energy use and anthropogenic inputs from traffic, etc.

14.6 External comfort

In hot climates, urban heat disrupts the use of outside space, although the extent of such disruptions remains disputed. Surveys of space use can be combined with physical measurements to help understand the relation between space use and outside environmental conditions. For example, three park areas in a new residential development in Hong Kong were studied and the results (Huang, Chen, et al., 2022) indicated a strong correlation of park use and a reduction in Universal Thermal Climate Index (UTCI). Of note is the importance of the provision of shaded areas.

The rise of 'big data' such as social media offers new opportunities for understanding external comfort. In a recent study, Hao et al. (2023) used Twitter data (tweets) to study park attendance under the influence of hot weather (Figure 14.5(a) and (b)). Some 20,000 tweets geo-coded within major parks were obtained in Hong Kong over a period of three years. Field studies were conducted in parallel in a large park covering the hot and cool seasons and some 40,000 attendances were recorded over three months. Both the 'small' and 'big data' were analysed and compared to each other. Findings suggest (Figure 14.5(c) and (d)) that a 1°C increase in temperature was

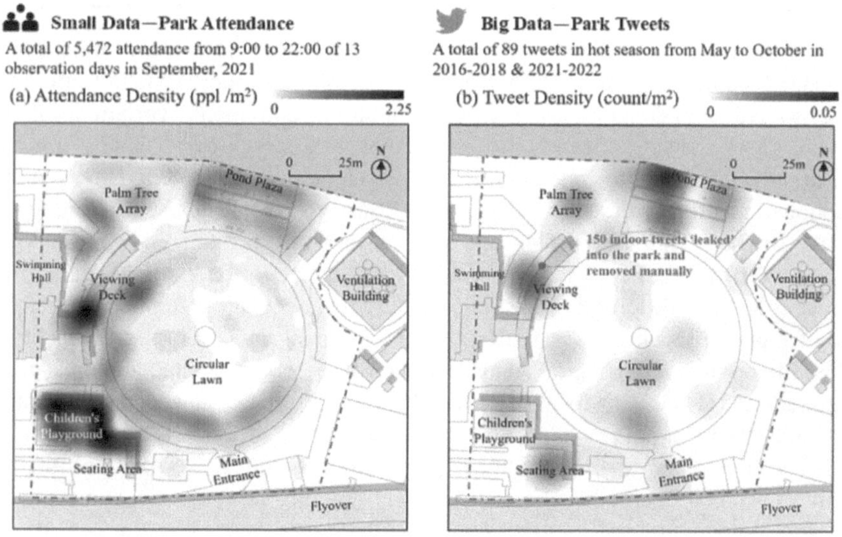

Figure 14.5 (a) The distribution of park attendance in Sun Yat-sen Park, Hong Kong; (b) the distribution of geo-coded tweets in Sun Yat-sen Park, Hong Kong; (c) park attendance and temperature (d) park tweets and temperature

Source: (Hao et al., 2023).

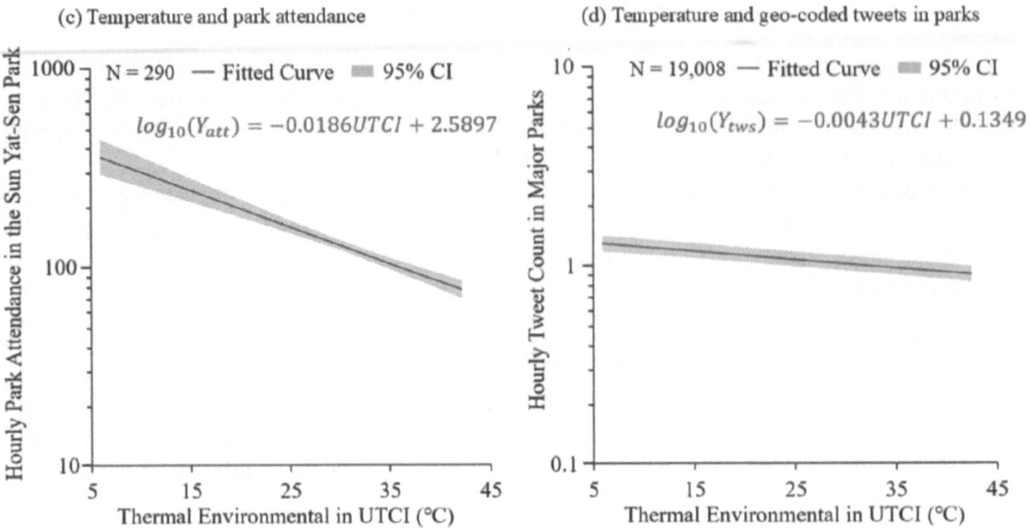

Figure 14.5 (Continued)

associated with a 4% drop in park attendance and a 1% drop in park tweets. Compared with other biometeorological indicators, the UTCI can better predict self-reported thermal sensations. This study contributes to methodologies and new evidence for the study of behaviours and thermal adaptations in an outdoor space, and geo-coded tweets can serve as a powerful data source.

14.7 Urban building energy use and external temperature simulation

The impact of the built form, together with anthropogenic heat gains from traffic and building HVAC exhaust, determines external urban environmental conditions. Existing building energy models are limited in accounting for combined temporal and spatial variations of the urban microclimate, which may significantly modify a building's energy performance in high-density cities. The urban building energy and climate (UrBEC) model, developed by Huang et al. (2020), directly coupled an urban microclimate model (UMM) and a building energy model (HTB2) (Jones et al., 2013) (Figure 14.6). This can be used to assess the time varying energy performance of a cluster of buildings and the combined heat gains to the external space from direct and reflected solar radiation, traffic, and the exhaust from HVAC systems. Simulation results have been evaluated by comparison with field measurement data collected from the Sai Ying Pun neighbourhood in Hong Kong on a summer and winter day. Predicted and measured air and surface temperature at the four locations were found to be in reasonable agreement. Simulation results indicate an average of 1–3°C of temperature rise in street canyons compared with the ambient air in summer. Street-level air is predicted to be 0.6°C warmer than that at higher levels (20m +). Anthropogenic heat from traffic and building HVAC exhaust are the dominant contributors to temperature rise in street canyons in summer, exceeding the contribution from urban surfaces. The building cooling demand is predicted to increase up to 15% in summer due to the warming effect in street canyons. The UrBEC model runs significantly faster than current CFD-based approaches. Therefore, the model has the potential to support early-stage design and planning decisions in a dense city.

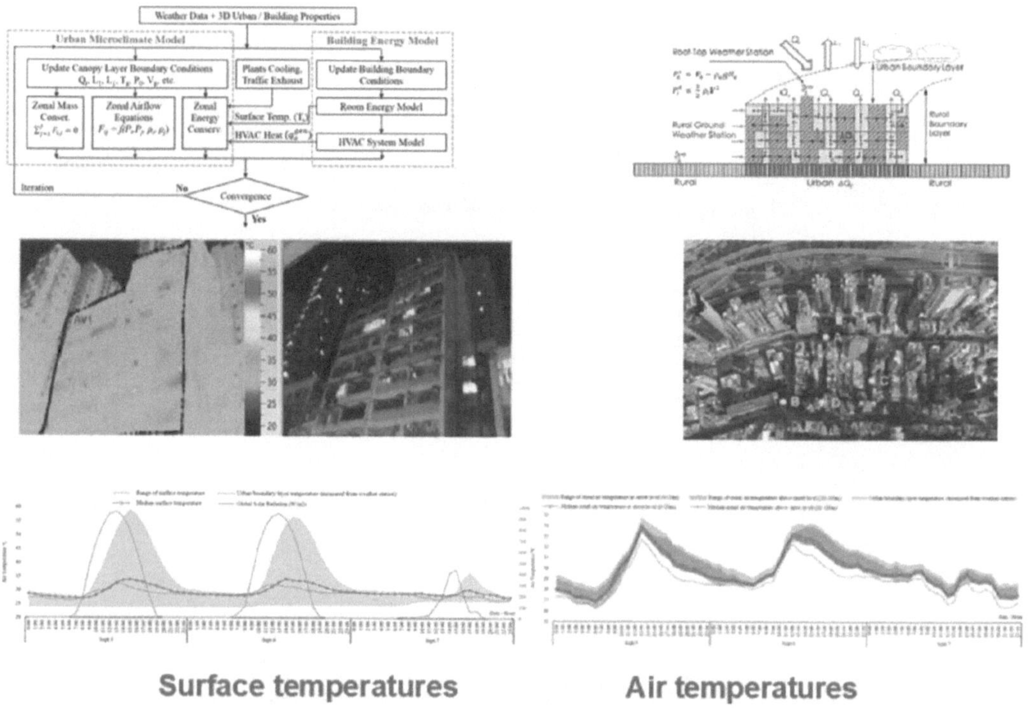

Figure 14.6 Urban modelling of microclimate and building energy in Hong Kong (Huang et al., 2020)

Urban greenery, notably trees, shrubs, groundcover, and rooftop gardens, have been widely promoted in municipal policies and planning practices for their cooling benefits. Tree canopies provide shading, which can lower surface temperatures by 11–25°C compared with unshaded surfaces while also reducing building energy used for cooling (Akbari et al., 1997). Foliage and moist soil in planting beds absorb latent heat via evapotranspiration, a process that lowers surface and air temperatures and purportedly reduces human heat stress during hot seasons. Combining shading and evapotranspiration, ground-based greenery has been predicted to reduce peak temperatures by 2–9°C (Alexandri & Jones, 2008; Wong et al., 2021). However, questions remain as to whether greenery can effectively reduce pedestrian heat stress in a high-density city, more so than alternative measures such as breeze enhancement, water spraying, and anthropogenic heat discharge management. The urban greenery and built environment (UGBE) model was developed to simulate the evapotranspiration of greenery in a high-density urban context, whilst also considering the anthropogenic exhaust heat from traffic and buildings, as well as shading and breezeways resulting from surrounding buildings. The UGBE model incorporates a green sub-model developed to assess the latent heat and moisture transfer by plant evapotranspiration, which is then coupled with an existing urban building and energy and climate (UrBEC) model (Huang et al., 2020), which models the time-varying energy exchanges of buildings and surrounding external spaces. The outputs are hourly urban microclimate conditions, i.e., air temperature, wind speed, humidity, and human heat stress, at the scale of streets and external open spaces (Figure 14.7). Simulated results have been compared with field measurement conducted in Shek Tong Tsui, Hong Kong under hot and humid weather conditions. Sensitivity studies have been conducted to simulate the cooling performances for a variety of greenery coverage and layout

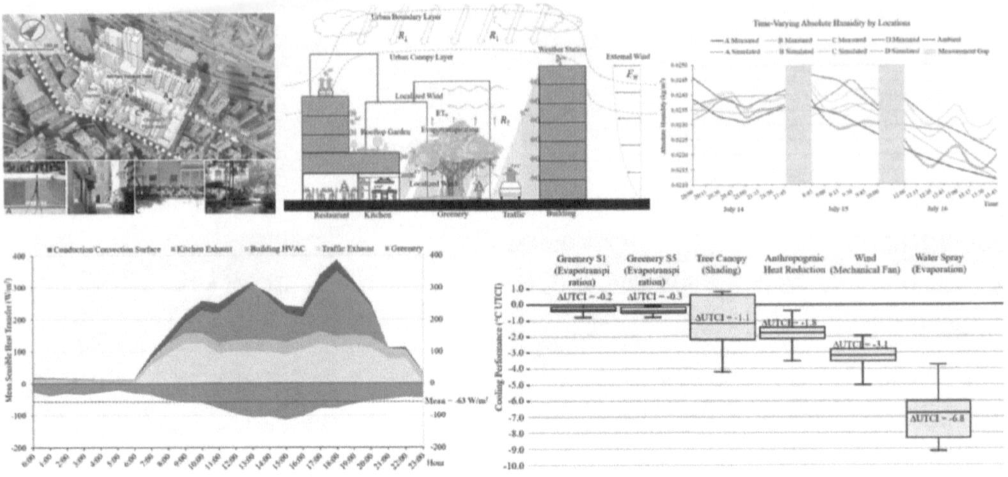

Figure 14.7 Simulation of reducing urban temperatures using greenery in Hong Kong (Huang, Hao, et al., 2022)

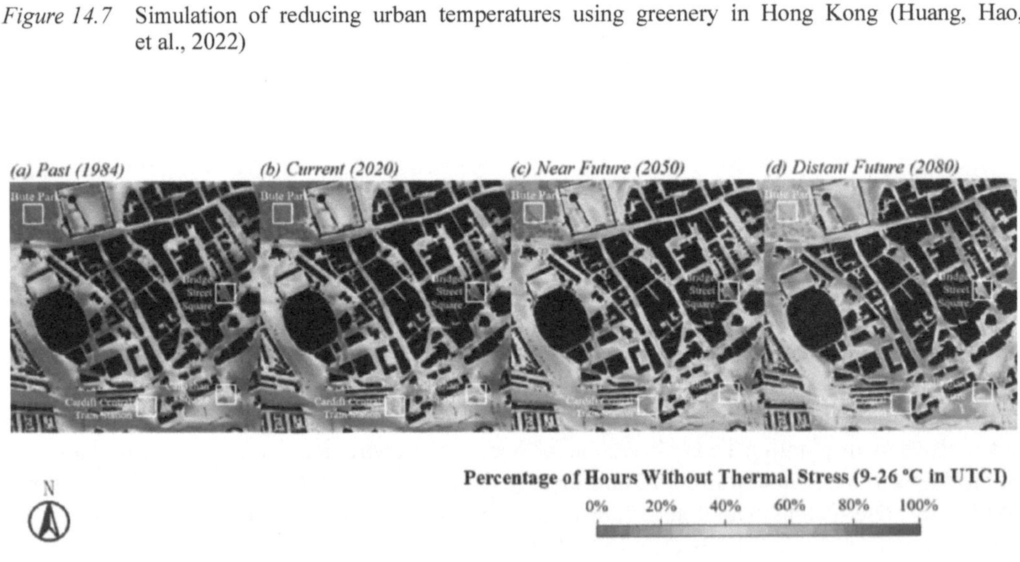

Figure 14.8 The percentage of daytime hours considered thermally stress-free in Cardiff during the two-week hot period, from August 24 to September 6, for 1) 1984 (past), 2) 2020 (present), 3) 2050 (near future), and 4) 2080 (distant future) (Huang et al., 2024)

conditions. By covering 40% of the site with greenery, a practical limit, the expected air temperature and UTCI reductions were 0.3°C, lower than previous estimates due to limited sunlight and ground-level surfaces for planting; the cooling benefits of greenery have been predicted to be higher in dry climates and lower in humid ones (Huang, Hao, et al., 2022). In a high-density city, plants converted sensible heat into latent gains at a slower rate than the anthropogenic exhaust heat. Alternative strategies, such as tree canopy, management of anthropogenic heat discharges, breeze enhancement, and water-spray were predicted to further help to cool the city by 1.1°C, 1.8°C 3.1°C, and 6.8°C, respectively (Figure 14.8).

14.8 Future urban temperatures

Recently, there has been greater recognition of heatwave risk to human health and well-being in the UK. However, there is a lack of understanding of heat stress at the pedestrian level and insufficient estimation of heat stress during future heatwaves, especially in urban situations. Computer simulation was used to assess heat stress as experienced by pedestrians in three Welsh cities, Cardiff, Newport, and Wrexham, during heatwaves (Huang et al., 2024). The aim was to identify the extent of pedestrian heat stress in both current and future heatwave events and to suggest implications for urban planning and design practice. A computerised simulation model combining first principles and empirical equations was applied to quantify the equivalent temperature at the ground level. The model considered localised radiant temperature, wind speed, air temperature, as well as the metabolic rate and clothing insulation of occupants. Simulated results (Figure 14.8) were partially evaluated using field measurement data obtained from Cardiff and from the land surface temperature data obtained from Landsat satellite thermography. The model was used to simulate conditions during past, present, and future heatwaves. Results suggest that peak heat stress is expected to increase by 4°C in UTCI equivalent temperature in 2080, compared with those recorded in the recent past. Mitigation measures will be essential to reduce future heat stress in cities and towns. The breadth of streetscape interventions and factors to be considered in urban layout include, (a) aspect and wind flows, (b) green and blue infrastructure, (c) choice of trees and artificial shading, (d) choice of both artificial surface materials and vegetation cover. The results have significant implications for local authorities, town planning, and landscape practice.

14.9 Breezeways

Breezeways are considered to be an important aspect of urban design because they not only exhaust pollutants and heat but also provide a certain degree of comfort and coolness. A new development area in Hong Kong was considered in relation to improving breezeways during the planning process. Various adjustments were made to the site planning to increase breezeways and improve outside comfort (Huang et al., 2019). These included streamlining buildings towards the prevailing wind directions during the hot seasons to enhance air ventilation, utilising tall buildings to shade public open spaces, and, for the remaining ground areas with excessive direct solar radiation, covering these areas with generous tree and shading canopies. The alignment of trees is also towards the prevailing wind to reduce the drag and to enhance airflow. A significant reduction of some 1.0°C in UTCI was achieved using the previous strategies.

14.10 Urban energy use

Building energy models can now be operated at an urban scale, simultaneously predicting energy demand of multiple buildings as described in the UrBEC model earlier. The example in Figure 14.9(a) presents the energy analysis of a new development proposed for Tianjin, including heating, cooling, lighting, fan power, and small power. Figure 14.9(b) presents a simulation result for a development in Hong Kong including energy demand and the potential for generating electricity from solar PV, suggesting that PV panels installed on building roofs and facades in a high-density neighbourhoods can meet the predicted building cooling demand, although not the entirety of the building energy use (Huang et al., 2019).

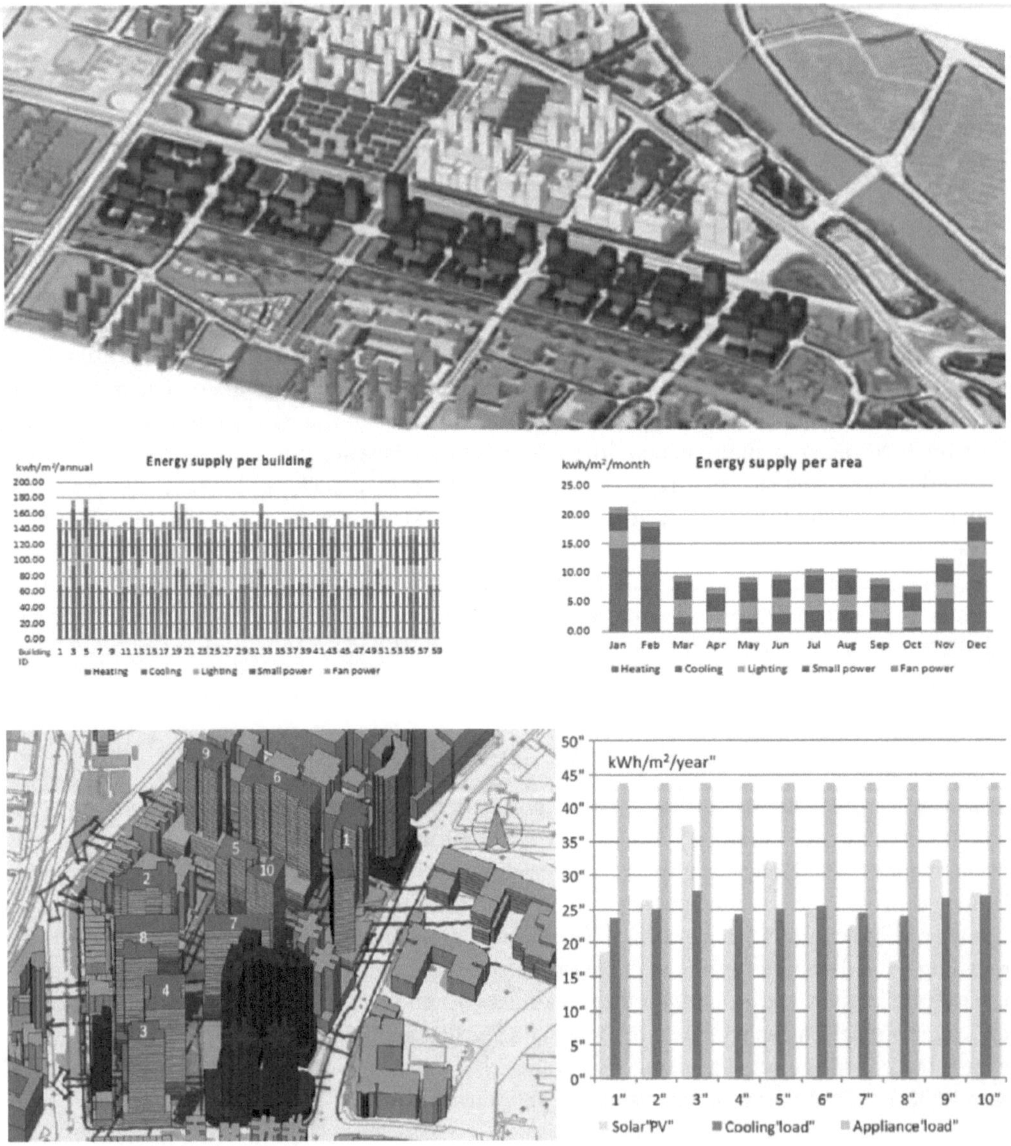

Figure 14.9 (a) Energy simulation for an urban block in Tianjin, China (Jones, 2019); (b) Energy simulation for an urban block in Hong Kong including the potential for collecting solar energy from PV (Huang et al., 2019)

14.11 Infectious disease

The spread of infectious disease, such as SARS and COVID-19, have had a catastrophic impact on urban living in recent years. Urban modelling has now been extended to include the external spread of infectious disease. Studies in Hong Kong have indicated that there is evidence of the spread of disease from one apartment to another via external paths (Huang et al., 2021). The modelling has been tested against case data for SARS, a close relative of the SARS-CoV-2 virus. This spread of

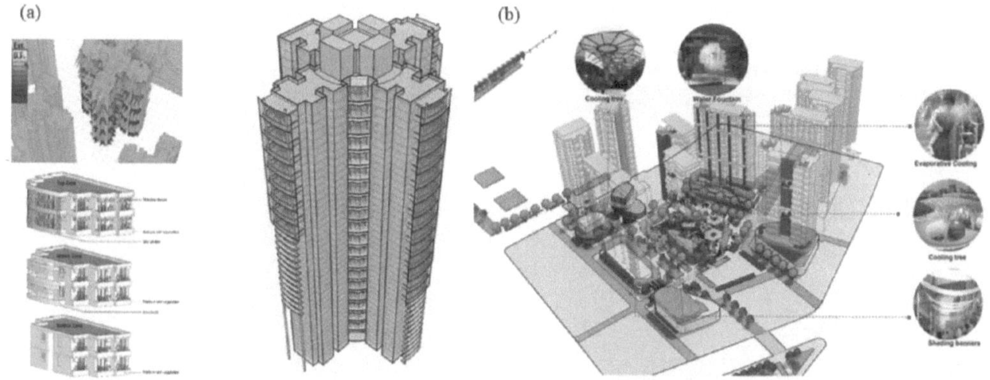

Figure 14.10 (a) Building design within the context of urban neighbours; (b) Innovations in microclimate control devices to improve outdoor comfort in To Kwa Wan, Hong Kong

infectious disease through external routes has implications on how buildings are ventilated and the role of external 'fresh air'. It contributes to the argument for whole-building mechanical ventilation with fresh air supply and exhaust at high level, away from potentially infected areas.

14.12 Urban design approaches

Clearly there is a need to consider current approaches to urban design. At a building scale the impact between adjacent buildings is important in relation to both internal and external environments. In addition, building façade design needs to respond to direction and height and the impact of sun, daylight, breezeways, noise, and pollution (Figure 14.10(a)). Building ventilation systems should be designed to access fresh air and exhaust stale air in a more considerate way in relation to outdoor air quality, perhaps including more centrally based HVAC systems. External comfort is important especially in high density cities in hot climates. New developments should consider breezeways and other innovative measures, such as shading canopies, mechanical air ventilation, evaporative cooling (i.e., the spraying of fine water droplets which can lower the ambient air temperature), and radiative cooling, the use of cool surfaces to improve outdoor comfort. An example has been shown in an urban renewal development in To Kwa Wan, Hong Kong (Figure 14.10 (b)).

14.13 Urban eco master planning

The concept of an Ecocity master-planning needs to be as much about process as product (Figure 14.11(a)), with quantitative and qualitative goals and evaluation at each stage. To achieve this, we must provide a clear vision and leadership across all levels, including government, industry, third sector, and communities. We all need to think differently and take ownership of the problem and immediately start to take action. We cannot change our cities overnight to be Ecocities, but we can work upwards, from individual projects, ensuring that they each contribute to providing clean and healthy environments that are accessible and affordable to all city dwellers.

14.14 Triple diamond

Sustainable urban design is a developing subject and needs to be underpinned by a strong theoretical framework. The double diamond design approach developed by the UK Design Council (2024) can be adapted for use in sustainable urban design (Figure 14.11(b)). The problem and solution

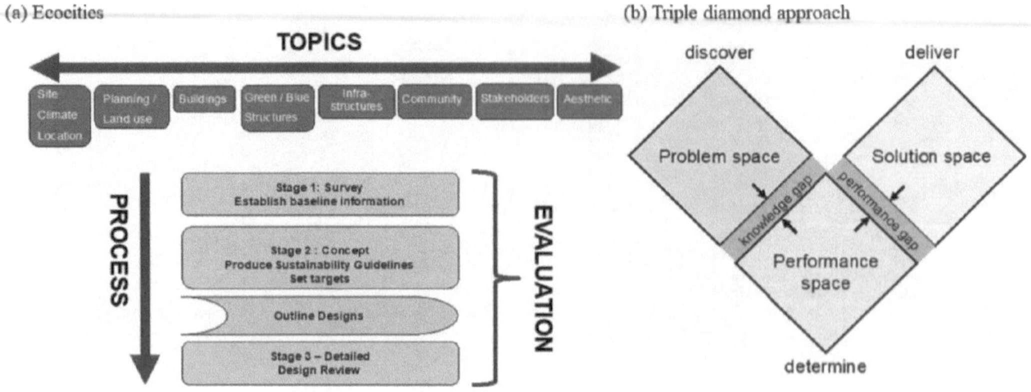

Figure 14.11 (a) Ecocities require sustainable master planning, which is multi-faceted, carried out through a number of stages with evaluation at each stage. (b) Triple diamond approach to the design of sustainable built environment in a high-density context

spaces can be extended to include a 'performance space'. This triple diamond approach can relate to the knowledge gap and performance gap, both of which need to be addressed to improve sustainable design in practice. The 'performance gap' is often used to describe why buildings do not perform as predicted. The 'knowledge gap' can be used to inform weaknesses in our current predictions and how our modelling might be improved.

14.15 Closing remarks

We need to rethink our approach to urban design with sustainability at its heart, recognising that sustainability is as much about process as it is about product. Sustainable urban design needs to be more holistic, combining top-down and bottom-up approaches to satisfy both government targets and end user's needs

Zero-carbon performance is the goal, with some developments achieving an energy neutral or even an energy positive performance, whilst others will rely on an increasingly zero-carbon grid supply. The control of future energy systems will be more centred at the local development scale than the grid, with developments becoming 'prosumers' based on the integration of generation, storage and energy use at the end-user level. In the future we also expect our definition of zero carbon to include embodied carbon, and tools to account for this are already becoming increasingly used at the design stage.

The concept of the 'city as a building' and the 'internet of energy' follows a holistic systems-based approach where there is integration of all spatial types and infrastructures in a robust, effective, and efficient way whilst allowing a flexibility of solutions.

For this to be effective we need simulation tools for predicting energy use, environmental conditions, etc., with the focus on demand reduction and the provision of clean, efficient, and effective supply relating to energy, water, air, and construction materials. We need to improve our understanding of the process of sustainable urban design and construction. In these processes, we can adopt simulation tools and conduct continuous tests according to the performance in use.

References

Aibar, E.; Bijker, W. E. (1997), Constructing a city: The cerdà plan for the extension of Barcelona, Science, Technology, & Human Values Vol. 22, No. 1, pp. 3-30

Akbari, H., Kurn, D. M., Bretz, S. E., & Hanford, J. W. (1997). Peak power and cooling energy savings of shade trees. *Energy and Buildings*, *25(2)*, 139–148. https://dx.doi.org/10.1016/s0378-7788(96)01003-1

Alexandri, E., & Jones, P. (2008). Temperature decreases in an urban canyon due to green walls and green roofs in diverse climates. *Building and Environment*, *43(4)*, 480–493. https://doi.org/10.1016/j.buildenv.2006.10.055

Hao, T., Chang, H., Liang, S., Jones, P., Chan, P. W., Li, L., & Huang, J. (2023). Heat and park attendance: Evidence from "small data" and "big data" in Hong Kong. *Building and Environment*, *234*, 110123. https://doi.org/10.1016/j.buildenv.2023.110123

Huang, J., Chen, Y., Jones, P., & Hao, T. (2022). Heat stress and outdoor activities in open spaces of public housing estates in Hong Kong: A perspective of the elderly community. *Indoor and Built Environment*, *31(6)*, 1447–1463. https://doi.org/10.1177/1420326X20950448

Huang, J., Hao, T., Hou, S. S., & Jones, P. (2019). Simulation-informed urban design: Improving urban microclimate in real-world practice in a high density city. *IOP Conference Series: Earth and Environmental Science*, 012047. https://doi.org/10.1088/1755-1315/329/1/012047

Huang, J., Hao, T., Wang, Y., & Jones, P. (2022). A street-scale simulation model for the cooling performance of urban greenery: Evidence from a high-density city. *Sustainable Cities and Society*, 103908. https://doi.org/10.1016/J.SCS.2022.103908

Huang, J., Jones, P., Zhang, A., Hou, S. S., Hang, J., Spengler, J. D., & Jones, P. (2021). Outdoor airborne transmission of coronavirus among apartments in high-density cities. *Frontiers in Built Environment*, *7*, 48. https://doi.org/10.3389/fbuil.2021.666923

Huang, J., Jones, P., Zhang, A., Peng, R., Li, X., & Chan, P. (2020). Urban Building Energy and Climate (UrBEC) simulation: Example application and field evaluation in Sai Ying Pun, Hong Kong. *Energy and Buildings*, *207*, 109580. https://doi.org/10.1016/J.ENBUILD.2019.109580

Huang, J., Tang, X., Jones, P., Hao, T., Tundokova, R., Walmsley, C., Lannon, S., Frost, P., & Jackson, J. (2024). Mapping pedestrian heat stress in current and future heatwaves in Cardiff, Newport, and Wrexham in Wales, UK. *Building and Environment*, 111168. https://doi.org/10.1016/j.buildenv.2024.111168

Jones, P. (2019). Can we achieve an energy positive built environment? (p. 34). University of Bath.

Jones, P. (2021). Forward: Setting the scene, ecocities eds Phillip Jones and Shan Shan Hou, pages 1–10, published by Gwyrdd.

Jones, P. 2017. A 'smart' bottom-up whole systems approach to a zero carbon built environment. Building Research & Information 46(5), pp. 566-577.

Jones, P.; Hou, S.S. (2021), Ecocities, Lighting Source UK, Ltd. Milton Keyes, UK

Jones, P., Lannon, S., Li, X., Bassett, T., & Waldron, D. (2013). Intensive building energy simulation at early design stage. *Proceedings of BS2013: 13th Conference of International Building Performance Simulation Association* https://doi.org/10.26868/25222708.2013.1299

The UK Design Council. (2024). Framework for Innovation. Available at: https://www.designcouncil.org.uk/our-resources/framework-for-innovation/ (Accessed: January 1, 2024)

UN, 2018, The world's cities in 2018, Available at https://www.un.org/development/desa/pd/ Accessed Jan. 1, 2024Zou, F.; Sun, X.; Song, K.; Yan, J. (2021). Tianjin Eco-city sponge city construction example. *Ecocities* eds Phillip Jones and Shan Shan Hou, pages 237–238, Published by Gwyrdd.

Wong, N. H., Tan, C. L., Kolokotsa, D. D., & Takebayashi, H. (2021). Greenery as a mitigation and adaptation strategy to urban heat. *Nature Reviews Earth & Environment*, *2(3)*, 166–181. https://doi.org/10.1038/s43017-020-00129-5

15

EXPLORING THE POTENTIALS AND OBSTACLES OF CITY INFORMATION MODELLING

Junhao Wu[1], Feng Qiu[2] and Ling Ma[3]

[1]PhD candidate, Dept. of Construction Management, National Center of Technology Innovation for Digital Construction, International Joint Research Laboratory of Smart Construction, School of Civil and Hydraulic Engineering, Huazhong University of Science and Technology, Wuhan Hubei 430074, China. E-mail: wujunhao@hust.edu.cn

[2]Master's candidate, Dept. of Construction Management, National Center of Technology Innovation for Digital Construction, International Joint Research Laboratory of Smart Construction, School of Civil and Hydraulic Engineering, Huazhong University of Science and Technology, Wuhan Hubei 430074, China. E-mail: qiufengchn@hust.edu.cn

[3]Professor, Dept. of Construction Management, National Center of Technology Innovation for Digital Construction, International Joint Research Laboratory of Smart Construction, School of Civil and Hydraulic Engineering, Huazhong University of Science and Technology, Wuhan Hubei 430074, China. E-mail: ling_ma@hust.edu.cn

15.1 Introduction

In an era marked by rapid urbanization and technological advancement, the concept of smart cities has emerged as a beacon of innovative urban management. Smart cities are the epitome of the seamless fusion between digital technology and urban ecosystems. This fusion represents a substantial shift in the approach to planning, delivering, and maintaining city services and infrastructure. At the heart of this shift lies the concept of city information modelling (CIM), a concept that is used for virtualizing the urban landscape. CIM has evolved primarily from the progress of building information modelling (BIM) and geographic information systems (GIS). CIM was first mentioned by Khemlani at Autodesk University in 2007 and later defined by Gil and Almeida (2011) at the eCAADe conference in 2011 as an expansion of the BIM framework to encompass urban design and scale BIM to the urban level. Xu et al. (2014) provided a technical implementation of CIM that inherits the functions of BIM and GIS, offering comprehensive management capabilities for digital cities. Thompson et al. (2016) described CIM as an interdisciplinary approach using urban data to manage land, property, and environmental resources to achieve sustainable and liveable cities. Chen et al. (2018) emphasized CIM's inclusion of spatial data and virtual representations of urban objects, while Wang et al. (2020) viewed CIM as a means for the collaborative development of physical and digital twin cities. CIM is seen as an interoperable platform for urban modelling and forecasting by Liu et al. (2022) and as an integration of cutting-edge information technologies by Zhang et al. (2022), providing improved urban governance and support

DOI: 10.1201/9781003383840-16

Table 15.1 Characteristics of CIM in the literature

Reference	*Characteristics*
(Gil and Almeida, 2011)	• Urban design • AutoCAD Map 3D • PostGIS
(Xu et al., 2014)	• Semantic mapping • BIM+GIS • Land planning
(Cureton and Hartley, 2023)	• Integration of digital models • Analysing workflows • Intelligent interaction • Public feedback
(Xue et al., 2021)	• GIS • Semantic enrichment • Smart city • Intelligent building
(Wang et al., 2020)	• GIS+BIM+IOT • CIM+
(Chen et al., 2018)	• Automatic reconstruction of CIM • LiDAR point cloud data • Level of model definition (LoD)

for smart cities. Last, Al Jurdi et al. (2023) highlighted CIM as a 3D model of city structure and interrelated layers, visualizing citizen feedback to improve their city. Overall, CIM represents a comprehensive and multidimensional approach to urban management and modelling, integrating various technologies to enhance efficiency and sustainability. Table 15.1 lists diverse definitions of CIM's characteristics given in scholarly literature.

CIM represents a cutting-edge paradigm in urban planning and management, holding the promise of transforming cities into smarter, more efficient, and more liveable spaces. Despite its burgeoning application across various sectors, there exists a significant gap in understanding the integrated use of CIM within urban contexts. This book chapter provides a comprehensive exploration of CIM, beginning with a systematic literature review that identifies and examines research areas and practical implementations of CIM. It then delves into the various challenges that have arisen during the development of CIM, offering a critical analysis of the obstacles faced by practitioners and theorists alike.

15.2 Methodology

This study follows the Preferred Reporting Items for Systematic Reviews and Meta-Analyses (PRISMA) guidelines to examine the adoption of CIM. Content was collected from journal articles, book chapters, and conference proceedings. Utilizing a targeted keyword search strategy using terms such as "CIM," "City," and "City information models," a total of 117 papers spanning the years 2009 to 2023 were retrieved from the Scopus database. Figure 15.1 delineates the tripartite methodology employed in this research, which encompasses the stages of identification, screening, and inclusion phases, ensuring a systematic and replicable approach to the literature search process.

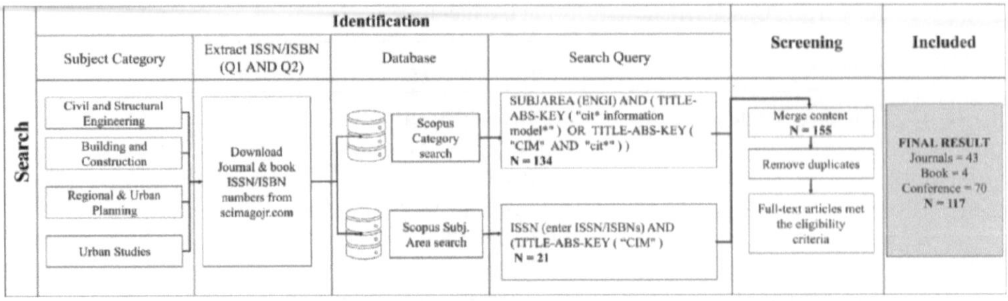

Figure 15.1 Procedure for obtaining scholarly references using Scopus

Five relevant review articles on CIM were identified among the 117 pieces of work. Xue et al. (2021) have provided an analytical framework to assess the evolution of semantic enrichment in BIM and CIM. Their work juxtaposes the last decade's research trends and prospects. Liu et al (2021) have delved into the transformative potential of blockchain technology when integrated with BIM within smart city infrastructures, particularly in enhancing building sustainability across construction lifecycle phases. Xu et al. (2021) contribute to the discourse by delineating the core segments of current CIM research, including data collection, integration, and visualization, while evaluating prevalent CIM platforms and their application across urban life cycles. Souza and Bueno (2022) offer an examination of CIM's various functions and tools, presenting case studies and conceptual frameworks that pave the way for future research and practical application. Finally, Omrany et al. (2022) have identified key areas and challenges of CIM implementation, proposing strategies to overcome obstacles and promote CIM's role in effective urban management. Together, these studies underscore the dynamic intersection of technology and urban development, charting a course for future innovation in the field.

Based on existing research, we offer the following clarifications about CIM. First, CIM is envisaged as a central repository, including geometric, semantic, and geospatial data, which is instrumental in the development of smart cities. Second, it should possess robust capabilities for processing multi-source heterogeneous data, enabling storage, computation, and analysis of diverse datasets (Cai et al., 2021; Del Duca et al., 2022). Third, CIM must support dynamic data updating, seamlessly integrating inputs from various IoT devices (Qian and Leng, 2021; Zhang et al., 2021). Last, the model should offer varying Levels of Detail (LoD) to accurately represent multi-scale urban indicators, catering to a spectrum of user requirements (Lu et al., 2020).

In essence, CIM serves as a sophisticated platform for comprehensive urban information management within a three-dimensional digital framework, merging spatial data with real-time IoT sensing, thereby fostering advanced urban governance strategies. The following section elaborates on research clusters of CIM applications and related technical components drawn from the 117 scholarly articles.

15.3 Results

15.3.1 Technical components

CIM represents a sophisticated integration of multi-dimensional data, encompassing subterranean and surface details, indoor and outdoor perspectives, and temporal urban data spanning past, present, and future.

This intricate web of urban intelligence is constructed within a 3D digital framework, presenting pivotal support for endeavours in urban planning, construction, management, and service provision. Fundamental to CIM's technological infrastructure are GIS, BIM, and IoT. The versatility and interactive capabilities of real-time 3D engines have cemented their status as the preferred medium for visualizing CIM data.

As the field evolves, an array of emergent technologies, such as blockchain, visual programming, information, and communication technology, along with advanced data acquisition methods like remote sensing, tilt photography, and point cloud modelling, are being integrated. Table 15.2 offers a detailed exposition of the technologies pertinent to CIM and identifies the industry's frontrunners.

Table 15.2 CIM-related technologies and major authoring tools

Name	*Characteristics*	*Major Authoring Tools*	*Result*
GIS	GIS is a computer system for capturing, storing, checking, and displaying data related to positions on Earth's surface.	Esri ArcGIS, QGIS, SuperMap, MapGIS	
BIM	BIM models are digital representations of the physical and functional characteristics of building objects (International Organization for Standardization, 2016).	Autodesk Revit, Rhinoceros, Bentley Advancing Infrastructure, Glodon 广联达	
IOT	IoT is a system that determines the interconnections among devices, machinery, and digital equipment using a universally unique identifier (UUID) and the capacity to transmit data over a network (Al-Fuqaha et al., 2015).	Huawei, Microsoft, Qualcomm, 中国移动 China Mobile	
Blockchain	Blockchain is a decentralized, unchangeable distributed ledger designed to record transactions and track assets in a commercial network. It serves as a platform for storing information, executing transactions, performing functions, and building trust in an open, transparent environment (Angles-Tafalla et al., 2023).	amazon, IBM, AON, ANT GROUP	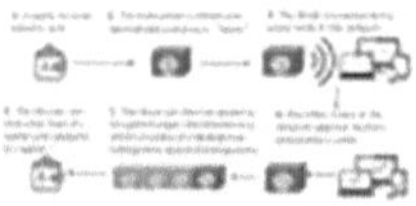
Real-time 3d engine	Real-time 3D engine technology facilitates the creation of life-like visuals and engaging experiences.	unity, Unreal Engine	
VP	Visual programming (VP) solves the challenge of providing computing resources to individuals with limited computer expertise by implementing visual (i.e., non-linear) representations during the programming process (Shu, 1999).	mendix, Zoho Creator, outsystems, Microsoft PowerApps	

(*Continued*)

Table 15.2 (Continued)

Name	*Characteristics*	*Major Authoring Tools*	*Result*
ICT	Information and communication technology (ICT) refers to industries and services that perform information processing and communication functions using electronic means (Chang et al., 2023).	Microsoft, hp, HUAWEI, cisco	
Data collection tools	Remote-sensing	 NASA	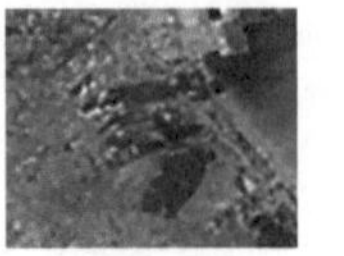
	Drone	 dji	
	LiDAR	 Trimble.	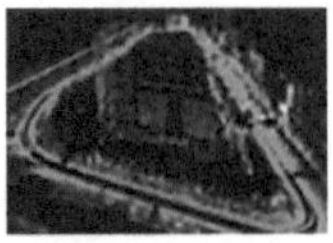

15.3.2 Research subject areas on CIM

The comprehensive literature analysis encompassed 22 application scenarios that are categorized into seven subject areas: (a) ontological study of CIM, (b) CIM model construction, (c) urban disaster management, (d) urban planning, (e) urban facility management, (f) sustainable urban management, (g) building virtual communities.

15.3.2.1 Ontological study of CIM

Zhang et al. (2022) emphasise that, for effective CIM data practices, there must be a comprehensive integration of various standards, including data, technical, management, and application standards. These standards underpin critical functions such as data collection, transmission, governance, and security, among others. Wang et al. (2020) discuss the development of CIM, which involves the organization and aggregation of data and the exploration of its application capabilities.

A pivotal aspect of CIM development is the establishment of a unified data architecture that synergizes BIM and GIS through standard representation formats like Industry Foundation Classes (IFC) and City Geography Markup Language (CityGML), as noted by Lu et al. (2020). CityGML is particularly noted for its ability to store and exchange semantic information about the urban built environment across various applications, making it suitable for modelling 3D outdoor variables. IFC, on the other hand, is geared towards the AEC industry, providing comprehensive physical and functional attributes of buildings, as Xu et al (2014) points out. Nonetheless, the challenge lies in the integration of diverse attribute data within CIM, which is crucial for data-driven analysis and decision-making processes.

To overcome this, researchers have proposed various data fusion methods. Ontology-based (Loo and Wong, 2023), semantic-based (Cai et al., 2021), and geometry-based (Chen et al., 2018) data fusion techniques have been explored to merge and express the vast array of data attributes within CIM. Shi et al (2023) have notably advanced this field by researching the geometric transformation from IFC to CityGML, establishing an extensive geometric ontology for multiple levels of detail, and integrating BIM, GIS, and IoT data through the creation of a multi-scale CIM ontology.

These advancements underscore the ongoing efforts to refine the data integration process within CIM, thereby enhancing its potential as a tool for comprehensive urban analysis and planning.

15.3.2.2 CIM model construction

Chen et al (2018) advanced the efficiency of CIM model construction by integrating LiDAR-derived geometric information with architectural knowledge, facilitating automatic reconstruction with a focus on detailed roof features. Complementing this, Xue et al. (2020) and Tan et al. (2019) contributed to the nuanced identification of roof albedo and elements, respectively. Zhang et al. (2021) streamlined the creation of 3D BIM models from 2D drawings, expediting the modelling process. To combat the lack of semantic information, Xue et al. (2019) introduced a derivative-free optimization method to generate semantically enriched BIM models from 3D point clouds, and Li et al. (2022) further proposed using deep migration learning to estimate architectural features in CIM. Addressing model light weighting, Chen et al. (2021) developed a semantic-based algorithm for efficiently scheduling scenes, thus minimizing computational demands.

Additionally, Xu et al (2021) suggested leveraging topological optimization to simplify city model reconstructions, substantially reducing data storage requirements. These advancements collectively represent a leap forward in the precision, efficiency, and practicality of CIM model construction.

15.3.2.3 Urban disaster management

Gu et al. (2023) introduced a wind damage assessment methodology using CIM that evaluates the potential threat to building complexes, including the effects of the surrounding vegetation. Further expanding CIM's utility, Lu et al. (2020) devised a multi-hazard simulation framework that addresses risks from earthquakes, fires, and wind, focusing on both individual structures and broader urban landscapes.

For visualizing disaster risks, Del Duca et al. (2022) created a 3D flood risk map along the Tahoe coastline, utilizing the CIM database. Additionally, Argenziano et al. (2018) enhanced the understanding of urban scale radiometric data through surveys in Pompeii, aiding in the formulation of effective rescue and response strategies. CIM's comprehensive data integration facilitates meticulous rescue route planning during fires (Xu et al., 2014) and supports targeted post-disaster restoration (Zhang et al., 2022).

Ultimately, CIM's role in simulating and visualizing urban risks not only prepares city managers for impending disasters but also fortifies the city's resilience and promotes public safety, while streamlining emergency and restoration operations.

15.3.2.4 Urban planning

The imperative of urban planning is to forge the path toward sustainable smart cities, a vision where CIM emerges as an indispensable tool. CIM's multifaceted applications range from crafting intricate 3D urban maps to the meticulous orchestration of city operations and extending to the innovative realm of urban design (Thompson et al., 2016). These maps not only encapsulate the physicality of cityscapes but also become repositories of rich, layered data – melding demographics, environmental metrics, and socioeconomic factors into a cohesive, dynamic model (Gao et al., 2023). Such integration is pivotal for urban data management, as it enhances decision-making by synthesizing both real-time and historical data (Wang, 2021).

Furthermore, CIM's prowess in granular city management is evident, as it amalgamates static urban layouts with the dynamic inputs from IoT devices, streamlining processes like heritage tourism governance (Liu et al., 2023). This precision extends to design, where CIM underpins parametric and generative methodologies, incorporating public input and enabling the construction of an adaptive infrastructure management framework.

According to Todor et al. (2022), in the domain of generative planning, CIM is a canvas for urban design creativity, aligning spatial syntax with sensory and aesthetic dimensions. It fosters a synthesis of urban morphology with sustainable design principles, IoT integration, and ecosystem considerations, ultimately elevating the sustainability quotient of smart cities.

Moreover, CIM serves as a catalyst for design automation, leveraging AI to refine design quality and operational efficiency, thus shaping the future of urban environments (Stojanovski et al., 2020).

15.3.2.5 Urban facility management

According to Loo and Wong (2023), CIM brings into consideration the dynamic interplay between the building and its urban environment. Zhang et al. (2022) highlight CIM's capability to produce detailed 3D spatial models that integrate geometric, physical, and structural data, which, when coupled with smart sensor network (SSN) technology, can potentially obviate the need for extensive rewiring in building control systems. Furthermore, the synergy of CIM with digital innovations like blockchain could revolutionize construction waste management by enhancing the sorting and reuse of materials, thereby extracting value from waste as identified by Liu et al. (2022). CIM's comprehensive data not only aids in property value assessment but also factors in landscape aesthetics, sunlight access, and other valuation parameters, as per underscoring the precision of CIM in evaluating window quality through detailed 3D texturing, enabling a unique assessment for each unit, thereby encapsulating the principle of "one house, one view, one assessment."

15.3.2.6 Sustainable urban management

By fusing CIM architecture with smart city design principles, Gao et al. (2023) devised robust strategies for sustainable urban development. CIM's utility extends to the direct appraisal of ISO sustainability indicators, offering a comprehensive view of a city's performance across economic, social, and environmental sectors (Dantas et al., 2019). It uniquely captures the interplay between urban spaces and human activity, fostering sustainable growth in management, construction, and service provision (Gao et al., 2023).

Moreover, CIM quantifies a city's sustainable capacity, considering both natural elements like green space and energy usage and constructed features such as infrastructure and resource distribution (Salles et al., 2023). This positions CIM as an indispensable instrument for achieving urban energy efficiency and conservation targets (Schaufler and Schwimmer, 2020). CIM also acts as a conduit for citizen engagement, allowing for the incorporation of public input into the urban enhancement process, thus elevating the standard of living (Al Jurdi et al., 2023).

Additionally, CIM provides an interactive platform for community participation in urban planning, fostering a collaborative approach to city design (Zingariello, 2021). It also aids in understanding the impact of urban policies on residential areas, with an eye on scientific and educational implications (Ungureanu, 2019).

15.3.2.7 Building virtual communities

Najafi et al (2023) have pioneered a virtual CIM environment to study how non-expert users interact with community design and policymaking processes. Similarly, Cureton and Hartley (2023) have innovated with the Lancaster urban information model (LCIM), a prototype that transforms urban planning data into a gamified, virtual experience, thereby facilitating public discourse on urban development proposals. Huang et al (2022) went a step further to integrate BIM, CIM, blockchain, and other data technologies to create a meta-universe that chronicles the dynamic nature of urban and architectural evolution.

At its core, the CIM platform is instrumental in creating digital twin cities, serving as a critical framework for the assimilation of data capture, management, and integration toward the realization of digital urban centres. Wang et al (2020) emphasise that the interconnection of the CIM platform with intelligent systems across various sectors is essential to achieving the vision of smart city construction, where CIM and business intelligence systems operate in synergy.

15.3.3 Challenges in the development of CIM

The deployment of CIM stands as a transformative prospect for the advancement of smart cities. It promises to streamline infrastructure management, optimize resource allocation, and foster sustainable development. Despite these potential benefits, the adoption of CIM faces significant hurdles, including integration complexities, data privacy concerns, and the need for standardized protocols. These challenges highlight the imperative for concerted research efforts to develop robust solutions that can facilitate the seamless integration of CIM into the smart city paradigm, ensuring its full potential is realized for the betterment of urban living.

15.3.3.1 Reluctance towards CIM application

Despite the considerable benefits that CIM offers for enhancing urban planning and management, there remains a notable reluctance among stakeholders to fully embrace this technology. This hesitancy can be attributed to a multitude of factors. Often, leadership may not be fully cognizant of the advantages CIM can provide, leading to a lack of advocacy at the decision-making level. There may also be a deficit of confidence in the reliability of the results yielded by CIM applications, coupled with an absence of mandates for its use in contractual obligations. Furthermore, the promotion of CIM utilization suffers from insufficient investment, which in turn leads to a lack of comprehensive training for prospective users. The challenge is compounded by a general shortfall in collaboration between partners in implementing CIM and a pervasive gap in specialized knowledge required for its effective application. Addressing these issues is crucial for unlocking the full potential of CIM in transforming urban environments.

15.3.3.2 Data quality

The efficacy of CIM is fundamentally contingent upon the integration of data that is not only accessible but also precise and coherent. However, the integrity of CIM's analytical insights and final outputs is susceptible to compromise by prevalent data quality issues, including inaccuracies, omissions, extraneous data, redundancies, and obsolescence. The construction of comprehensive 3D urban models mandates the assimilation of extensive geometrical, spatial, and semantic data about cityscapes. This data is procured from a diverse spectrum of sources, encompassing

GIS data from remote sensing technologies, community-generated content, scholarly expertise, and official land records. Given the disparate methodologies and origins of urban data collection, upholding data quality presents a significant challenge. Such lapses in data quality can adversely affect the reliability and authenticity of the insights derived from CIM analyses.

15.3.3.3 Computing resources and storage inefficiency

In the realm of CIM, the volume and complexity of data present formidable challenges to existing computing and storage infrastructures. To ensure efficient and secure data management, there is a pressing need for advanced hardware and software solutions. The execution of detailed simulation analyses, which scrutinize millions of city elements like buildings and infrastructure, demands great computing power that is currently within the ability of cutting-edge supercomputers. Additionally, the surge of big data in urban modelling exacerbates the difficulties of data storage management, necessitating the development of innovative data management architectures. These architectures must be robust enough to handle the geospatial intricacies, diverse sources, and evolving nature of urban data. Furthermore, to overcome interoperability hurdles and enhance project delivery, it is imperative to consider additional dimensions such as semantic understanding and legal frameworks.

15.3.3.4 Standardized workflow

Efficient implementation of CIM is paramount to the advancement of smart city initiatives. Despite the availability of numerous standards and tools designed to facilitate CIM deployment, the industry continues to grapple with the challenge of creating a unified workflow. This workflow must clearly define the prerequisites for applying CIM and standardize the data inputs and associated procedures within the built environment. The establishment of a comprehensive standardized process is critical to improving the efficiency, quality, and transparency of CIM, ultimately enabling stakeholders to make more informed decisions and better manage urban development.

15.3.3.5 Synergy between all parties involved

The advancement of CIM is contingent upon the synergistic efforts of a diverse array of urban stakeholders, including policymakers, planners, designers, and citizens. While the field is buoyed by the burgeoning contributions of numerous research entities, evidenced by an array of innovative tools, methods, and databases, the harmonization of disparate objectives and perspectives remains a formidable challenge. A collaborative framework must be established to bridge the gaps between the varied interests and capabilities of all involved parties. Only through such collaborative endeavours can the full potential of CIM be realized, fostering urban environments that are sustainable, resilient, and conducive to an enhanced quality of life for all inhabitants.

15.3.3.6 Cybersecurity and intellectual property

The integration of CIM within the built environment necessitates stringent attention to intellectual property and cybersecurity concerns. As CIM systems facilitate extensive data exchange among stakeholders, they inherently elevate the risk of unauthorized access and potential copyright violations.

The collaborative nature of CIM, involving diverse experts, complicates the delineation of intellectual property rights, posing challenges in the licensing of CIM models and replication of results. Consequently, robust legal frameworks and advanced cybersecurity measures are imperative to safeguard proprietary information and maintain the integrity of the CIM process.

15.3.3.7 Scalability of CIM models

Traditional technical components have historically concentrated on the tangible infrastructure of urban environments, inadvertently marginalizing the complex tapestry of human dynamics that animate these spaces. This reductionist approach can result in a "trapped in the matrix" effect, where the essence of urban life is not fully captured. Furthermore, scalability emerges as a formidable obstacle in CIM, particularly when aligning models with the expansive and often indeterminate nature of real-world urban areas. Ignoring the global context in which cities operate can lead to oversimplified models that fail to accurately reflect the intricate interdependencies of urban systems and their broader regional, national, and international linkages.

15.4 Conclusion

The primary aim of this study was to delve into the application of CIM within the realm of the built environment. Through the examination of 117 scholarly works, the research identified seven principal areas where CIM has been effectively researched and employed. The findings underscore CIM's significant impact on advancing decision-making capabilities across these diverse fields. However, the widespread adoption of CIM is curtailed by challenges such as limited enthusiasm for its use, data quality issues, computational constraints, the absence of standardized workflows, stakeholder coordination, cybersecurity risks, and model scalability.

To facilitate the integration of CIM into urban management, awareness of CIM's practical benefits must be heightened through targeted educational campaigns and workshops. The education system bears the responsibility of cultivating skilled CIM modelers to assist decision-makers throughout project development. The emergence of the CIM modeler as a new urban management profession is anticipated. Moreover, the research advocates for the mandatory inclusion of CIM in urban project contracts to ensure its use in enhancing collaboration and decision-making. Incentives for companies that adopt CIM could further promote best practices within the sector.

Additionally, the evolution of CIM is closely tied to advancements in computer science, which will empower CIM models to handle vast urban datasets effectively. Establishing standardized data protocols is essential for ensuring data quality and facilitating inter-system sharing. Protecting the intellectual property of CIM models is also critical, necessitating the development of appropriate contractual templates. The broad potential of CIM necessitates extensive research to substantiate its benefits and explore its integration with lifecycle thinking, thereby reinforcing its role as a transformative tool in urban management.

References

A. Al-Fuqaha, M. Guizani, M. Mohammadi, M. Aledhari, M. Ayyash, 2015. Internet of things: A survey on enabling technologies, protocols, and applications. IEEE Commun. Surv. Tutor. 17, 2347–2376. https://doi.org/10.1109/COMST.2015.2444095

Al Jurdi, M., Wehbe R., Mroueh H., 2023. Integration of citizens' feelings and feedback into the city information modeling environment. Sustain. Cities Soc. 99. https://doi.org/10.1016/j.scs.2023.104971

Angles-Tafalla, C., Viejo, A., Castella-Roca, J., Mut-Puigserver, M., Magdalena Payeras-Capella, M., 2023. Security and privacy in a blockchain-powered access control system for low emission zones. IEEE Trans. Intell. Transp. Syst. 24, 580–595. https://doi.org/10.1109/TITS.2022.3211659

Argenziano, P., Avella, A., Albanese, S., 2018. Building materials, ionizing radiation, and HBIM: A case study from Pompei (Italy). Buildings 8, 18. https://doi.org/10.3390/buildings8020018

Cai, Y., Huang, H., Wang, K., Zhang, C., Fan, L., Guo, F., 2021. Selecting optimal combination of data channels for semantic segmentation in City Information Modelling (CIM). Remote Sens. 13. https://doi.org/10.3390/rs13071367

Chang, C.M., Salinas, G.T., Gamero, T.S., Schroeder, S., Vélez Canchanya, M.A., Mahnaz, S.L., 2023. An infrastructure management humanistic approach for smart cities development, evolution, and sustainability †. Infrastructures 8. https://doi.org/10.3390/infrastructures8090127

Chen, K., Lu, W., Xue, F., Tang, P., Li, L.H., 2018. Automatic building information model reconstruction in high-density urban areas: Augmenting multi-source data with architectural knowledge. Autom. Constr. 93, 22–34. https://doi.org/10.1016/j.autcon.2018.05.009

Chen, Q., Chen, J., Huang, W., 2021. Visualizing large-scale building information modeling models within indoor and outdoor environments using a semantics-based method. ISPRS Int. J. Geo-Inf. 10, 756. https://doi.org/10.3390/ijgi10110756

Cureton, P., Hartley, E., 2023. City Information Models (CIMs) as precursors for Urban Digital Twins (UDTs): A case study of Lancaster. Front. Built Environ. 9, 1048510. https://doi.org/10.3389/fbuil.2023.1048510

Dantas, H.S., Sousa, J.M.M.S., Melo, H.C., 2019. The importance of City Information Modeling (CIM) for cities' sustainability, in: IOP Conf. Ser. Earth Environ. Sci. Presented at the IOP Conference Series: Earth and Environmental Science, Institute of Physics Publishing. https://doi.org/10.1088/1755-1315/225/1/012074

Del Duca, G., Rocha G., Orszt, M., Mateus, L., 2022. A preliminary contribution towards a risk-based model for flood management planning using BIM: A case study of Lisbon. Sensors 22. https://doi.org/10.3390/s22197456

Gao, C., Wang, F., Hu, X., Martinez, J., 2023. Research on sustainable design of smart cities based on the internet of things and ecosystems. Sustainability 15, 6546. https://doi.org/10.3390/su15086546

Gil, J., Almeida, J., 2011. The backbone of a City Information Model (CIM): Implementing a spatial data model for urban design. Presented at the eCAADe 2011 : Respecting Fragile Places, Ljubljana, Slovenia, pp. 143–151. https://doi.org/10.52842/conf.ecaade.2011.143

Gu, D., Shuai, Q., Wang, Y., 2023. CIM-powered physics-based assessment of wind damages to building clusters considering trees. Dev. Built Environ. 15. https://doi.org/10.1016/j.dibe.2023.100178

Huang, H., Zeng, X., Zhao, L., Qiu, C., Wu, H., Fan, L., 2022. Fusion of building information modeling and blockchain for metaverse: A survey. IEEE Open J. Comput. Soc. 3, 195–207. https://doi.org/10.1109/OJCS.2022.3206494

International Organization for Standardization, 2016. ISO 29481-1:2016, Building information models – Information delivery manual – Part 1: Methodology and format. Eid.10.1109/2-s2.0-85145616016

Li, M., Xue, F., Wu, Y., Yeh, A.G.O., 2022. A room with a view: Automatic assessment of window views for high-rise high-density areas using city information models and deep transfer learning. Landsc. Urban Plan. 226, 104505. https://doi.org/10.1016/j.landurbplan.2022.104505

Liu, Z., Chi, Z., Osmani, M., Demian, P., 2021. Blockchain and Building Information Management (BIM) for sustainable building development within the context of smart cities. Sustainability 13, 2090. https://doi.org/10.3390/su13042090

Liu, Z., Wu, T., Wang, F., Osmani, M., Demian, P., 2022. Blockchain enhanced construction waste information management: A conceptual framework. Sustain. Switz. 14. https://doi.org/10.3390/su141912145

Liu, Z., Zhang, M., Osmani, M., 2023. Building Information Modelling (BIM) driven sustainable cultural heritage tourism. Buildings 13. https://doi.org/10.3390/buildings13081925

Loo, B.P.Y., Wong, R.W.M., 2023. Towards a conceptual framework of using technology to support smart construction: The case of Modular Integrated Construction (MiC). Buildings 13. https://doi.org/10.3390/buildings13020372

Lu, X., Gu, D., Xu, Z., Xiong, C., Tian, Y., 2020. CIM-powered multi-hazard simulation framework covering both individual buildings and urban areas. Sustainability 12, 5059. https://doi.org/10.3390/su12125059

Najafi, P., Mohammadi, M., van Wesemael, P., Le Blanc, P.M., 2023. A user-centred virtual city information model for inclusive community design: State-of-art. Cities 134. https://doi.org/10.1016/j.cities.2023.104203

Omrany, H., Ghaffarianhoseini, Amirhosein, Ghaffarianhoseini, Ali, Clements-Croome, D.J., 2022. The uptake of City Information Modelling (CIM): A comprehensive review of current implementations, challenges and future outlook. Smart Sustain. Built Environ. https://doi.org/10.1108/SASBE-06-2022-0116

Qian, Y., Leng, J., 2021. CIM-based modeling and simulating technology roadmap for maintaining and managing Chinese rural traditional residential dwellings. J. Build. Eng. 44, 103248. https://doi.org/10.1016/j.jobe.2021.103248

Salles, A., Salati, M., Bragança, L., 2023. Analyzing the feasibility of integrating urban sustainability assessment indicators with City Information Modelling (CIM). Appl. Syst. Innov. 6. https://doi.org/10.3390/asi6020045

Schaufler, C., Schwimmer, E., 2020. City information modeling – An expedient tool for developing sustainable, responsive and resilient cities?, in: H. W., A. H., L. T., P. F., I. K., K. M., C. F. (Eds.), IOP Conf. Ser. Earth Environ. Sci. Presented at the IOP Conference Series: Earth and Environmental Science, IOP Publishing Ltd. https://doi.org/10.1088/1755-1315/588/3/032005

Shi, J., Pan, Z., Jiang, L., Zhai, X., 2023. An ontology-based methodology to establish city information model of digital twin city by merging BIM, GIS and IoT. Adv. Eng. Inform. 57. https://doi.org/10.1016/j.aei.2023.102114

Shu, N.C., 1999. Visual programming: Perspectives and approaches. IBM Syst. J. 38, 199–221. https://doi.org/10.1147/sj.382.0199

Souza, L., Bueno, C., 2022. City Information Modelling as a support decision tool for planning and management of cities: A systematic literature review and bibliometric analysis. Build. Environ. 207, 108403. https://doi.org/10.1016/j.buildenv.2021.108403

Stojanovski, T., Partanen, J., Samuels, I., Sanders, P., Peters, C., 2020. Viewpoint: City Information Modelling (CIM) and digitizing urban design practices. Built Environ. 46, 637–646. https://doi.org/10.2148/BENV.46.4.637

Tan, T., Chen, K., Lu, W., Xue, F., 2019. Semantic enrichment for rooftop modeling using aerial LiDAR reflectance, in: 2019 IEEE International Conference on Signal Processing, Communications and Computing (ICSPCC). Presented at the 2019 IEEE International Conference on Signal Processing, Communications and Computing (ICSPCC), IEEE, Dalian, China, pp. 1–4. https://doi.org/10.1109/ICSPCC46631.2019.8960769

Thompson, E.M., Greenhalgh, P., Muldoon-Smith, K., Charlton, J., Dolník, M., 2016. Planners in the future city: Using city information modelling to support planners as market actors. Urban Plan. 1, 79–94. https://doi.org/10.17645/up.v1i1.556

Todor, S., Lefosse, D., Torres, M., Samuels, I., Zhang, H., Zojaji, S., Peters, C., 2022. Convexity and imageability convex maps and urban (Space) envelopes, in: A. van N., R.E. de K. (Eds.), Proc. Int. Space Syntax Symp., SSS. Presented at the Proceedings 13th International Space Syntax Symposium, SSS 2022, Western Norway University of Applied Sciences (HVL). https://www.scopus.com/inward/record.uri?eid=2-s2.0-85145616016&partnerID=40&md5=c32e74eae0d383b389099c00aca4197f

Ungureanu, T., 2019. The potential of City Information Modeling (CIM) in understanding and learning from the impact of urban regulations on residential areas in Romania, in: I. R. (Ed.), eLearning Softw. Educ. Conf. Presented at the eLearning and Software for Education Conference, National Defence University – Carol I Printing House, pp. 422–428. https://doi.org/10.12753/2066-026X-19-056

Wang, P., 2021. Bridging human mobility and urban growth. Nat. Comput. Sci. 1, 778–779. https://doi.org/10.1038/s43588-021-00166-0

Wang, Z., Jiang, H., Zhang, W., Liu, L., 2020. The problem analysis and solution suggestion in the process of city information model construction, in: 2020 4th International Conference on Smart Grid and Smart Cities (ICSGSC). Presented at the 2020 4th International Conference on Smart Grid and Smart Cities (ICSGSC), IEEE, Osaka, Japan, pp. 109–112. https://doi.org/10.1109/ICSGSC50906.2020.9248544

Xu S., Mo W., Luan L., Tong R., Liu T., 2021. Optimization of distribution network topology based on CIM model, in: IET. Conf. Proc. Presented at the IET Conference Proceedings, Institution of Engineering and Technology, pp. 822–827. https://doi.org/10.1049/icp.2021.2334

Xu, Xun, Ding, L., Luo, H., Ma, L., 2014. From building information modeling to city information modeling. J. Inf. Technol. Constr. 292–307.

F. Xue, W. Lu, K. Chen, A. Zetkulic, 2019. From semantic segmentation to semantic registration: derivative-free optimization-based approach for automatic generation of semantically rich as-built building information models from 3D point clouds. J. Comput. Civ. Eng. 33. https://doi.org/10.1061/(ASCE)CP.1943-5487.0000839

Xue, F., Lu, W., Tan, T., Chen, K., 2020. Semantic enrichment of city information models with lidar-based rooftop albedo, in: K., P., C.S., C., I., G., G., G., F., G. (Eds.), Sustain. Build. Struct.: Build. Sustain. Tomorrow – Proc. Int. Conf. Sustain. Build. Struct. Presented at the 2nd International Conference in Sustainable Buildings and Structures, ICSBS 2019, CRC Press/Balkema, pp. 207–212. https://doi.org/10.1201/9781003000716-27

Xue, F., Wu, L., Lu, W., 2021. Semantic enrichment of building and city information models: A ten-year review. Adv. Eng. Inform. 47, 101245. https://doi.org/10.1016/j.aei.2020.101245

Zhang, C., Zou, Y., Dimyadi, J., 2021. A systematic review of automated BIM modelling for existing buildings from 2D documentation, in: C., F., T., L., I., B. (Eds.), Proc. Int. Symp. Autom. Robot. Constr. Presented at the 38th International Symposium on Automation and Robotics in Construction, ISARC 2021, International Association for Automation and Robotics in Construction (IAARC), pp. 220–226. https://doi.org/10.22260/ISARC2021/0032

Zhang, D., Li, Z., Sun, W., Zhu, X., Ge, H., 2022. Data analysis of building microrenovation based on city information modeling and smart sensor networks. J. Sens. 2022, 1–11. https://doi.org/10.1155/2022/4384795

Zingariello I., 2021. From BIM to CIM: A new instrument for urban planners and a new bottom-up planning process, in: D. L.R., R. P. (Eds.), Lect. Notes Civ. Eng. Presented at the Lecture Notes in Civil Engineering, Springer Science and Business Media Deutschland GmbH, pp. 189–193. https://doi.org/10.1007/978-3-030-68824-0_20

16
MOBILITY-AS-A-SERVICE (MAAS) AND THE BUILT ENVIRONMENT

Qiao Si[1], *Anthony G.O. Yeh*[1], *Guan Huang*[1] *and Yang Yue*[2]

[1] Department of Urban Planning and Design, The University of Hong Kong (siqiao@connect.hku.hk, hdxugoy@hku.hk, guanhuang@connect.hku.hk)

[2] Department of Urban Informatics, Shenzhen University (yueyang@szu.edu.cn)

16.1 Introduction

The demand for urban travel is growing rapidly, leading to increased congestion, energy consumption, and carbon emissions in urban areas (Kenworthy 2003). Mobility-as-a-Service (MaaS) has emerged as a potential solution to this problem by providing users with a seamless and integrated travel experience that allows them to conveniently access and use different modes of transportation seamlessly through a single digital platform with a united payment or monthly subscription bundle (Jittrapirom et al. 2017). Previous studies of MaaS, as well as other related smart mobility services, are largely dependent on assessing the potential changes in travel preferences and travel behaviors, which shows the complicated mechanism of MaaS competing with or supplementing current public transit, sustainable travel modes, and private cars (e.g., Alyavina, Nikitas and Njoya 2020; Alonso-González et al. 2020; Rongen et al. 2022; Lopez-Carreiro et al. 2021; Kim et al. 2021; Lopez-Carreiro, Monzon, and Lopez-Lambas 2021; Qiao and Yeh 2023a); interactions between the popularity of smart mobilities and urban environment (e.g., Qiao and Yeh 2021; Qiao, Huang, and Yeh 2022; Bian, Qiao, and Yeh 2023); and consequent impacts on social-spatial justice (e.g., Bian and Qiao 2024; Qiao, Zhang, and Yeh 2023; Qiao and Yeh 2023b; Qiao 2022). For instance, connecting active transport of shared bikes or e-scooters to rail stations would expand the service area of urban rail systems (Shaheen and Chan 2016).

However, the use of MaaS is still in its early stage, and there is a general lack of discussion on the potential impacts on the built environment systematically. Thus it is a need to clarify the concept of MaaS and examine its potential contribution to the built environment. This could help to guide future urban planning and management towards sustainability and social inclusion (Qiao, Huang, and Yeh 2022).

This chapter starts by distinguishing MaaS from similar travel systems and defining the current research paradigm in Section 2. Section 3 aims to provide a comprehensive review of the use of smart technologies in the operation of MaaS. Section 4 will then discuss the potential contributions of MaaS to the built environment in a smart way, including accessibility improvements, energy consumption reductions, and carbon emissions reductions. By providing an

DOI: 10.1201/9781003383840-17

overview of the state-of-the-art research on MaaS and its potential contributions to the built environment, this chapter aims to better understand this emerging concept and its role in creating sustainable smart cities.

16.2 Clarifying MaaS and other concepts

MaaS is relatively new and is still evolving rapidly. In past discussions of MaaS, multiple concepts and definitions have often been intertwined and partially overlapped. Thus, it is essential to clarify what the MaaS concept is and how to distinguish it from similar related travel systems, including multimodal travel (MT), Mobility-on-Demand (MOD), and shared mobility (SM). While these systems share some similarities with MaaS, there are also vital differences in the scope of services provided and the level of integration of various travel modes.

Mobility-as-a-Service (MaaS). As shown in Figure 16.1, MaaS refers to an innovative transportation concept that seeks to integrate various forms of transport services into a single mobility platform to provide seamless door-to-door mobility service on demand (Hietanen 2014). It emphasizes a unified payment process or monthly subscribe bundle for integrating different transport modes and services under one digital platform to break the barriers of integrating various mobility services. By offering a range of transport options under one platform, such as public transit, bike-sharing, and ride-hailing services, MaaS aims to increase the share of

Figure 16.1 Similarities and differences of multiple concepts

public transport by providing users with a seamless, convenient, and personalized travel experience directly from origin to destination. This approach encourages the use of mobility services instead of owning private vehicles, ultimately reducing traffic congestion and promoting sustainable urban development.

Multimodal travel (MT). MT refers to the free use of multiple modes of transportation, such as public transit, taxis, ride-hailing services, and micro-mobility options such as shared bikes and e-scooters during a single trip within a certain time period (Kuhnimhof, Chlond, and Von Der Ruhren 2006). With the goal of providing a more efficient, cost-effective, and environmentally friendly way to travel, MaaS typically integrates MT as the core component, which helps to optimize and maximize the use of existing transportation services.

Mobility-on-Demand (MoD). MoD is a user-centric transportation model that focuses on providing real-time, flexible, and personalized mobility solutions to meet the diverse demands of travelers, rather than relying on a fixed schedule or owning the transport assets (Campisi et al. 2021). With the help of advanced technologies like mobile applications and data analytics, MoD services allow users to access various on-demand transport services, such as ride-hailing, ridesharing, and bike-sharing, whenever and wherever they need, without owning the transport assets. This on-demand approach not only enhances the efficiency and convenience of urban transportation but also promotes the shift towards shared mobility and sustainable transport systems. In virtue of timely service response according to online order requests, MaaS platforms typically offer real-time information on transportation options, allowing users to book and pay for transport services, including MoD services, through a single digital platform in a short time in advance of a trip (Shaheen and Cohen 2020).

Shared mobility (SM). SM refers to transportation services that allow users to share vehicles such as cars or bikes, as well as trips such as rides among users, either simultaneously or one after another. Examples of shared mobility services include car-sharing, bike-sharing, and ridesharing. Traditional public transport modes like metro and bus are also considered as a kind of shared mobility service (Board 2017). SM aims to maximize the utilization of transportation resources and promote more sustainable and efficient transportation options. MaaS also integrates on-demand bus service as a core travel choice in suburban or low-density areas to connect to the rail system in the city center, which helps balance between the costs and benefits of implementing public transport services (Hensher 2017).

While each of these concepts has unique aspects, they all share a common goal of promoting sustainable, efficient, and user-centric transportation systems. By integrating MT and leveraging advanced technologies of MoD and SM, the current paradigm of the study of MaaS on urban built environments focuses on the understanding of the impacts of these emerging concepts under the umbrella of the integrated various aspects of urban life, such as transportation efficiency (Shaheen and Chan 2016; Wong, Hensher, and Mulley 2020), sustainability (Albatayneh et al. 2020; Alyavina, Nikitas, and Njoya 2020), land use (Rantasila 2015), and social equity (Chen and Sheldon 2016; Pangbourne et al. 2020; Becker et al. 2020). It also explores ways to optimize the integration of different transportation modes and services to create more accessible, resilient, and sustainable urban transport systems. Researchers in this field are particularly interested in examining the potential of MaaS to promote a shift away from private vehicle ownership towards more sustainable and efficient forms of mobility (Kamargianni et al. 2018; Ho, Mulley, and Hensher 2020; Polydoropoulou et al. 2020), as well as the role of urban planning and policy in facilitating this transition (Pangbourne et al. 2018; Audouin and Finger 2018; Hirschhorn et al. 2019).

16.3 Smart technologies in the operation of MaaS

MaaS operates on a digital platform, such as a smartphone app, to support an integrated way for planning multimodal trips and payment. In the operation of a MaaS platform, the user inputs their destinations for journey planning, and the platform plans, optimizes, and provides various trip options, considering factors such as travel time, cost, and transfer convenience. The operation of MaaS service involves multiple smart technologies to guarantee the positive experience of passengers, efficiency of the services, and outcomes of the transportation systems.

16.3.1 Dynamic dispatching of fleet

Dynamic dispatching is one of the most crucial algorithms in the transportation system, which aims at optimizing the allocation of the fleet and improving the efficiency of the transportation system. The basic concept of dynamic dispatching is to allocate and relocate the transportation resources responsively, such as vehicles and drivers, to match the spatiotemporally dynamic transportation demand in real time (Gendreau and Potvin 1998). For instance, the dynamic dispatching of for-hire vehicles, repositioning of shared bikes, and scheduling of subways and busses. Through advanced prediction or real-time monitoring of travel demand and vehicle supply, the dynamic dispatching turns the matching problem into an operation problem subjected to a series of constraints, such as the capacity and fleet size of vehicles, accepted waiting time of passengers, maximum road speed, and minimum departure interval (Pillac et al. 2013). Drivers/vehicles are dispatched to serve passengers to meet one or several objectives, such as maximizing served passengers and occupancy and/or minimizing fleet size and passengers' waiting time.

The optimization nature of dynamic dispatching determines its potential benefits to the urban environment from two aspects. On the one hand, it assigns drivers to passengers with the optimal combination and route so that unnecessary deadheading and detouring time for vehicles can be avoided, and excessive cars can be reduced. For example, Vazifeh et al. (2018) proposed an optimal vehicle dispatching and empty vehicle repositioning algorithm and found a 30% reduction in vehicle fleet size without leading to any trip delay. Then, the reduction in vehicle fleet size can consequentially contribute to the reduction in car emissions (Yu et al. 2020). On the other hand, the improvement in the experience of passengers, such as the waiting time reduction, will reduce the cost of using these public transportation modes – such as subway, bus, bike, and for-hire vehicles – and decrease private car trips and their related emissions (Washbrook, Haider, and Jaccard 2006; Salon 2009).

In recent years, the advancement of information communication technologies (ICTs), big data, and artificial intelligence has brought further development of dynamic dispatching. For the for-hire vehicle, the prevalence of mobile phones and ICTs enables dynamic dispatching to capture the real-time spatiotemporal heterogeneous travel demand more accurately through the mobile phone app (Alemi et al. 2018), gradually transforming traditional cruising services to demand responsive services that efficiently fulfill the travel demand. Meanwhile, the advancement in big data and artificial intelligence algorithms empowers dynamic dispatching to predict short-term demand and better plan the vehicles and drivers to work out optimal dispatching (Ke et al. 2017; Yao et al. 2018; Qin et al. 2021). In terms of shared bikes and scooters, dynamic repositioning strategies are developed based on the accurate prediction algorithm, benefiting from the massive bike/scooter demand big data with spatial and temporal tags (Legros 2019; Ham et al. 2021). Due to the rebalancing of bikes and scooters, they are more accessible during the rush period, and they won't excessively accumulate around hot destinations such as subway stations and crowd out the public space at other times (Liu et al. 2020). For the subway, due to the advancement in the algorithms, studies can not only accurately predict regular demand but also forecast the demand under special events (Li et al. 2017; Chen et al. 2019). Based on the prediction, subway schedule and crowd control measurements can be adjusted and improved to ensure service performance. For MaaS, its service performance can be improved by implementing dynamic dispatching in each of its component services, such as ride-hailing, ride-splitting, taxi, bike-sharing, bus, and subway. However, challenges also come from better dynamic dispatching vehicles to coordinate the connections between different modes in MaaS, which has not been discussed in depth. For example, Huang, Kockelman, and Garikapati (2022) found that 87% of passengers can be delivered to their stations in time using a dynamic pooling algorithm tightly coordinated with train arrivals, while uncoordinated assignments deliver just 58%.

16.3.2 Models of balance between supply and demand

The mismatch between demand and supply can result in the inefficiency of the transportation system, as an over-intensive demand will greatly increase the waiting time of passengers. In contrast, an excessive supply will lead to vehicle waste. The determinants of the mismatch between demand and supply are mainly from two aspects. On the one hand, the concentrated commuting activities in the morning and afternoon and the leisure activities in the evening lead to a sudden increase in travel demand, significantly exceeding the capacity of the supply side. To overcome the great pressure from the sudden change in demand, efforts are made to manage and distribute the demand. First, one core concept of MaaS is to take public transit as the backbone transportation mode (Arias-Molinares and García-Palomares 2020; Kamargianni and Melinda 2018). Compared to other transportation modes, public transit has a higher capacity to carry more passengers, higher flexibility to increase the frequency to alleviate the long wait time, and, therefore, better performance to meet the intensive travel demand. However, historically, the planning process of public transit has not considered equity (Welch and Mishra 2013), which may lead to the accessibility equity between communities with different characteristics, such as income (Liu et al. 2022). Meanwhile, the shock of COVID-19 shows the vulnerability of public transit (Liu, Miller, and Scheff 2020). Then, dynamic pricing in ride-sourcing is found to reduce the excessive travel demand or increase the number of drivers for ride-sourcing during peak time to alleviate the mismatch. For example, Castillo, Knoepfle, and Weyl (2017) found that a surge in prices during peak time can help to reduce the excessive demand for ride-sourcing and avoid the wild-goose-chase (WGC) of vehicles. Meanwhile, Chen and Sheldon (2016) observed that dynamic fares help increase the payment to drivers during peak time so that there will be more drivers to balance the mismatch between demand and supply. However, the adoption of dynamic pricing induces an equity problem, which may hinder low-income groups from using the shared mobility service (Giorgione, Ciari, and Viti 2020). Lastly, MaaS contributes to the balance between demand and supply by providing a series of transportation modes. For example, shared bike, scooter, and ride-sourcing acting as a feeder service helps to solve the first/last mile problem of public transit, extending the catchment area of public transit (Shaheen and Chan 2016) and concentrating these demands to public transit, which alleviates the mismatching between demand and supply of other services. Additionally, when these feeder transportation modes are mainly utilized as the supportive transportation mode in MaaS, their occupied time for each trip will be shorter since they do not need to fulfill the entire trip, which can be regarded as an increase in supply for these transportation modes (Ke et al. 2020).

On the other hand, although the supply is sufficient, there can also be a mismatch between demand and supply because of spatial mismatch (Tang et al. 2019). The spatial mismatch between demand and supply is mainly solved by spatially reallocating transportation resources, such as vehicles, bikes, and scooters. For example, the supply distribution can be better optimized and dispatched by the aforementioned dynamic dispatching algorithm by accurately capturing the current demand and predicting the future demand and the optimal vehicle dispatch. Xu et al. (2020) proposed an online driver repositioning system, which was found to significantly help drivers find passengers faster and increase the income of drivers by 2%. For shared bikes and scooters, a manual vehicle reposition is necessary because the bikes and scooters often encroach on the public spaces around public transit stations during the morning peak hour. Still, it is typically hard to find an available vehicle during the evening peak hour (Xu et al. 2019). To solve this problem, heuristic and reinforcement learning are commonly used (Forma, Raviv, and Tzur 2015; Ho and Szeto 2014; Ghosh et al. 2017).

16.3.3 Smart traffic management system in MaaS

MaaS is not a simple bundling of a series of transportation modes but an efficient integration of these modes. Therefore, MaaS requires a smart traffic management system to not only optimize the efficiency of each component mode but also to coordinate the integration between them so that MaaS can achieve its expected sustainable environmental impact on urban transportation systems. The modern transportation service is smart for robust data analytic support. The data collection system comprises the foundation of the MaaS smart traffic management system. Regarding data collection systems, traditional coils, cameras, and radar collect the basic traffic flow information at the road intersection or curbside to reflect the travel intensity. Meanwhile, other sensors also help collect real-time data from each transportation mode to improve the efficiency of MaaS. For instance, the precise spatial location of vehicles provided by the GPS terminal helps with a precise estimation of bus arrival time (Lin and Zeng 1999), contributing important information to the seamless transportation modes transfer in MaaS.

Additionally, GPS data is also used for asset management for car-sharing and bike-sharing in MaaS to detect whether the vehicle is returned within the designated virtual geofencing, which is also possible via Bluetooth or Wi-Fi transmitter (Qiao, Huang, and Yeh 2022). In addition to the data collected by the sensor, the transportation services in MaaS also keep generating extensive and massive service data, such as the trip order data and trajectory data of ride-hailing, bike-sharing, and car-sharing and the smart card data of subway and bus. These data are crucial for transportation analytics. For example, the trip ridership data of ride-hailing and bike-sharing with spatial and temporal stamps is the most fundamental data for trip demand prediction (Ke et al. 2017; Yao et al. 2018; Liang, Huang, and Zhao 2022), which is the prerequisite to dynamic dispatching. Besides, they are also quite useful for analyzing travel behaviour, environmental impact, and human activity (Henao and Marshall 2019; Tirachini 2020; Pelletier, Trépanier, and Morency 2011) to better develop the MaaS service.

A smart analytic system is another important component of smart traffic management in MaaS, which aims to improve service efficiency. By utilizing the MaaS data, the smart analytic system of MaaS can provide several services. On the one hand, by integrating the schedules of different transportation modes, a route planning service helps the user to plan a customized itinerary consisting of multiple services, which have been implemented on many navigations and MaaS apps. Meanwhile, the arrival time estimation service is provided by the location information of the GPS terminal and prediction of the artificial intelligent algorithm (Chen 2018; Wang, Fu, and Ye 2018; Yu, Lam, and

Tam 2011) to reduce passengers' uncertainty and anxiety during waiting for the transportation service and give confidence to the user in intermodal mobility (Barreto, Amaral, and Baltazar 2020). On the other hand, the smart analytic system works to accurately predict the spatiotemporally dynamic traffic demand and optimize the distribution and fleet size of vehicles to coordinate the demand-supply balance of component modes, facilitating the mode complementary and avoiding the mode competition to ensure the efficiency of MaaS (Arias-Molinares and García-Palomares 2020).

16.4 Potential contributions of MaaS to the built environment

Implementing MaaS has the potential to contribute to the built environment in several ways. One potential contribution is accessibility improvements, as MaaS can provide more efficient and affordable transportation options for people with disabilities and those living in underserved areas. Another potential contribution is through energy consumption reductions, as MaaS can encourage the use of more sustainable modes of transportation and reduce reliance on personal vehicles. MaaS can also contribute to carbon emissions reductions by promoting the use of cleaner transportation options – such as electric vehicles – and reducing the need for personal vehicles. Meanwhile, spatial heterogeneity is an important factor that may affect the overall carbon emission and complement competing effects diverse in suburbs and city centers.

16.4.1 Accessibility improvements through MaaS

Transportation accessibility refers to the easiness of reaching desired destinations and services using various transportation modes, which ensures equitable and efficient mobility for individuals and communities (Morris, Dumble, and Wigan 1979). Improving transportation accessibility has the potential to enhance quality of life, promote economic development, and reduce social disparities, such as increased social inclusion, better opportunities for disabled populations, and reduced environmental impacts by encouraging sustainable travel choices (Church and Marston 2003; Farrington and Farrington 2005; Kwok and Yeh 2004).

Despite the great importance of accessibility, it is challenging to achieve high accessibility, especially in suburban areas. On the one hand, due to the poor planning of urban areas, the development of public transportation often falls behind urban sprawl, and the transportation infrastructures are poorly designed to facilitate passenger accessibility (Silva and Larsson 2018). On the other hand, due to the limitation of transportation resources and consideration of service efficiency, the suburbs often have poorer accessibility. For example, due to a relatively lower travel demand in the suburban areas, the public transit service area typically will not cover these areas, and the idle ride-hailing, taxis, and bikes are not prone to be repositioned to these areas (Clewlow and Mishra 2017).

MaaS integrates different transportation modes into a unified service accessible via a single mobile phone app and plays a critical role in promoting inclusivity by addressing the specific mobility needs of various user groups. For example, prior studies found the accessibility of suburbs and rural areas can benefit from the implementation of MaaS because the for-hire vehicle, bike, or bus service in MaaS can act as a feeder to connect these distant areas to the accessible public transit network (Barreto, Amaral, and Baltazar 2018; Eckhardt et al. 2018; Hasselwander et al. 2022). Additionally, Campisi et al. (2021) found that implementing MaaS may help serve elderly and poor residents. One successful MaaS project that has significantly improved accessibility is "Whim" in Helsinki, Finland, which has made it easier for users to navigate the city by integrating public transit, taxi services, car rentals, and bike-sharing into a single app (Weinreich and Riihelä 2019).

16.4.2 Public/parking space savings through MaaS

Transport takes up 35–50% of urban land through roads, parking, and associated infrastructure (Rodrigue 2020). The increasing number of private cars exerts great pressure on public spaces, including green spaces, pedestrian areas, and other community amenities. MaaS has the potential to alleviate parking congestion by offering comparable or even superior mobility services (Hesselgren, Sjöman, and Pernestål 2020).

Notably, most of the private cars used for commuting are parked idle. MaaS could help to reduce private car usage on the road by providing shared mobility services and optimized travel routes, especially with ride-sharing services. The number of cars and, thereby, the road and parking space, has been saved (Ströhle, Flath, and Gärttner 2019). It has been estimated that only 40% of taxis would be needed to serve all the trips in New York if an optimal algorithm was implemented (Vazifeh et al. 2018). A similar study in Shenzhen showed that the car fleet needs merely 17% of current private cars without delaying passengers' pick-up times, and around 1 million parking spaces can be released. In particular, 0.13 million cars are enough in working days and 0.1 million cars for weekends to meet existing travel demands, considering 350 million car ownership in this city (Yue et al. 2022). This would fundamentally reduce the demands of private cars and free up valuable public space that would otherwise be used for parking lots or garages.

16.4.3 Energy consumption and carbon emission reductions through MaaS

As reported by the International Energy Agency (IEA), private cars and vans were responsible for more than 25% of global oil use (IEA 2023). MaaS can help to reduce energy consumption and carbon emissions by offering shared mobility options, optimizing travel routes, and promoting the use of electric vehicles.

MaaS promotes the use of shared mobility, which helps reduce the number of vehicles on the road and reduces energy consumption and carbon emissions. The Organization for Economic Co-operation and Development reports estimate that the wide adoption of shared mobility can lead to a 30% reduction in carbon emissions in urban areas by 2050 (ITF 2019). Bike-sharing provides an alternative mode of transportation for short trips typically made by car, leading to decreased traffic congestion and energy consumption (Shaheen et al. 2011; Shaheen, Cohen, and Martin 2013). Existing studies have found that bike-sharing could reduce motor vehicle mileage in high bike substitution rates cities (Fishman, Washington, and Haworth 2014). Ride and car sharing also provide opportunities for reducing energy consumption and carbon emissions with low costs (Akimoto, Sano, and Oda 2022; Cai et al. 2019). It was found that the global marginal Co_2 abatement cost in 2050 for the 2°C target could be greatly reduced under the ride-sharing and car-sharing scenario (Akimoto, Sano, and Oda 2022). However, the impacts of shared mobility on car use usually vary in city characteristics, such as bike/car dependency rate, population density, and infrastructure condition (Kurisu et al. 2023).

MaaS platforms seamlessly integrate various transportation modes and provide users with optimized routes and options, greatly improving transport efficiency. Besides, MaaS platforms can analyze data from various sources, such as traffic patterns, weather conditions, and user preferences, to provide a personalized travel plan (Butler, Yigitcanlar, and Paz 2021). By avoiding congested areas and selecting the shortest paths, MaaS can minimize the time and distance traveled, thereby reducing energy consumption and emissions. Reducing the distance and time needed to reach charging stations through MaaS also helps reduce energy consumption. The service that provides convenient access to charging stations for electric vehicles is also known as charging as a service (CaaS) (Zhang et al. 2020). Existing research has found that an optimal CaaS system could reduce the waiting time and save the travel distance to visit a charging station (Guo et al. 2020). By integrating with renewable energy generation and storage systems, CaaS can further reduce reliance on non-renewable energy.

Another critical way that MaaS helps reduce energy consumption and carbon emissions is by incentivizing the use of electric vehicle fleets, which are more energy efficient than traditional gasoline-powered vehicles (Wu et al. 2015; Albatayneh et al. 2020). If electricity can be generated from renewable energy, the wide adoption of electric vehicles would significantly reduce fossil fuel consumption in transport in the long run, contributing to a more sustainable energy system. To achieve the goal of net zero emissions, it is crucial to promote the low-carbon transformation of electric vehicles throughout their entire lifecycle in the future. This transition would not only lead to a significant reduction in carbon emissions in the transport sector but contribute to a more sustainable energy system.

16.5 Future developments

MaaS has the potential to play a significant role in changing the built environment by integrating different transport modes and services under one digital platform using smart technologies. By providing a more seamless and efficient transportation system, one potential impact of MaaS on the built environment is increased accessibility, especially for suburbs and rural areas. Another potential impact of MaaS on the built environment is reduced energy consumption and carbon emissions. By promoting public transport and shared mobility options, MaaS can help reduce the number of cars on the road, which in turn can reduce greenhouse gas emissions. Additionally, MaaS can facilitate using low-carbon transport options such as electric vehicles and bikes. MaaS can also reduce the need for car ownership and usage in the city center, which in turn can lead to a reduction in the demand for parking spaces and road infrastructure. Reducing demand for cars and car-related infrastructure can free up valuable space in cities, which can then be repurposed for other uses.

The impact of the current MaaS on the transport system has only been studied in a preliminary form, and there is room for further work here, especially in rural areas. With the wider use of MaaS, how traffic flow will develop and, if necessary, how traffic routes will be adjusted and, ultimately, how the urban built environment will be changed are questions that must be studied more deeply, especially through simulation. We can only find out the real impacts of MaaS on the built environment by observing the real changes in traffic flows and transport usage when it is fully implemented.

References

Akimoto, K., F. Sano, and J. Oda. 2022. Impacts of ride and car-sharing associated with fully autonomous cars on global energy consumptions and carbon dioxide emissions. *Technological Forecasting and Social Change* 174.

Albatayneh, A., M. N. Assaf, D. Alterman, and M. Jaradat. 2020. Comparison of the overall energy efficiency for internal combustion engine vehicles and electric vehicles. *Rigas Tehniskas Universitates Zinatniskie Raksti* 24 (1):669–680.

Alemi, F., G. Circella, S. Handy, and P. Mokhtarian. 2018. What influences travelers to use Uber? Exploring the factors affecting the adoption of on-demand ride services in California. *Travel Behaviour and Society* 13:88–104.

Alonso-González, M. J., S. Hoogendoorn-Lanser, N. van Oort, O. Cats, and S. Hoogendoorn. 2020. Drivers and barriers in adopting Mobility as a Service (MaaS) – A latent class cluster analysis of attitudes. *Transportation Research Part A: Policy and Practice* 132:378–401.

Alyavina, E., A. Nikitas, and E. T. Njoya. 2020. Mobility as a service and sustainable travel behaviour: A thematic analysis study. *Transportation Research Part F: Traffic Psychology and Behaviour* 73:362–381.

Arias-Molinares, D., and J. C. García-Palomares. 2020. The Ws of MaaS: Understanding mobility as a service fromaliterature review. *IATSS Research* 44 (3):253–263.

Audouin, M., and M. Finger. 2018. The development of mobility-as-a-service in the Helsinki metropolitan area: A multi-level governance analysis. *Research in Transportation Business & Management* 27:24–35.

Barreto, L., A. Amaral, and S. Baltazar. 2018. Mobility as a Service (MaaS) in rural regions: An overview. Paper read at 2018 International Conference on Intelligent Systems (IS).

Barreto, L., A. Amaral, and S. Baltazar. 2020. Mobility in the era of digitalization: Thinking Mobility as a Service (MaaS). In Jardim-Goncalves, R., Sgurev, V., Jotsov, V., and Kacprzyk, J. (eds.), *Intelligent systems: Theory, research and innovation in applications. studies in computational intelligence*, vol 864: Springer, Cham. https://doi.org/10.1007/978-3-030-38704-4_12

Becker, H., M. Balac, F. Ciari, and K. W. Axhausen. 2020. Assessing the welfare impacts of shared Mobility and Mobility as a Service (MaaS). *Transportation Research Part A: Policy and Practice* 131:228–243.

Bian, F., and S. Qiao. 2024. Challenging the gender neutrality of on-demand mobility platforms. *Annals of the American Association of Geographers*: 1–23.

Bian, F., S. Qiao, and A. G.-O. Yeh. 2023. Mobility resilience: Transit-oriented development, ride-hailing, and car ownership. *Transportation Research Part D: Transport and Environment* 123:103924.

Board, T. R. 2023. *TCRP Report 188: Shared Mobility and the Transformation of Public Transit* 2017 [cited Oct 14, 2023]. Available from https://www.trb.org/Main/Blurbs/174653.aspx.

Butler, L., T. Yigitcanlar, and A. Paz. 2021. Barriers and risks of Mobility-as-a-Service (MaaS) adoption in cities: A systematic review of the literature. *Cities* 109:103036.

Cai, H., X. Wang, P. Adriaens, and M. Xu. 2019. Environmental benefits of taxi ride sharing in Beijing. *Energy* 174 (May 1):503–508.

Campisi, T., C. Garau, G. Acampa, F. Maltinti, A. Canale, and M. Coni. 2021. Developing flexible mobility on-demand in the era of mobility as a service: An overview of the italian context before and after pandemic. Paper read at International Conference on Computational Science and Its Applications.

Castillo, J. C., D. Knoepfle, and G. Weyl. 2017. Surge pricing solves the wild goose chase. Paper read at Proceedings of the 2017 ACM Conference on Economics and Computation.

Chen, C.-H. 2018. An arrival time prediction method for bus system. *IEEE Internet of Things Journal* 5 (5):4231–4232.

Chen, E., Z. Ye, C. Wang, and M. Xu. 2019. Subway passenger flow prediction for special events using smart card data. *IEEE Transactions on Intelligent Transportation Systems* 21 (3):1109–1120.

Chen, M. K., and M. Sheldon. 2016. Dynamic pricing in a labor market: Surge pricing and flexible work on the Uber platform. *Ec* 16:455.

Church, R. L., and J. R. Marston. 2003. Measuring accessibility for people with a disability. *Geographical Analysis* 35 (1):83–96.

Clewlow, R. R., and G. S. Mishra. 2017. *Disruptive transportation: The adoption, utilization, and impacts of ride-hailing in the United States*. UC Davis: Institute of Transportation Studies. Available from https://escholarship.org/uc/item/82w2z91j

Eckhardt, J., L. Nykänen, A. Aapaoja, and P. Niemi. 2018. MaaS in rural areas-case Finland. *Research in Transportation Business & Management* 27:75–83.

Farrington, J., and C. Farrington. 2005. Rural accessibility, social inclusion and social justice: towards conceptualisation. *Journal of Transport Geography* 13 (1):1–12.

Fishman, E., S. Washington, and N. Haworth. 2014. Bike share's impact on car use: Evidence from the United States, Great Britain, and Australia. *Transportation Research Part D: Transport and Environment* 31:13–20.

Forma, I. A., T. Raviv, and M. Tzur. 2015. A 3-step math heuristic for the static repositioning problem in bike-sharing systems. *Transportation Research Part B: Methodological* 71:230–247.

Gendreau, M., and J.-Y. Potvin. 1998. Dynamic vehicle routing and dispatching. In *Fleet management and logistics*, 115–126. Springer.

Ghosh, S., P. Varakantham, Y. Adulyasak, and P. Jaillet. 2017. Dynamic repositioning to reduce lost demand in bike sharing systems. *Journal of Artificial Intelligence Research* 58:387–430.

Giorgione, G., F. Ciari, and F. Viti. 2020. Dynamic pricing on round-trip carsharing services: travel behavior and equity impact analysis through an agent-based simulation. *Sustainability* 12 (17):6727.

Guo, S., X. Qian, and J. Liu. 2020. Charging-as-a-Service: On-demand battery delivery for light-duty electric vehicles for mobility service. *arXiv preprint.* https://doi.org/10.48550/arXiv.2011.10665

Ham, S. W., J.-H. Cho, S. Park, and D.-K. Kim. 2021. Spatiotemporal demand prediction model for e-scooter sharing services with latent feature and deep learning. *Transportation Research Record* 2675 (11):34–43.

Hasselwander, M., J. F. Bigotte, A. P. Antunes, and R. G. Sigua. 2022. Towards sustainable transport in developing countries: Preliminary findings on the demand for mobility-as-a-service (MaaS) in metro Manila. *Transportation Research Part A: Policy and Practice* 155:501–518.

Henao, A., and W. E. Marshall. 2019. The impact of ride-hailing on vehicle miles traveled. *Transportation* 46 (6):2173–2194.

Hensher, D. A. 2017. Future bus transport contracts under a mobility as a service (MaaS) regime in the digital age: Are they likely to change? *Transportation Research Part A: Policy and Practice* 98:86–96.

Hesselgren, M., M. Sjöman, and A. Pernestål. 2020. Understanding user practices in mobility service systems: Results from studying large scale corporate MaaS in practice. *Travel Behaviour and Society* 21:318–327.

Hietanen, S. 2014. Mobility as a service. *The New Transport Model* 12 (2):2–4.

Hirschhorn, F., A. Paulsson, C. H. Sørensen, and W. Veeneman. 2019. Public transport regimes and mobility as a service: Governance approaches in Amsterdam, Birmingham, and Helsinki. *Transportation Research Part A: Policy and Practice* 130:178–191.

Ho, C. Q., C. Mulley, and D. A. Hensher. 2020. Public preferences for mobility as a service: Insights from stated preference surveys. *Transportation Research Part A: Policy and Practice* 131:70–90.

Ho, S. C., and W. Szeto. 2014. Solving a static repositioning problem in bike-sharing systems using iterated tabu search. *Transportation Research Part E: Logistics and Transportation Review* 69:180–198.

Huang, Y., K. M. Kockelman, and V. Garikapati. 2022. Shared automated vehicle fleet operations for first-mile last-mile transit connections with dynamic pooling. *Computers, Environment and Urban Systems* 92:101730.

IEA. 2023. *Oil 2023.* IEA, Paris. https://www.iea.org/reports/oil-2023, Licence: CC BY 4.0.

ITF. 2019. *ITF transport outlook 2019.* OECD Publishing, Paris. https://doi.org/10.1787/transp_outlook-en-2019-en.

Jittrapirom, P., V. Caiati, A. M. Feneri, S. Ebrahimigharehbaghi, M. J. Alonso-González, and J. Narayan. 2017. Mobility as a service: A critical review of definitions, assessments of schemes, and key challenges. *Urban Planning* 2 (2):13–25.

Kamargianni, M., M. Matyas, W. Li, and J. Muscat. 2018. *Londoners' attitudes towards car-ownership and Mobility-as-a-Service: Impact assessment and opportunities that lie ahead.* MaaSLab – UCL Energy Institute for Transport for London.

Kamargianni, M., and M. Melinda. 2018. Exploring individual preferences for mobility as a service plans: A mixed methods approach. *MaaSLab Working Paper* (Series Paper No.18–02).

Ke, J., H. Yang, X. Li, H. Wang, and J. Ye. 2020. Pricing and equilibrium in on-demand ride-pooling markets. *Transportation Research Part B: Methodological* 139:411–431.

Ke, J. T., H. Y. Zheng, H. Yang, and X. Q. Chen. 2017. Short-term forecasting of passenger demand under on-demand ride services: A spatio-temporal deep learning approach. *Transportation Research Part C-Emerging Technologies* 85:591–608.

Kenworthy, J. R. 2003. Transport energy use and greenhouse gases in urban passengertransport systems: A study of 84 global cities. Paper Read at International sustainability conference.

Kim, Y., E.-J. Kim, S. Jang, and D.-K. Kim. 2021. A comparative analysis of the users of private cars and public transportation for intermodal options under Mobility-as-a-Service in Seoul. *Travel Behaviour and Society* 24:68–80.

Kuhnimhof, T., B. Chlond, and S. Von Der Ruhren. 2006. Users of transport modes and multimodal travel behavior: Steps toward understanding travelers' options and choices. *Transportation Research Record* 1985 (1):40–48.

Kurisu, K., K. Tsuji, J. Nakatani, and Y. Moriguchi. 2023. What are important factors to determine CO_2 reduction by car sharing? Simulation of car-sharing impact in cities with different car dependencies considering variable uncertainty. *Resources, Conservation and Recycling* 193.

Kwok, R. C., and A. G. Yeh. 2004. The use of modal accessibility gap as an indicator for sustainable transport development. *Environment and Planning A* 36 (5):921–936.

Legros, B. 2019. Dynamic repositioning strategy in a bike-sharing system; how to prioritize and how to rebalance a bike station. *European Journal of Operational Research* 272 (2):740–753.

Li, Y., X. Wang, S. Sun, X. Ma, and G. Lu. 2017. Forecasting short-term subway passenger flow under special events scenarios using multiscale radial basis function networks. *Transportation Research Part C: Emerging Technologies* 77:306–328.

Liang, Y., G. Huang, and Z. Zhao. 2022. Joint demand prediction for multimodal systems: A multi-task multi-relational spatiotemporal graph neural network approach. *Transportation Research Part C: Emerging Technologies* 140:103731.

Lin, W.-H., and J. Zeng. 1999. Experimental study of real-time bus arrival time prediction with GPS data. *Transportation Research Record* 1666 (1):101–109.

Liu, D., M.-P. Kwan, J. Huang, Z. Kan, Y. Song, and X. Li. 2022. Analyzing income-based inequality in transit nodal accessibility. *Travel Behaviour and Society* 27:57–64.

Liu, L., H. J. Miller, and J. Scheff. 2020. The impacts of COVID-19 pandemic on public transit demand in the United States. *PLoS One* 15 (11):e0242476.

Liu, Y., Y. Ji, T. Feng, and Z. Shi. 2020. Use frequency of metro – Bikeshare integration: Evidence from Nanjing, China. *Sustainability* 12 (4):1426.

Lopez-Carreiro, I., A. Monzon, D. Lois, and M. E. Lopez-Lambas. 2021. Are travellers willing to adopt MaaS? Exploring attitudinal and personality factors in the case of Madrid, Spain. *Travel Behaviour and Society* 25:246–261.

Lopez-Carreiro, I., A. Monzon, and M. E. Lopez-Lambas. 2021. Comparison of the willingness to adopt MaaS in Madrid (Spain) and Randstad (The Netherlands) metropolitan areas. *Transportation Research Part A: Policy and Practice* 152:275–294.

Morris, J. M., P. L. Dumble, and M. R. Wigan. 1979. Accessibility indicators for transport planning. *Transportation Research Part A: General* 13 (2):91–109.

Pangbourne, K., D. Stead, M. Mladenović, and D. Milakis. 2018. The case of mobility as a service: A critical reflection on challenges for urban transport and mobility governance. In *Governance of the smart mobility transition*, 33–48: Emerald Publishing Limited.

Pangbourne, K., M. N. Mladenović, D. Stead, and D. Milakis. 2020. Questioning mobility as a service: Unanticipated implications for society and governance. *Transportation Research Part A: Policy and Practice* 131:35–49.

Pelletier, M.-P., M. Trépanier, and C. Morency. 2011. Smart card data use in public transit: A literature review. *Transportation Research Part C: Emerging Technologies* 19 (4):557–568.

Pillac, V., M. Gendreau, C. Guéret, and A. L. Medaglia. 2013. A review of dynamic vehicle routing problems. *European Journal of Operational Research* 225 (1):1–11.

Polydoropoulou, A., I. Tsouros, I. Pagoni, and A. Tsirimpa. 2020. Exploring individual preferences and willingness to pay for mobility as a service. *Transportation Research Record* 2674 (11):152–164.

Qiao, S. 2022. Understanding ride-hailing and inventing future transit: Pathways to spatial justice. *HKU Theses Online (HKUTO)*.

Qiao, S., and A. G.-O. Yeh. 2021. Is ride-hailing a valuable means of transport in newly developed areas under TOD-oriented urbanization in China? Evidence from Chengdu City. *Journal of Transport Geography* 96:103183.

Qiao, S., and A. G.-O. Yeh. 2023a. Is ride-hailing competing or complementing public transport? A perspective from affordability. *Transportation Research Part D: Transport and Environment* 114:103533.

Qiao, S., and A. G.-O. Yeh. 2023b. Mobility-on-demand public transport toward spatial justice: Shared mobility or Mobility as a Service. *Transportation Research Part D: Transport and Environment* 123:103916.

Qiao, S., G. Huang, and A. G.-O. Yeh. 2022. Mobility as a Service and urban infrastructure: From concept to practice. *Transactions in Urban Data, Science, and Technology* 1 (1–2):16–36.

Qiao, S., M. Zhang, and A. G.-O. Yeh. 2023. Mind the gender gap in ride-hailing from the demand side. *Journal of Transport Geography* 107:103531.

Qin, G., Q. Luo, Y. Yin, J. Sun, and J. Ye. 2021. Optimizing matching time intervals for ride-hailing services using reinforcement learning. *Transportation Research Part C: Emerging Technologies*, 129:103239.

Rantasila, K. 2015. The impact of mobility as a service concept to land use in Finnish context. Paper read at 2015 International Conference on Sustainable Mobility Applications, Renewables and Technology (SMART), at Kuwait, Kuwait.

Rodrigue, J. P. 2020. *The geography of transport systems*. Routledge.

Rongen, T., T. Tillema, J. Arts, M. J. Alonso-González, and J.-J. Witte. 2022. An analysis of the mobility hub concept in the Netherlands: Historical lessons for its implementation. *Journal of Transport Geography* 104:103419.

Salon, D. 2009. Neighborhoods, cars, and commuting in New York city: A discrete choice approach. *Transportation Research Part A: Policy and Practice* 43 (2):180–196.
Shaheen, S., and N. Chan. 2016. Mobility and the sharing economy: Potential to facilitate the first-and last-mile public transit connections. *Built Environment* 42 (4):573–588.
Shaheen, S., and A. Cohen. 2020. Mobility on Demand (MOD) and Mobility as a Service (MaaS): Early understanding of shared mobility impacts and public transit partnerships. In *Demand for emerging transportation systems*, 37–59: Elsevier.
Shaheen, S. A., A. P. Cohen, and E. W. Martin. 2013. Public bikesharing in North America: early operator understanding and emerging trends. *Transportation Research Record* 2387 (1):83–92.
Shaheen, S. A., H. Zhang, E. Martin, and S. Guzman. 2011. China's Hangzhou public bicycle: understanding early adoption and behavioral response to bikesharing. *Transportation Research Record* 2247 (1):33–41.
Silva, C., and A. Larsson. 2018. Challenges for accessibility planning and research in the context of sustainable mobility. Paper read at International Transport Forum Discussion Paper.
Ströhle, P., C. M. Flath, and J. Gärttner. 2019. Leveraging customer flexibility for car-sharing fleet optimization. *Transportation Science* 53 (1):42–61.
Tang, J., Y. Zhu, Y. Huang, Z.-R. Peng, and Z. Wang. 2019. Identification and interpretation of spatial – Temporal mismatch between taxi demand and supply using global positioning system data. *Journal of Intelligent Transportation Systems* 23 (4):403–415.
Tirachini, A. 2020. Ride-hailing, travel behaviour and sustainable mobility: An international review. *Transportation* 47 (4):2011–2047.
Vazifeh, M. M., P. Santi, G. Resta, S. H. Strogatz, and C. Ratti. 2018. Addressing the minimum fleet problem in on-demand urban mobility. *Nature* 557 (7706):534–538.
Wang, Z., K. Fu, and J. Ye. 2018. Learning to estimate the travel time. Paper read at Proceedings of the 24th ACM SIGKDD International Conference on Knowledge Discovery & Data Mining.
Washbrook, K., W. Haider, and M. Jaccard. 2006. Estimating commuter mode choice: A discrete choice analysis of the impact of road pricing and parking charges. *Transportation* 33:621–639.
Weinreich, M., and A. Riihelä. 2019. Insights from the world's first MOBILITY-as-a-Service (MaaS) system. Paper read at Proceedings from the Annual Transport Conference at Aalborg University.
Welch, T. F., and S. Mishra. 2013. A measure of equity for public transit connectivity. *Journal of Transport Geography* 33:29–41.
Wong, Y. Z., D. A. Hensher, and C. Mulley. 2020. Mobility as a service (MaaS): Charting a future context. *Transportation Research Part A: Policy and Practice* 131:5–19.
Wu, X., D. Freese, A. Cabrera, and W. A. Kitch. 2015. Electric vehicles' energy consumption measurement and estimation. *Transportation Research Part D: Transport and Environment* 34:52–67.
Xu, D., Y. Bian, J. Rong, J. Wang, and B. Yin. 2019. Study on clustering of free-floating bike-sharing parking time series in Beijing subway stations. *Sustainability* 11 (19):5439.
Xu, Z., C. Men, P. Li, B. Jin, G. Li, Y. Yang, C. Liu, B. Wang, and X. Qie. 2020. When recommender systems meet fleet management: practical study in online driver repositioning system. Paper read at Proceedings of The Web Conference 2020.
Yao, H., F. Wu, J. Ke, X. Tang, Y. Jia, S. Lu, P. Gong, J. Ye, and Z. Li. 2018. Deep multi-view spatial-temporal network for taxi demand prediction. Paper read at Proceedings of the AAAI Conference on Artificial Intelligence, at New Orleans, Lousiana, USA.
Yu, B., W. H. Lam, and M. L. Tam. 2011. Bus arrival time prediction at bus stop with multiple routes. *Transportation Research Part C: Emerging Technologies* 19 (6):1157–1170.
Yu, Q., H. Zhang, W. Li, X. Song, D. Yang, and R. Shibasaki. 2020. Mobile phone GPS data in urban customized bus: Dynamic line design and emission reduction potentials analysis. *Journal of Cleaner Production* 272:122471.
Yue, Y., Xie, S., Wang, Y., Jin, Y., & Li, Q. 2022. How Many Private Cars Can Be Reduced by the On-Demand Car-Sharing Services at a Metropolitan Scale?. Available at SSRN 4048913. https://dx.doi.org/10.2139/ssrn.4048913
Zhang, X., Y. Cao, L. Peng, J. Li, N. Ahmad, and S. Yu. 2020. Mobile charging as a service: A reservation-based approach. *IEEE Transactions on Automation Science and Engineering* 17 (4):1976–1988.

17

INTEGRATED SMART CURB MANAGEMENT IN URBAN BUILT ENVIRONMENTS ADDRESSING HUMAN DYNAMICS AND DISRUPTIVE TECHNOLOGIES

Yan Wang[1]

[1]Assistant Professor, Dept. of Urban and Regional Planning and Florida Institute for Built Environment Resilience, University of Florida, P.O. Box 115706, Gainesville, FL 32611. Email: yanw@ufl.edu

17.1 Introduction

Cities around the world are experiencing a period of significant advancements in electric vehicles (EVs), autonomous vehicles (AVs), and the technology found inside vehicles like radars, cameras, sensors, computing systems, communication systems, the global positioning system (GPS), mobile devices, digital maps, and various driving-assistance apps (National Academies of Sciences & Medicine, 2020). However, the infrastructure and overall environment in which these vehicles operate have not kept up with these advancements (U.S. Department of Transportation, 2021). The introduction of new vehicle technologies and on-demand mobility services (such as carsharing, bikesharing, ridesharing, and microtransit) has particularly put pressure on the curbside areas in urban centers. These curbside areas are now facing competition for space due to activities like pick-ups and drop-offs, food and package delivery, freight loading, scooter and bike parking, and EV charging (Brown et al., 2020; Butrina et al., 2020; Curb Control, n.d.; Curbside Management TPP, n.d.; Mitman et al., 2018).

Curbside areas serve as a crucial connection point between road traffic and pedestrian walkways in urban communities. They provide access for people to move between vehicles and different destinations, and they also house infrastructure for various transportation options (Curb Control, n.d.). Figure 17.1 demonstrates a curbside parking area at UF Innovate Center, City of Gainesville, FL, USA. However, the lack of coordination and the absence of smart operating systems in these environments have resulted in increased congestion (Weinberger et al., 2020), safety concerns (Dumbaugh & Li, 2010), and higher greenhouse gas emissions related to transportation (K. Zhang & Batterman, 2013; Aftab et al., 2020).

Therefore, this book chapter aims to review existing research and practices that address the planning and management innovations of smart curb environments. This will involve the integration of vehicles, humans (passengers and pedestrians), mobile devices/apps, and traditional infrastructure,

DOI: 10.1201/9781003383840-18

Figure 17.1 Curbside at UF Innovate Center, City of Gainesville, FL, USA

as well as new physical and cyber infrastructures that emerge with the introduction of EVs, mobility on demand (MODs), and artificial intelligence (AI). These innovations have the potential to revolutionize the utilization of curb space by vehicles and people, to ensure fair and convenient access to social infrastructure (such as local businesses, entertainment and education centers, medical services, and parks). They also seek to enhance curb productivity for charging and pick-ups and drop-offs (PUDOs), while simultaneously reducing greenhouse gas (GHG) emissions for the well-being of communities and promoting sustainability. These advancements can also help alleviate congestion and crowding issues, ultimately enhancing the livability of community residents.

17.2 Literature review

Curbside management has received increasing attention from academia and industry due to the changes that we have discussed in the introduction. The field has gained momentum since 2018 according to statistics generated by Web of Science (Shoup, 2006). The fast-growing literature body covers research areas across environmental science, engineering, transportation, computer science, urban studies, automation, and behavioral science among others. However, curbside management is not a new topic at all because extant studies have focused on the management of on-street parking of private vehicles (Shoup, 2006).

The review of this chapter will focus on the "smart" components of the curb, curb planning, and management across four cutting-edge but also cross-cutting topics, including (a) an overview of smart curbside management, (b) developing smart curb digital twin (focused on new data collection), (c) spatial planning of curb spaces across the macro-built environment, and (d) dynamic curbside management in micro/localized built environment. These largely cover many data-driven empirical studies and technical developments at the intersection of curb environment and artificial intelligence. The review progresses with curb management in more fine-grained space (macro-built environment vs. localized environment) and time (longer-term planning vs. short-term or even real-time management) units. It focuses on studies that intended to address the more intensified and diversified curb uses rather than topics like curb lanes for bus/stops or bicycles, curbsides at airports, and pure curbside parking.

17.2.1 Smart curbside management

Spatial planning of curbside locations addresses the relative longer-term allocation of curbside spaces with static functions. However, there is a need to create more flex zones at curb areas to address diverse demands that may be impeded by static curb functional zones. Generally speaking, curbside management concerns the "planning, measuring, managing, allocating, and enforcement of the curb uses and users by a government agency" to achieve the optimization goals of mobility and safety across user groups (IPMI Research and Innovation Task Force, 2021). The recent report (National Academies of Sciences & Medicine, 2022) published by National Academies has identified four key components of curb management, namely, (a) context (e.g., adjacent land uses), (b) uses and users of the curb, (c) time, and (d) space. The chapter will take two types of curbside locations in the City of Gainesville, FL as illustration examples. Curbside, which is located in the downtown area nearby restaurants, needs to prioritize customer access, food/supply deliveries, and accommodate pick-up and drop-offs. In comparison, curbsides within the campus needs to allow the parking of vehicles with different permits (faculty, staff, or students), EV charging, e-scooter/e-bike parking, and occasionally reserved parking for guests or construction vehicles. The designation of curb function may vary depending on the time of day, day of the week, and adjacent land use, among other considerations.

17.2.2 Smart Curb Digital Twin

Digitalizing physical facilities and environments, as well as capturing dynamic uses, is a crucial aspect of creating Smart Curb Digital Twins. Though the concept of smart cities Digital Twin has been adopted in different engineering management domains (Taylor et al., 2021), limited efforts have been devoted to digitizing the curb environment. Only some researchers have recently explored methods to achieve this objective. Hao and Wang (2022) addressed the lack of digital inventories of curb environments by proposing a data-driven framework that utilizes computer vision and Google Street View (GSV) imagery. They employed a state-of-the-art semantic segmentation model, DeepLab V3, pre-trained on the CityScapes dataset, to detect curb facilities from GSV images. The Inverse Perspective Mapping (IPM) technique was then utilized to estimate the spatial location of each detected facility, followed by spatial processing to aggregate and filter the estimation results. Their framework was successfully demonstrated in the Innovation District of Gainesville, FL and provides an intelligent approach to inventorying curbs, particularly beneficial for resource-limited small and medium-sized cities. Salazar-Miranda et al. (2023) proposed a real-time framework to measure on-street pedestrian activities and transportation modes. The framework combines machine learning and computer vision methods to identify various forms of street activity from images collected in real time by a moving vehicle. It trains the street activity detection model using Google Street View (GSV) and MS COCO images. This method can be used to distinguish between different types of pedestrian activity and other curb uses, such as the number of cars and micro-mobility modes in a street, and it provides an accurate and detailed way to study street use. In addition to the image-based approach, studies have utilized emerging "big" data from various sensors to understand and track curb usage patterns. For example, Wang et al. (2022) have integrated longitudinal parking transaction and violation records (obtained from the city office), place visitation data (GPS-based), and geospatial data of parking assets (open geospatial data from government websites) to retrieve different curb use patterns and conflicts across time and space.

17.2.3 Spatial planning of curb spaces and locations across macro built environment

Spatial allocation of curb zones tends to be ad-hoc or incremental. Sometimes, curb zones are designated due to new needs, but there have not been useful tools to help the systematic planning of curb zones over the years to adapt to new transportation demands and to minimize congestion/delays

(Butrina et al., 2020; Salewski et al., 2012). The ad-hoc planning approach cannot address curb management problems systematically and may cause other issues, for example, meeting the needs of certain types of curb uses but missing others or solving the problem at one curb but burdening nearby ones (Hao et al., 2023). Some recent work has proposed data-driven or simulation-based methods to allocate curbs and/or their functions across space. Most of the research has focused on a specific context/community value (e.g. pandemic) (Mandhan & Gregg, 2023), type of curb use such as PUDOs (Wang et al., 2022), EV charging (Guo & Wang, 2023), specific regulation/policy (e.g. dwell time, speed limit, pricing), or mobility type.

Here are a few recent works to address the specific aspects of spatial planning of curbs. Mandhan and Gregg (2023) conducted comparative case studies of curbside cafes in Toronto and Chicago to evaluate how two cities with different legacies of regulation surrounding curbside parking spaces reacted and enabled the adaptation of curbside parking spaces during the pandemic. A particular discussion focuses on transferring curbside space from the public realm into the private during the pandemic. They qualitatively analyzed the regulation aspects of curbside function adaptation influenced by the prioritized **community value** to support urban economics in crises for curb management. Wang et al. (2022) intended to address the increasing curbside uses of PUDO activities due to Mobility-on-Demand (MoD) services and proposed a data-driven agent-based simulation approach to plan designated PUDO zones in limited public curbside spaces. They conducted a case study for five street blocks in urban cores of the City of Gainesville, Florida, United States, based on longitudinal parking transaction and violation records, place visitation data, and geospatial data of parking assets. The results show that **temporary PUDO zones** should be designated at all investigated blocks during peak hours even when the MoD market penetration rate is low (i.e., 10%), which helps mitigate the occurrences of competing use events. Permanent PUDO zones should be designated for the two busiest blocks when the MoD market penetration rates increase to 30% or 50%. Sensitivity analyses suggest that the designated PUDO zones can mitigate curbside stresses more effectively when regulating MoD users' PUDO dwell time (e.g., within one minute). This research aims to contribute to data-driven public asset managerial decision-making and strategies in smart cities and benefits more accessible and sustainable living environments in urban cores.

Pochowski et al. (2022) developed an estimated demand-based curb allocation tool that allocates curb space with demand constraints using an integer linear programming approach. The tool creates an output for the number of feet of curbside space necessary to accommodate the demand across different transportation modes, including ride-hailing services, commercial loading, on-street parking, transit stops, micro-mobility uses, and non-transportation uses such as parklets or "streateries" (listed based on priority). It also provides some practical methods to estimate and maximize both societal values and economic values across curb use types. However, the research has simplified the societal value as the number of person trips related to the five transportation modes and the number of people served for the non-transportation use. Due to their limitations of economic data to take both direct and indirect curb use economics, the economic value was measured by direct economic benefits in consumer spending. Additionally, the curb space demand was mainly estimated based on some sociodemographic factors using regression models, which cannot capture the dynamic demands in real situations that may respond to different variables such as pricing and time limit.

Machado-León et al. (2023) investigated the dwell time of over 6,000 passenger vehicles' PUDO passengers in Seattle and found this key metric of curb management can be significantly affected by the number of passenger maneuvers, location and time of day, and traffic and operation management factors. They also assessed the effects of two curb management strategies on dwell time. They found that adding PUDO zones, together with geofencing, was found to be related to faster PUDO operations and improved productivity at the curb. However, the effect is heterogeneous, and more observation data needs to be obtained for different street conditions (e.g. location of stop sign, split of vehicle types) and traffic congestion levels. The research also did not fully consider the co-use situation when PUDO competes with other curb uses. Also, Liu et al. (2023) explored optimal curbside pricing for managing ride-hailing PUDO with a sensitivity analysis-based method to minimize the total social cost of a bi-modal regional traffic network among all trips. The pricing would influence travelers' modal choices and parking location choices. The pricing strategy is a bi-level optimization with the upper level to minimize total social costs for all travelers in the network, and the lower level is the bi-modal user equilibrium.

A very recent study (Hao et al., 2023) has proposed to integrate spatiotemporal deep learning into scenario planning to anticipate possible diverse curb uses across time and space. They have developed a place-specific graph-based deep learning approach, i.e., MultiGCN-LSTM, to capture the spatial, temporal, and semantic dependencies between curb spaces, functions, regulations (e.g., meter time limits/rates), built-environment features (e.g. nearby mobility facilities, POI, pedestrian/auto accessibility), and curbside parking for the City of Gainesville. The model was also calibrated to the City of San Francisco, which had a different curb management challenge in the downtown area. The study demonstrated multi-task learning with data collected for two types of curb uses, i.e., curbside parking and docked bike-sharing, from San Francisco, providing an integrated understanding of diverse curb uses. The model aims to capture the spatial, temporal, and semantic dependencies between curb spaces, curb functions, parking regulations (e.g., meter time limits/rates), and built-environment features (e.g., nearby mobility facilities, points of interest, pedestrian/auto accessibility) to predict curbside parking. However, there are opportunities for future work to address the following: (a) the model has not been calibrated to anticipate more diverse curb uses across time and space. New data collection of diverse curb uses such as micro-mobility, EV charging, docked bike-sharing, commercial loading, and PUDO can further enhance calibration; (b) the model did not assess the on-road traffic impact of curb allocation; (c) changing weather and construction conditions (e.g., road and nearby buildings) were excluded, despite their influence on curb uses; and (d) the graph was built upon existing curbside parking zones without considering changed/new curb locations. Overall, the study presents a novel approach integrating deep learning into scenario planning for diverse curb uses, but there are areas for future improvement and validation in real-world settings.

17.2.4 Dynamic curbside management for micro and localized built environment

In addition to the research that focused on spatial planning (i.e., static) reviewed in the previous section, the increasingly intensified and diversified curb use demands have driven more dynamic curbside management to allocate regulations to distinct curb spaces dynamically for optimized curb uses (National Academies of Sciences & Medicine, 2022). Such management of dynamic curbside spaces has become increasingly important to optimize curb productivity while considering community values in micro/localized built environments. Flexible regulations and time limitations are essential in dynamically allocating curb spaces based on a data-driven understanding of existing curb uses. Leveraging community values such as livability, accessibility, economic vitality, equity, safety, air quality, and congestion management is also crucial for effective curb management (National Academies of Sciences & Medicine, 2022). Innovative technologies have also been explored to facilitate real-time adjustments and accommodate diverse parking behaviors.

Particularly, AI-enabled technologies, such as reinforcement learning (RL), have been proposed to address real-time online curbside parking space assignments, especially for short-term parking demands such as PUDO.

For instance, Ye et al. (2022) focused on effectively managing the fixed capacity of curb space to accommodate a mixed flow of parking behaviors across PUDOs, docking, and on-street parking. They utilized a partially learned deep deterministic policy gradient (DDPG) algorithm to train a model that approximates sequential dispatching decisions addressing whether to accept the requests and where to accommodate them. The learning objectives are to maximize the rates of parking service and curbside occupancy. The RL model successfully distinguished various types of parking demands and prioritized PUDOs and docking requests. The model was proposed under the context of an intelligent transport system with a particular focus on shared autonomous vehicles that carry out the three types of demands; each has a different parking duration assumed at 1.5 minutes, 10 minutes, and 60 minutes. However, the real-world situation can be complex as vehicles have different automation levels with various types of sensors and un-unified parking durations across use types. Different curb environments may also require distinct RL models to consistently improve the policy. Community values such as the safety of pedestrians and passengers, equity, and urgency level, among other local socioeconomic factors, may also need to be leveraged in the optimization objectives. Additionally, for some destinations, several close-by candidate curbside areas can be coordinated in the assignment process using multi-agent RL.

Zhang et al. (2022) focused on the context of coexisting connected vehicles (CV) and non-connected vehicles (NCV) to propose a multi-agent deep reinforcement learning framework to generate efficient online parking assignment strategies with partial observations of parking demand. The two agents include one that measures the impact of NCVs and the other for exploring the parking characteristics of CVs. The research presented a fine-designed DRL formulation that includes agent appointment, state description, action selection, and a reward function. They designed an efficient multi-agent learning algorithm (QMIXm) with a modified exploitation strategy for directing exploration based on the nature of the problem to improve the efficiency of network training. The model was not developed specifically for curbside uses but has an implication for curbs in similar vehicle connectivity contexts; however, the unique challenges with diverse/intensified curb uses have not been addressed in existing model architecture.

Yin et al. (2022) addressed curb space management from the perspective of infrastructure autonomy, which has distinguished curbside space management from the parking problem. It developed and extended an RL model-advantage actor-critic (A2C algorithm) – for a curbside space agent that leverages partial CV observation of traffic state and controls the physical allocation of space to CV vs. non-CV. The reward that an agent would receive was set as a minus number of reroutes for all vehicles in all curbs. The model allows for (a) multiple types of vehicles that compete for space and (b) mixed vehicle connectivity scenarios. Its performance was validated at an isolated blockface with synthetic parking demand. The research mainly differentiated two types of curb parking vehicles: short stops (e.g., food delivery, ride-sourcing) vs. longer-duration parking (e.g., commuters and loading trucks). It assumes that only commercial delivery vehicles are connected. Again, the real-world nuanced curb use challenges have not been fully addressed in this context and it also does not consider any community values. Technically, the model can be further calibrated with more complex curb demand patterns and address across-curb coordination using multi-agent RL. It can also include more realistic traffic and curb geometrics to simulate the curb use dynamics. Future research should consider these factors and develop comprehensive models that account for nuanced curb demand patterns, multi-agent coordination, realistic traffic, and curb geometrics to simulate curb use dynamics effectively.

17.3 Smart curb planning and management framework

Based on the existing curb management challenges and the literature review, this chapter summarizes three sets of research and management tasks for building a Smart Curb Management framework leveraging big data and artificial intelligence in the following subsections. Before proposing smart management solutions, the first step is to fully understand the existing curb environments, their diverse uses, any conflicting or competing usages, users and behavioral patterns, local regulations, and policies, among others.

Task 1: developing a localized curb environment involves investigating the challenges associated with multi-faceted curb uses, as well as safety, congestion, and accessibility issues. The data collection may leverage existing resources and data from local government, transportation agencies, and data companies. Community engagement, social sensing, and citizen surveys are also key to identifying curbside management challenges and needs. This research task aims to address the gaps in the existing literature by leveraging heterogeneous data analytics, modern sensors, and AI algorithms to create new curb management and operational strategies.

Task 2: building a framework for community-wide curbside environments focuses on distinct types of curbside uses such as EV charging. For example, the research task combines cyberinfrastructure (including sensors, networking, data centers, and mobile apps) with physical infrastructure (such as available charging stations) to coordinate EV charging requests. The objective is to minimize travel and waiting time, alleviate congestion, reduce environmental impact, and maximize curb productivity and accessibility for all users. Similar approaches can be developed for other curb uses, such as pick-up and drop-off activities and micro-mobility parking.

Task 3: establishing a real-time computing and analytics platform to support the curb management tasks. This involves designing an edge-cloud configuration that ensures responsiveness and scalability. Additionally, this research task may investigate data, human, and vehicle privacy issues. The goal is to develop and experiment with novel algorithms and computing systems that can effectively support operations in both localized and community-wide curb environments. Potential management solutions and strategies may include reservations, parking-space co-use, designating flex zones, geo-fencing, and floating parking coordination among others.

Overall, this smart curb planning and management framework aims to address the challenges of dynamic curb management and optimize curb productivity while considering community values, safety, congestion, and accessibility. By leveraging advanced technologies and data-driven approaches, the framework strives to create efficient and sustainable curb environments.

17.4 Prospects and challenges of the respective field

The nature of the problem of building smart curb environments and coordinating curb space uses is intrinsically social and technological. Future research could include synergistic components in urban planning, social science, engineering, and computer science to address community-specific curb challenges. This chapter has summarized several key prospects for interdisciplinary efforts leveraging new data and techniques:

Equity and commuting well-being in smart curb management. Curb use equity and commuting well-being are essential aspects of curb management that require careful consideration. While equity is recognized as an important community value in curb planning and management, there is a significant gap in quantitative studies addressing equity and how it should be integrated into decision-making processes, particularly regarding use priorities. The question of who should have the privilege to use limited curb space and for what purpose remains unresolved, and there is a need to establish criteria for measuring equity in curb allocation. Traditionally, equity in curb use has focused on addressing the needs of vulnerable populations, such as individuals with disabilities, through the provision of reserved parking spaces. However, it is important to redefine equity at different time resolutions. At a finer time scale, such as real-time, equity considerations can be expanded to accommodate more nuanced situations and diverse needs. For instance, the parking needs of a parent with toddlers in the vehicle may require specific accommodations that contribute to their overall commuting well-being. Introducing new metrics to measure equity is also crucial. In addition to traditional measures of accessibility and fairness, metrics such as time poverty, stress levels, and other commuting well-being indicators (Giurge et al., 2020; Chatterjee et al., 2020) can provide valuable insights into the equitable distribution of curb resources. These metrics can capture the impact of curb management decisions on individuals' overall quality of life and well-being during their commuting experiences. To achieve equity in curb use, it is imperative to engage diverse stakeholders and communities in the planning and decision-making processes. This collaborative approach ensures that the voices and needs of various user groups are considered and integrated into curb management strategies. By embracing equity as a core principle in curb planning, we can create inclusive and accessible urban environments that prioritize the well-being and experiences of all individuals, leading to more equitable and sustainable transportation systems.

Integrating human behavioral science. Integrating human behavioral science into curb management presents both prospects and challenges for the field. Research has revealed the difficulty of modifying commuting behavior, which is habitual and deeply ingrained in individuals' daily routines (Kristal & Whillans, 2020). Similarly, curb parking behaviors, encompassing cruising-for-parking and parking-choice behaviors, exhibit patterns that are influenced by various structural, institutional, environmental, and cognitive factors. Understanding these behaviors systematically is essential to inform the development of smart environment strategies that can adapt to differentiated patterns and needs. The prospects lie in the potential to leverage human behavioral science to shape and influence curb parking behaviors. By comprehensively understanding the underlying factors driving these behaviors, policymakers and urban planners can design interventions that encourage desirable parking practices. For instance, applying theories such as the "theory of planned behavior" can help assess the effectiveness of interventions aimed at changing perceptions and attitudes toward curb parking. Additionally, evaluating the impact of new curb management technologies on behavior can provide valuable insights into their effectiveness and potential for widespread adoption. By systematically understanding and influencing curb parking behaviors, the smart environment can adapt to differentiated patterns and needs. However, addressing challenges related to equity measurement, defining use priorities, and evaluating the behavioral impact of new technologies is crucial for advancing the field. Future research can explore these prospects and challenges to create more equitable, efficient, and user-centered curb management practices that contribute to the overall well-being and sustainability of urban environments.

Agile curb management considers evolving contextual factors, including changing environmental conditions such as weather and construction activities. These factors have a significant impact on curb availability and usage. To effectively manage dynamic curbs, it is important to create a dynamic digital curb inventory that leverages diverse information sources and sensors. This digital inventory enables smart management by providing real-time information on curb availability and utilization. Weather conditions play a role in people's parking behavior, and integrating patterns mined from past data can be valuable in developing a management system that addresses specific needs during adverse weather conditions like heavy rains and extreme heat. By analyzing historical data, the system can anticipate and respond to parking demands and adjust curb allocations accordingly. Furthermore, construction activities, including road closures and nearby construction sites, have a direct impact on curb availability and may require temporary adjustments in curb management. The agile curb management approach recognizes the importance of monitoring and adapting to these changing contextual factors to optimize curb utilization and address emerging challenges. By considering evolving environmental factors and integrating data-driven insights, agile curb management systems can dynamically allocate and manage curbs to meet changing demands, optimize utilization, and enhance overall efficiency.

Expanding both macro- and micro-spatial scale research. Expanding research on both macro and micro-spatial scales is essential for comprehensive curb management. Existing studies have predominantly focused on either the community level, such as an entire city or the local level, such as a specific neighborhood block. However, it is crucial to integrate both scales in practical curb management, as curbs are inherently part of an urban system while also being influenced by local contexts. Managing curbsides in downtown environments, for example, has implications for the overall traffic flow and transportation network of the entire city. The attractiveness of different destinations such as perceived environmental safety, local regulations such as pricing schemes, and other micro-environmental factors significantly shape the usage and effectiveness of curbs. Therefore, when developing strategies to manage curbside functions and uses, it is important to test these strategies at both macro and micro levels to ensure their robustness and sustainability. By considering the macro-level impact on urban systems and the micro-level influence of local contexts, curb management strategies can be designed and implemented with a holistic approach. This integrated perspective enables a better understanding of the interconnectedness between curbs and their surrounding environments, leading to a more effective and informed decision-making processes. Ultimately, expanding research across both macro and micro scales enhances the ability to develop comprehensive and contextually appropriate curb management strategies that optimize curb utilization, improve traffic flow, and support sustainable urban development.

Growing convergent socio-technical research. The field of curb management presents a socio-technical challenge that requires socio-technical solutions. Existing studies in the realm of curb management have been somewhat fragmented, primarily focusing on either planning or engineering aspects. However, to effectively address the complex challenges of curb management, integrated solutions that encompass both social and technical considerations are desired. One example of an emerging approach in curb management is the use of reinforcement learning, which has shown promise in real-time transportation management, including intelligent curbside management. However, it is important to note that the "reward" design in these reinforcement learning models has typically been focused solely on transportation objectives, such as optimizing traffic flow or reducing congestion. Social objectives, such as commuter well-being and reducing commuter stress, have often been overlooked. To advance the field of curb management, there is a need to integrate more community values and social objectives in the design of rewards for intelligent systems based on reinforcement learning. By incorporating a broader range of objectives, including community well-being, equity, and environmental sustainability, we can develop more comprehensive and socially responsive solutions. Convergent socio-technical research in curb management can bridge the gap between the planning and engineering disciplines and foster collaboration among stakeholders with diverse expertise. By bringing together experts from transportation planning, urban design, sociology, psychology, and other relevant fields, we can develop holistic and contextually sensitive approaches to curb management. This interdisciplinary approach will enable us to better understand the social and technical dimensions of curb management and design intelligent systems that not only optimize transportation outcomes but also promote community well-being, reduce inequalities, and enhance the overall quality of life in urban environments.

Striking a balance: smart systems and pricing in curb management policy. When examining curb management strategies, it is essential to consider the balance between smart systems and pricing mechanisms, among other regulations. Though smart curb management systems, including the aforementioned advanced technologies and data-driven approaches, offer valuable tools for optimizing curb use and enhancing efficiency, it may not be feasible to replace pricing mechanisms entirely. Pricing mechanisms, such as parking fees or congestion pricing, have long been recognized as effective tools to manage curb demand and promote more efficient use of limited space. By adjusting prices based on demand and availability, pricing mechanisms can incentivize behavior changes, reduce congestion, and generate revenue for infrastructure investment or other public purposes. On the other hand, smart systems leverage technology, data analytics, and automation to optimize curb functions, improve accessibility, and enhance the overall user experience. Smart systems can provide valuable insights into real-time demand patterns, support efficient use of curb space, and enable effective enforcement of regulations. Rather than replacing pricing mechanisms with smart systems, a more balanced approach would involve integrating both strategies. Pricing mechanisms can still play a crucial role in managing curb demand and encouraging behavior changes, while smart systems can enhance the implementation and effectiveness of these pricing mechanisms. For example, smart systems can facilitate dynamic pricing adjustments based on real-time demand, support seamless payment methods, and enable efficient enforcement and compliance monitoring. Furthermore, it is important to consider that curb management policies should align with broader transportation and urban policy objectives. The goal is not solely to optimize curb use but also to promote equity, accessibility, sustainability, and other community values. Therefore, a comprehensive policy lens should be applied to evaluate the trade-offs and impacts of different strategies, considering the specific context, local priorities, and diverse needs of various stakeholders.

Anticipating the future: addressing the challenges of self-driving vehicles in curb management. As we delve into the realm of self-driving vehicles, it becomes evident that the existing body of user-inspired studies is not sufficient to tackle the challenges that lie ahead. It is imperative to adopt a forward-looking approach and conduct proactive research to proactively plan for revolutionary changes. This forward-looking research involves envisioning different future transportation scenarios that take into account various factors, including the levels of autonomous vehicle (AV) deployments, the evolving market penetration of electric vehicles, the rise of shared mobility services, and other variables that shape the transportation landscape. By considering these dynamic variables, researchers can explore potential implications and devise strategies to optimize curb management in anticipation of the future. For instance, they can examine how different levels of AV deployments might impact curb demand and usage patterns, taking into account factors such as changes in parking behavior, pick-up and drop-off dynamics, and the need for charging infrastructure. Additionally, understanding the evolving market penetration of electric vehicles and shared mobility services will be crucial in determining the required curb infrastructure and developing efficient allocation strategies. To address these future challenges, researchers can employ scenario-based planning to simulate and analyze different transportation futures. These scenarios can encompass a wide range of possibilities, from gradual AV adoption to full-scale autonomous fleets and consider factors such as regulatory frameworks, consumer preferences, and technological advancements. By exploring these scenarios, researchers can gain insights into potential outcomes, identify potential bottlenecks or opportunities, and devise proactive curb management strategies that can accommodate and optimize the operations of self-driving vehicles.

17.5 Conclusion

The adoption of smart and agile curb management using data-driven approaches has the potential to revolutionize our urban environments, transforming the way we utilize and optimize our curbsides. By harnessing the power of diverse data sources, advanced analytics techniques, and emerging technologies, we can unlock new possibilities for enhancing efficiency, accessibility, and sustainability in our cities. The integrated approach allows us to gain a deeper understanding of curb use patterns, anticipate future challenges brought by self-driving vehicles and changing transportation scenarios, and proactively plan for revolutionized changes. It also enables us to embrace flexibility and responsiveness, dynamically allocating curb spaces and adapting to evolving contextual factors, such as weather conditions and construction events. However, successful curb management goes beyond technological advancements; it requires multi-disciplinary collaboration, stakeholder engagement, and the integration of community values and equity considerations. By embracing a holistic and forward-looking perspective, we can create vibrant, people-centric urban spaces that promote seamless mobility, support sustainable transportation choices, and enhance the overall livability of our cities. This chapter of this handbook serves as a comprehensive guide, providing insights, best practices, and practical strategies for implementing smart and agile curb management, ultimately shaping the future of our urban environments.

References

Aftab, K., Kulkarni, P., Shergold, I., Jones, M., Dogramadzi, M., Carnelli, P., & Sooriyabandara, M. (2020). Reducing parking space search time and environmental impacts: A technology driven smart parking case study. *IEEE Technology and Society Magazine*, *39*(3), 62–75.

Brown, A., Klein, N. J., Thigpen, C., & Williams, N. (2020). Impeding access: The frequency and characteristics of improper scooter, bike, and car parking. *Transportation Research Interdisciplinary Perspectives*, *4*, 100099.

Butrina, P., Le Vine, S., Henao, A., Sperling, J., & Young, S. E. (2020). Municipal adaptation to changing curbside demands: Exploratory findings from semi-structured interviews with ten US cities. *Transport Policy*, *92*, 1–7.

Chatterjee, K., Chng, S., Clark, B., Davis, A., De Vos, J., Ettema, D., Handy, S., Martin, A., & Reardon, L. (2020). Commuting and wellbeing: A critical overview of the literature with implications for policy and future research. *Transport Reviews*, *40*(1), 5–34.

Curb Control. (n.d.). *American Planning Association*. Retrieved February 5, 2021, from https://www.planning.org/planning/2019/jun/curbcontrol/

Curbside Management TPP – Google Search. (n.d.). Retrieved February 8, 2021, from https://www.google.com/search?rlz=1C1GCEA_enUS838US838&ei=8TEhYJGGBNC4tAaAiYewCw&q=curbside+management+TPP&oq=curbside+management+TPP&gs_lcp=CgZwc3ktYWIQAzIFCAAQzQIyBQgAEM0C-OgcIABBHELADOgsILhDHARCvARCTAjoCCAA6BggAEBYQHjoFCCEQoAFQw2Vyim9g2nBoAXACeACAAWSIAdwDkgEDNC4xmAEAoAEBqgEHZ3dzLXdpesgBCMABAQ&sclient=psy-ab&ved=0ahUKEwjRwrKiqNruAhVQHM0KHYDEAbYQ4dUDCA0&uact=5

Dumbaugh, E., & Li, W. (2010). Designing for the safety of pedestrians, cyclists, and motorists in urban environments. *Journal of the American Planning Association*, *77*(1), 69–88.

Giurge, L. M., Whillans, A. V., & West, C. (2020). Why time poverty matters for individuals, organisations and nations. *Nature Human Behaviour*, *4*(10), 993–1003.

Guo, Z., & Wang, Y. (2023) Anticipatory planning for equitable and productive curbside electric vehicle charging stations. *Sustainable Cities and Society*, 104962.

Hao, H., & Wang, Y. (2022). Smart curb digital twin: Inventorying curb environments using computer vision and street imagery. *IEEE Journal of Radio Frequency Identification*, *7*, 168–172.

Hao, H., Wang, Y., Du, L., & Chen, S. (2023). Enabling smart curb management with spatiotemporal deep learning. *Computers, Environment and Urban Systems*, *99*, 101914.

IPMI Research and Innovation Task Force. (2021). *Curbside management: Glossary of terms*. International Parking and Mobility Institute.

Kristal, A. S., & Whillans, A. V. (2020). What we can learn from five naturalistic field experiments that failed to shift commuter behaviour. *Nature Human Behaviour*, *4*(2), 169–176.

Liu, J., Ma, W., & Qian, S. (2023). Optimal curbside pricing for managing ride-hailing pick-ups and drop-offs. *Transportation Research Part C: Emerging Technologies*, *146*, 103960.

Machado-León, J. L., MacKenzie, D., & Goodchild, A. (2023). An empirical analysis of passenger vehicle dwell time and curb management strategies for ride-hailing pick-up/drop-off operations. *Transportation*, 1–27.

Mandhan, S., & Gregg, K. (2023). Managing the curb – Public space and use of curbside cafes during the Coronavirus pandemic. *Cities*, *132*, 104070.

Mitman, M. F., Davis, S., Armet, I. B., & Knopf, E. (2018). *Curbside management practitioners guide*. Institute of Transportation Engineers (ITE). https://trid.trb.org/View/1568721

National Academies of Sciences & Medicine. (2020). *Foreseeing the impact of transformational technologies on land use and transportation*. National Academies of Sciences & Medicine.

National Academies of Sciences & Medicine. (2022). *Dynamic curbside management: Keeping pace with new and emerging mobility and technology in the public right-of-way, part 1: Dynamic Curbside management guide and part 2: Conduct of research report*. National Academies of Sciences & Medicine.

Pochowski, A. L., Crim, S., Liu, L., Trask, L., Sherman Baker, C., Woodworth, S., & Klion, J. (2022). Solving the curb space puzzle through the development of a curb space allocation tool. *Transportation Research Record*, *2676*(10), 601–621.

Salazar-Miranda, A., Zhang, F., Sun, M., Leoni, P., Duarte, F., & Ratti, C. (2023). Smart curbs: Measuring street activities in real-time using computer vision. *Landscape and Urban Planning*, *234*, 104715.

Salewski, A. J., Buckley, S., & Weinberger, R. (2012). Regulating curb space: Developing a framework to understand and improve curbside management. (Compendium of Papers). *91st Annual Meeting of the Transportation Research Board, Washington, DC*.

Shoup, D. C. (2006). Cruising for parking. *Transport Policy*, *13*(6), 479–486.

Taylor, J. E., Bennett, G., & Mohammadi, N. (2021). Engineering smarter cities with smart city digital twins. *Journal of Management in Engineering*, *37*(6).

US Department of Transportation. (2019). *Preparing for the future of transportation: Automated vehicles 3.0 (AV 3.0)*. Retrieved February 5, 2021, from https://www.transportation.gov/av/3/preparing-future-transportation-automated-vehicles-3.

Wang, Y., Hao, H., & Wang, C. (2022). Preparing urban curbside for increasing mobility-on-demand using data-driven agent-based simulation: Case study of city of Gainesville, Florida. *Journal of Management in Engineering*, *38*(3), 05022004.

Weinberger, R. R., Millard-Ball, A., & Hampshire, R. C. (2020). Parking search caused congestion: Where's all the fuss? *Transportation Research Part C: Emerging Technologies*, *120*, 102781.

Ye, Q., Feng, Y., Qiu, J., Stettler, M., & Angeloudis, P. (2022). Approximate optimum curbside utilisation for Pick-Up and Drop-Off (PUDO) and parking demands using reinforcement learning. In *2022 IEEE 25th International Conference on Intelligent Transportation Systems (ITSC)*, 2628–2633.

Yin, S., Cui, Z., & Wang, Y. (2022). Reinforcement learning for curbside space management with infrastructure autonomy and mixed vehicle connectivity. In *2022 IEEE 25th International Conference on Intelligent Transportation Systems (ITSC)*, 3276–3282.

Zhang, K., & Batterman, S. (2013). Air pollution and health risks due to vehicle traffic. *Science of the Total Environment*, *450*, 307–316.

Zhang, X., Zhao, C., Liao, F., Li, X., & Du, Y. (2022). Online parking assignment in an environment of partially connected vehicles: A multi-agent deep reinforcement learning approach. *Transportation Research Part C: Emerging Technologies*, *138*, 103624.

18

DIGITAL TECHNOLOGIES FOR CLIMATE-ADAPTIVE SMART URBAN TRANSPORTATION SYSTEMS

Jinying Xu[1], Kristen MacAskill[2], Asaf Tzachor[3] and Catherine Richards[4]

[1]Marie Skłodowska – Curie Fellow, Future Roads Programme, Department of Engineering, University of Cambridge, Cambridge, CB3 0FA, UK. jx314@cam.ac.uk

[2]Associate Professor, Department of Engineering, University of Cambridge, Cambridge, CB2 1PZ, UK. kam71@eng.cam.ac.uk

[3]Associate Professor, Reichman University, Herzliya, Israel. at875@cam.ac.uk

[4]Research Associate, The Centre for the Study of Existential Risk, University of Cambridge, Cambridge, CB2 1SB, UK. cer76@cam.ac.uk

18.1 Introduction

An urban transportation system (UTS) is a complex system of interdependent systems (Ahmed & Dey, 2020; Zhou et al., 2019a). Being one of the most critical infrastructures to support daily human activities, today's UTS is facing intractable challenges including congestion, parking issues, and inefficient traffic management, as well as air and sound pollution. These challenges become overwhelming in the face of hazards and disasters whose probability and impacts are exacerbated by climate change (Kurth et al., 2020; Zhou et al., 2019a). UTSs in developed economies have been through significant development in the 20th and early 21st centuries. Design assumptions have often been based on historical climate conditions. The infrastructure is vulnerable to floods, hurricanes, droughts, extreme heats, and other extreme weather events, and this exposure is changing, often for the worse (Markolf et al., 2019). Meanwhile, the transport sector is criticized as a major energy consumer (consuming 25% of final energy) and environment polluter, responsible for nearly 40% of end-user emissions (International Energy Agency, 2022) which are fuelling climate change. To evade the dilemma, there have been urgent appeals that cities and UTSs should take immediate climate mitigation (greenhouse gas emission reduction) and adaptation (adapting to a changing climate) actions (Bai et al., 2018; Markolf et al., 2019). While it is widely agreed that adaptation and mitigation should be considered at parity (Sharifi, 2021), many cities are taking climate mitigation measures but few have emphasized climate adaptation yet, especially in the transportation sector (Reckien et al., 2018; Sharifi, 2021). However, before we might reach net-zero targets, the historically existing and newly built infrastructure still need to adapt to the

DOI: 10.1201/9781003383840-19

climate change impacts. This article will shed light on the less attended but imminent climate adaptation issues in UTS, which requires the ability to prepare for and respond to weather events, while planning for possible changes in what might be experienced in the future.

How to plan and manage a complex UTS to enhance its adaptability and resilience for the uncertainty and non-stationarity of climate change stressors requires significant research efforts (Markolf et al., 2019; Zawieska & Pieriegud, 2018), especially when considering the interactions between UTS and uncertain human behaviours in extreme weather events. Such investigations require big data (Donovan & Work, 2017a; Zhu et al., 2019), sophisticated models (Ahmed & Dey, 2020), and supercomputing (Ali et al., 2022). The emerging and growing efforts in smart cities, smart transportation, and intelligent transportation systems provide promising vision and technical feasibility to explore climate adaptation in smart UTSs. By integrating new methodologies of data collection and analysis, information management, decision-making, and system control (Wang et al., 2019a; Zawieska & Pieriegud, 2018), a UTS can play a crucial role in climate adaptability toward resilience, reliability, and low-carbon/zero-carbon cities. This chapter also identifies the challenges to be addressed.

18.2 Literature review to map digital technologies for climate-adaptive smart UTS

18.2.1 The value of digital technologies

Urban areas are developing rapidly from a technological viewpoint, and innovative digital technologies create new possibilities for smart urban transportation management (Zawieska & Pieriegud, 2018). High expectations and huge investments have been attached to digital technologies for our future UTS in the context of climate change and other challenges. The most attractive value of digital technologies is that they enable the collection and analysis of larger quantities of data in less time. Data is the fuel driving most smart systems; the amount and quality of data directly determine the performance of system intelligence. In the absence of reliable and adequate data, it is impossible to fully assess the risks associated with hazards and to accurately analyse the feasibility of possible interventions.

The Internet of Things (IoT), weaved by satellites, cameras, sensors (remote, attached, and embedded), communicators, and smart actuators, can capture and control transport activities, generating tremendous data about the fabrics and dynamics of transport systems (Nižetić et al., 2020a). IoT and modelling technologies empower the digital twinning of UTS as a cyber-physical system. Artificial intelligence (AI) can then dig out the behaviour patterns from the big data to enable smart transportation planning, control, and prediction through the cyber-physical system (Wang et al., 2019a). Smart navigation maps and autonomous driving benefit from improved IoT, big data, AI, and geographic information system (GIS) capabilities. These digital technologies, together with the data-driven methods, can also inform climate adaptation interventions by analysing road traffic accidents to decrease casualties; predicting traffic flow in extreme weather events to better plan transportation services and evacuations; optimizing resource utilization to lower carbon; and supporting predictive maintenance to reduce traffic interruptions (Zhou et al., 2019a; Zhu et al., 2019).

Digital technologies can also contribute to the quantitative investigation of human transport behaviours in extreme weather events. Humans are much more complex and difficult to measure than the transport system itself, especially in extreme situations (Zhou et al., 2019a). Uncertainty about the potential impacts of various hazards, added to the unpredictability of human reaction to extreme situations, challenges the resilience and reliability of current human-centred transport systems and calls for the implementation of digital technologies to foster more efficient and accurate data-driven methods (Markolf et al., 2019). Existing relevant research mainly relies on traffic data (e.g., taxi GPS data, smart card data, sensor data of vehicle speeds and densities, or social media data) (Zhu et al., 2019) to speculate the resilience of transport systems in extreme events (Donovan & Work, 2017a). This tends to be at the level of sketching overall system performances at the macro level and does not address behavioural modes at a micro level. Micro-level behaviour data collection and analysis are critical for understanding human behaviour and associated transport demand in extreme events (Pantic et al., 2006), which are the prerequisite for explicit and precise climate adaptation interventions. Analysing travel data collected from smart vehicles, mobile applications, locating devices, and surveillance equipment using big data analysis and AI could assist in painting transportation flows at a micro level; meanwhile, blockchain technology for data encryption could help alleviate privacy concerns.

18.2.2 The requirements of climate-adaptive smart UTS

A UTS is a system of systems knitted with human factors. Using digital technologies to collect well-rounded data including human factors is essential to exploring how to reach transportation goals such as equal availability, effective connectivity, economic reachability, robust reliability, and, more importantly, quick response to disruptions and adaptation to climate challenges (Sterchi et al., 2019). Achieving these goals requires massive data and decision-support information of the UTS status quo, real-time seamless integration between the convenient public transportation subsystems, the handy micro-mobility subsystems, and the flexible intermodal transit subsystems (Oeschger et al., 2020). Each of these subsystems of a UTS have varied strengths and weaknesses when it comes to climate adaptation. Public transportation (e.g., planes, trains, metros, and buses) is more robust against adverse events, more effective for disaster absorption, better for emergency access, and quicker and easier in evacuation (Sharifi, 2021). Micro-mobility subsystems (e.g., walkability, neighbourhood accessibility by scooters and bicycles) can reduce accessibility inequalities and strengthen coping capacity of societies in the first/last mile transport during extreme events. Intermodal transit subsystems include integrated metro and bus service network, as well as intermodal freight transport network, etc., sharing and transferring traffic demand from one mode to another if disruption happens on one mode (Zhou et al., 2019a). Digital technologies can enable real-time information collection and updates at different interdependent subsystems, as well as sharing information among subsystems for their seamless integration. For example, global positioning system (GPS) for estimated arrival time prediction and transport tool locating, near field communication (NFC) for contactless facility access and payment, and smart mobile apps for real-time transit planning and transfer.

18.2.3 Mapping digital technologies for climate-adaptative smart urban transportation systems

This section introduces a map of climate-adaptative smart UTS requirements and key supporting digital technologies. The technology map is summarized in Figure 18.1. The structure follows the data supply chain from data generation and acquisition to data visualization. The section focuses on the broad categories of supporting digital technologies and their specific application to supporting climate-adaptive requirements.

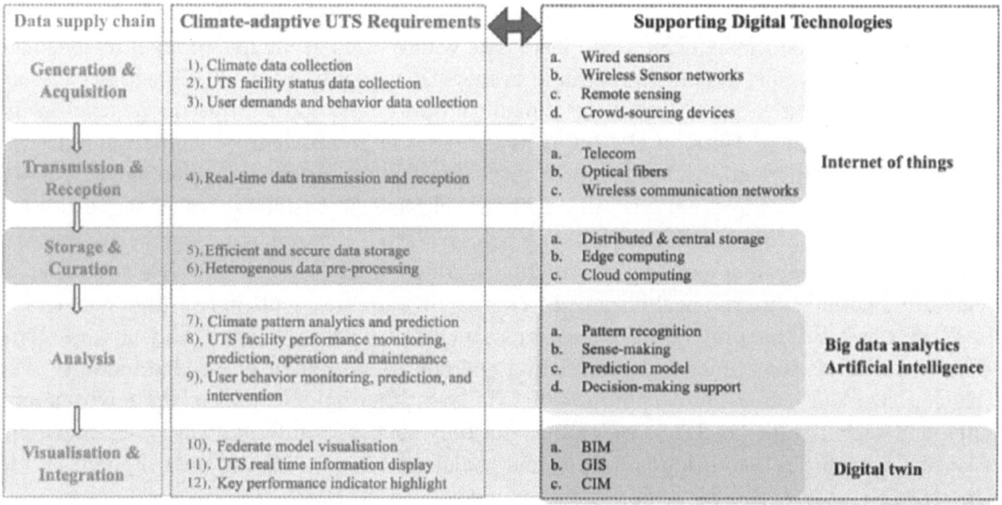

Figure 18.1 Mapping climate-adaptive smart UTS requirements and digital technologies

18.2.3.1 IoT and data supply

IoT can help to generate, acquire, transmit, receive, store, and pre-process data. They have a perception layer made of sensors, actuators, controllers, etc., a transmission layer to transmit the data from perception nodes to information processing units, and an application layer to manage and provide applications globally (Gupta & Quamara, 2020). The IoT provides the platform via which useful data can be sourced for improving the operation of UTSs. In considering climate adaptation there are three core categories of data that might be obtained via the IoT: data about the infrastructure, about the users, and about the climate. Data on the infrastructure network itself might be collected via remote sensing devices, sensor networks, smart vehicles, etc. Data about user behaviour might be collected via sensor networks and smartphones. Climate models are another key source – giving an indication of the probability and severity of future hazards. The volume and quality of data related to facilities, users, and climate can influence the comprehensiveness and accuracy of hazard risk assessments. It can also shape the level of sophistication of analysis of possible interventions in a range of events/climate scenarios.

Climate data is critical for climate adaptation plans and actions in UTS. High resolution satellites can capture global and regional climate dynamics. Various sensors can collect temperature, humidity, wind speed, pressure, concentrations of particulate matters (PM10/PM2.5), CO_2 and other gas levels, etc. Such climate-related data will help to better track weather conditions, air quality, and climate change dynamics.

The UTS facility data that IoT can directly or indirectly collect include the temporal-spatial data of transportation infrastructure (e.g., roads, highways, rails, traffic signs, and lights), other relevant infrastructure (e.g., pipelines, cables, power stations, fuel stations), and vehicles (user vehicles and maintenance vehicles). These various forms of data plot the fabrics and dynamics of UTS subsystems so that we can better monitor the structures and system performance of the UTS (Nižetić et al., 2020b).

Applying IoTs to collect data of human factors is crucial to empower user-centred climate adaptation in UTS. IoTs can enable real-time human factor collection and updates. In existing studies, taxi global positioning system (GPS) data, smart card data, sensor data of vehicle speeds and densities, or social media (Zhu et al., 2019) are widely used at the macro level to speculate the resilience of transport systems in extreme events (Donovan & Work, 2017b). IoTs can contribute to the quantitative investigation of human transport behaviours at the micro level in the extreme weather events (Pantic et al., 2006), which are the prerequisite of explicit and precise climate adaptation interventions. For example, GPS data can be used for real-time human locating and tracking to identify travel patterns, to predict travel plans, to customize travel suggestions in extreme weather, to locate humans and arrange help in rescue. Smartphone usage behaviour data can support extreme weather warning, information sharing among families and friends, evacuation guidance and notification, and health status assessment in extreme events in real-time manners.

IoTs are capable of not only collecting various data but also data pre-processing, storage, communication, display, sometimes edge computing and control with their embedded modules. The decreasing size and increasing capabilities of IoTs boost the digitalization of the environment, facility, and social entities of UTS, which become fully sensed in a digital way. These capabilities lay the fundamentals to bridge physical and social worlds of UTS and their digital twin; to empower big data, AI, and automation; and to unlock the potential in smart UTS confronting climate risks.

18.2.3.2 Big data and pattern recognition

Data collected from the IoT, such as from smart vehicles, smart cards, mobile applications, locating devices, satellites, sensors, and surveillance equipment (Zhu et al., 2019), becomes big data featured by volume, variety, veracity, velocity, and value (5V) (Khan et al., 2021; Xu et al., 2020). Other sources of data collected from survey and modelling on climate and the UTS resilience to climate change, as well as probability and severity of future hazards (Mehryar et al., 2022), also add the 5V of big data. When fused and analysed properly, big data can release huge value to assist UTS to deal with CCA.

Big data can identify hidden patterns that small data cannot, especially in global big cities with complex and scalable UTS. The insights from hidden patterns of climate change, infrastructure health and performance, and transport behaviour can facilitate traffic monitoring and prediction, facility anomaly detection, operation efficiency, and safety improvement, transportation and travel route planning, transportation facility management (Khan et al., 2021; Zhu et al., 2019), resource-based or risk-based efficient decision-making (Land et al., 2021), and assessing climate resilience of transport network (Stamos et al., 2015). With the accumulation of volume, variety, veracity, and velocity, big data may produce new values for UTS. For example, by analysing transport patterns and public health (Koehl, 2021), transportation network companies, and road congestion (Diao et al., 2021), new insights about UTS efficiency will emerge. These emergent insights will help us better understand UTS and make better plans and interventions for climate adaptation.

18.2.3.3 AI and rapid decision-making

On top of IoT, big data, and digital twin, AI can then make sense of the data collected from various sources in heterogenous forms, thereafter collect context and situation awareness, and finally make informed decisions. AI largely include machine learning, deep learning, natural language processing, and computer vision (Xu et al., 2019). Technologies in AI have been advancing rapidly to support the digitalization and intelligent systems development in all areas including UTS. The unstoppable trend is that future UTS will be more and more reliant on AI. AI can be used to dig out the behaviour patterns of transportation systems; to enable smart transportation planning, control, and prediction (Abduljabbar et al., 2019); to simulate interventions for the climate adaptation through road traffic accident analysis to decrease casualties; to predict traffic flow in extreme weather events to better plan transportation services and evacuations; to optimize transportation resources utilization for lower carbon emission; and to reduce traffic interruptions by predictive maintenance of transportation systems (Wang & Ng, 2022; Wang et al., 2019b; Zhou et al., 2019b; Zhu et al., 2019).

The climate-adaptative UTS is not a linear single-objective local optimization problem but a complex multi-modal multi-objective dynamic global optimization. It must borrow the power of strong AI which is fuelled by IoT, big data, and digital twins. Sense-making and decision-making are the two key potentials of AI. Sense-making, very much equivalent to explainable AI (XAI), is a bit more advanced than black-box pattern recognition because it aims to help users make sense of the AI model. The explainability determines the interpretability of the correlation and causation in the AI model. The climate adaptation actions in UTS are highly reliant on the situated sense-making process by humans (Shin, 2021). AI models need to be aware of the context and situation to suggest sensible and appropriate decisions for efficient climate adaptation instead of maladaptation which can create long-term lock-in of vulnerability and risks. After sense-making, AI can help with decision-making by the inclusion of principles, priorities, and cost-benefit considerations. Our society is still at a very early stage of developing AI for climate-adaptive UTS; the efforts shall be continued and improved.

18.2.3.4 Digital twin and data visualization

Rising IoT and digital modelling technologies facilitate the digital twinning of UTS. A digital twin of UTS is a realistic up-to-date digital representation of UTS's physical and functional properties (Bolton et al., 2018; Sacks et al., 2020). With (near) real-time data updating, digital twin fuses various data from the cyber, physical, and social worlds in a federated digital model. The fused digital twin empowers interoperability among systems, high-dimension data visualization, and quantitative analytics in the cyber world. Sacks et al. (2020) provide an outline of the concept of the digital twin in construction and the associated information system workflow, outlining the value that can be added to managing the construction process through integrating this workflow with BIM and AI to create a data-centric process for construction management. This concept could be extended to a different systems-level problem: creating a more data-centric process for improving the capacity to analyse, predict, intervene, and adapt to extreme climate events.

Currently, building information modelling (BIM) and city information modelling (CIM) are typical technologies under the umbrella of digital twins. BIM for transportation infrastructure is widely used for planning, design, project scheduling, construction management, maintenance, structural health monitoring, as-built data registration and documentation, renovation and rehabilitation, and behaviour modelling and prediction across the lifecycle of roads and highways, bridges, railways, etc. (Costin et al., 2018). CIM includes not only infrastructure but also urban elements and environments. Transportation and urban climate modelling are the applications that can harness the power of CIM (Xue et al., 2021). Both BIM and CIM are umbrella technologies that can be integrated with other technologies such as IoT, GIS, big data, and AI.

The digital twin for climate adaptation in UTS would ideally require a grand digital replica to visualize what happened, is happening, and will happen in the physical and social worlds to better understand, analyse, and predict climate change, its impact on UTS, and adaptation plans. The grand digital twin consists of three digital twins: an urban climate digital twin to gain context awareness of urban climate at long-term, short-term, and up-to-date scales; a transportation infrastructure digital twin to gain situation awareness of its status and performance; and a transportation activity digital twin to monitor the social physics of transportation behaviours. Put together, the three digital twins mirror the UTS and its climate context in a digital and quantitative way so that visualized key information can be easily communicated to decision-makers and the mass public for equity, diversity, and inclusion in UTS planning for climate adaptation.

Another requirement to create digital twins is the timeliness of information. Time is life and money, especially in extreme climate events. The timely information update in the digital twin can enable well-informed quick decisions about emergency measures, rescue actions, evacuation guidance, damage assessment, and post-disaster recovery arrangements.

18.3 Climate-adaptive smart UTS deployment

The climate-adaptative UTS needs to address the aforementioned challenges by making the most out of the advanced digital technologies to adapt to climate change and achieve resilience, reliability, and low-carbon/net-zero targets.

18.3.1 Framework and deployment principles

A deployment framework of the climate-adaptative UTS is shown in Figure 18.2. It visualizes a future integrated transport system and illustrates how technologies could support it. The deployment framework encompasses nine key principles that span across strategic urban planning, transportation planning, and technical operation levels. Each principle contributes to the overall effectiveness of the framework in creating a resilient, sustainable, and climate-adaptive urban transportation system.

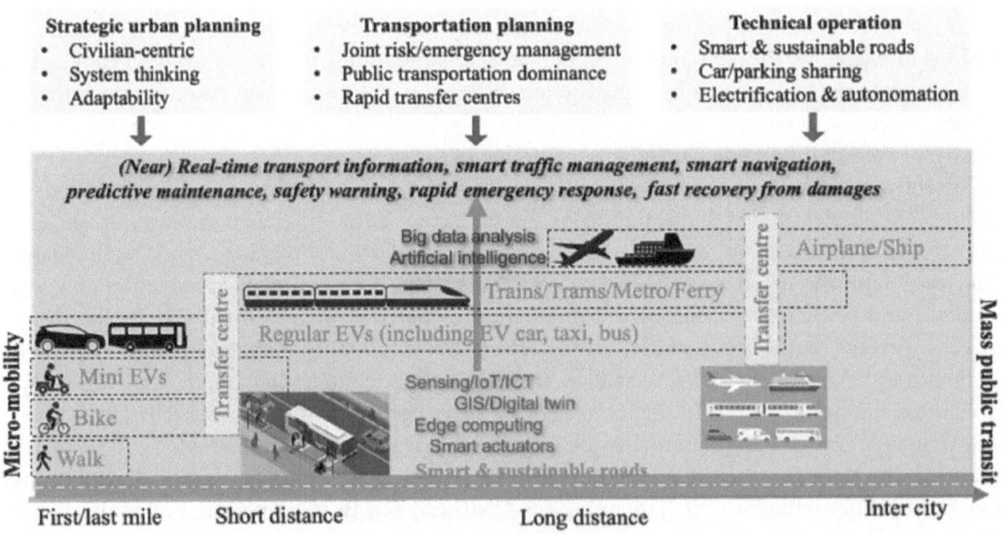

Figure 18.2 Climate-adaptative urban transportation system deployment framework

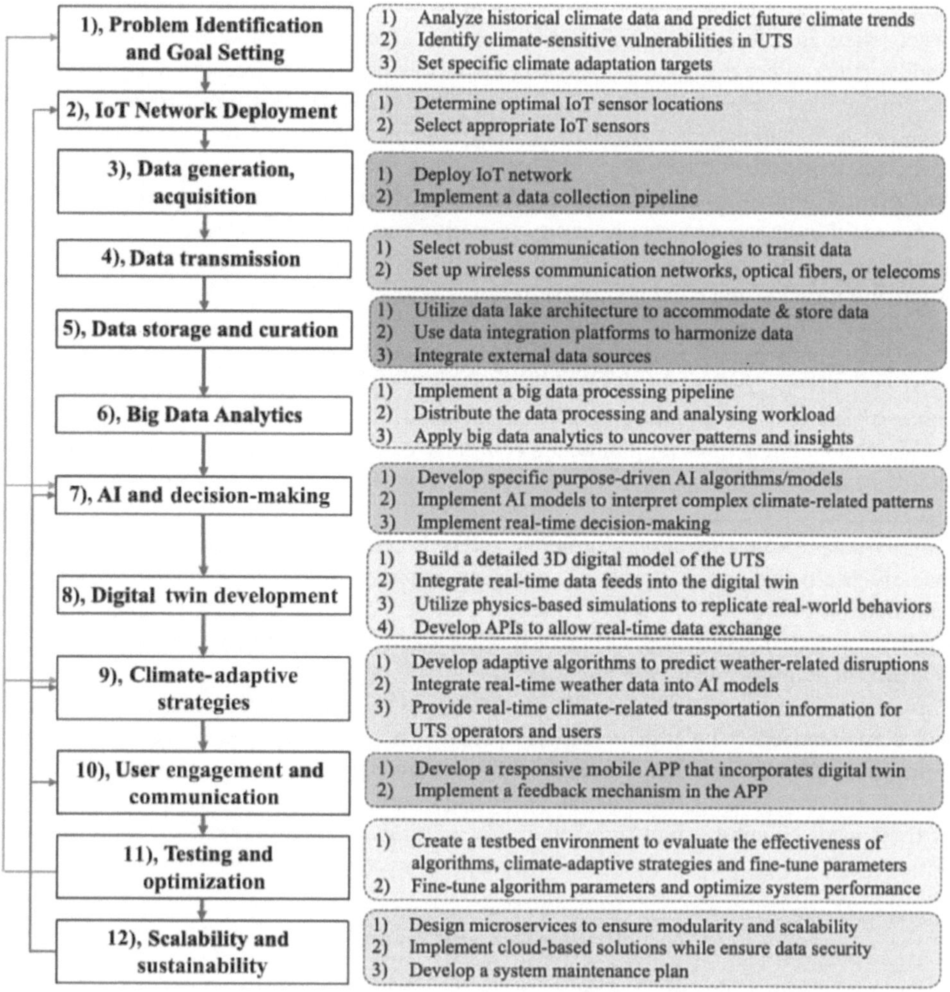

Figure 18.3 Process diagram of digital technology deployment for climate-adaptative UTS

18.3.1.1 Principles

At the very top of the framework are the nine principles at three levels of planning and operation. These principles are the authors' proposition, informed by reviewing literature of UTSs and smart cities. They can give strategic guidance for cities when they are developing their own UTS to address the urgent need for sustainable, resilient, and efficient transportation systems in the face of climate change.

Strategical urban planning level

1) *Civilian-centric*: putting civilians at the heart of the transportation system is crucial. Prioritizing their needs, safety, and convenience ensures that the system serves the community effectively. By understanding the behaviours and preferences of the citizens, the framework can create transportation solutions that are both practical and socially acceptable

2) *System thinking*: UTS is a system of systems with complexity and dynamic changes; a holistic system view is a foremost principle to ensure the interconnected systems can synergically work together to provide robust urban services at the highest resource efficiency. Understanding how different aspects interact with one another helps in identifying potential synergies, trade-offs, and unintended consequences. This principle aids in designing a coherent and interconnected transportation network
3) *Adaptability*: climate change introduces uncertainty, making adaptability a crucial factor. The UTS should be designed with the ability to evolve and respond to changing climate patterns and user demands. Incorporating flexibility in strategic urban planning allows for adjustments as new information and conditions emerge

Transportation planning level

4) *Joint risk/emergency management*: different levels of risk and emergency management organizations and facilities working jointly and coordinatively can help cities proactively prepare for and respond to extreme weather events, minimizing the impact on transportation operations and public safety while optimizing resource scheduling and allocation
5) *Public transportation dominance*: efficient and well-connected public transportation networks incentivize citizens to use alternatives to private vehicles, ultimately reducing congestion and emissions. It is more efficient, cost-effective, and energy-saving under the prerequisite that public transportation provides sufficient services to ensure availability, connectivity, and convenience in mobility needs with attention to EDI
6) *Rapid transfer centres*: strategic placement of rapid transfer centres enhances the efficiency and convenience of the transportation system. These centres should be close to public transport and vehicle-sharing stops to offer secure parking, different transport tool transit, and EV chargers. These hubs encourage seamless transitions between different modes of transport, reducing travel times and encouraging the use of sustainable transportation options.

Technical operation level

7) *Smart and sustainable roads*: roads are the fundamental infrastructure in UTS. Infusing roads with smart technology, such as sensors and real-time traffic management systems, optimizes traffic flow and reduces congestion. Sustainable road design considers factors like permeable surfaces and proper drainage, mitigating the impact of extreme weather events. Roads should be designed and improved to support the smart and sustainable transportation services by using smart and green materials, integrating digital technologies, supporting EV and autonomous driving, and being linked to smart traffic management systems.
8) *Car/parking sharing*: sharing of vehicles and parking spaces should be highly recommended and prioritized to minimize private cars. This reduces congestion, lowers emissions, and optimizes land use by reducing the need for extensive parking infrastructure.
9) *Electrification and automation*: transitioning to electric vehicles (EVs) and integrating automation technologies can significantly reduce carbon emissions and enhance safety, while automation of driving and maintenance of vehicles can improve traffic flow, reduce accidents, and optimize energy usage, allowing for more resilience especially under extreme events where humans may not be able to work.

18.3.1.2 The framework architecture

One significant part of the future UTS is the combination of smart and sustainable roads, varied transport tools for different distances of travel, and the transfer centres for low-carbon transportation. They are planned and built under the guidance of the previous nine principles. Another significant part of the deployment framework is comprised of digital technologies to empower resilient, reliable, and low-carbon applications. Remote, attached, and embedded sensing devices, IoT, and Information and Communication Technologies (ICT) networks are used to collect data in a (near) real-time fashion. GIS and digital twin are applied for information integration, visualization, and analysis. Edge computing can help with processing data on local equipment at the edge of IoT networks in a decentralized manner and thus minimize massive data security and privacy issues (Cao et al., 2020).

The physical UTS, its digital twin, the sensing, communicating, computing, and controlling equipment, and the users, together form a cyber-physical-social system (CPSS) (Xu et al., 2019). Within the CPSS, big data analysis and AI can enable many resilient, reliable, and low-carbon applications. For example, (near) real-time transport information updating and sharing, smart traffic management (e.g., context-based traffic light/flow control, traffic warning), and smart navigation based on real-time transport information and user preference can enable reliable services from UTS. The following services can be enabled by climate-adaptive smart UTS: (a) predictive maintenance based on road conditions and traffic pattern to avoid disruption and severe damage in climate events; (b) safety warning based on road conditions under extreme events; (c) rapid emergency response by analysing emergency level and location information and allocating rescue resources automatically in climate disasters; and (d) fast damage recovery actions based on damage estimation and quick repair planning after climate disasters.

Future UTS using the deployment framework will (a) be more environment-friendly and user-friendly at the same time by encouraging and prioritizing public transportation, shared vehicles and parking, and EV; (b) have the ability to sense, perceive, and anticipate changes/disruptions (Markolf et al., 2019) and take quick actions, which is very significant for complex systems under climate change stressors; and (c) have the ability to adapt to urban development, human needs, and technology evolution. However, the detailed action plan to implement the nine principles and the deployment framework requires further collaborative research from different sectors at all levels.

18.3.2 Digital technology deployment plan for climate-adaptative UTS

Developing a climate-adaptive urban transportation system that incorporates IoT, digital twin, big data, and AI technologies requires a comprehensive deployment plan. Figure 18.3 is a step-by-step process diagram with technical details. The key guidance points for each step will be articulated in the following section.

1) Problem identification and goal setting

First and foremost, it is pivotal to identify the problem to solve and set the goal to be focused on. In the context of a climate-adaptive UTS, efforts should be shaped with climate scientists, urban planners, and transportation experts to identify key climate-related challenges and climate-sensitive vulnerabilities in UTS (e.g., extreme weather, pollution). By analysing historical data and predicting future trends in both climate and UTS aspects, we need to set specific climate adaptation goals (e.g., net-zero in construction and maintenance by 2040, net-zero in operation by 2050), enhancing efficiency (e.g., achieving a 15% increase in public transit ridership by 2030) and improving overall transportation experience (e.g., decreasing travel time during extreme weather events by 20%)

2) Infrastructure and IoT deployment

As discussed in Section 3, IoT is significant for data collection. To get the right data from the right place in a massive UTS, we need to first perform a comprehensive assessment of the urban landscape to determine the optimal locations for the establishment of a IoT network across the UTS, including roads, traffic signals, vehicles, and public transit stations. The optimal locations will be selected based on traffic patterns (congestion, bottleneck), pollution hotspots and levels, and weather impact zones. Appropriate IoT sensors should be selected based on their accuracy, reliability, and ability to withstand environmental conditions for each location and their function requirements, which is based on the problem identified and goals set. Sensors including temperature and humidity sensors, air quality monitors, traffic cameras, and vehicle status trackers can be employed to measure parameters such as temperature, humidity, air quality, traffic flow, and vehicle status, respectively. To transmit data from sensors to the central repository is as substantial as data collection. Therefore, it is necessary to deploy robust communication technologies to ensure stable connectivity through a combination of wireless technologies (e.g., 5G, LoRaWAN) to transmit data from sensors to the central repository and continuous data transmission even in the case of network failures

3) Data generation and acquisition

After deploying the IoT network, it is suggested to implement a data collection pipeline that includes data generation and acquisition, transmission, storage, and curation stages that can handle real-time data streams from various IoT sensors. The data collection pipeline is necessary to streamline the process of intaking data from various sources with various formats. UTS management parties need to make sure the IoT network is properly set up and turned on to generate and acquire data. Sensors should have appropriate sampling rates to capture real-time changes while not overwhelming the data processing infrastructure. In areas with high sensor density, we can implement edge computing capabilities for real-time analysis and decision-making locally at the edge of the network to reduce latency and offload the central infrastructure

4) Data transmission

After deploying the IoT network, the next unneglectable crucial step is to transmit data from IoT devices to the data storage. The data communication technologies should be evaluated based on factors like data rate, range, power consumption, reliability, and cost. Given the density and distribution of IoT devices in UTS, hybrid communication network typology is suggested. To cover various communication needs, we can consider a combination of technologies within the system, especially wireless communication networks using technologies such as Wi-Fi, cellular (3G, 4G, 5G), Bluetooth, Zigbee, and LoRaWAN. In some harsh environments or areas with high data traffic, wired communication technologies such as Ethernet, optical fibres, and telecoms can also be used. When setting up the communication networks, the access points, routers, and gateways should be deployed strategically to ensure optimal coverage. Communication protocols that support low power consumption and reliable data transmission, such as MQTT or CoAP, should be chosen

5) Data storage and curation

Acknowledging that there will be a large volume of data coming every second, we need to utilize data lake architecture to accommodate various data formats and allow for scalable storage. More importantly, data integration platforms should be developed to harmonize data from different sources into a unified format for analysis to ensure consistency and compatibility. External data sources, such as weather forecasts, transit systems and schedules, micro-mobility services, and historical traffic data should also be integrated with the IoT data for decision-making purposes later. While overall it's suggested to store data in a central data repository for easy data access, it is efficient to utilize edge computing to process critical data locally for faster responses. Raw sensor data may contain noise, outliers, and missing values and thus needs to be pre-processed using techniques like data imputation, filtering, and outlier detection to clean the data and ensure its quality and reliability

6) Big data analytics

With a vast amount of data collected and integrated in digital twins, we can implement big data analytics tools to process and analyse the data. A big data processing pipeline to handle the influx of data from sensors and other sources is needed before the big data analytics. Due to the volume, velocity, variety, and veracity of big data, it is better to implement distributed computing frameworks like Apache Spark, as well as data partitioning and sharding techniques to distribute the data processing workload and analyse the data in parallel. Afterwards, semi-processed data from partitioning and sharding will be further explored using big data analytics techniques such as clustering, time-series analysis, and neural networks to uncover patterns and insights such as anomaly detection, predictive modelling, and traffic pattern recognition

7) AI and decision-making

On top of big data analytics which analyses existing data for pattern recognition, advanced AI models and/or algorithms using a combination of supervised, unsupervised, and reinforced machine learning and deep learning techniques can be developed to assist real-time decision-making. The difference between big data analytics and AI is that the latter requires the former to learn and evolve to enable machines to simulate human intelligence. In AI, we can leverage machine learning frameworks like TensorFlow and PyTorch to build models for advanced data analysis, such as identifying vehicle types, plate numbers, and driver health conditions from traffic camera feeds. AI has much more potential to understand and interpret complex situations, such as climate-related data patterns, to develop adaptive algorithms that can predict weather-related disruptions and traffic congestions. Furthermore, it can enable automatic and optimized interventions, for example, adaptively adjusting traffic signals, opening or closing roads based on emergent situations, dynamically allocating resources based on changing conditions, and optimizing vehicle routes and schedules according to urgent needs in climate disasters. A combination of AI-driven algorithms and rule-based systems can facilitate real-time decision-making in complex and rapidly changing situations in climate events for UTS operators and users. Reinforcement learning can then be applied to continuously improve decision-making based on outcomes and feedbacks

8) Digital twin development

Digital twins should reflect real-world conditions and update in real-time information based on big data analytics and AI. With the enormous real-time data from the IoT network, we can then create digital twins of the UTS by collaborating with architects, engineers, and GIS specialists to build a detailed 3D digital model of the urban transportation infrastructure, including roads, public transit stations, and micro-mobility hubs. The 3D digital model is not a digital twin until integrating real-time data feeds from IoT sensors. BIM and CIM are the main technologies to be used for building 3D digital models. Extra information (e.g., time series information, cost, etc.) can be added to extend the dimensions of the 3D models. For accurate representation, some physics engines and advanced physics-based simulations should be unutilized to replicate real-world behaviours accurately of the entire transportation ecosystem, including vehicles, infrastructure, and user behaviours. We should also develop APIs to allow real-time data exchange between the digital twin and smart UTS applications, as well as micro service units

9) Climate-adaptive strategies

The insights and lessons learnt from big data analytics and AI can be formulated into climate-adaptive strategies that better adapt UTS to climate events. The strategies should be supported by dynamic route optimization algorithms that consider weather and traffic conditions to minimize energy consumption. They can be embedded in risk assessment and alert systems that can trigger automatic adaptation and responses. Such strategies include automatic rerouting or scheduling changes for UTS services based on predicted weather patterns to optimize energy consumption and minimize exposure to hazards; automatic maintenance; repair of UTS according to the damage; automatic alert; and evacuation guidance to users at different locations. The specific strategies should be adjusted according to the severity level and impact scope of ongoing and imminent weather events

10) User engagement and communication

The main purpose of developing the climate-adaptive UTS is to limit the impacts of climate events and prepare them for possible disasters. Therefore, users should be closely engaged in the development of the system and the relevant information should be effectively communicated to them. It is highly suggested to develop a responsive mobile application with user-friendly interfaces and real-time updates on transportation options, delays, and climate-related alerts. In the app, interactive maps and visualizations from the digital twin can be incorporated to help users understand the impact of their transportation choices on climate and efficiency, therefore, to engage users and help them make informed decisions. Meanwhile, APIs to provide users with accurate information about available transportation options, including public transit, micro-mobility services, and real-time traffic conditions, should be used to provide interface for other functional development. It is significant to include a feedback mechanism in the app to gather user input and suggestions for the continuous improvement of user experience with the UTS

11) Testing and optimization

It is necessary to conduct extensive testing of the integrated system in simulation environments before deploying it in the real world and to continuously optimize algorithms and strategies based on performance metrics and user feedback. To do this, we need to (a) create a testbed environment that emulates real-world scenarios for validating AI algorithms and climate-adaptive strategies; (b) implement A/B testing methodologies to compare the performance of different algorithms in controlled settings; (c) conduct controlled experiments to evaluate the effectiveness of climate-adaptive strategies and fine-tune parameters; (d) use historical data and simulation results to fine-tune algorithm parameters and optimize system performance; and (e) establish a feedback loop to conduct regular reviews and updates to the system based on user feedback, emerging technologies, and changing climate conditions

12) Scalability and sustainability

UTS is always evolving; the smart climate-adaptive digital UTS should be designed with microservices to ensure modularity and scalability to accommodate the addition of new sensors, data sources, and transportation modes as the system expands. Containerization using technologies like Docker and orchestration with Kubernetes to efficiently manage system components and automatic scaling mechanisms to handle increased sensor deployment and user traffic is highly recommended. Cloud-based solutions should be implemented for scalability, while ensuring data privacy and security measures are in place. It's also necessary to develop a maintenance plan that considers the long-term system sustainability including regular sensor calibration and maintenance, software updates, and security audits. Also, for the scalability and sustainability of the system, collaboration among local governments, transportation agencies, environmental organizations, and technology vendors to ensure seamless integration of the system into existing infrastructure, ensure regulatory compliance, and secure necessary permits for sensor deployment is necessary.

By following this deployment plan and leveraging IoT, digital twin, big data, and AI technologies, a climate-adaptive urban transportation system can be developed to enhance both the efficiency of transportation and the resilience of the city's infrastructure to climate-related challenges. It should be noted that each deployment plan should be tailored to the specific characteristics of the city and its transportation ecosystem. Flexibility, adaptability, collaboration, and responsiveness to real-world challenges are crucial throughout the entire deployment process of a climate-adaptive urban transportation system.

18.4 Challenges ahead

Before we design a deployment framework of resilient, reliable, and low-carbon UTS with its key driving digital technologies, it is necessary to identify the challenges that should be addressed. The challenges can be generally divided into three categories: technical, policy, and ethical.

18.4.1 Technical challenges

The main technical challenge is the multiplied engineering system integration complexity (Helmrich & Chester, 2022). First, since transportation system is a system of interconnected systems (e.g., road, traffic signs and lights, water, electricity, petroleum, ICT), the failure of one system will cause cascading disruption for other systems. However, the planning and management of different systems in practice usually lack synergy with others. Second, high interdependence among components in a single transport subsystem can also lead to cascading failures when a critical point fails. Third, different stages of the prolonged lifecycle of a UTS are also siloficated and inconsistent, resulting in ineffectiveness, abuse of available resources, and fragmentation. Fourth, non-stationarity population and demographic shifts, uncertain human behaviour/response during extreme events, evolution of technology, and people's preferences bring more uncertainty and difficulty to the UTS. For example, the emergence of autonomous vehicles is likely to result in several (albeit uncertain) changes to our transportation infrastructure (Markolf et al., 2019). Added together, the complexity and uncertainty of the UTS exponentially increase and become overwhelming for its plan and management in a resilient and reliable way. A digital twin is likely a potential solution to integrating, visualizing, modelling, analysing, monitoring, and controlling the complex system of systems. Also, edge computing can help to decentralize and lighten the computing of UTS central control system, allowing for fast response times, freeing up bandwidth, and sensitive data privacy.

18.4.2 Policy challenges

The resilient and reliable governance of low-carbon UTS remains a worldwide significant challenge for policymakers at different levels. An adequate UTS policy must incorporate a number of interrelated and disconnected or even mutually exclusive components, yet it has to have clear ownership, governance, and regulation (Bolton et al., 2018). There are many governance questions to be considered by policymakers, for example, how to develop local smart UTS strategies to create added value for society, environment, and economy; how to incentive companies, organizations, and citizens to willingly shift to low-carbon UTS; how to promote new technology development to effectively adapt to climate threats; how to develop in-situ upgrade policy to support more sustainable UTS; and how to govern data ethics while using open data for smart applications. Policymaking to address these challenges takes a long time to study the prolonged lifecycle of UTS and the complex network of components involved. It needs to reach solid anchoring and common understanding, to design concrete measures and make the necessary adaptations to standards and regulations, to pilot testing the risks and mistakes, and to put the new policy into implementation while continuing to monitor its effects and impacts. It requires constructive cooperation with many different actors from both the public and private sectors. It needs systematic, flexible, and agile planning with the consideration of city development, people's changing requirements, technology upgrades, extreme event adaptations, and public good.

18.4.3 Ethical challenges

Public good includes improved services and benefits for users under proper ethical principles. Although the development of a smart system is for the purpose of better services, how to avoid the violation of privacy, human rights, and cybersecurity has been widely debated among all sectors of society. For example, using a camera to capture behavioural data and using GPS to track location are heatedly criticized as the violation of privacy. There is a spectrum of measures dealing with privacy concerns, ranging from refusing any data acquisition from users to collecting all data even without consent (Xu et al., 2022). A negative refusal of collaboration is criticized as dragging the leg of technology development; however, ignoring privacy is against common ethical rules. The practice of general data protection regulation (GDPR) in the European Union is a good attempt to allow users to decide how their privacy is protected. However, GDPR is also criticized for causing less transparency and disclosure (Zaeem & Barber, 2021) and explainability challenges to automated decision-making systems and customized recommendation systems (Sovrano et al., 2021). Security of data, personnel, sensitive physical assets, commercial interests, intellectual property, and cyberspace is another major ethical concern (Bolton et al., 2018). Meanwhile, security concerns should be balanced against data openness and value creation. A third critical concern is the user equity, diversity, and inclusion (EDI) in the planning, design, and management of UTS. How a UTS meets the needs of all population groups is a question awaiting more attention. Civilian-centric principles should be applied when dealing with these ethical challenges. Pseudonymization and anonymization of personally identifiable information and encryption technologies like blockchain could also help at the tactical level.

18.5 Conclusion

With the rapid technology progression, city outspread, and demand shifts, existing UTS are becoming more complex and interconnected while less resilient to exacerbated climate changes (Markolf et al., 2019). This article focuses on how UTS could be planned and managed for climate adaptation, presenting a view on how the development and adoption of technology will support this. It first discussed the value of digital technologies in data collection, integration, visualization, and analysis of UTS. Then a future UTS deployment framework is proposed with nine principles at strategic, planning, and technical levels, a physical layer of multiple transport tools integration, and a digital layer to sense, perceive, and anticipate changes and disruptions for climate adaptation actions. The deployment framework conceptually presents a UTS into a cyber-physical-social system to efficiently facilitate resilience, reliability, and low carbon. It provides a feasible pathway toward data-informed urban governance and sustainable development (Acuto, 2018). Furthermore, a step-by-step deployment plan is introduced to guide cities to plan and deploy the climate-adaptive smart UTS. Finally, the major challenges include technical complexity of multiplied engineering system integration, policy challenges of governance and planning, and ethical challenges of privacy and security. Future research is still needed to properly implement and validate the deployment framework to combat the urban transportation challenges.

References

Abduljabbar, R., Dia, H., Liyanage, S., & Bagloee, S. A. (2019). Applications of artificial intelligence in transport: An overview. In *Sustainability (Switzerland)* (Vol. 11, Issue 1). MDPI. https://doi.org/10.3390/su11010189

Acuto, M. (2018). Global science for city policy It is time for a global reform of science advice to cities. In Science (Vol. 359, Issue 6372, pp. 165–166). *American Association for the Advancement of Science*. https://doi.org/10.1126/science.aao2728

Ahmed, S., & Dey, K. (2020). Resilience modeling concepts in transportation systems: A comprehensive review based on mode, and modeling techniques. *Journal of Infrastructure Preservation and Resilience*, 1(1). https://doi.org/10.1186/s43065-020-00008-9

Ali, Z. H., Zaki, J. F., & El-Rashidy, N. (2022). Dynamic urban evaluation routing protocol for enhanced vehicle ad hoc networks. *Journal of Supercomputing*. https://doi.org/10.1007/s11227-022-04877-7

Bai, X., Dawson, R. J., Ürge-Vorsatz, D., Delgado, G. C., Barau, A. S., Dhakal, S., Dodman, D., Leonardsen, L., Masson-Delmotte, V., Roberts, D. C., & Schultz, S. (2018). Six research priorities for cities and climate change. *Nature*, 555, 23–25. https://doi.org/10.1038/d41586-018-02409-z

Bolton, A., Butler, L., Dabson, I., Enzer, M., Evans, M., Fenemore, T., Harradence, F., Keaney, E., Kemp, A., Luck, A., Pawsey, N., Saville, S., Schooling, J., Sharp, M., Smith, T., Tennison, J., Whyte, J., Wilson, A., & Makri, C. (2018). The Gemini principles: Guiding values for the national digital twin and information management framework. Centre for Digital Built Britain, University of Cambridge.

Cao, K., Liu, Y., Meng, G., & Sun, Q. (2020). An overview on edge computing research. In *IEEE Access* (Vol. 8, pp. 85714–85728). Institute of Electrical and Electronics Engineers Inc. https://doi.org/10.1109/ACCESS.2020.2991734

Costin, A., Adibfar, A., Hu, H., & Chen, S. S. (2018). Building Information Modeling (BIM) for transportation infrastructure – Literature review, applications, challenges, and recommendations. *Automation in Construction*, 94, 257–281. https://doi.org/10.1016/j.autcon.2018.07.001

Diao, M., Kong, H., & Zhao, J. (2021). Impacts of transportation network companies on urban mobility. *Nature Sustainability*, 4(6), 494–500. https://doi.org/10.1038/s41893-020-00678-z

Donovan, B., & Work, D. B. (2017a). Empirically quantifying city-scale transportation system resilience to extreme events. *Transportation Research Part C: Emerging Technologies*, 79, 333–346. https://doi.org/10.1016/j.trc.2017.03.002

Donovan, B., & Work, D. B. (2017b). Empirically quantifying city-scale transportation system resilience to extreme events. *Transportation Research Part C: Emerging Technologies*, 79, 333–346. https://doi.org/10.1016/j.trc.2017.03.002

Gupta, B. B., & Quamara, M. (2020). An overview of Internet of Things (IoT): Architectural aspects, challenges, and protocols. *Concurrency and Computation: Practice and Experience*, 32(21), e4946.

Helmrich, A. M., & Chester, M. V. (2022). Reconciling complexity and deep uncertainty in infrastructure design for climate adaptation. *Sustainable and Resilient Infrastructure*, 7(2), 83–99. https://doi.org/10.1080/23789689.2019.1708179

International Energy Agency. (2022). World energy outlook 2022. https://iea.blob.core.windows.net/assets/830fe099-5530-48f2-a7c1-11f35d510983/WorldEnergyOutlook2022.pdf

Khan, S., Nazir, S., García-Magariño, I., & Hussain, A. (2021). Deep learning-based urban big data fusion in smart cities: Towards traffic monitoring and flow-preserving fusion. *Computers & Electrical Engineering*, 89, 106906. https://doi.org/10.1016/j.compeleceng.2020.106906

Kochl, A. (2021). Urban transport and COVID-19: Challenges and prospects in low- and middle-income countries. *Cities and Health*, 5(sup1), S185–S190. https://doi.org/10.1080/23748834.2020.1791410

Kurth, M., Kozlowski, W., Ganin, A., Mersky, A., Leung, B., Dykes, J., Kitsak, M., & Linkov, I. (2020). Lack of resilience in transportation networks: Economic implications. *Transportation Research Part D: Transport and Environment*, 86. https://doi.org/10.1016/j.trd.2020.102419

Land, C., Andersson, S. E., Elmqvist, T., Mayor, U., Lin, B. B., Ossola, A., Alberti, M., Andersson, E., Bai, X., Dobbs, C., Elmqvist, T., Evans, K. L., Frantzeskaki, N., Fuller, R. A., Gaston, K. J., Haase, D., Yung Jim, C., Konijnendijk, C., Nagendra, H., . . . Yok Tan, P. (2021). Integrating solutions to adapt cities for climate change. In Viewpoint Lancet Planet Health. www.thelancet.com/

Markolf, S. A., Hoehne, C., Fraser, A., Chester, M. V., & Underwood, B. S. (2019). Transportation resilience to climate change and extreme weather events – Beyond risk and robustness. *Transport Policy*, 74, 174–186. https://doi.org/10.1016/j.tranpol.2018.11.003

Mehryar, S., Sasson, I., & Surminski, S. (2022). Supporting urban adaptation to climate change: What role can resilience measurement tools play? *Urban Climate*, 41. https://doi.org/10.1016/j.uclim.2021.101047

Nižetić, S., Šolić, P., López-de-Ipiña González-de-Artaza, D., & Patrono, L. (2020a). Internet of Things (IoT): Opportunities, issues and challenges towards a smart and sustainable future. *Journal of Cleaner Production*, 274. https://doi.org/10.1016/j.jclepro.2020.122877

Nižetić, S., Šolić, P., López-de-Ipiña González-de-Artaza, D., & Patrono, L. (2020b). Internet of Things (IoT): Opportunities, issues and challenges towards a smart and sustainable future. *Journal of Cleaner Production*, 274. https://doi.org/10.1016/j.jclepro.2020.122877

Oeschger, G., Carroll, P., & Caulfield, B. (2020). Micromobility and public transport integration: The current state of knowledge. *Transportation Research Part D: Transport and Environment*, 89. https://doi.org/10.1016/j.trd.2020.102628

Pantic, M., Pentland, A., Nijholt, A., & Huang, T. (2006). Human computing and machine understanding of human behavior: A survey. ICMI'06: 8th International Conference on Multimodal Interfaces, Conference Proceeding, 239–248. https://doi.org/10.1145/1180995.1181044

Reckien, D., Salvia, M., Heidrich, O., Church, J. M., Pietrapertosa, F., de Gregorio-Hurtado, S., D'Alonzo, V., Foley, A., Simoes, S. G., Krkoška Lorencová, E., Orru, H., Orru, K., Wejs, A., Flacke, J., Olazabal, M., Geneletti, D., Feliu, E., Vasilie, S., Nador, C., . . . Dawson, R. (2018). How are cities planning to respond to climate change? Assessment of local climate plans from 885 cities in the EU-28. *Journal of Cleaner Production*, 191, 207–219. https://doi.org/10.1016/j.jclepro.2018.03.220

Sacks, R., Brilakis, I., Pikas, E., Xie, H. S., & Girolami, M. (2020). Construction with digital twin information systems. *Data-Centric Engineering*, 1, e14. https://doi.org/10.1017/dce.2020.16

Sharifi, A. (2021). Co-benefits and synergies between urban climate change mitigation and adaptation measures: A literature review. In Science of the Total Environment (Vol. 750). Elsevier B.V. https://doi.org/10.1016/j.scitotenv.2020.141642

Shin, D. (2021). The effects of explainability and causability on perception, trust, and acceptance: Implications for explainable AI. *International Journal of Human-Computer Studies*, 146, 102551. https://doi.org/10.1016/j.ijhcs.2020.102551

Sovrano, F., Vitali, F., & Palmirani, M. (2018, December). Making things explainable vs explaining: Requirements and challenges under the GDPR. In *International Workshop on AI Approaches to the Complexity of Legal Systems* (pp. 169–182). Cham: Springer International Publishing. https://doi.org/10.1007/978-3-030-89811-3_12

Stamos, I., Mitsakis, E., Salanova, J. M., & Aifadopoulou, G. (2015). Impact assessment of extreme weather events on transport networks: A data-driven approach. *Transportation Research Part D: Transport and Environment*, 34, 168–178. https://doi.org/10.1016/j.trd.2014.11.002

Sterchi, M., Faverjon, C., Sarasua, C., Vargas, M. E., Berezowski, J., Bernstein, A., Grütter, R., & Nathues, H. (2019). The pig transport network in Switzerland: Structure, patterns, and implications for the transmission of infectious diseases between animal holdings. P*LoS One*, 14(5). https://doi.org/10.1371/journal.pone.0217974

Wang, T., & Ng, A. K. Y. (2022). Responding to the barriers in climate adaptation planning among transport systems: Insights from the case of the port of Montreal. *International Journal of Sustainable Transportation*, 16(10), 942–956. https://doi.org/10.1080/15568318.2021.1960450

Wang, Y., Zhang, D., Liu, Y., Dai, B., & Lee, L. H. (2019a). Enhancing transportation systems via deep learning: A survey. In *Transportation Research Part C: Emerging Technologies* (Vol. 99, pp. 144–163). Elsevier Ltd. https://doi.org/10.1016/j.trc.2018.12.004

Wang, Y., Zhang, D., Liu, Y., Dai, B., & Lee, L. H. (2019b). Enhancing transportation systems via deep learning: A survey. In *Transportation Research Part C: Emerging Technologies* (Vol. 99, pp. 144–163). Elsevier Ltd. https://doi.org/10.1016/j.trc.2018.12.004

Xu, J., Lu, W., Wu, L., Lou, J., & Li, X. (2022). Balancing privacy and occupational safety and health in construction: A blockchain-enabled P-OSH deployment framework. *Safety Science*, 154. https://doi.org/10.1016/j.ssci.2022.105860

Xu, J., Lu, W., Xue, F., & Chen, K. (2019). 'Cognitive facility management': Definition, system architecture, and example scenario. *Automation in Construction*, 107, 102922. https://doi.org/10.1016/j.autcon.2019.102922

Xu, J., Lu, W., Ye, M., Xue, F., Zhang, X., & Lee, B. F. P. (2020). Is the private sector more efficient? Big data analytics of construction waste management sectoral efficiency. *Resources, Conservation and Recycling*, 155. https://doi.org/10.1016/j.resconrec.2019.104674

Xue, F., Wu, L., & Lu, W. (2021). Semantic enrichment of building and city information models: A ten-year review. *Advanced Engineering Informatics*, 47, 101245. https://doi.org/10.1016/j.aei.2020.101245

Zaeem, R. N., & Barber, K. S. (2021). The effect of the GDPR on privacy policies. *ACM Transactions on Management Information Systems*, 12(1). https://doi.org/10.1145/3389685

Zawieska, J., & Pieriegud, J. (2018). Smart city as a tool for sustainable mobility and transport decarbonisation. *Transport Policy*, 63, 39–50. https://doi.org/10.1016/j.tranpol.2017.11.004

Zhou, Y., Wang, J., & Yang, H. (2019a). Resilience of transportation systems: Concepts and comprehensive review. *IEEE Transactions on Intelligent Transportation Systems*, 20(12), 4262–4276. https://doi.org/10.1109/TITS.2018.2883766

Zhou, Y., Wang, J., & Yang, H. (2019b). Resilience of transportation systems: Concepts and comprehensive review. *IEEE Transactions on Intelligent Transportation Systems*, 20(12), 4262–4276. https://doi.org/10.1109/TITS.2018.2883766

Zhu, L., Yu, F. R., Wang, Y., Ning, B., & Tang, T. (2019). Big data analytics in intelligent transportation systems: A survey. In *IEEE Transactions on Intelligent Transportation Systems* (Vol. 20, Issue 1, pp. 383–398). Institute of Electrical and Electronics Engineers Inc. https://doi.org/10.1109/TITS.2018.2815678

19

BLOCKCHAIN FOR SMART BUILT ENVIRONMENT

Current status, challenges, and prospects

Liupengfei Wu[1], *Weisheng Lu*[2], *Jinying Xu*[3], *Jinfeng Lou*[4] *and Liang Yuan*[5]

[1]Research Assistant Professor, Department of Real Estate and Construction, Faculty of Architecture, The University of Hong Kong, Pokfulam Road, Hong Kong SAR. Email: liupengfeiwu@connect.hku.hk

[2]Chair Professor, Department of Real Estate and Construction, Faculty of Architecture, The University of Hong Kong, Pokfulam Road, Hong Kong SAR. http://orcid.org/. Email: wilsonlu@hku.hk

[3]Marie Curie Future Roads Fellow, Department of Engineering, University of Cambridge, UK. Email: jx314@cam.ac.uk

[4]Carnegie Bosch Institute fellow, Department of Civil and Environmental Engineering, Carnegie Mellon University, 5000 Forbes Ave, Pittsburgh, USA. Email: jinfengl@andrew.cmu.edu

[5]PhD Candidate, Department of Real Estate and Construction, Faculty of Architecture, The University of Hong Kong, Pokfulam Road, Hong Kong SAR. Email: u3007096@connect.hku.hk

19.1 Introduction

The built environment has an impact on every aspect of our daily lives (Lu & Anumba, 2022). This includes the structures people reside in, the systems that supply citizens with water and electricity, and the transportation systems residents use to move around (United States Environmental Protection Agency [USEPA], 2023). These man-made or altered constructions provide us with spaces for living, working, and leisure. At present, poor productivity is identified as a significant factor in the lack of success in the built environment (Farmer, 2016). On the other hand, it is also a significant area for possible enhancement, as per McKinsey Global Institute, which indicates a worldwide productivity shortfall of $1.6 trillion that can be addressed by advancing the industry's performance (Agarwal et al., 2016). The built environment faces a major obstacle in its modernization due to its lack of ability to adopt technological advancements, which is in stark contrast to the successes observed in the logistics, automotive, and mechanical engineering sectors (Li et al., 2019). In this regard, stakeholders across the built environment are constantly looking for ways to address challenges.

The role of technology is crucial in tackling challenges in both the overall economy and the smart built environment, in particular. The use of computers, the Internet, and advancements in information and communication technologies (ICTs) have contributed to an improvement in

DOI: 10.1201/9781003383840-20

information flow and communication (Dixit et al., 2021). This has resulted in the development of a culture of trust, honesty, and shared values, which has led to increased productivity in the smart built environment (Orgut et al., 2020). Despite this, new challenges continue to arise such as an aging workforce (Kamardeen & Hasan, 2022) and high cost (Navaratnam et al., 2022). The advent of the fourth industrial revolution and the associated digital transformation, commonly referred to as Industry 4.0, appears to offer unparalleled opportunities for facilitating advancements in the built environment (Ghobakhloo, 2020). Thus, the way people live and work is being fundamentally transformed by the digital revolution, which utilizes various technologies such as the Internet of Things (IoT), cloud computing, smart sensors, advanced robotics, big data analytics, and three-dimensional (3D) printing (Wang et al., 2020). Blockchain has also been hailed as a disruptive technology for Industry 4.0 and has sparked strong interest in the smart built environment (Wu et al., 2022a).

Advances in blockchain are increasingly investigated as one of the constituents in the digital transformation of the smart built environment and its response to the aforementioned challenges (Wu et al., 2024). To support its functionality, blockchain relies on three key components: (a) distributed ledgers, (b) consensus mechanisms, and (c) cryptographic algorithms (Wu et al., 2022a). When combined, these components allow transaction data to be collected, validated, shared, and stored with minimal risk of unauthorized alteration (Lu et al., 2022). The widely propagated benefits of blockchain, therefore, include reinforced transparency, immutability, traceability, efficiency, democratic decision-making, trust, and reduced costs (Turk & Klinc, 2017; Lu et al., 2021a). Since 2021, an extensive literature review has become feasible due to the substantial amount of research conducted on blockchain, which has contributed to a significant body of knowledge in the field (Scott et al., 2021). However, limited attempts have been made to systematically discuss the current status, challenges, and prospects of blockchain in the smart-built environment. Considering that blockchain is a rapidly evolving technology, it is necessary to have an up-to-date and comprehensive review of it.

The overarching aim of this chapter is to analyze the current state of blockchain in the built environment on which to support its adoption in the built environment. Three objectives are established to achieve this: (a) to provide a consolidated view of the applications explored in the smart built environment, which could potentially support the disruption of the industry, by presenting the state-of-the-art and literature review on blockchain; (b) to identify the challenges of blockchain research in the smart built environment; (c) to summarize the prospects of blockchain research in the smart built environment.

19.2 Research status of blockchain for smart built environment

A literature review was first conducted to identify the current applications of blockchain in the built environment. The built environment can be influenced by various aspects such as energy, transport, the construction industry, smart homes, smart government, smart cities, and the sharing economy, which play crucial roles in creating, adapting, and supporting its evolution. As a result, the scope of this chapter extends to provide a comprehensive understanding of how blockchain technology can contribute to the development of the built environment, encompassing all its aspects.

19.2.1 Summary of blockchain applications in the smart-built environment

The various emerging applications of blockchain in the smart-built environment have been identified and categorized into seven streams, which are summarized in the following sections.

19.2.1.1 Smart energy

The potential applications of blockchain in smart energy systems have generated significant research interest in recent years. Smart energy aims to optimize the utilization of affordable and sustainable energy sources (Siddiquee et al., 2022). By enabling peer-to-peer energy generation and consumption, blockchain has the potential to enhance the resilience and efficiency of the energy industry (Hassan et al., 2019). Incorporating blockchain technology into urban ecosystems can optimize energy efficiency and improve energy resource management (Siddiquee et al., 2022). Also, employing blockchain to regulate energy distribution and transformation can foster transparency in energy transactions. By acting as a reliable communication backbone, blockchain can also simplify and secure peer-to-peer energy trade transactions within the energy network (Hassan et al., 2019). Furthermore, data generated by energy management systems can be safely stored in the ledgers of blockchain (Guan et al., 2021).

19.2.1.2 Smart transportation

Smart transportation aims to enhance the accessibility of contemporary transportation systems that are efficient for users and sustainable for the smart city and the environment at large (Wazid et al., 2022). Effective management of transportation networks is one of the primary challenges that every city encounters (Yuan & Wang, 2016). To enable ongoing real-time monitoring of vehicles and passengers, IoT devices and systems can be combined with blockchain technology (Humayun et al., 2020). Several blockchain applications have been proposed to assist transportation decision-makers in optimizing routing methods and timetables (Li et al., 2023), anticipating diverse passenger needs (Wang & Zhang, 2020), and achieving greater efficiency in terms of environmental and sustainability objectives (Wazid et al., 2022). In addition, with blockchain technology, users can securely pay for transportation services within an efficient transportation network (Wazid et al., 2022).

19.2.1.3 Construction management

According to prior research, several blockchain applications have been shown to be helpful in various stages of construction projects. The value of blockchain in the design stage lies in facilitating collaborative design and automatic code compliance checking (Tao et al., 2021). As per the findings, supply chain management (SCM), progress payment management, and construction quality management (CQM) emerged as the top three topics during the construction stage. Regarding SCM, various frameworks and prototypes based on blockchain have been created to ensure real-time delivery of precast components or prefinished modules and prevent conflicts by improving information sharing from manufacturing to assembly. Readers are suggested to follow Lu Weisheng's team, which has published several excellent studies in this field (e.g., Lu et al., 2021b; Wu et al., 2022b, 2022c). Lu et al. (2021a) noted that the effectiveness of blockchain technology might be restricted when data is sourced from a single centralized or low-quality source. To tackle this limitation, they proposed a solution that combines smart construction objects-enabled blockchain oracles and signature techniques to ensure the authenticity and integrity of information in construction (Lu et al., 2023).

Besides, Han Luo and Moumita Das contributed significantly to blockchain-based construction process payment. For example, Luo et al. (2019) proposed a smart contract-based blockchain framework for construction payment automation. Das et al., 2020) presented a blockchain-based framework to secure interim payments in construction projects. Furthermore, Hamledari and Fischer (2021) have successfully integrated blockchain technology with other advanced technologies, such as unmanned aerial vehicle-based progress monitoring, to complete the construction progress payment automation. Another important topic in blockchains is CQM. Blockchain allows quality information to be recorded immutably, transparently, and traceably. Thus, several studies developed blockchain-based quality information management systems in construction (e.g., Sheng et al., 2020; Zhong et al., 2020). Only a few blockchain studies focused on facility management and maintenance in the operation stage. For example, Li et al. (2022) proposed a comprehensive solution that integrates blockchain, building information modeling (BIM), and IoT to enhance the traceability of maintenance records for built assets.

This solution automates numerous maintenance operations through smart contracts, such as tendering, work completion approvals, and payments.

19.2.1.4 Smart homes

Blockchain-based smart home applications are another vital research stream. For example, Moniruzzaman et al. (2020) proposed a blockchain-based system architecture to achieve a decentralized, transparent, and secured smart home data marketplace. Also, the emerging field of third-party access control of smart homes presents promising opportunities for applying blockchain technology. For instance, Dorri et al. (2017) highlighted that blockchains can be leveraged to implement access control on policy headers, reinforcing security in the media access control layer when transmitting data in a blockchain-based smart home. Additional research endeavors have investigated utilizing blockchain technology in decentralized homecare systems to safeguard data transmission between homecare and healthcare service providers (Tantidham & Aung, 2019), establish a repository of falling records for elderly individuals (Rupasinghe et al., 2019), and develop a patient-centric access control system (Dwivedi et al., 2019). Like the construction progress payment, several blockchain-based solutions have been proposed to realize the automated utility payment system (e.g., Zhong et al., 2019).

19.2.1.5 Smart government

Numerous blockchain applications have been found for smart government. Several studies proposed blockchain-based e-voting systems to reduce time, cost, and fraud associated with electoral voting (e.g., Pawlak et al., 2018). Blockchain use cases are also applicable in tax collection (Yayman, 2021), identity management (Shobanadevi et al., 2021), benefits distribution (Li et al., 2019), property and land registries (Themistocleous, 2018), local and national digital currencies (Lu et al., 2021b), government records management (O'Dair & Beaven, 2017), and regulatory compliance checking (Lu & Xu, 2017). Also, China is currently exploring using blockchain technology to verify the reliability of individuals' data (Hou, 2017). This authentication process enables the data to be utilized in an enterprise credit system, facilitating mutual information exchange between individuals and enterprises (Hou, 2017). This exchange empowers both parties to make more informed decisions.

19.2.1.6 Smart cities

Blockchain has been explored for numerous smart city service systems. Smart city service systems require fast and secure communication between city authorities to function effectively (Moniruzzaman et al., 2020). However, malicious attackers constantly eavesdropping on communication channels pose various security threats that can put these services at high risk (Moniruzzaman et al., 2020). To mitigate these threats and facilitate smart city services, blockchain-based systems could be beneficial. For example, Sharma and Park (2018) adopted blockchain for secured and privacy-enhanced smart city services. Furthermore, blockchain is actively explored as a data-sharing platform for smart cities due to its distributed ledgers and decentralized network (e.g., Michelin et al., 2018). Further blockchain use cases in smart cities include monitoring population growth (Li et al., 2019), maintenance of health records (Attaran, 2022), and granting access to relevant parties (Sookhak et al., 2021).

19.2.1.7 Sharing economy

The progress in blockchain technology has also made the sharing economy much easier and more accessible (Huckle et al., 2016). Several studies discussed the use of blockchain to enhance trust in the sharing economy (e.g., Hawlitschek et al., 2018). The reason is that blockchain enables individuals to access real-time and transparent data about the availability of resources, empowering them to make more informed choices. With the recent development of blockchain technology, the opportunity also exists to create myriad sharing applications, such as peer-to-peer automatic payment mechanisms (Toorajipour et al., 2022), foreign exchange platforms (Moşteanu & Flocea, 2023), cross-border payment systems (Hongmei, 2021), digital rights management, and cultural heritage (Trček, 2022).

19.3 Challenges of blockchain research in the smart-built environment

While the potential of blockchain is widely recognized, several technical and non-technical challenges prevent the smart-built environment from realizing these promises in real-world projects.

19.3.1 Technical challenges

19.3.1.1 Blockchain network topology

In the built environment, it is customary to establish a centralized governance system where the government oversees activities, such as construction activities. The primary objective of this oversight is to ensure that built environment work meets the required standards of quality, safety, progress, and compliance independently (Tuuli et al., 2010). The government's obligation to supervise projects, including those owned by private entities, is driven by concerns related to the public interest, as described by Li et al. (2019). This supervision is governed by various mechanisms, such as national standards, ordinances, codes, and professional codes of conduct, as outlined by Recarte and Jaselskis (1993). To ensure compliance with these regulations, specific government supervision units have been established, such as Hong Kong's buildings department. These units are responsible for granting activity permits and conducting inspections as projects progress.

Thus, a challenging question must be answered before a blockchain is developed to avoid pointless blockchain projects: "what kinds of blockchain network topology should be adopted?" The decentralization of control and decision-making from a central entity to a distributed network through a consensus mechanism is why blockchain technology is deemed anti-authoritarian. Nevertheless, if blockchain is to be employed in conjunction with government supervision units, it is vital to identify a network topology that can achieve a balance between centralized and decentralized governance.

19.3.1.2 Data privacy

The transparency of blockchain technology may give rise to data privacy concerns. Although blockchain guarantees that all transactions are recorded and visible to network participants, this openness can compromise sensitive information that individuals or organizations may want to keep confidential. The challenge is to strike a balance between transparency and privacy in blockchain systems. While the technology provides a high degree of transparency, it is essential to implement adequate measures to safeguard sensitive data (e.g., payment data). This may involve utilizing encryption techniques, pseudonymization, or permissioned blockchains that limit access to authorized users. For example, Li et al. (2021) proposed a two-layer adaptive blockchain-based supervision model for off-site modular housing production. The proposed solution has a unique two-layer structure that allows it to benefit from avoiding any tampering with operation records on the main blockchain. Additionally, it can encourage participants to publish their operation records promptly on the sidechain without any privacy leaks. By incorporating privacy-enhancing features, blockchain can mitigate the risks linked with data privacy and allow individuals and organizations to maintain a certain level of confidentiality while still witnessing the inherent transparency and security offered by the technology.

19.3.1.3 Interoperability issues

Numerous blockchain systems have been created for various applications in the built environment, as outlined earlier. These systems were suggested for particular purposes, indicating that there may be variations in blockchain architectures and system participants. As an illustration, Ethereum was frequently utilized in developing blockchain systems for cross-border payment in construction projects (Das et al., 2020). On the other hand, Wu et al. (2022b) developed a prototype blockchain-based system based on the Hyperledger Fabric architecture to enhance the accuracy of information sharing in on-site modular construction assembly. Thus, stakeholders may have to manage multiple blockchain systems for various objectives in a single construction project, leading to potential interoperability concerns. Therefore, it is important to consider interoperability between blockchain systems and other existing systems to support different applications.

19.3.2 Non-technical challenges

19.3.2.1 Resistance to change

For a significant period, the built environment has relied on information systems that operate on centralized architectures. During the implementation of such systems, encountering resistance is a common occurrence. One primary reason for this is the uncertainty surrounding the impact that blockchain will have on the built environment. Additionally, if the economic benefits of blockchain-enabled solutions are not clear to industry professionals, they may hesitate to embrace the technology. Furthermore, the creation and maintenance of blockchain-enabled systems require specialized expertise, which is currently lacking in the built environment sector. These factors contribute to the industry's reluctance to adopt this new technology. To overcome this opposition, it is crucial to demonstrate the advantages of blockchain technology and provide successful implementation examples, as demonstrated by Lu et al. (2023). Exploring cost-sharing models and establishing platforms that encourage collaboration and knowledge exchange among industry stakeholders can also help disseminate best practices and facilitate change. As the built environment becomes more informed about the benefits of blockchain technology and witnesses its increasing adoption in other industries, it may gradually realize the transformative potential of blockchain.

19.3.2.2 Additional costs

One key factor that hinders the widespread adoption of blockchain technology is the significant financial investment required to implement blockchain solutions. According to Brody et al. (2019), the cost of implementing a blockchain solution varies based on several factors, such as the user scenario, type of blockchain, complexity, blockchain platform, and other technology stacks. In general, implementing a blockchain involves several steps, including consulting, design, development, quality assurance, third-party services, and maintenance costs, as outlined by LeewayHertz (n.d.). Allocating funds for investment in emerging technologies such as blockchain may pose a challenge for the built environment, especially the construction industry, known for its low profit margins and cost-sensitive nature. Construction companies' decision-makers prioritize urgent project requirements and might be reluctant to invest in unproven blockchain solutions. Until enterprise blockchain data becomes more accessible, applications are standardized, and pricing becomes consistent, any cost forecasting model will heavily depend on assumptions and benchmarking data (Brody et al., 2019). Convincing evidence of the return on investment and long-term benefits is necessary before committing substantial financial resources. One of the significant benefits of blockchain technology is its ability to foster collaboration among stakeholders. Therefore, when considering blockchain implementation, it is essential to determine which stakeholders need to contribute to the cost and in what proportion.

19.3.2.3 Lack of regulatory clarity

There is no well-established policy environment for applying blockchain technology in the built environment, including relevant regulations, laws, policies, and standards (Li et al., 2019). One instance is the absence of clear policies on how current regulations apply to blockchain systems, particularly concerning data privacy protection, intellectual property rights, and responsibility for fraudulent or modified data. Furthermore, as blockchain technology progresses and its usage in the built environment grows, regulatory frameworks may find it challenging to keep up with these swift advancements. For instance, there is a need for new policies and regulations to govern the use of blockchain in construction contracts and payment processes. The lack of well-defined guidelines and regulatory frameworks instigates a state of indeterminacy among industry stakeholders, impeding their propensity to adopt blockchain solutions. To address the challenge of regulatory uncertainty, it is crucial to foster cooperation and open conversations among industry stakeholders, regulatory entities, and policymakers.

19.4 Prospects of blockchain research in the smart built environment

The built environment stands to undergo a profound transformation through the utilization of blockchain technology. With its decentralized nature, ability to trace and track information, and immutable and transparent attributes, blockchain is well-positioned to make a significant impact in the smart built environment. This subsection delves into an exploration of the notable prospects that blockchain holds in the built environment.

19.4.1 The adoption of blockchain smart contracts

A smart contract is a set of instructions programmed into a blockchain that can automatically enforce an agreement when specific conditions are fulfilled (Zheng et al., 2020). It consists of two main components: predetermined conditions and corresponding actions to be taken when those conditions are met (Wu et al., 2022b). The terms of the smart contract are executed on the blockchain using a computerized protocol that automatically carries out the instructions (Zheng et al., 2020). Once a smart contract is established on a blockchain network, the rules for execution cannot be altered (Li et al., 2021). This allows for the digitization and automation of business processes, reducing the need for human involvement, especially from intermediaries.

Smart contracts have several potential applications in the built environment. Ye et al. (2022) reviewed the potential applications of smart contracts in the built environment. These potential applications include but are not limited to: information exchange (Tao et al., 2021), document change recording (Xue & Lu, 2020), contract administration and payment management (Das et al., 2020), provenance tracking (Zhong et al., 2020), compliance checking for quality (Sheng et al., 2020) and site safety (Xu et al., 2022), trust building (Lu et al., 2021a), information sharing (Wu et al., 2022b), and facility monitoring (Moniruzzaman et al., 2020). Thus, future research should explore these potential applications enabled by smart contracts.

19.4.2 Blockchain with token-based incentives

With the development of the information age, the efficiency of information resource sharing on the blockchain platform has not yet been fully released, and the willingness of participants to share information needs to be improved. Therefore, a blockchain with a token-based incentive mechanism is necessary for users in the smart building environment. A token is a digital asset that is created and managed via a blockchain platform in the realm of blockchain technology. Tokens can be transferred using smart contracts and can fulfil various objectives, such as motivating users to share information or operating as a type of currency within a defined ecosystem (Lu & Wu, 2024). There are several benefits to using token-based incentives to encourage users to share information on a blockchain platform. One of the primary advantages is an upsurge in information sharing. By giving tokens as incentives, users are motivated to share information more openly and frequently, collecting more information and providing managers with a more comprehensive information set for analysis and decision-making. Another benefit is better information quality. Tokens encourage users to provide precise and dependable information. The incentive mechanism motivates users to invest time and effort to guarantee the quality of the shared information, which leads to more precise insights and better decision-making. Increased participation and engagement may also be a potential benefit. The token-based incentive mechanism fosters a sense of participation among blockchain users. They are motivated to actively contribute to the smart built environment by sharing information, enabling better collaboration. Additionally, token-based incentives can enhance trust and transparency. Using tokens and blockchain technology provides a transparent and auditable record of information-sharing activities. This transparency builds trust among users as they can see the fairness and integrity of the incentive mechanism, ensuring their contributions are reasonably recognized and rewarded. Therefore, future research can explore various incentive models for information sharing on blockchain platforms.

19.4.3 Integration of blockchain and other technologies

One area of focus for future research could be the effective integration of blockchain and other digital technologies. The integration of blockchain technology with BIM and IoT has the potential to address numerous challenges in the built environment. These include improving trust in collaborative efforts, streamlining payment processes, optimizing procurement, transforming construction workflows, and enhancing design information management. Future research can also explore the integration of blockchain with artificial intelligence (AI) and machine learning (ML). As an example, Shinde et al. (2020) suggested that integrating blockchain technology with AI and ML has the potential to address fraud and corruption issues in the conventional contract system within the built environment. However, it is essential to note that this research is mainly in the early stages of development. Furthermore, numerous other emerging technologies can be combined with blockchain, such as augmented reality, virtual reality, cloud computing, big data, and digital twins.

19.5 Conclusion

In conclusion, blockchain technology has emerged as a disruptive force for Industry 4.0 and the built environment. This chapter provides a comprehensive overview of the various emerging applications of blockchain in the smart built environment, classified into seven categories. However, several technical and non-technical challenges need to be addressed to ensure the successful implementation of blockchain in the built environment. Despite these challenges, the integration of blockchain with other digital technologies and the adoption of blockchain smart contracts and token-based incentives offer promising prospects for the technology in this industry. As blockchain adoption in the built environment is still in its infancy, the future holds great potential for this technology to revolutionize the industry and pave the way for a more connected and efficient smart built environment.

References

Agarwal, R., Chandrasekaran, S., & M. Sridhar. (2016). *Imagining construction's digital future*. Retrieved June 28, 2023 from https://www.mckinsey.com/business-functions/operations/our-insights/imagining-constructions-digital-future

Attaran, M. (2022). Blockchain technology in healthcare: Challenges and opportunities. *International Journal of Healthcare Management*, *15*(1), 70–83.

Brody, P., Holmes, A., Wolfsohn, E., & Frechette, J. (2019). *Total cost of ownership for blockchain solutions*. Ernst & Young LLP. https://theblockchaintest.com/uploads/resources/EY%20-%20Total%20cost%20of%20ownership%20for%20blockchain%20solutions%20-%202019%20-%20Apr.pdf

Das, M., Luo, H., & Cheng, J. C. (2020). Securing interim payments in construction projects through a blockchain-based framework. *Automation in Construction*, *118*, 103284.

Dixit, S., Stefańska, A., Musiuk, A., & Singh, P. (2021). Study of enabling factors affecting the adoption of ICT in the Indian built environment sector. *Ain Shams Engineering Journal*, *12*(2), 2313–2319.

Dorri, A., Kanhere, S. S., Jurdak, R., & Gauravaram, P. (2017). Blockchain for IoT security and privacy: The case study of a smart home. In *2017 IEEE international conference on pervasive computing and communications workshops (PerCom workshops)* (pp. 618–623). IEEE.

Dwivedi, A. D., Srivastava, G., Dhar, S., & Singh, R. (2019). A decentralized privacy-preserving healthcare blockchain for IoT. *Sensors*, *19*(2), 326.

Farmer, M. (2016). The Farmer review of the UK construction labour model: modernise or die. *Construction Leadership Council*. Retrieved May 18, 2023, from http://www.constructionleadershipcouncil.co.uk/wp-content/uploads/2016/10/Farmer-Review.pdf

Ghobakhloo, M. (2020). Industry 4.0, digitization, and opportunities for sustainability. *Journal of Cleaner Production*, *252*, 119869.

Guan, Z., Lu, X., Yang, W., Wu, L., Wang, N., & Zhang, Z. (2021). Achieving efficient and Privacy-preserving energy trading based on blockchain and ABE in smart grid. *Journal of Parallel and Distributed Computing, 147*, 34–45.

Hamledari, H., & Fischer, M. (2021). Construction payment automation using blockchain-enabled smart contracts and robotic reality capture technologies. *Automation in Construction, 132*, 103926.

Hassan, N. U., Yuen, C., & Niyato, D. (2019). Blockchain technologies for smart energy systems: Fundamentals, challenges, and solutions. *IEEE Industrial Electronics Magazine, 13*(4), 106–118.

Hawlitschek, F., Notheisen, B., & Teubner, T. (2018). The limits of trust-free systems: A literature review on blockchain technology and trust in the sharing economy. *Electronic Commerce Research and Applications, 29*, 50–63.

Hongmei, Z. (2021). A cross-border e-commerce approach based on blockchain technology. *Mobile Information Systems, 2021*, 1–10.

Hou, H. (2017). The application of blockchain technology in E-government in China. In *2017 26th International Conference on Computer Communication and Networks (ICCCN)* (pp. 1–4). IEEE.

Huckle, S., Bhattacharya, R., White, M., & Beloff, N. (2016). Internet of things, blockchain and shared economy applications. *Procedia Computer Science, 98*, 461–466.

Humayun, M., Jhanjhi, N. Z., Hamid, B., & Ahmed, G. (2020). Emerging smart logistics and transportation using IoT and blockchain. *IEEE Internet of Things Magazine, 3*(2), 58–62.

Kamardeen, I., & Hasan, A. (2022). Occupational health and safety implications of an aging workforce in the Australian construction industry. *Journal of Construction Engineering and Management, 148*(10), 04022112.

LeewayHertz. (n.d.). *How to determine the cost of blockchain implementation?* Retrieved July 1, 2021, from https://www.leewayhertz.com/cost-of-blockchain-implementation/

Li, J., Greenwood, D., & Kassem, M. (2019). Blockchain in the built environment and construction industry: A systematic review, conceptual models and practical use cases. *Automation in Construction, 102*, 288–307.

Li, T., Xiong, X., Zheng, G., Li, Y., & Tolba, A. (2023). A blockchain-based shared bus service scheduling and management system. *Sustainability, 15*(16), 12516.

Li, X., Lu, W., Xue, F., Wu, L., Zhao, R., Lou, J., & Xu, J. (2022). Blockchain-enabled IoT-BIM platform for supply chain management in modular construction. *Journal of Construction Engineering and Management, 148*(2), 04021195.

Li, X., Wu, L., Zhao, R., Lu, W., & Xue, F. (2021). Two-layer adaptive blockchain-based supervision model for off-site modular housing production. *Computers in Industry, 128*, 103437.

Lu, Q., & Xu, X. (2017). Adaptable blockchain-based systems: A case study for product traceability. *IEEE Software, 34*(6), 21–27.

Lu, W., & Anumba, C. J. (Eds.). (2022). *Research companion to building information modeling*. Edward Elgar Publishing.

Lu, W., Lou, J., & Wu, L. (2023). Combining smart construction objects – Enabled blockchain oracles and signature techniques to ensure information authentication and integrity in construction. *Journal of Computing in Civil Engineering, 37*(6), 04023031.

Lu, W., & Wu, L. (2024). A blockchain-based deployment framework for protecting building design intellectual property rights in collaborative digital environments. *Computers in Industry, 159*, 104098.

Lu, W., Wu, L., & Xue, F. (2022). Blockchain technology for projects: A multicriteria decision matrix. *Project Management Journal, 53*(1), 84–99.

Lu, W., Wu, L., & Zhao, R. (2021a). Rebuilding trust in the construction industry: A blockchain-based deployment framework. *International Journal of Construction Management*, 1–12.

Lu, W., Wu, L., Zhao, R., Li, X., & Xue, F. (2021b). Blockchain technology for governmental supervision of construction work: Learning from digital currency electronic payment systems. *Journal of Construction Engineering and Management, 147*(10), 04021122.

Luo, H., Das, M., Wang, J., & Cheng, J. C. (2019). Construction payment automation through smart contract-based blockchain framework. In *ISARC: Proceedings of the international symposium on automation and robotics in construction* (Vol. 36, pp. 1254–1260). IAARC Publications.

Michelin, R. A., Dorri, A., Steger, M., Lunardi, R. C., Kanhere, S. S., Jurdak, R., & Zorzo, A. F. (2018). Speedy chain: A framework for decoupling data from blockchain for smart cities. In *Proceedings of the 15th EAI international conference on mobile and ubiquitous systems: Computing, networking and services* (pp. 145–154).

Moniruzzaman, M., Khezr, S., Yassine, A., & Benlamri, R. (2020). Blockchain for smart homes: Review of current trends and research challenges. *Computers & Electrical Engineering*, *83*, 106585.

Moşteanu, N. R., & Flocea, T. (2023). Revolutionizing foreign exchange market: A critical analysis of blockchain's opportunities and challenges. *European Journal of Theoretical and Applied Sciences*, *1*(5), 33–44.

Navaratnam, S., Satheeskumar, A., Zhang, G., Nguyen, K., Venkatesan, S., & Poologanathan, K. (2022). The challenges confronting the growth of sustainable prefabricated building construction in Australia: Construction industry views. *Journal of Building Engineering*, *48*, 103935.

O'Dair, M., & Beaven, Z. (2017). The networked record industry: How blockchain technology could transform the record industry. *Strategic Change*, *26*(5), 471–480.

Orgut, R. E., Batouli, M., Zhu, J., Mostafavi, A., & Jaselskis, E. J. (2020). Critical factors for improving reliability of project control metrics throughout project life cycle. *Journal of Management in Engineering*, *36*(1), 04019033.

Pawlak, M., Poniszewska-Marańda, A., & Kryvinska, N. (2018). Towards the intelligent agents for blockchain e-voting system. *Procedia Computer Science*, *141*, 239–246.

Recarte Suazo, G. A., & Jaselskis, E. J. (1993). Comparison of construction safety codes in United States and Honduras. *Journal of Construction Engineering and Management*, *119*(3), 560–572.

Rupasinghe, T., Burstein, F., Rudolph, C., & Strange, S. (2019). Towards a blockchain based fall prediction model for aged care. In *Proceedings of the Australasian computer science week multiconference* (pp. 1–10).

Scott, D. J., Broyd, T., & Ma, L. (2021). Exploratory literature review of blockchain in the construction industry. *Automation in Construction*, *132*, 103914.

Sharma, P. K., & Park, J. H. (2018). Blockchain based hybrid network architecture for the smart city. *Future Generation Computer Systems*, *86*, 650–655.

Sheng, D., Ding, L., Zhong, B., Love, P. E., Luo, H., & Chen, J. (2020). Construction quality information management with blockchains. *Automation in Construction*, *120*, 103373.

Shinde, R., Nilakhe, O., Pondkule, P., Karche, D., & Shendage, P. (2020). Enhanced road construction process with machine learning and blockchain technology. In *2020 international conference on industry 4.0 technology (I4Tech)* (pp. 207–210). IEEE.

Shobanadevi, A., Tharewal, S., Soni, M., Kumar, D. D., Khan, I. R., & Kumar, P. (2021). Novel identity management system using smart blockchain technology. *International Journal of System Assurance Engineering and Management*, 1–10. Siddiquee, S. S., Khan, M. M. H., Al-Ismail, F. S., Ullah, A., Alam, M. S., & Ahmed, H. (2022). Blockchain applications in smart sustainable city context – A systematic mapping study. *Energy Reports*, *8*, 162–169.

Sookhak, M., Jabbarpour, M. R., Safa, N. S., & Yu, F. R. (2021). Blockchain and smart contract for access control in healthcare: A survey, issues and challenges, and open issues. *Journal of Network and Computer Applications*, *178*, 102950.

Tantidham, T., & Aung, Y. N. (2019). Emergency service for smart home system using ethereum blockchain: System and architecture. In *2019 IEEE international conference on pervasive computing and communications workshops (PerCom Workshops)* (pp. 888–893). IEEE.

Tao, X., Das, M., Liu, Y., & Cheng, J. C. (2021). Distributed common data environment using blockchain and interplanetary file system for secure BIM-based collaborative design. *Automation in Construction*, *130*, 103851.

Themistocleous, M. (2018). Blockchain technology and land registry. *Cyprus Review*, *30*(2), 195–202.

Toorajipour, R., Oghazi, P., Sohrabpour, V., Patel, P. C., & Mostaghel, R. (2022). Block by block: A blockchain-based peer-to-peer business transaction for international trade. *Technological Forecasting and Social Change*, *180*, 121714.

Trček, D. (2022). Cultural heritage preservation by using blockchain technologies. *Heritage Science*, *10*(1), 6.

Turk, Ž., & Klinc, R. (2017). Potentials of blockchain technology for construction management. *Procedia Engineering*, *196*, 638–645.

Tuuli, M. M., Rowlinson, S., & Koh, T. Y. (2010). Dynamics of control in construction project teams. *Construction Management and Economics*, *28*(2), 189–202. United States Environmental Protection Agency (USEPA). (2023). *Basic information about the built environment*. Retrieved March 8, 2023, from https://www.epa.gov/smm/basic-information-about-built-environment

Wang, D., & Zhang, X. (2020). Secure ride-sharing services based on a consortium blockchain. *IEEE Internet of Things Journal*, *8*(4), 2976–2991.

Wang, M., Wang, C. C., Sepasgozar, S., & Zlatanova, S. (2020). A systematic review of digital technology adoption in off-site construction: Current status and future direction towards industry 4.0. *Buildings, 10*(11), 204.

Wazid, M., Bera, B., Das, A. K., Mohanty, S. P., & Jo, M. (2022). Fortifying smart transportation security through public blockchain. *IEEE Internet of Things Journal, 9*(17), 16532–16545.

Wu, H., Zhang, P., Li, H., Zhong, B., Fung, I. W., & Lee, Y. Y. R. (2022a). Blockchain technology in the construction industry: Current status, challenges, and future directions. *Journal of Construction Engineering and Management, 148*(10), 03122007.

Wu, L., Li, X., Zhao, R., Lu, W., Xu, J., & Xue, F. (2022c). A blockchain-based model with an incentive mechanism for cross-border logistics supervision and data sharing in modular construction. *Journal of Cleaner Production, 375*, 133460.

Wu, L., Lu, W., & Chen, C. (2024). Compliance checking for cross-border construction logistics clearance using blockchain smart contracts and oracles. *International Journal of Logistics Research and Applications*, 1–35.

Xu, J., Lu, W., Wu, L., Lou, J., & Li, X. (2022). Balancing privacy and occupational safety and health in construction: A blockchain-enabled P-OSH deployment framework. *Safety Science, 154*, 105860.

Wu, L., Lu, W., Zhao, R., Xu, J., Li, X., & Xue, F. (2022b). Using blockchain to improve information sharing accuracy in the onsite assembly of modular construction. *Journal of Management in Engineering, 38*(3), 04022014. Xue, F., & Lu, W. (2020). A semantic differential transaction approach to minimizing information redundancy for BIM and blockchain integration. *Automation in Construction, 118*, 103270.

Yayman, D. (2021). Blockchain in taxation. *Journal of Accounting and Finance, 21*(4), 140–155.

Ye, X., Zeng, N., & König, M. (2022). Systematic literature review on smart contracts in the construction industry: Potentials, benefits, and challenges. *Frontiers of Engineering Management, 9*(2), 196–213.

Yuan, Y., & Wang, F. Y. (2016). Towards blockchain-based intelligent transportation systems. In *2016 IEEE 19th international conference on intelligent transportation systems (ITSC)* (pp. 2663–2668). IEEE.

Zheng, Z., Xie, S., Dai, H. N., Chen, W., Chen, X., Weng, J., and Imran, M. (2020). An overview on smart contracts: Challenges, advances and platforms. *Future Generation Computer Systems, 105*, 475–491.

Zhong, B., Wu, H., Ding, L., Luo, H., Luo, Y., and Pan, X. (2020). Hyperledger fabric-based consortium blockchain for construction quality information management. *Frontiers of Engineering Management, 7*(4), 512–527.

Zhong, L., Wu, Q., Xie, J., Li, J., & Qin, B. (2019). A secure versatile light payment system based on blockchain. *Future Generation Computer Systems, 93*, 327–337.

20

EFFECTS OF SIGNAGE COLOR ON HUMAN PSYCHOPHYSIOLOGICAL STATES AND WAYFINDING PERFORMANCE IN URBAN UNDERGROUND SPACES WITH DIFFERENT NOISE LEVELS

A combined EEG and VR experiment

*Isabelle Y.S. Chan[1], Hao Chen[2] and Pei Ma[3]**

[1]ASSISTANT PROFESSOR, DEPARTMENT OF REAL ESTATE AND CONSTRUCTION, THE UNIVERSITY OF HONG KONG, HONG KONG, PR CHINA. EMAIL: IYSCHAN@HKU.HK

[2]PHD CANDIDATE, DEPARTMENT OF REAL ESTATE AND CONSTRUCTION, THE UNIVERSITY OF HONG KONG, HONG KONG, PR CHINA. EMAIL: HAOC@CONNECT.HKU.HK

[3]*POST-DOCTORAL FELLOW, DEPARTMENT OF REAL ESTATE AND CONSTRUCTION, THE UNIVERSITY OF HONG KONG, HONG KONG, PR CHINA (CORRESPONDING AUTHOR). EMAIL: PEIMA@CONNECT.HKU.HK

20.1 Introduction

Rapid urbanization resulting from population growth has become an increasingly significant concern worldwide. According to a recent report by the United Nations, 55% of the global population currently resides in urban areas, and this figure is projected to rise to 68% by 2050 (Department of Economic and Social Affairs, 2018). This trend is particularly pronounced in cities with high population densities, which have resorted to the concept of the "vertical city" involving the construction of massive high-rise buildings. While this approach has demonstrated the efficacy of vertical development (Al-Kodmany, 2018), it is not without its drawbacks, such as the accumulation of air pollutants and obstruction of natural light and breezes (Ding, 2013; Chan & Liu, 2018). Therefore, the optimal utilization of urban underground space (UUS) has become imperative. UUS has notable differences in spatial design and environmental quality compared to aboveground spaces, presenting a range of challenges and risks to UUS users. For example, the enclosed nature of UUS limits users' visual connectivity with outdoor landmarks or greenery, impeding their sense of orientation,

DOI: 10.1201/9781003383840-21

navigation abilities, perception of environmental control, and psychological well-being and potentially impacting their physical health and emergency response (Liu, et al., 2020).

Existing empirical studies on building users' wayfinding and psychophysiological states have primarily focused on visual stimuli, such as signage (Vigolo, et al., 2020). For instance, Kalantari et al. (2022) demonstrated the integrated effects of signage design style, interior color, and architectural features on wayfinding performance. Rodrigues et al. (2019) summarized nine signage design practices from a literature review, covering aspects such as text formatting, information hierarchy, density, terminology usage, and pictograms. Additionally, Shi et al. (2020) validated the impact of signage illuminance level, color, height, and observation angle on wayfinding performance. However, individuals' visual attention may vary under different noise levels. For instance, Coutrot et al. (2012) indicated that non-spatial sound can affect visual attention in terms of eye position, fixation duration, and saccade amplitude. Liu et al. (2020) discovered that sound level affects the visibility of visual elements, paralleling changes in visual attention due to variations in sound levels.

While previous studies have offered some insights into the association between environmental noise and the visual functions of building occupants, there remains a significant gap in comprehensive research that investigates the simultaneous effects of signage color and noise on the wayfinding abilities and psychophysiological states of UUS users. Moreover, there is a lack of studies that approach these combined audio-visual effects from a neuroscience perspective. Electroencephalography (EEG) has proven to be an effective tool in uncovering human cognitive mechanisms under different built environments (Lee, et al., 2022; Bower, et al., 2023). By employing a combined EEG-Virtual Reality (VR) approach, this study aims to investigate the influence of different signage color systems (visual stimulus) and noise levels (audio stimulus) on psychophysiological states (mental fatigue, mental workload, alertness, anxiety) and wayfinding performance (task accomplishment time) of UUS users.

20.2 Literature Review

20.2.1 Wayfinding in Urban Underground Spaces

Despite the growing interest in underground spaces in recent years, their public acceptance has historically been low due to the absence of natural ventilation, sunlight, and a connection with nature (Chan & chen, 2023). The enclosed nature of underground spaces often leads to disorientation, making it challenging for occupants to navigate effectively. This lack of effective navigation negatively impacts their well-being, resulting in feelings of frustration, irritation, anxiety, and stress (Chan & Chen, 2023; Barati, et al., 2021). Therefore, it is crucial to prioritize the design of wayfinding systems in underground spaces (van der Hoeven & van Nes, 2014).

Existing studies on wayfinding in underground spaces tend to focus on the influence of space design, evacuation facilities, and lighting on navigation in evacuations. For instance, Mossberg et al. (Mossberg, et al., 2021) evaluated the impact of evacuation system design on elevator usage in a deep underground subway station. They developed four scenarios, including a baseline scenario, advanced design (with more informative voice alarms and green lighting), smartphone guidance, and a countdown timer. The results indicated that both the advanced design and smartphone guidance improved elevator usage behaviors. Similarly, Wang et al. (2021) examined the effects of various factors such as the number of evacuees, minimal group impact, and pre-action time settings on evacuation performance. Their findings demonstrated how different settings, including crowd congestion, group effects, and pre-action time, can impact evacuation efficiency during emergencies. Also, fire characteristics and alarm designs have been found to affect evacuation performance during emergencies (Wang, et al., 2021). Additionally, Wang, et al. (2022) examined the influence of natural lighting presence and intensity on human wayfinding performance in underground spaces during emergencies. While the presence of natural lighting affected wayfinding efficiency, the intensity of lighting did not significantly impact performance. Overall, a large number of studies on underground wayfinding have focused on the impact of architectural design, natural lighting, and evacuation system design on wayfinding and evacuation performance. However, in UUS where architectural design and natural lighting are limited, signage systems play an essential role in facilitating wayfinding.

In fact, signage has gained increasing attention as a valuable tool to provide information to occupants during the wayfinding process. For instance, Kubota, et al. (2021) conducted a study to evaluate the effectiveness of signage installation location (distance and interaction angle) and signage arrow design on direction compliance behaviors. The findings validated the effectiveness of interaction angle and arrow design pointing to the route center in promoting direction compliance. In a subsequent study, Kubota et al. (2022) investigated the effects of signage placement (closeness to the route and interaction angle) and types (floor signage, one-face signage on the ceiling, and four-face signage on the ceiling) on occupants' direction compliance behaviors. Moreover, Kostakos, et al. (2021) explored the effectiveness of a signage system with different assistive lights on human psychophysiological reactions (measured by heart rate) and wayfinding performance in a burning parking lot. These studies provide initial insights into the vital role that signage plays in wayfinding within underground spaces. However, in reality, underground users are affected by various environmental stimuli during the wayfinding process, with noise being one of the most significant influencing factors in enclosed UUS.

20.2.2 Visual-auditory stimulus from environmental psychology perspective

The field of environmental psychology and neuroarchitecture has demonstrated that different architectural designs can elicit a range of emotional and physiological responses in humans, which can be measured using neuroscience tools (Lee, et al., 2022). While there is a growing body of literature on the relationship between building environments and associated health consequences, most of the research has focused on aboveground spaces. The studies of neuroarchitecture in underground environments are still in their early infancy, despite the fact that the impact of indoor environments on health can be more severe in UUS. When perceiving the environment, humans are exposed to multisensory stimuli, including visual, auditory, and olfactory stimuli, which elicit different responses. This study specifically investigates the psychophysiological impact of visual and auditory stimuli on wayfinding performance of UUS users.

20.2.2.1 Signage color (visual stimuli)

Signage serves as a means of conveying practical information to building occupants and providing guidance on directions, identification, safety, and so on (Ding, et al., 2023). The effectiveness of the occupant-signage interaction is determined by various factors such as signage color (Ding, et al., 2023). For instance, Ding et al. (2023) conducted a study using eye tracking to evaluate four aspects of evacuation signage design, including color (red vs. green), placement location (high vs. low), graphics (running man vs. arrow), and flashing settings (flashing vs. no flashing). Their findings indicated that low-altitude signs with a red color and running man graphics achieved the highest evacuation performance. Moreover, Shi et al. (2022) examined the effects of signage design, specifically color and direction, on attention bias and perceptual processing. They found that red signage resulted in longer perceptual processing times, while green signage attracted stronger attention. Chen et al. (2020) evaluated the physiological, eye-tracking, and escape performance in different color combinations of signage design in underground subway stations. Their results demonstrated that signage with a combination of green and black achieved the best evacuation escape performance, low cognitive load, and high search efficiency. Similarly, Kalantari et al. (2022) found that a color pattern highlighting destinations and increasing contrast of signage against the overall environment effectively enhanced individuals' wayfinding in healthcare environments. While there have been a growing number of studies investigating the effects of signage color on occupants' wayfinding, the results have been inconsistent. More importantly, these existing studies have not taken into consideration the simultaneous effects of other stimuli, such as environmental noise, on the wayfinding process.

20.2.2.2 Impacts of signage under different environmental noise levels (visual-auditory stimulus)

Acoustic environments are capable of generating powerful reactions in human beings, influencing how they interact with and perceive their surroundings. Previous studies have recognized the psychophysiological effects of soundscapes, which refers to human perception of the acoustic environment (Erfanian, et al., 2019; Algargoosh, et al., 2022). For instance, Lee et al. (2020) examined the impact of different background sounds on cognitive performance, satisfaction, preference, and physiological responses in a controlled open-plan office. The results showed that white noise and office noise helped maintain a certain level of alertness and improve cognitive performance. Furthermore, previous studies have shown that environmental noise is closely related to various cognitive functioning, behaviors, and health of building occupants (Hahad, et al., 2022). While there have been studies investigating the respective impacts of visual and auditory stimuli on building occupants separately, studies mimicking real-life situations and investigating their simultaneous effects are rare. Liu et al. (2023) conducted a stress-induction-reduction experiment to examine the effect of visual, auditory, and visual-auditory stimuli on stress reduction. The results indicated that visual and visual-auditory stimuli had a more significant stress-reducing effect compared to auditory stimuli, as evidenced by physiological sensors and behavioral data. Algargoosh et al. (2022) investigated the impacts of audiovisual stimuli in a worship space and found that the acoustic environment and audiovisual congruency amplified the intensity of emotional impact, which varied depending on the building's acoustic environment. These limited studies highlight the importance of understanding the effects of both signage colors and background noise levels on the wayfinding performance of UUS users.

20.2.3 Conceptual framework

The extensive review of existing literature reveals two significant gaps: (a) a lack of comprehensive studies investigating the combined effects of signage color and noise levels on UUS users' wayfinding performance and (b) a lack of understanding regarding the mentioned combined effects on psychophysiological states of UUS users. Hence, this study aims to investigate the impact of signage color (visual stimuli) and background noise (auditory stimuli) on the wayfinding performance and psychophysiological states of UUS occupants. These psychophysiological states include mental fatigue, mental workload, alertness, and anxiety, as illustrated in Figure 20.1.

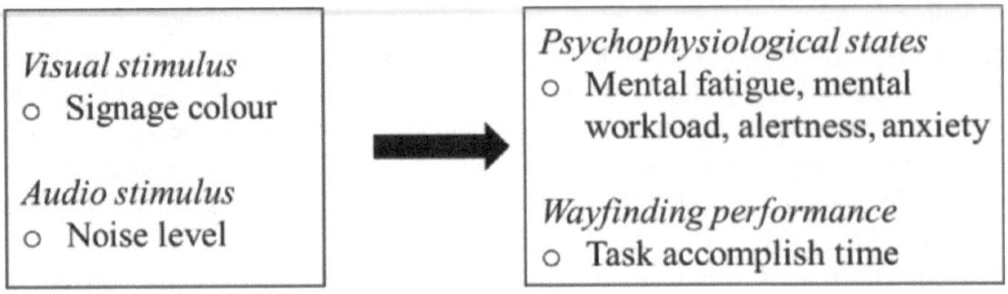

Figure 20.1 Conceptual framework

20.3 Research methodology

20.3.1 Virtual reality experiment

The VR technology provides researchers with the opportunity to explore the influence of different environmental design settings on human performance, encompassing physical, psychophysiological, and overall performance. In view of its advantages in ecological validity, experimental control, safety, and cost-effectiveness, immersive VR technologies have been widely adopted in building environment and wayfinding research in recent decades (Wang, et al., 2022). To investigate the effects of signage color and noise levels on occupants' psychophysiological states and wayfinding performance, four virtual underground subway station models were developed in this study, with varying signage color systems and noise levels.

20.3.2 Experiment design

The VR experiment focused on an underground subway station, a widely utilized urban underground space worldwide. The choice of a subway station as the virtual model was motivated by the significance of complex circulation design and ambient noise in influencing passengers' wayfinding behaviors. By simulating a subway station environment, the study aimed to investigate the impact of these factors on participants' navigational performance. The 3D virtual model of the subway station was developed using 3D Max software and imported into Unity to create a VR testbed. A control panel was designed to allow the selection of different signage systems and background noise levels for different groups of participants, ensuring controlled experimental conditions.

Two signage design systems were developed for the study: black-and-white and color-coded. The black-and-white system lacked color differentiation across spatial information and architectural features, while the color-coded system utilized different colors to distinguish names and areas of space in maps, signage, and architectural features. This allowed for a comparison of the effectiveness of these two signage color systems. Regarding the background noise, two levels were implemented in the virtual model: a low level at 55 dB and a medium level at 65 dB. These noise levels were set based on the Environmental Impact Assessment Ordinance in Hong Kong (Environmental Protection Department). To ensure participant safety and avoid any risks of hearing impairment, scenes with high noise levels were excluded from the study. The background noise included common sounds found in subway stations, such as talking noise, human movement noise, and mechanical noise. To control for the potential impact of actual sounds in the physical experiment environment, the actual sound level of the room was measured and regulated during the study.

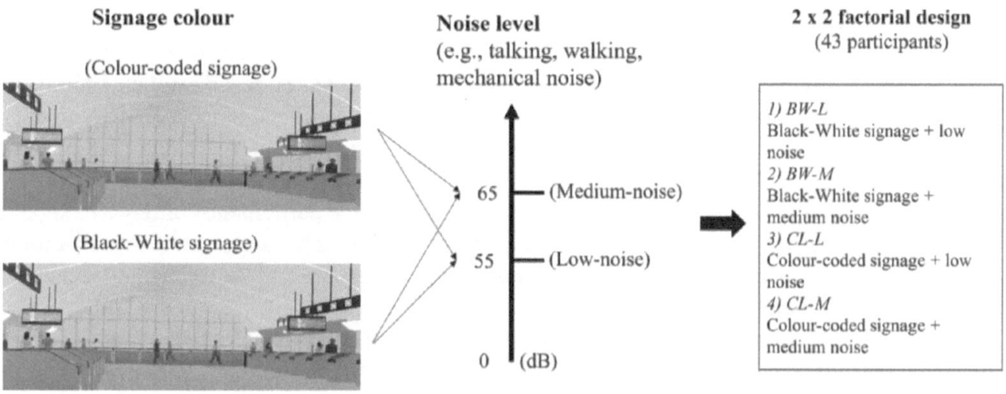

Figure 20.2 Four VR scenarios

Four experimental scenarios were developed, each representing a distinct combination of wayfinding systems (black-and-white vs. color-coded) and background noise levels (low vs. medium). To prevent the carry-over effect, a between-group approach was employed, where participants were randomly assigned to one of the four experimental scenarios, as illustrated in Figure 20.2. This approach ensured that each participant experienced only one specific combination of signage color systems and background noise levels:

- Scenario 1 – black-and-white signage system + low noise level (BW-L): signage in the environment displayed white text on a black background. The background noise level was set at 55dB, representing a low-noise environment
- Scenario 2 – black-and-white signage system + medium noise level (BW-M): signage in the environment featured white text on a black background. The background noise level was set at 65dB, representing a medium-noise environment
- Scenario 3 – color-coded signage + low noise (CL-L): different signage was assigned with different colors, aligning with those of the architectural colors adopted in respective areas. Additionally, the background noise level was set at 55dB, indicating a low-noise environment
- Scenario 4 – color-coded signage + medium noise (CL-M): similar to the CL-L scenario, different signage was assigned with different colors, aligning with those of the architectural colors adopted in respective areas. Additionally, the background noise level was set at 65dB, representing a medium-noise environment.

20.3.3 Measurements of psychophysiological states and wayfinding performance

In extant studies, questionnaires are commonly used to assess individuals' emotional states in response to stimuli. Examples of such questionnaires include the Positive and Negative Affect Schedules (Thompson, 2007) and the Profile of Mood States (Yokoyama, et al., 1990). However, these questionnaire surveys have limitations due to their subjective nature and the potential for the "Hawthorne effect," where participants may alter their behaviors or responses due to being observed. Consequently, there has been growing interest in objectively evaluating human psychophysiological states using neuroscience and neuroimaging technologies, particularly in the field of the built environment (Li, et al., 2020). This study employed electroencephalography (EEG) to objectively measure participants' psychophysiological states based on their brainwave signals. The Emotiv EPOC 14-channel wireless EEG headset was utilized to measure participants' psychophysiological states throughout the experiment. EEG enables the recording and analysis of electrical activity in the brain, providing insights into various cognitive and emotional processes. By utilizing EEG, this study sought to decode and analyze participants' brainwave signals to obtain objective measures of their psychophysiological states. This approach provides a more direct and precise evaluation of human emotional states in response to experimental scenarios, overcoming the limitations of subjective questionnaire surveys.

The collected EEG data underwent preprocessing using EEGLAB, which consisted of three main steps: (a) high-low pass filtering: the EEG signals were filtered to retain brainwave signals within the frequency range of 2–39 Hz. This filtering step aided in removing unwanted noise and artifacts outside the relevant frequency range; (b) Independent Component Analysis (ICA) and manual artifact removal: ICA was applied to separate the EEG signal into independent components. These components were then visually inspected, and any unwanted artifacts, such as those related to blinking or eye movement, were manually identified and removed from the data; (c) calculation of relative power spectral density: the relative power spectral density of different frequency bands was calculated using validated equations. In this study, the frequency bands of interest included delta (δ: 0.5–4 Hz), relative theta (θ: 4–8 Hz), relative alpha (α: 8–13 Hz), sensory-motor rhythm (RSMR: 13–15 Hz), relative mid beta (β: 15–20 Hz), and relative beta (RB: 13–30 Hz) (Kim, et al., 2020). These frequency bands are associated with diverse brainwave activities and can provide insights into participants' alertness, mental workload, mental fatigue, and anxiety levels.

The brainwave equations derived from previous studies are summarized in Table 20.1 (Kim, et al., 2020; Mathersul, et al., 2008). These equations facilitate the measurement of alertness, mental workload, mental fatigue, and anxiety based on the relative power spectral density values obtained from the EEG data. The details of these equations can be referred to in the corresponding cited sources. Additionally, wayfinding performance, which served as the experiment outcome, was assessed by measuring the time each participant took to complete the wayfinding task. This measurement provides an objective evaluation of participants' efficiency in navigating in the virtual subway environment.

Table 20.1 Brainwave equations for different psychophysiological indicators

Indicators	*Equations*	*Ref*
Alertness	$\frac{RSMR_{Frontal} + RMB_{Frontal}}{RT_{Frontal}}$	(Kim, et al., 2020)
Mental workload	$\frac{RB_{Frontal}}{RT_{Frontal} + RA_{Frontal}}$	(Kim, et al., 2020)
Mental fatigue	$\frac{RB_{Frontal}}{RA_{Frontal}}$	(Kim, et al., 2020)
Anxiety	$In(RA)\vert_{F3} - In(RA)\vert_{F4}$	(Mathersul, et al., 2008)

20.3.4 Experiment procedures

Participants were recruited for the VR experiment through methods such as posters, social networks, and emails. A total of 43 participants engaged in the study and were randomly assigned to the four scenarios, with 11 participants in the BW-L scenario, 11 in the BW-M scenario, ten in the CL-L scenario, and 11 in the CL-M scenario. The majority of participants were aged 18–29 (93%), followed by a smaller group within the age range of 30–39 (7%).

After confirming that the EEG and VR headset connections were functioning properly, the experiment commenced with a scene set in a subway environment, i.e., T0 as shown in Figure 20.3. Participants had the opportunity to explore and familiarize themselves with the VR environment by moving around. Subsequently, they were informed about the tasks to be performed. The participants were told that they were in route to an important job interview and were already running late. They needed to navigate through the underground subway station to reach the interview venue as quickly as possible. A timer was displayed throughout the wayfinding process to emphasize the time pressure. When participants pressed the "start" button, the timer began counting down, marking this time point as "T1." Upon locating the correct exit, a time point labeled as "T2" was recorded, signifying the end of the experiment. The experiment process is illustrated in Figure 20.3.

20.4 Results

The combined effects of color-coded wayfinding systems and noise levels on human psychophysiological states and wayfinding performance are depicted in Figures 20.4 and 20.5. In general, participants using color-coded wayfinding systems displayed increased levels of alertness, mental workload, and mental fatigue, irrespective of the noise levels. Surprisingly, their wayfinding performance was also inferior compared to those using black-and-white systems, regardless of noise level.

Beta brainwaves indicate an individual's attention and consciousness levels, alpha brainwaves represent relaxation and calmness levels, and theta brainwaves signify deep relaxation. These variables served as key measurements of alertness, mental workload, and mental fatigue (refer to

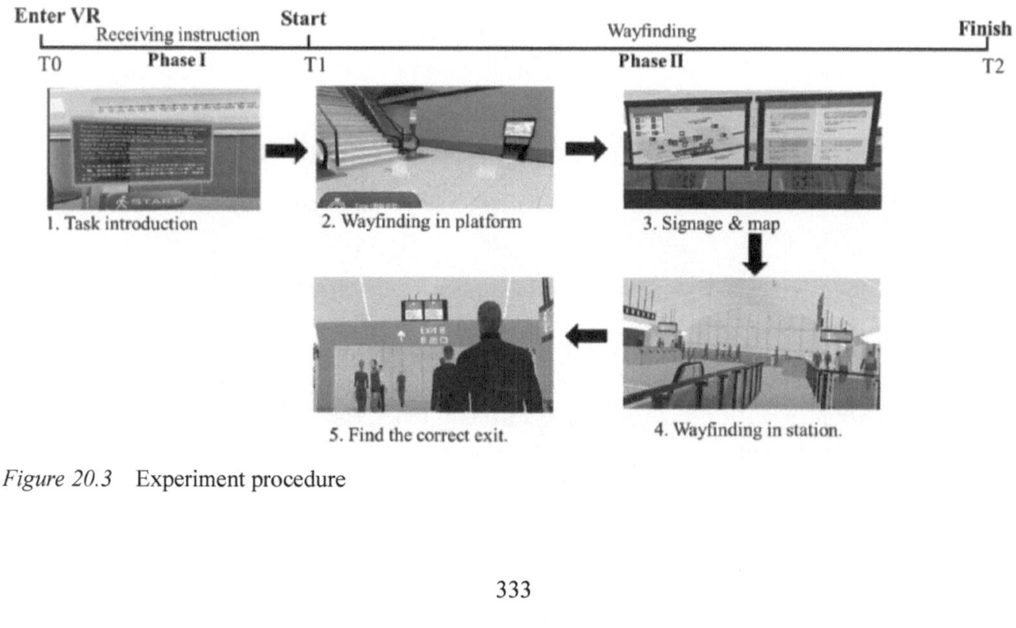

Figure 20.3 Experiment procedure

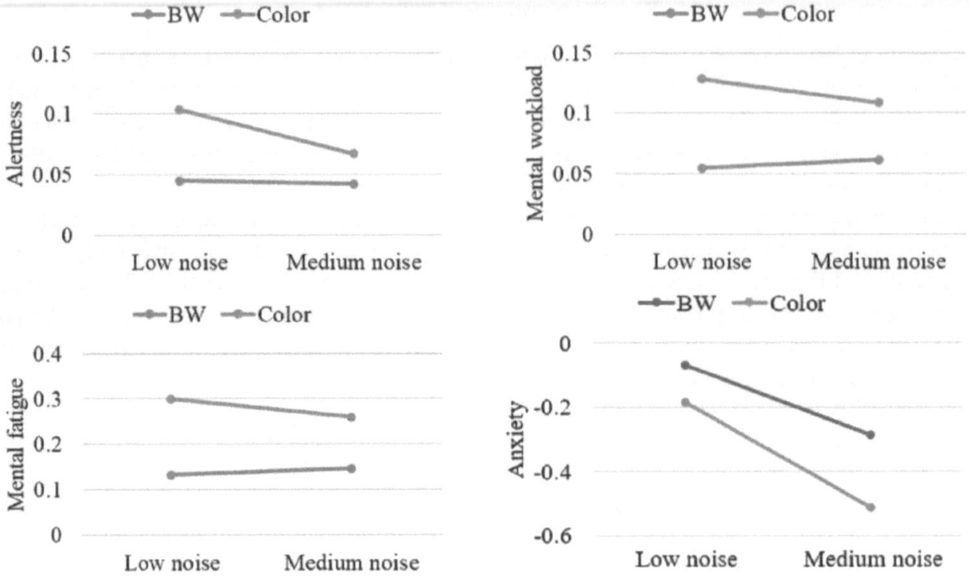

Figure 20.4 Combined effects of signage color and noise levels on human psychophysiological states

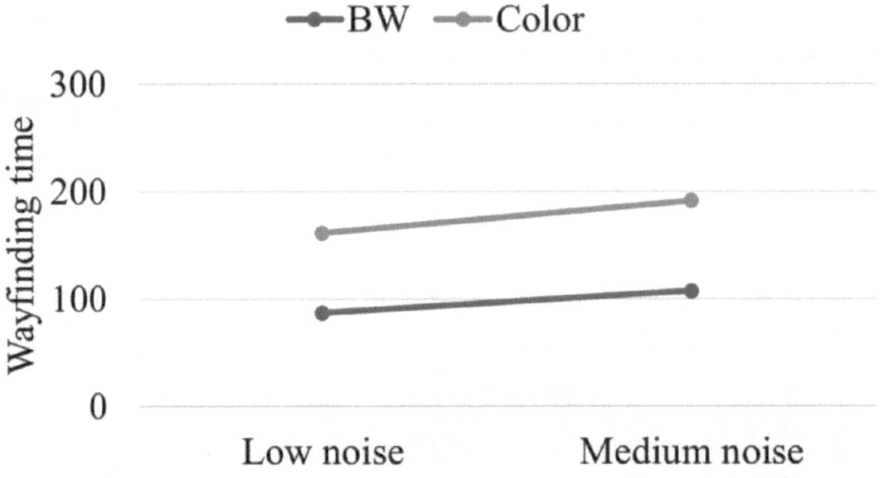

Figure 20.5 Combined effects of signage color and noise levels on wayfinding performance

Table 20.1), which were primarily determined by higher levels of beta and lower levels of alpha and theta brainwaves.

The color-coded wayfinding system employed in this study exposed participants to a combination of stimulations created by four different colors. Previous studies have indicated the varying effects of color tones on human arousal. For instance, warm color tones, such as red and orange, can stimulate an increase in beta brainwaves due to their exciting and positive effects on active feelings (e.g., (Yoto, et al., 2007)). In contrast, cold color tones, such as black and white in this case, may inhabit individuals' beta brainwaves and increase alpha waves due to their relaxing and calming effects (e.g., (Yoto, et al., 2007; Ueda, et al., 2004; Shimagami, 1991)). This study, drawing from the most commonly adopted color themes in underground zoning designs, adopted the colors red, orange, green, and blue in the color-coded system. Compared with the black-and-white scenario, the color-coded signage system tends to evoke higher brain activation and arousal, resulting in elevated levels of alertness, mental workload, and even mental fatigue.

For anxiety levels, these are indicated by the difference between left and right frontal alpha activities. A negative value indicates that frontal right alpha activities are higher than frontal left alpha activities, which happens when individuals are exposed to stressful stimuli (Giannakakis, et al., 2022). In some situations, higher frontal right alpha activities also signify a greater tendency for withdrawal behaviors. The results indicated that participants in the black-and-white groups experienced lower levels of stress than those in the color-coded groups, regardless of noise levels. This echoes some previous studies that find the color black reduces anxiety behaviors (de Abreu et al., 2020).

Interestingly, despite the increased levels of alertness, participants using the color-coded wayfinding system took longer to complete the wayfinding task compared to those using the black-and-white system. In fact, previous studies have suggested that cold color tones, through enhancing one's task concentration, were associated with improved task results and better memory (Llinares, et al., 2021). Furthermore, there is also empirical evidence from existing studies that warm color tones enhance performance in simple tasks only, while cold color tones improve performance in difficult tasks (Xia, 2016). Navigating in an enclosed underground environment could be regarded as a challenging task. The results indicated that a colorful underground environment might cause distractions during wayfinding, negatively impacting individuals' wayfinding efficiency.

Furthermore, the results indicated that under noisier environments, both color-coded and black-and-white groups took longer to complete the wayfinding task. This could be attributed to the increased level of acoustic distractions. For the black-and-white groups, a higher noise level led to elevated levels of mental workload and fatigue. This could be an indication of increased stimulation and information loading caused by simultaneous visual and auditory stimuli. For the color-coded groups, a higher noise level resulted in decreased levels of alertness, mental workload, and fatigue. This could be an indication of overstimulation and information overload. Further studies are recommended to explore the potential non-linear impacts of audiovisual stimuli on UUS users' wayfinding and psychophysiological states in greater depth.

20.5 Discussion and conclusion

Unlike their aboveground counterparts, UUS users encounter unique challenges in wayfinding due to the absence of visual connectivity with outdoor landmarks or green spaces. Additionally, the enclosed nature of UUS poses difficulties in managing noise levels. While previous studies have separately examined the effects of noise levels and signage colors on human cognitive performance, little is known about their combined impact on wayfinding and psychophysiological states in UUS. This study investigated the effect of signage colors (black and white vs. color coded) and noise levels (low vs. medium) on wayfinding performance, alertness, mental workload, mental fatigue, and anxiety levels of UUS users. By conducting a combined EEG-VR experiment, the following key findings were observed:

1) Alertness, mental workload, mental fatigue, and anxiety levels of participants using the color-coded wayfinding system were higher under both low and medium noise levels
2) Wayfinding performance of participants using the black-and-white wayfinding system was better than that of their counterparts using the color-coded system under both low and medium noise levels

3) Mental workload and fatigue levels of participants in the color-coded wayfinding system were lower in noisier underground environments
4) Mental workload and fatigue levels of users in the black-and-white wayfinding system were higher in noisier underground environments.

In complex UUS designs, the use of multiple colors becomes necessary to define or differentiate different zones. However, the increased number of colors and the mixture of color temperatures may result in mixed stimulating effects and information overload for users, causing negative impacts on UUS users' psychophysiological states and wayfinding performance. Although this study found empirical evidence that underground occupants' wayfinding performance and physiological states were better in black-and-white signage systems and that color systems may overstimulate occupants during their wayfinding process, it may be too early to draw practical recommendations at this stage. Further studies are recommended to delve into the effects of different color combinations and tones on users' cognitive and psychophysiological states. In addition to the current background noise, further studies can consider other sources of sound, such as music and natural sounds.

Acknowledgment

This research was funded by the Early Career Scheme (No. 27203319), the General Research Fund (17203920), the Research Grant Council, and HKSAR.

References

Algargoosh, A., Soleimani, B., O'Modhrain, S., and Navvab, M. "The impact of the acoustic environment on human emotion and experience: A case study of worship spaces," *Building Acoustics,* vol. 29, no. 1, pp. 85–106, 2022/03/01 2022, doi: 10.1177/1351010X211068850.

Al-Kodmany, K. The vertical city: a sustainable development model. WIT Press, 2018.

Barati, N., Hahsemi Zadegan, S. A., and Kasravi, R. "The role of survey details for wayfinding problem in complex pedestrian underground interchange with poor architectural configuration," *Tunnelling and Underground Space Technology,* vol. 108, p. 103718, 2021/02/01/ 2021, doi: 10.1016/j.tust.2020.103718.

Bower, I. S., Hill, A. T., and Enticott, P. G. "Functional brain connectivity during exposure to the scale and color of interior built environments," *Human Brain Mapping,* vol. 44, no. 2, pp. 447–457, 2023, doi: 10.1002/hbm.26061.

Chan, I. Y., and Chen, H. "Towards an integrative analysis of underground environment and human health: A survey and field measurement approach," *Engineering, Construction, and Architectural Management,* 2023, doi: 10.1108/ECAM-12-2021-1120.

Chan, I. Y. S., and Chen, H. "Lessons learned from the COVID-19 pandemic: A multigroup structural equation modelling of underground space environment and users' health," *Buildings*, vol. 13, no. 5, doi: 10.3390/buildings13051321.

Chan, I. Y. S., and Liu, A. M. M. "Effects of neighborhood building density, height, greenspace, and cleanliness on indoor environment and health of building occupants," *Building and Environment,* vol. 145, pp. 213–222, 2018/11/01/ 2018, doi: 10.1016/j.buildenv.2018.06.028.

Chen, N., Zhao M., Gao, K., and Zhao, J. "The physiological experimental study on the effect of different color of safety signs on a virtual subway fire escape – An exploratory case study of Zijing Mountain subway station," *International Journal of Environmental Research and Public Health,* vol. 17, no. 16, doi: 10.3390/ijerph17165903.

Coutrot, A., Guyader, N., Ionescu, G., and Caplier, A. "Influence of soundtrack on eye movements during video exploration," *Journal of Eye Movement Research,* vol. 5, no. 4, p. 2, 2012.

de Abreu, M. S., *et al.* "The impact of housing environment color on zebrafish anxiety-like behavioral and physiological (cortisol) responses," *General and Comparative Endocrinology,* vol. 294, pp. 113499–113499, 2020, doi: 10.1016/j.ygcen.2020.113499.

Department of Economic and Social Affairs, "68% of the world population projected to live in urban areas by 2050, says UN," in *United Nations*, ed: United Nations, 2018.
Ding, C. "Building height restrictions, land development and economic costs," *Land Use Policy,* vol. 30, no. 1, pp. 485–495, 2013/01/01/ 2013, doi: 10.1016/j.landusepol.2012.04.016.
Ding, N., Ma, Y., Fan, Z., and Shi, J. "Test the effectiveness of building safety guidance signs in a T-junction corridor based on eye movement data," *Journal of Safety Science and Resilience,* vol. 4, no. 2, pp. 123–129, 2023/06/01/ 2023, doi: 10.1016/j.jnlssr.2022.11.001.
Environmental Protection Department. "A guide to the EIA ordinance." HKSAR. Accessed https://www.epd.gov.hk/eia/english/guid/ordinance/guide7_4-8.html.
Erfanian, M., Mitchell, A. J., Kang, J., and Aletta, F. "The psychophysiological implications of soundscape: A systematic review of empirical literature and a research agenda," *International Journal of Environmental Research and Public Health,* vol. 16, no. 19, doi: 10.3390/ijerph16193533.
Giannakakis, G., Grigoriadis, D., Giannakaki, K., Simantiraki, O., Roniotis, A., and Tsiknakis, M. "Review on psychological stress detection using biosignals," *IEEE Transactions on Affective Computing,* vol. 13, no. 1, pp. 440–460, 2022, doi: 10.1109/TAFFC.2019.2927337.
Hahad, O., *et al. "Cerebral* consequences of environmental noise exposure," *Environment International,* vol. 165, p. 107306, 2022/07/01/ 2022, doi: 10.1016/j.envint.2022.107306.
Kalantari, S., *et al.* "Evaluating the impacts of color, graphics, and architectural features on wayfinding in healthcare settings using EEG data and virtual response testing," *Journal of Environmental Psychology,* vol. 79, p. 101744, 2022/02/01/ 2022, doi: 10.1016/j.jenvp.2021.101744.
Kim, H., Hong, T., Kim, J., and Yeom, S. "A psychophysiological effect of indoor thermal condition on college students' learning performance through EEG measurement," *Building and Environment,* vol. 184, p. 107223, 2020.
Kostakos, P., Alavesa, P., Korkiakoski, M., Marques, M. M., Lobo, V., and Duarte, F. "Wired to exit: Exploring the effects of wayfinding affordances in underground facilities using virtual reality," *Simulation & Gaming,* vol. 52, no. 2, pp. 107–131, 2021, doi: 10.1177/1046878120944567.
Kubota, J., Sano, T., and Ronchi, E. "Assessing the compliance with the direction indicated by emergency evacuation signage," *Safety Science,* vol. 138, p. 105210, 2021/06/01/ 2021, doi: 10.1016/j.ssci.2021.105210.
Kubota, J., Sano, T., and Ronchi, E. "The impact of people-signage interaction on way-finding evacuation behaviour," *Fire Safety Journal,* vol. 142, p. 104023, 2024, doi: 10.1016/j.firesaf.2023.104023.
Lee, Y., Nelson, E. C., Flynn, J., and Jackman, J. S. "Exploring soundscaping options for the cognitive environment in an open-plan office," *Building Acoustics,* vol. 27, no. 3, pp. 185–202, 2020/09/01 2020, doi: 10.1177/1351010X20909464.
Lee, S., Shin, W., and Park, E. J. "Implications of neuroarchitecture for the experience of the built environment: A scoping review," (in English), *Archnet-IJAR,* Article 2022, doi: 10.1108/ARCH-09-2021-0249.
Li, J., Jin, Y., Lu, S., Wu, W., and Wang, P. "Building environment information and human perceptual feedback collected through a combined virtual reality (VR) and electroencephalogram (EEG) method," *Energy and Buildings,* vol. 224, p. 110259, 2020/10/01/ 2020, doi: 10.1016/j.enbuild.2020.110259.
Liu, C., Kang, J., and Xie, H. "Effect of sound on visual attention in large railway stations: A case study of St. Pancras railway station in London," *Building and Environment,* vol. 185, p. 107177, 2020/11/01/ 2020, doi: 10.1016/j.buildenv.2020.107177.
Liu, G., *et al.* "Stress recovery at home: effects of the indoor visual and auditory stimuli in buildings," *Building and Environment,* vol. 244, p. 110752, 2023/10/01/ 2023, doi: 10.1016/j.buildenv.2023.110752.
Llinares, C., Higuera-Trujillo, J. L., and Serra, J. "Cold and warm coloured classrooms. Effects on students' attention and memory measured through psychological and neurophysiological responses," *Building and Environment,* vol. 196, p. 107726, 2021, doi: 10.1016/j.buildenv.2021.107726.
Mathersul, D., Williams, L. M., Hopkinson, P. J., and Kemp, A. H. "Investigating models of affect: Relationships among EEG alpha asymmetry, depression, and anxiety," *Emotion,* vol. 8, no. 4, p. 560, 2008.
Mossberg, A., Nilsson, D., and Wahlqvist, J. "Evacuation elevators in an underground metro station: A virtual reality evacuation experiment," *Fire Safety Journal,* vol. 120, p. 103091, 2021/03/01/ 2021, doi: 10.1016/j.firesaf.2020.103091.
Rodrigues, R., Coelho, R., and Tavares, J. M. R. "Healthcare signage design: a review on recommendations for effective signing systems," *HERD: Health Environments Research & Design Journal,* vol. 12, no. 3, pp. 45–65, 2019.
Shi, J., Ding, N., and Jiang, F. "The influence of color and direction on the perceptual processing of standard evacuation signs and the effect of attention bias," *Fire Safety Journal,* vol. 132, p. 103638, 2022/09/01/ 2022, doi: 10.1016/j.firesaf.2022.103638.

Shi, Y., Zhang, Y., Wang, T., Li, C., and Yuan, S. "The effects of ambient illumination, color combination, sign height, and observation angle on the legibility of wayfinding signs in metro stations," *Sustainability,* vol. 12, no. 10, 2020, doi: 10.3390/su12104133.

Shimagami, K. "Change in environmental image and state of mind and body by color light," *Jpn J Hyg,* vol. 46, p. 135, 1991.

Thompson, E. R. "Development and validation of an internationally reliable short-form of the positive and negative affect schedule (PANAS)," *Journal of Cross-cultural Psychology,* vol. 38, no. 2, pp. 227–242, 2007.

Ueda, Y., Hayashi, K., Kuroiwa, K., Miyoshi, N., Kashiba, H., and Takeda, D. "Consciousness and recognition of five colors: Using functional-MRI and brain wave measurements (International conference on mind body science: Physical and physiological approach joint with the eighteenth symposium on life information science)," *Journal of International Society of Life Information Science,* vol. 22, no. 2, pp. 366–371, 2004.

van der Hoeven, F., and van Nes, A. "Improving the design of urban underground space in metro stations using the space syntax methodology," *Tunnelling and Underground Space Technology,* vol. 40, pp. 64–74, 2014/02/01/ 2014, doi: 10.1016/j.tust.2013.09.007.

Vigolo, V., Bonfanti, A., Sallaku, R., and Douglas, J. "The effect of signage and emotions on satisfaction with the servicescape: An empirical investigation in a healthcare service setting," *Psychology & Marketing,* vol. 37, no. 3, pp. 408–417, 2020.

Wang, D., Liang, S., Chen, B., and Wu, C. "Investigation on the impacts of natural lighting on occupants' wayfinding behavior during emergency evacuation in underground space," *Energy and Buildings,* vol. 255, p. 111613, 2022/01/15/ 2022, doi: 10.1016/j.enbuild.2021.111613.

Wang, D., Yang, Y., Zhou, T., and Yang, F. "An investigation of fire evacuation performance in irregular underground commercial building affected by multiple parameters," *Journal of Building Engineering,* vol. 37, p. 102146, 2021/05/01/ 2021, doi: 10.1016/j.jobe.2021.102146.

Wang, D., Zhou, T., and Li, X. "Impacts of environment and individual factors on human premovement time in underground commercial buildings in China: A virtual reality – based study," *ASCE-ASME Journal of Risk and Uncertainty in Engineering Systems, Part A: Civil Engineering,* vol. 7, no. 1, p. 04020056, 2021/03/01 2021, doi: 10.1061/AJRUA6.0001112.

Xia, T., Song, L., Wang, T. T., Tan, L., and Mo, L. "Exploring the effect of red and blue on cognitive task performances," *Frontiers in Psychology,* vol. 7, pp. 784–784, 2016, doi: 10.3389/fpsyg.2016.00784.

Yokoyama, K., Araki, S., Kawakami, N., and Tkakeshita, T. "Production of the Japanese edition of profile of mood states (POMS): assessment of reliability and validity," *[Nihon koshu eisei zasshi] Japanese Journal of Public Health,* vol. 37, no. 11, pp. 913–918, 1990.

Yoto, A., Katsuura, T., Iwanaga, K., and Shimomura, Y. "Effects of object color stimuli on human brain activities in perception and attention referred to EEG alpha band response," *Journal of Physiological Anthropology,* vol. 26, no. 3, pp. 373–379, 2007.

CONCLUSION

Weisheng Lu and Chimay J. Anumba

Over the past two decades or so, smart technologies have witnessed an unprecedented development and created a sea of change in our society, economy, business, education, and recreation through to our daily necessities. They have also profoundly disrupted the built environment, which refers to the human-made surroundings that provide the setting for human activity, including homes, buildings, zoning, streets, sidewalks, open spaces, transportation options, and other infrastructure. These smart technologies are still developing at a rapid pace that can go beyond the imagination of any visionary. With such technologies to enable autonomous and smart control of immediate environments, work processes, and the natural and industrial systems, the built environment is also experiencing a transformation in its design, construction, and operation in a bid to increase efficiency and achieve overall sustainability or, in other words, to achieve a smart built environment (SBE).

This book is a collection of 20 chapters from renowned researchers and teams in the domain of SBE around the world. It is encouraging to see that researchers have explored the latest smart technologies in various aspects of SBE, including big data, artificial intelligence (AI), building information modeling (BIM), digital twins (DTs), blockchain, Internet of Things (IoT), and, most recently, large language models such as ChatGPT and many others. The associated insurgence of big data, computing power, and smart algorithms has fostered a virtuous circle of smart technology development. SBE is becoming the very frontier for implementing these smart technologies, and, in turn, for fostering the continuous development of the smart technologies to deliver societal benefits. It is also noteworthy that there are several interdisciplinary collaborations amongst the different disciplines and across the stages of the smart built environment. The 19 chapters range from design to construction, operation, smart living and conservation stages and cover building, community, and city scales. They also include both theoretical and practical contributions. The contributors are all geographically dispersed with a few of them being collaborations across different institutions.

It can be seen from the 20 chapters that SBE domains have been exploited and advanced, particularly by tapping into the enormous opportunities as enabled by the smart technologies. As expected at the conception of the book, the contributors have created an invaluable resource for those seeking to make smarter and more sustainable development decisions in the built environment sector. Readers will find these studies inspiring and helpful. Further developments are

DOI: 10.1201/9781003383840-22

expected to be undertaken in concert with stakeholders (such as city administrators, developers, constructors, technology developers, policymakers, and the general public) to ensure that research efforts deliver real impacts that leverage the potential of SBE. It can be seen from the book that the contributions from "smart living and operation" are fewer than those from "design and construction" and "smart city" fields. This is a growth area that will garner more attention in the near future.

The 20 chapters in the book represent only a small proportion of the work in the field of smart built environment. They are intended to stimulate further research, development, and implementation in this field.

INDEX

Note: Page numbers in *italics* indicate figures, **bold** indicate tables in the text, and references following "n" refer notes.

Abanda, F. H. 142
active exoskeletons 120, **121**, *121*, 127
activity-based models 221, 237; city digital twins and 227–228; constraint-based models 221, 228, 230–231; development of 224; econometric models 221, 228, 231–234, *232*, *233*; micro-simulation models for 224; rule-based model 221, 234–236, *235*; single-facet models 221, 228–230; time geography in 224, 225–227, 230–231; travel demand model 236
activity pattern 220, 224–225; cultural environment in 222; effects of 222–223; influence of urban environment on 221; interpersonal interactions among residents in 222; natural environment in 222; *see also* activity-based models
adaptability 38, 42, 56, 59, 61–62, 124, 153, 196, 297, 304, 309
Adaptive Gaussian Cloud Transformation (A-GCT) 97
Adegoke, E. I. 18
advanced work packaging (AWP) 17
agent-based modelling (ABM) 18, 22, 224–225
agile curb management 291–292, 294
Akanbi, T. 138
Akhavian, R. 18
ALBATROSS model 230, 234–236
Albreshne, A. 18
Alemi, M. M. **123**
Al-Jumaily, A. A. 123
Al Jurdi, M. 259
Alzraiee, H. 138
Anam, K. 123
Antwi-Afari, M. F. **126**
AnyBody Modelling System 86
Apache Jena framework 153
ARCBERT model 77
ArchiGAN model 41
Arentze, T. 230
Argenziano, P. 263
Arisha, A. 164
artificial intelligence (AI) 10, 67, 138, 193, 197, 264, 276, 285, 297, 300, 305, 339; algorithms 96; blockchain technology with 322; building management through 175; climate-adaptive smart UTS 300–301, 307–308; compatibility 182, 187; data-driven 39; enabled technologies 289; generative 39, 42; with hybrid method 13; integrating 178–179, 187; integrating techniques 79; medical 164; model 18; natural language processing in 64–65, 67; ontology, defined 133; *see also specific models*
Aryal, A. 85
audio stimulus *see* noise levels
augmented reality (AR) 163, 168–171, *168*, 182, 322
Autodesk Revit software 113
Autodesk Tandem 13, 181–182, **183–186**, 187
Auto-GPT 78
autonomous vehicles (AVs) 284, 293
autoregressive language models 73

Babalola, A. 124
Baltrusch, S. J. **123**, **124**
Bances, E. **126**
Behzadan, A. H. 18
Benachio, G. L. F. 18

Bennett, S. T. **125**, **126**
bi-directional coordination 195, 197–199, 205, 207, 212, *213*, 215–216
big data 64, 249, 266, 274, 286, 290, 297, 300, 301, 309, 322, 339; analytics 41, 196, 298, 305, 307–308, 315; climate-adaptive smart UTS 300; harnessing 39; incorporating 67; SBE and 39
bike-sharing 272, 273, 275–277, 278, 288
BIM-Revit 202
biomechanical indicators 84, 86–88, 89, **90**, 97–98
Bjornsson, H. C. 18
blockchain technology **261**, 298, 315, 317–320, 339; additional costs 320; with artificial intelligence 322; with building information modelling 260, 322; components 315; construction management 316–317; data privacy 319; effectiveness of 316; incorporating into urban ecosystems 316; Internet of Things with 315–317, 322; interoperability issues 319; lack of regulatory clarity 320; with machine learning 322; network topology 318; non-technical challenges 319–320; and other technologies integration 322; resistance to change 319–320; sharing economy 315, 318; smart cities 315, 317–318; smart contract 321; smart energy 316; smart government 315, 317; smart homes 315, 317; smart transportation 316; technical challenges 318–319; with token-based incentives 321
Bloomberg-GPT 77
Borg CR10 scale 84
Borg rate of perceived exertion (RPE) scale 84
Boschert, S. 195
bottom-up approach 55, 224, 244, *245*, 256
breezeways 253
Brilakis, I. 197
Brody, P. 320
Bueno, C. 260
Bughin, J. 197
building automation systems (BAS) 13, 194, 198–199
building energy model (HTB2) 250
building information modeling (BIM) 4, 9, 14, 17, 65, 105, 116, 137, 169, 197, 199, 202, 317, 339; 3D models 263, 307; -based digital construction 106–107, 115; -based digital fabrication 115; -based prototype system 115; BIM-R platform 142; in building lifecycle 174–175; city information modelling and 260–261, **261**, 262, 265; classification of 176–177; collaborating format (BCF) 177; in design and construction practice 176; digital twins and 5, 11, *11*, 106–107; goal of 142; incomplete or incorrect 138, 142; information extraction 135; integration 179, 182, 187; integration of blockchain technology with 260, 322; Internet of Things integration 107; levels of detail 176–177; levels of development 5; libraries 35; -oriented tools 135; platforms 108, 112–113, *113*, 115; -related databases 135; -related sources 135; -RFID management 107; semi-automated reasoning system 142; for transportation infrastructure 301
building lifecycle 181–182, 187; building information modeling in 174–175; digital twins in 174–175; Internet of Things in 174–175
buildingSMART Data Dictionary (bSDD) 69

Calastri, C. 234
Campisi, T. 277
Caneparo, L. 143
capacity constraints 76, 226, 230
carbon emission 221, 222, 271, 277, 278–279, 301, 304
cardiovascular indicators 85
Castaldini, F. 195
Castillo, J. C. 275
center of area (CoA) method 93
Chapin, F. S. 224
charging as a service (CaaS) 278–279
ChatGPT 39, 41–42, 68, 69–71, 79; as knowledge base 69, *69*; large-scale language model 73–75, *73*, 79, 339; limitation 77; zero-shot prediction 75, **75**, *75*
ChatLaw 77
Chatterjee, A. 216
Chen, K. 258, 263
Chen, M. K. 275
Chen, S. **127**
Cheng, J. C. P. 9
chi-squared automatic interaction detector (CHAID)-based method 236
Cho, Y. K. 118
Chua, C. K. 206
Church of Colònia Güell 35, 37
city as a building *246*, 247, 256
city digital twins (CDTs) 227, 237, 265
City Geography Markup Language (CityGML) 262–263
city information modelling (CIM) 258–259, 301, 307; advancement of 266; challenges 265; characteristics of **259**; clarifications about 260; computing resources and storage inefficiency 266; cybersecurity and intellectual property 266–267; data quality 265–266; efficacy of 265; model construction 263; ontological study of 262–263; PRISMA guidelines to 259; reluctance towards application 265; scalability of 267; standardized workflow 266; sustainable urban management 264; technical components of 260–261, **261–262**; urban disaster management 262, 263; urban facility management 264; urban planning 263–264; virtual communities 265

civilian-centric principles 303, 310
climate-adaptive smart UTS 296–297; adaptability 304; ai and decision-making 307; artificial intelligence 300–301; big data 300, 307; car/parking sharing 304; civilian-centric 303; data generation and acquisition 306; data storage and curation 306–307; data supply 299–300; data transmission 306; data visualization 301–302; decision-making 300–301; deployment framework 302, *302–303*; digital technologies 297–302; digital technology deployment plan for 305–309; digital twin 301–302; digital twin development 307–308; electrification and automation 304; ethical challenges 310; infrastructure and IoT deployment 306; Internet of Things 299–300; joint risk/emergency management 304; pattern recognition 300; policy challenges 309–310; principles 303; problem identification and goal setting 305; public transportation dominance 304; rapid transfer centres 304; requirements of 298; scalability and sustainability 308–309; smart and sustainable roads 304; strategical urban planning 303–304; strategies 308; system thinking 304; technical challenges 309; technical operation 304–305; testing and optimization 308; transportation planning 304; user engagement and communication 308
cloud computing 196, 315, 322
cloud model (CM) 92–94, 98, 261
CO_2 sensors 163, 164–167, 171
cognitive digital twins (CDTs) 10, 13–14
Combinatorial Algorithm for Rescheduling Lists and Activities (CARLA) 231
communication networks 6, 194, 195–196, 203, 207, 215, 306
community value 287–290, 292–294
comparison matrix 178, 179, 181–182, **183–186**, 187
complex survey method 108, 109, 116
comprehensive utility-based system of activity-travel scheduling options modelling (CUSTOM) 232
computational process models *see* rule-based models
computer-aided facility management (CAFM) 199–200
computer numerical control cutting 47, 105
connected vehicles (CV) 289
Constantinescu, C. 122
constraint-based models 221, 228, 230–231
Construction 4.0 104
construction management 98, 158, 261, 301, 316–317; multi-prime method 48; procurement route 51; software 152
construction manager at risk method 48
Construction Motion Data Library (CML) 92, 93, 96
contact-sensor-based methods 88–89
convolutional neural network (CNN) 153
coupling constraints 226, 230, 231
Cox proportional hazard model 229
credibility 18, 32
Cureton, P. 265
cyber-physical-social system (CPSS) 305, 311
cyber-physical systems (CPS): characteristics 6; convergence 6; digital twins and 7–8
cybersecurity 6, 32, 266–267, 310

D3M model 41
Dahmen, C. 122
Daiwa House Group 115
D'Amico, R. D. 14
data acquisition 153, 175, 182, 190, 261, 310
data analytic techniques 203
data-based digital twins 12, 13
data collection methods *see specific methods*
data-driven generative DfX 38; big data 39; challenges 40; optimization across disciplines 41; prospects 41; stages 40; workflow 40, *40*
data integration module 153–154
data transmission 306, 317
Davis, K. G. 123
DaySim 232–233, *233*
De Bock, S. **123**
decision-support tools 31
deep deterministic policy gradient (DDPG) algorithm 289
deep learning approach 39–41, 288, 300, 307
Delbert, C. M. 212
Del Duca, G. 263
Deloitte 196
de Looze, M. P. 119
demand and supply, mismatch between 275–276
design-bid-build (D-B-B) method 48, 51, 106
Design-Build (D-B) method 48–49, 51, 53
de Soto, B. G. 32
de Vries, A. W. **127**
DFAB house project 52, *52*; DFAB implementation to 53; DFAB techniques in **54**; Project delivery to 53–55
dialogue understanding *see* spoken language understanding
diffusion model 40–41
diffusion of innovation (DOI) theory 121
digital construction 107, 116; BIM-based 106–107, 115; defined 104–105; implementation framework of 109, *110*; project 113–114; stakeholders 111; terminologies and concepts related to 108; Zurich Switzerland 105, 111, 116
digital design 104, 115; capabilities 169; integrated process 105–106, 112, *112*; mock-ups *170*

digital fabrication 107, 116; BIM-based 115; conventional manual construction vs. 106; defined 104–105; design modeling information for *109*; digital twin for 106; implementation framework of 109, *110*; information 109–111, 112; information integration 106; manager 106; methods 105; and prerequisites 105–106; project 112; terminologies and concepts related to 108; West Coast of the United States 105, 111, 112, 116; *see also* integrating digital fabrication
digital mock-ups 168, *169*, 206; digital 169, 170, *170*; physical 163, *163*, 169, 170–171
digital model 8, 53, 190, 194, 196, 201–203, 208; 3D 307; advanced 169; federated 301; geometric 198, *198*; integrated 163, 207; of physical system 7, *7*
digital platform 201, 272, 274, 279; collaborative 162; integrating 163; single 271, 273
digital shadow 7, *7*
digital technologies *see specific technologies*
digital twins (DTs) 4, 104, 162, 167–169, 180, 190–191, 220, 308–309, 322, 339; action-based 171; approaches to implementing 196; benefits 8, *8*; building information modeling and 5, 11, *11*, 106–107; in building lifecycle 174–175; challenges 9; city 227, 237, 265; classifications 9–10; climate-adaptive smart UTS 301–302; cognitive 10, 13–14; commercial modeling 13; components 190, *191*, 195–196; conceptual model of architecture *176*; connection between real and virtual ophthalmology *163*; in construction industry 197–198; continuum 177, *177*; control group in 180; cyber-physical systems and 7–8; data-based 12; data formats for 177; defined 194–195; development group in **178**, 179, **179**, 180, 182, 187; for digital fabrication 106; digital forms to *7*; enabling technologies 201–203, *202*; essential components to create building 4, *5*; evaluation criteria to analyze platforms 179–181, 182; existence 5, 9, 10; fused 301; gaming engines for 13; Gemini Principles for 10, *10*; for healthcare facilities management 199–201, 191–194, 198 (*see also* healthcare facilities management); history 194; incremental 163; innovators group in 180, 182; integrated 164; Internet of Things for 7, 12, 13; iterative 162–163, 171; management group in **178**, 180, **181**, 182, 187; maturity 9, *9*; methods for designing and implementing 12–13; operational 9; in ophthalmology healthcare 162–164, 170–171; in other industries 196, **197**; parts 7–8, 195–196; platforms 13, 179, 181–182, **183–186**, 187; process of delivering *165*; prototype system (*see* prototyping); sensor-actuator connections in 180, 182, **184**, 185; simulation 10, 13; and smart cities 11–12, **12**; in smart cities 4, 11, **12**, 14; Smart Curb 285, 286; status 9, 10; steps to implementing 196; strategy 111; success 6; system-based 12–13; transportation activity 302; transportation infrastructure 302; Universal Scene Description and 178; urban 221; urban climate 301–302; user identification 178, **178**; of UTS (*see* climate-adaptive smart UTS); visualization (*see* visualization)
Dijst, M. 231
discrete event simulation (DES) 18, 163, 167–169, *167*, 170–171
domain-specific language model 36, 77
Dorri, A. 317
Draper, J. 215
drone 97, **262**
dynamic curbside management 285, 288
dynamic dispatching 274–275, 276
dynamic scheduling 143, 152–158

Easwaramoorthy, M. 192
ecocity master-planning 255
econometric models 221, 228, 231–234, *232*, *233*
electric vehicles (EVs) 284–285, 305; challenges 228; charging 228, 284, 286–288, 290, 304; driving 304; plug-in 236
electrocardiography (ECG) 85
electrodermal activity (EDA) 85
electroencephalography (EEG) 85, 327, 332–333, 335
electronic indicators 85, 89
Ellegård, K. 226
emotional fatigue 84
Enckell, M. 26
energy consumption 36–38, 221, 223, 271, 278–279, 308
energy efficiency 20, 29, 31, 36, 38, 248, 264, 279, 316
energy supply 244–245, 247
energy use 250–252, *251*, 253
engineering, procurement, and construction (EPC) 56–57
Epic Games Unreal Engine 181–182, **183–186**, 187
ergonomic assessment method *92*; body-segment-level 91; for construction workers *92*; directions for advancing 96; heuristic data-driven 84, 91; joint-level 90–93, 96, 98; vision-based 89, 98; whole-body assessment 85–86, 88, 91, 93
ergonomic indicators 84, 123; contact-sensor-based methods 88–89; data collection methods and **90**; directions for 97–98; facilitating optimal PFA 89; non-contact-sensor-based methods 88–89, 97; scales 88, 89–91, 96, 98; systematic

observation 88; types of methods employed 88; *see also* ergonomic assessment method
existence digital twin 5, 9, 10
exoskeletons 119; active 120, *121*, **121**, 127; classification 120, **121**; in construction industry 124–127; examples **121**; full body support 120, **121**; in healthcare 122; lower body support 120, **121**; in manufacturing industry 122; in other industries 123, **123**; passive 120, *121*, **121**, 127; powered 120, *121*, **121**, 127; research directions 127–128; rigid 120, **121**; semi-active 120, **121**; soft 120, **121**; theories to 121–122; types 120, **121**, **123–124**, **125–127**; upper body support 120, **121**
explainable AI (XAI) 301
external comfort 249, 255

fatigue: in construction 83; defined 84; emotional 84; mental 84; physical 84, 89 (*see also* physical fatigue prevention); types 84
Fethke, N. B. 123
Fine, C. H. 49–50, 52, 58–59, 62
Fischer, M. 317
forecasting evolutionary activity-travel of households and their environmental repercussions (FEATHERS) 236
Forner-Cordero, A. **124**
Fraser Tower, Singapore 197–198
front-end planning 17
Fuerst, F. 226
fuzzy reasoning mechanism 93, 97, 98

Gao, C. 264
garbage in, garbage out mechanism 42
Garikapati, V. 275
Gaudi, A. 35, 37
Gemini Principles 10, *10*
general data protection regulation (GDPR) 310
generative adversarial network (GAN) 40–41
generative design 36–37; computer-aided 36; for excellence (*see* data-driven generative DfX; rule-based generative DfX); reasons for flourishing of 35–36; tools 35
geographic information system (GIS) 261–263, **261**, 266, 297, 301, 305, 307
Gil, J. 258
Glaessgen, E. H. 195
global positioning system (GPS) 276, 284, 286, 298, 300, 310
Golabchi, A. 125, **125**
Golparvar-Fard, M. 32
Google Street View (GSV) image 286
Goulias, K. G. 229
Grasshopper software 37, 106
Grau, D. 18
Gregg, K. 287
Grieves, M. 171, 194
ground reaction forces (GRF) 87
Gu, D. 263
Gutierrez, N. 124

Habib, K. M. N. 229
Hafezi, M. H. 232
Hägerstrand, T. 224, 230
Hambrick, D. 216
Hamledari, H. 317
Han, K. K. 32
Hao, H. 286
Hao, T. 249
Hartley, E. 265
healthcare facilities management (HFM): digital twin for 191–194, 198–201, *200*, *202*, 203, 209, 212, 214–216; issues 198–199; system architecture 200–201, *200*
heart rate (HR) 85
heart rate variability (HRV) 85
Hemminger, S. 208
Heuristics Gaussian Cloud Transformation (H-GCT) 91–94, 96, 98
high-technology healthcare building 163, 164, 171; *see also* digital twins
Hobolink 203, 207, 209, *212*, 216
Horni, A. 230
how-what-who principles 115
Huang, H. 265
Huang, Y. 275
Huigen, P. P. 231
human behavioral science 291
human-in-the-loop methods 78
Hwang, J. **124**

indoor air quality (IAQ) sensor 201, 202, 205–209, *206*, **208**, 212–213, 215
Industry 4.0 (Fourth Industrial Revolution) 64, 104, 107, 315, 322
Industry Foundation Classes (IFC codes) 17, 69, 107, 136, 177–178, 182, 262–263
Inertial Measurement Units (IMUs) 88
infectious disease, spread of 220, 244, 248–249, 254–255
Information and communication technology (ICT) **262**, 274, 305, 309, 314
information extraction 65, 135, 137–138, **139**, **141**, 142–143, **144–146**, **149–151**, 157
information sharing 110, 115, 116, 300, 316, 319, 321; on cloud BIM platform 112; digital construction 113; multi-server 98
institutional constraints 226, 230, 234
integrated project delivery (IPD) 48, 51, 53, 57–58, 60–61, 106, 112
integrated smart curb management 284–286; agile 291–292, 294; challenge 288, 290; digital twin

286; dynamic curbside management 288–289; equity and commuting well-being in 290–291; evolving contextual factors 291–292; integrating human behavioral science into 291; macro- and micro-spatial scale research for 292; pricing in 293; prospects and challenges 290–294; self-driving vehicles in 293–294; smart curb planning and management framework 290; socio-technical challenge 292–293; spatial planning of 286–288; strategies 287, 290
integrating digital fabrication (DFAB) 61–62; cross-case analysis 57–58; defined 46; degree of 47, *48*; DFAB house *52*, 53–55; full 47; hybrid 47; implementation of 46, 47, *48*; minimal 47; partial 47; project delivery methods to (*see* project delivery method); Tujia Pan-Museum *53*, 55–57
intellectual property 266–267, 310, 320
intelligent planning unit (IPU) 17; advantages of 20; application 21, 23, *24*; characteristics of 20; composite *24*; credibility and cybersecurity concerns 32; decision support tools for 31; defined 19–20; design 21–22; implementation in real world 31; interaction 24–25; network 21, 24–25, *24*; optimized electricity demand management case 28–29, *29*; physical entities 23; planning 21–22, *23*; proposed 18; rating system 30–31; refinement 25; replication 23; scale and number of entities *19*; sports centre design case 29–30, *30*; strategy 22; structural health monitoring (SHM) of bridges case 25–28
intent detection 68, *68*, 70, 72, 74
International Energy Agency (IEA) 278
Internet-of-Things (IoT) 6, 9, 32, 39, 152–153, 196, 247, 256, 297, 339; blockchain technology with 315–317, 322; building information modeling and 107; in building lifecycle 174–175; city information modelling and 261, **261**, 263–264; climate-adaptive smart UTS 299–300, 308; convergence 6; deployment 306; for digital twins 7, 12, 13; in intelligent planning unit 20, 24; people tracking 163, 164–167, 171; urban transportation systems 297, 299–301, 305–307, 309
Inverse Perspective Mapping (IPM) 286
Iranzo, S. **123**
Iterative digital twins 162–163, 171

Jaccard similarity algorithm 154
Jaselskis, E. J. 318
JavaScript Object Notation (JSON) 196
joint discrete-continuous model 229
Jorgensen, M. J. **124**

Karshenas, S. 142
Keen, M. 197
Key Item Method (KIM) 88
key performance indicators (KPIs) 10, 14, 18–20, 22, 195
Khosakitchalert, C. 142
Kim, J. 29
Kim, S. 118, 143, 236
Kitchenham, B. 120
Knoepfle, D. 275
knowledge modelling 135–138, **139–140**, 142, **144**, **147–151**
Kockelman, K. M. 275
Koo, C. 18, 20, 32
Kopp, V. **126**
Kotowski, S. E. 123
Kritzinger, W. 7, 9

Lancaster urban information model (LCIM) 265
LangChain 78
language models 64–65; autoregressive 73; as backbone models 65, 69, 71, 74–75, 77, 79; challenges 78–79; complexity of construction 78; decoder-only architectures 67, 73–74, *73*; demand for high-quality data 79; domain-specific 77; dynamic project life cycle 79; encoder-decoder architecture 40, 67, 73–74, *72*, 76; encoder-only architectures 67, 71–74, *71*; error analysis 76–77; human-in-the-loop methods 78; integration with domain ontology 77; as knowledge bases 65, 69, 70, 77, 79; large-scale 67–68, 73–79, *73*; multi-modal models 78; multi-scale 68; in natural language understanding tasks 72; scope of 67; small-scale 67, 69, 74–77, 79; trade-off between cost and effectiveness 79; utilization of external tools 78; *see also specific models*
large language models (LLMs) 42, 67, 339
large-scale language model 67–69, 73–74, *73*, 75, 77–79; *see also* ChatGPT
Le Corbusier style 40
Lee, S.-K. 143
Lenntorp, B. 231
LiDAR 97, **262**, 263
Li, J. 318
Li, M. 263
Likert scale 193, 213
Lin, X. 40
Liu, J. 288
Liu, K. 142
Liu, Y. 164
Liu, Z. 142, 258, 260, 264
Loo, B. P. Y. 264
Lu, X. 262, 263
Luo, H. 316

Ma, J. 229
Ma, Z. 142

Machado-León, J. L. 287
machine learning (ML) 13, 43, 67, 143, 197; blockchain technology with 322; building management through 175
Madni, A. M. 196
Mandhan, S. 287
Mantha, B. R. 32
mass customization 65, 66–67, 79, 115
matrix comparative analysis 178, 179, 181–182, **183–186**, 187
MediaPipe 89
mental fatigue 327, 329, 332–335, **332**; *see also* physical fatigue prevention
Meyer, R. **125**
micro-simulation models 224, 232
Miller, E. J. 229
Min, J. U. 18
misalignment relationships 50, 58–62
mixed reality (MR) 182
mobile devices 203
Mobility-as-a-Service (MaaS) 271; accessibility improvements through 277; balance between demand and supply 275–276; defined 272–273; dynamic dispatching on 274–275; energy consumption and carbon emission reductions through 278–279; operation of 274; public/parking space savings through 278; smart traffic management system in 276–277
Mobility-on-Demand (MOD) 272, 285, 287
model distillation 41
model of action spaces in time intervals and clusters (MASTIC) 231
modular construction 20, 58, 66–67, 319
modular integrated construction (MiC) project 41
modularity 46–47, 57, 308; defined 48; Fine's three dimensions of 58–59, 62; horizontal 50–51, **50**; integration-modularity *51*, 52, 58, 62; misalignment 61–62; modular complement relationship through decrease of *60*; modular complement relationship through increase of *60*; multi-dimensional *49*, 52, 59; process 55, 59; strategies 49; supply chain 48–49; vertical 50–51, **50**, 52
modularization 17
modular prototyping 169, 171
modus operandi 120, **121**, 128
Moher, D. 119
Moniruzzaman, M. 317
Moyon, A. **124**
MS COCO images 286
Mukherjee, A. 171
multi-agent learning algorithm (QMIXm) 289
MultiGCN-LSTM 288
multi-modal models 78
multimodal travel (MT) 272, 273
multinomial logit (MNL) model 224, 232
multiple discrete continuous extreme values (MDCEV) framework 234
multivariate multiple discrete-continuous probit (MDCP) model 233
muscle fatigue 84–85
MX Gateway 203, 207, 209

Najafi, P. 265
natural language generation (NLG) 65, 73
natural language processing (NLP) 64, 135–136, 142, 143, 300; in artificial intelligence 64–65, 67; in smart construction 65–66, 69–70, 74, 77, 79; tasks 68, 74
natural language understanding (NLU) 65, 66, 71–74
Neil, J. S. 212
Nepal, M. P. 142
neural networks 40, 67, 153, 307
Niknam, M. 142
"nine-nine-six" work system 223
Niu, Y. 18
noise levels 327, *334*; background 330–331, 336; human movement 330; low and medium 331, 335; mechanical 330; signage under 329; talking 330; unwanted 332
non-connected vehicles (NCV) 289
non-contact-sensor-based methods 88–89
NVIDIA Omniverse 13, 181–182, **183–186**, 187

Occupational Repetitive Actions (OCRA) 88
Ogunseiju, O. **126**
Okpala, I. 121, **126**
Omoniyi, A. **123**
Omrany, H. 260
ontology 143; applications 135, 136–138, **139**, **144–149**; -based framework 18; based on New Rules of Measurement 142; benefits 158; for BIM-related information extraction 135; city information modelling 262–263; constraint theory 137; crane operations 143, 152–157; creation 136–138, **139**, 142, **150–151**; defined 133; domain 71, 72, 77; functions 134–138; information extraction 65, 135, 137–138, 142–143, 157; integration 136–138, **140–141**; knowledge modelling 135–138, 142; model 109, 111; reasoning and conformance check 135–138, 142; for reasoning and conformance checks 136; regulatory ontology 136; resource planning (*see* resource planning); scheduling (*see* scheduling); technologies enabling implementation of 133–134; universal ontology 136
OpenPose 89, 94
OpenSim 86, *86*
operational digital twin 9
optimized electricity demand management 28–29, *29*

Otten, B. M. **127**
Ovako Working Posture Analysis System (OWAS) 88

Pan, W. 47, 58
Park, J. H. 318
passive exoskeletons 120, *121*, **121**, 127
pattern recognition 300, 301, 307
Pauwels, P. 69
PESASP model 231
physical asset 11, 13, 170, 175, 179, 190, 195–197, 201
physical fatigue prevention (PFA) 83; biomechanical indicators 84, 86–88; physiological indicators 84, 85; research directions for 96–98; significance of 84; subjective indicators 84–85; *see also* ergonomic assessment method; ergonomic indicators
physical twin (PT) 6, 9, 194; acquire state 7; digital 171; digital twins and 7–8
physiological indicators 84, 85, 89, **90**, 98
pick-ups and drop-offs (PUDOs) 285, 287–289
Pillai, M. V. **127**
Pinho, J. P. **124**
plug-in electric vehicles (PEVs) 236
Pochowski, A. L. 287
Poisson regression model 229
powered exoskeletons *see* active exoskeletons
product lifecycle management (PLM) 194–195
project delivery method 47–49, 54, 61; contractual 58; to DFAB house 53–55; DFAB implementation and 47, 52; horizontal integration 6, 50–52, **50**; horizontal modularity 50–52, **50**; inherent 57; innovative 17; integration-modularity in *51*, 52; to Tujia Pan museum 56–57; types 50–51, **50**, 52; vertical integration 6, 50–52, **50**, 106, 115; vertical modularity 50–51, **50**, 52
prototyping 191, 214–216; 3D monitoring of IAQ 205–206, *206*; actuation layer 201; application/digital platform layer 201; bi-directional coordination 195, 197–199, 205, 207, 212, *213*, 215–216; case study 192; communication layer 201; communication network 203; data analytic techniques 203; defined 206; development 193; device layer 201; DT enabling technologies 201, *202*; evaluation of 212–213, *214*; focus group for 193–194; integration of hardware and software of *207*; interviews 192; limitations 216; mobile devices 203; monitoring and controlling HVAC system 204–205, *205*; operations of 207–209, *209–212*; physical layer 201; scenario planning 192–193, *193*; sensing layer 201; sensors 202–203; steps for implementation 206–207; virtual design/modeling technologies 202
psychophysiological states 326; measurements of 331–332; psychophysiological indicators **332**; *see also* signage color; noise levels
public-private partnership method 48
public transportation 274, 277, 298, 304–305

Random Forest algorithm 41
random utility maximisation (RUM) 230, 232
Rao–Scott chi-squared tests 108
Rapid Entire Body Assessment (REBA) 88–94, 96, 98
rapid transfer centres 304
Rapid Upper Limb Assessment (RULA) 88, 89
RDF Schema (RDFS) 134
real-time 3D engine 261, **261**
real-time ray tracing 182, 187
reasoning and conformance check 135–138, **139**, 142, **144–146**, **148**
Recarte Suazo, G. A. 318
Red-Green-Blue (RGB) cameras 89, 93, 97
regulatory ontology 136
reinforcement learning (RL) 235, 276, 289, 292, 307
reinforcement learning from human feedback (RLHF) 68, 78
remote sensors 261, **262**, 266, 299
Ren, G. 142
Resource Description Framework (RDF) 134
resource planning 133, 137–138, 142, 152; benefits for 143, 158; crane 143; functions for utilising of ontology in 135; limitations 143, 158; literature review for **144–151**; non-physical 134; physical 134; static 157
rigid exoskeletons 120, **121**
risk and emergency management 304
robotics 46–47, 52–53, 62, 99, 120; advanced 315; arm cutting **56**; arms 56; assemblies 105–106; drilling 105; fabrication 55, 56; nailing 105; welding 105
Rosen, R. 195
rule-based generative DfX: advantages 38; application 37; challenges 38; stage 37–38; workflow 37–38, *37*
rule-based models 221, 234–236, *235*
Rule Interchange Format (RIF) 134

Sacks, R. 301
Salazar-Miranda, A. 286
Salvador Palau, A. 7
Saraswathy, V. 26
Sauter, E. 18
scaled model 154, *154*
scheduling 115, 133–138, *153*, 164, 235–236, 263, 274, 308; activities 225, 231, 232, 234; benefits for 138, 158; defined 137; drag-and-drop lift 152, 153–154; dynamic 143, 152–158; elements

134; functions for utilising of ontology in 135; limitations 158; literature review for **139–141**; project 301; rescheduling 222; resource 304; static 152
Schwartz, M. **124**
Schwatka, N. V. 118
self-driving vehicles 293–294
self-report methods 84–85, 89, **90**
semantic segmentation model 286
Semantic Web Rule Language (SWRL) 134, 153
Semantic web technology 133–134, 135, 152–153, 158
sequence labelling 65, 66, 68, 72
Settembre, N. **123**
Shabanpour, R. 229
shared mobility (SM) 272, 273, 275, 278–279, 293
sharing economy 315, 318
Sharma, P. K. 318
Sharmeen, F. 229
Sheldon, M. 275
Shi, J. 263
Shi, X. 18
Shinde, R. 322
signage color 328–329, *334*; black-and-white 330–331, 333–336; color-coded 330–331, 333–336
simplistic device models (SDM) 196
simulation digital twin 10
simulation engine 195, 202–203, 206–207, *208*, 209
simulation of travel/activity responses to complex household interactive logistic decisions (STARCHILD) 232
single-facet models 221, 228–230
slot filling 68, *68*, 72, 74
smart analytic system 276–277
smart and sustainable transportation 304–305
smart built environment (SBE) 339; blockchain for (*see* blockchain technology); generative design for excellence (dfx) for (*see* generative design); intelligent planning unit for (*see* intelligent planning unit)
smart cities 220, 265, 286–287, 297, 303; blockchain and 315, 316, 317–318; design principles 264; digital twins in 4, 11, **12**, 14; infrastructures 260; initiatives 266; sustainable 263, 272
smart configuration 65–68, 70, 74, 77, 79; *see also* language models
smart contract 316–317, 321, 322
smart decision-making, digital twins for *see* digital twins
smart energy 316
smart government 315, 317
smart homes 315, 317
smart sensor network (SSN) 264
smart traffic management system 276, 304
smart transportation 297, 301, 316
soft exoskeletons 120, **121**
Song, H. W. 26
Souza, L. 260
SPARQL 134, 157
spatial planning 285, 286–288
spoken language understanding (SLU) 64, 65, 68–69, *68*, 79
'sponge city' concept 247
sports centre design 29–30, *30*
Stargel, D. 195
status digital twin 9, 10
Staub-French, S. 142
stochastic programming (SP) approach 236
strategic urban planning 302, 303–304
structural equation model (SEM) 229
structural health monitoring (SHM) of bridges: IPU application (replication and physical entities) 27, *27*; IPU network (interaction and refinement) 28, *28*; IPU planning (design and strategy) 25–27, *26*
subjective indicators 84–85, 89, **90**, 98
supply chain modularity 48–49
surface electromyography (sEMG) 85, 87, *87*, 88
sustainable urban built environment 243, *244*; breezeways 253; city as building 247–248, *247–248*; design approaches 255, *255*; design tools 249; ecocity master-planning 255; energy supply 244–245, 247; energy use 250–252, *251*, 253; external comfort 249–250, *249*; infectious disease 254–255; temperature 250–252, *252*, 253; top-down bottom-up 244, *245*; triple diamond approach 255–256, *256*; zero-carbon performance 244–245, 256; *see also* activity-based models
sustainable urban management 262, 264
systematic observation method 88
system-based digital twins 12–13
system thinking 304

T5 model 67, 73–76, *75*, **75**, 79
Tao, F. 7, 11
technology acceptance model (TAM) 121, 122
Tee, R. 49, 58
temperature 165, 177, 205, 207, *210*, 249, 299, 306; air 215, 250–251, *251*, 252, 253; color 336; external 250; sensors 85, 202–203; skin 85; surface 250–251, *251*, 253; urban *252*, 253
temporary PUDO zones 287
text classification 65, 66, 68, 72
theory of planned behavior (TPB) 121, 122, 291
theory of reasoned action (TRA) 121–122
Thompson, E. M. 258
time geography 224, 225–227, 230–231
Timmermans, H. 230

Todor, S. 264
top-down approach 55, 163, 244, *245*, 256
transitional spaces *247*, 248
tripartite diagram 109, *110*, 116, 259
triple diamond approach 255–256, *256*
Tujia Pan-Museum, Chinese 53, *53*; DFAB implementation to 55; DFAB techniques in **56**; project delivery to 56–57
two-stage stochastic mixed-integer programming (TSMIP) model 236
TxtImg2Img model 41

ultra-wide band (UWB) technology 164–165
unified theory of acceptance and use of technology (UTAUT) 122
United States Bureau of Labor Statistics 118, 119
universal ontology 136
Universal Scene Description (USD) 13, 177–178, 182
Universal Thermal Climate Index (UTCI) 249–250, 252–253
Upasani, S. 123
urban building and energy and climate (UrBEC) model 250, 251, 253
urban built environment 220–221, 262, 267, 271, 273–274, 276–279, 284–285, 287, 290–294; activity-based models (*see* activity-based models); activity patterns, influence of 221–222; effects of activity patterns on 222–223; integrated smart curb management in (*see* integrated smart curb management); *see also* sustainable urban built environment
urban climate digital twin 301–302
urban climate modelling 301
urban disaster management 262, 263
urban ecosystems 316
urban facility management 262, 264
urban greenery and built environment (UGBE) model 251
urban intelligence 261
urban microclimate model (UMM) 250
urban planning 221, 253, 259–261, 262–265, 271, 273, 290; modernist 248; strategic 302, 303–304
urban underground space (UUS) 326–330, 335–336
user experience (UX) 137, 180, **183**, **185**, *214*, 293, 308
user interface (UI) 8, 13, *86*, 154, 180–182, **183**, **185**, 187, 201, 203, 208, *208*, 213
utility-maximising models *see* econometric models

Van Eck, J. R. 231
Vazifeh, M. M. 274
Vickers, J. 171, 194, 195
Vidakovic, V. 231
virtual models 13, 107, 128, 171, 190, 195–197, 200–201; 3D 330; dynamic 4; unified 175
virtual reality (VR) 40, 163, 168–171, *168*, 182, 322, 327, 330
visualization 14, 175, 179–182, **183**, 201, 203, 260, 305, 308, 310; 3D 4, 198–199, 207, 213, 215; capabilities 13, 180–181, 213; data 298, 301; effectiveness *214*; equipment 168; limiting 198; process 109, 111; tool 215
visual programming (VP) 106, 180, 182, 261, **261**
visual stimuli *see* signage color
Voordijk, H. 48–50, 59

Wang, Z. 258, 262, 265
wayfinding performance 330; black-and-white 335–336; challenges in 335; color-coded 333–336; combined effects of signage color and noise levels on *334*; in healthcare environments 329; measurements of 331–332; in urban underground spaces 327–329
Web Ontology Language (OWL) 134
Wegener, M. 226
Weyl, G. 275
Wong, R. W. M. 264
work-related musculoskeletal disorders (WMSDs) 83, 118–119, 122–124, 128
World Health Organization 118

XGBoost model 41
XR Studio at the Bartlett School of Architecture 168, *168*
Xu, S. 260, 263
Xu, Xun 258, 262
Xu, Z. 276
Xue, F. 260, 263

Yan, J. 8
Yan, T. **126**
Ye, Q. 289
Ye, X. 321
Yildiz, A. E. 142
Yin, S. 289

Zarinpoush, F. 192
zero-carbon performance 244–245, 256
Zhang, C. 263
Zhang, D. 258, 262, 264
Zhang, J. 138, 229
Zhang, X. 289
Zheng, J. 77
Zheng, L. 123
Zhu, Z. 124